INFRASTRUCTURE SYSTEMS

WILEY SERIES IN INFRASTRUCTURE MANAGEMENT AND DESIGN

Lonnie E. Haefner, Ph.D., P.E., Series Editor

INFRASTRUCTURE SYSTEMS: MECHANICS, DESIGN, AND ANALYSIS OF COMPONENTS

Demeter G. Fertis, Ph.D.

INFRASTRUCTURE SYSTEMS

Mechanics, Design, and Analysis of Components

Demeter G. Fertis

A WILEY-INTERSCIENCE PUBLICATION

JOHN WILEY & SONS, INC.

New York • Chichester • Weinheim • Brisbane • Singapore • Toronto

This text is printed on acid-free paper.

Copyright © 1997 by John Wiley & Sons, Inc.

All rights reserved. Published simultaneously in Canada.

Reproduction or translation of any part of this work beyond that permitted by Section 107 or 108 of the 1976 United States Copyright Act without the permission of the copyright owner is unlawful. Requests for permission or further information should be addressed to the Permissions Department, John Wiley & Sons, Inc., 605 Third Avenue, New York, NY 10158-0012.

This publication is designed to provide accurate and authoritative information in regard to the subject matter covered. It is sold with the understanding that the publisher is not engaged in rendering legal, accounting, or other professional services. If legal advice or other expert assistance is required, the services of a competent professional person should be sought.

Library of Congress Cataloging-in-Publication Data
Fertis, Demeter G.
   Infrastructure systems : mechanics, design, and analysis of components / Demeter G. Fertis.
      p. cm. — (Wiley series in infrastructure management and design)
   "A Wiley-Interscience publication."
   Includes index.
   ISBN 0-471-17907-8 (cloth : alk. paper)
   1. Structural analysis (Engineering)   2. Structural dynamics.
I. Title. II. Series.
TA645.F464 1997
624.1′71—dc21   97-53

Printed in the United States of America

10 9 8 7 6 5 4 3 2 1

To my family

CONTENTS

Preface ix

1 Infrastructural Fundamentals 1

    1.1   Introduction, 1
    1.2   Infrastructure Systems, 1
    1.3   Structural Elements and Loadings, 3
            1.3.1 Beams, Columns, and Beam-Columns, 4
            1.3.2 Trusses and Frames, 8
            1.3.3 Arches, 16
            1.3.4 Plates and Shells, 18
            1.3.5 Foundations and Footings, 19
    1.4   Equilibrium of a Particle, 21
    1.5   Equilibrium of a Rigid Body or Structure, 26
    1.6   Moment, Shear, and Axial Force Equations and Diagrams, 31
    1.7   Relations Between Load, Shear, and Bending Moment, 38
    1.8   Dynamic Aspects of Component Elements of an Infrastructure System, 47
            1.8.1 Dynamic Characteristics of Highway Bridges, 48
            1.8.2 Dynamic Bridge Studies by Fertis and Bender, 51
    1.9   The Earthquake Problem, 54
            1.9.1 Practical Design Considerations, 57
    1.10 The Los Angeles Earthquake, 61
           Problems, 62

2 Stress and Deformation 71

    2.1   Introduction, 71
    2.2   Fundamentals of Stress, Strain, and Deformation, 71
            2.2.1 Axially Loaded Members, 72
            2.2.2 Members Subjected to Torsional Moments, 77
    2.3   Members Subjected to Bending, 88
    2.4   Combined Loading, 105
            2.4.1 Bending Combined with Axial Loading, 105
            2.4.2 Members Subjected to Combined Bending, Torsion, and Axial Force, 109
    2.5   Shear Stresses Associated with the Bending of Members, 113

2.6 Maximum and Minimum Normal and Shear Stresses, 121
2.7 Mohr's Circle for Plane Stress, 128
Problems, 132

3 Infrastructural Dynamics 147

3.1 Introduction, 147
3.2 The Simplest Infrastructural Dynamics Problem, 147
    3.2.1 Practical Applications of the Spring-Mass System, 149
3.3 Solution of the Simplest Infrastructural Dynamics Problem, 157
    3.3.1 Free Undamped Vibration, 159
    3.3.2 Free Vibration with Viscous Damping, 162
3.4 Response Due to Harmonic Excitations, 169
    3.4.1 Closed-Form Solution of the Differential Equation, 170
3.5 Response Due to Force of a General Type, 178
    3.5.1 Closed-Form Solution of the Differential Equation, 178
    3.5.2 Numerical Solution of the Differential Equation, 185
3.6 Vibration of Highway Bridges, 194
3.7 Vibration Analysis of Continuous Highway Bridges, 200
    3.7.1 Stodola's Method and Iteration Procedure, 200
    3.7.2 Vibration of Highway Bridges Using Stodola's Method, 203
3.8 Practical Design for Earthquakes Using Available Codes, 208
Problems, 212

4 Advanced Methods and Problems of Infrastructural Analysis 219

4.1 Introduction, 219
4.2 Members of Arbitrary Stiffness Variations, 219
    4.2.1 Theory and Method of the Equivalent Systems, 220
    4.2.2 Approximate Method of the Equivalent Systems, 224
4.3 Equivalent Systems for Highway Bridges with Girders of Variable Depth, 230
4.4 Statically Indeterminate Variable Stiffness Members, 234
4.5 Moment-Area and Conjugate-Beam Methods, 248
    4.5.1 Moment-Area Method, 248
    4.5.2 Conjugate-Beam Method, 258
    4.5.3 Statically Indeterminate Members, 263
4.6 Moment Distribution Method, 266
4.7 Moment Distribution Method for Continuous Bridge Girders of Variable Depth, 273
4.8 The Three-Moment Equation, 277

4.9 Dynamic Response of Rigid Frames Composed
of Members of Variable Stiffness, 283
Problems, 292

5 Infrastructural Nonlinearities, Instability, and Inelastic Response 303

5.1 Introduction, 303
5.2 Infrastructural Nonlinear Response, 303
    5.2.1 The Elastica Theory, 304
    5.2.2 Pseudolinear Equivalent Systems of Constant Stiffness, 306
    5.2.3 Simplified Nonlinear Equivalent Systems of Constant Stiffness, 308
    5.2.4 Loading and Stiffness Dependence on the Geometry of Deformation, 309
5.3 Solution of Nonlinear Problems Using Pseudolinear Equivalent Systems, 311
5.4 Solution of Nonlinear Problems Using Simplified Nonlinear Equivalent Systems, 323
5.5 Infrastructural Instabilities, 326
    5.5.1 The Euler Column, 328
    5.5.2 Columns with Other Boundary Conditions, 331
    5.5.3 Eccentrically Loaded Columns, 331
5.6 Influence of the Axial Force on the Free Vibration and Flexibility of Columns or Beam-Columns, 336
5.7 Influence of the Axial Compressive Force on the Stability and Free Vibration of Elastically Supported Beam-Columns, 338
5.8 Suspension Bridges and the Collapse of the Tacoma Narrows Bridge, 340
    5.8.1 Brief Discussion on Fundamental Aspects of Suspension Bridges, 342
    5.8.2 More on the Disaster of the Tacoma Narrows Suspension Bridge, 344
    5.8.3 Other Failures and What We Learn from Them, 348
5.9 Inelastic Analysis and Design of Infrastructural Elements, 350
    5.9.1 Theory of Equivalent Systems for Inelastic Analysis and Reduced Modulus E_r, 351
5.10 Ultimate Design Loads Based on Inelastic Response, 364
    5.10.1 Ultimate Loads for Beams of Monel Material, 365
    5.10.2 Ultimate Loads for Mild Steel Beams, 371
    5.10.3 Ultimate Loads for Aluminum Alloy Beams, 377
Problems, 383

Answers to Selected Problems	579
References	595
Index	601

PREFACE

Today, more than ever before, the design and analysis of our structural and mechanical systems require basic knowledge and physical comprehension regarding the many aspects that constitute optimum structural and mechanical performance. The dynamic and vibration analysis of a highway bridge, for example, requires knowledge of the basic properties of the material, or materials, used to construct the bridge. Such properties are then used to perform the static and dynamic analysis, so that we can develop a bridge design that is safe and functional.

Such static and dynamic analysis, however, also requires knowledge and fundamental understanding of statics, mechanics, and structural mechanics methods; these methods must be used to compute the required static and dynamic deformations of the bridge, its frequencies of vibration, and also to draw conclusions as to how long a bridge span should be allowed to be in order to maintain its required rigidity for a safe design.

These are only a part of a group of studies that need to be performed in the design of a highway bridge. We can already see, however, that if we wish to perform the dynamic and vibration design and analysis of a bridge, we will need to comprehend all of the problems and methods that are needed for such a task.

The work in this text represents an integration of the many topics that are needed in the dynamic design and analysis of our structures. Since structures, highways, aircraft, and so on, are essential parts of our modern infrastructure, the total approach is integrated to the problems and concerns of our total infrastructure. We need to know the evolutionary aspects of our present and future infrastructure in order to be able to provide the structural, mechanical, and transportation systems, as well as other component elements of our infrastructure, in order to be able to secure its functional capability and provide the required services to the user, we the people.

On this principle, the book is intended to serve as a text in the Departments of Civil Engineering and Mechanics for junior and senior undergraduate courses in structural dynamics and structural analysis. For a 15-week course, the instructor may cover completely the first four chapters, as well as the sixth chapter. As an alternative, the instructor could develop a syllabus including a selection of sections throughout the text that are most appropriate to the level of his or her class.

The text includes also advanced material, making it suitable for a graduate course in structural dynamics or structural analysis. For such a course, the first two chapters serve as supporting material to fully comprehend the more advanced material in the following chapters of the book.

The book may be used also as a text in infrastructure curricula that deal with the static and dynamic design and analysis of infrastructural components such as buildings, bridges, mechanical systems, and so on. Aeronautical and mechanical engineers will find this book very interesting, particularly the material covered in Chapters 4 and 5. These chapters represent new methods and theories that have been developed and tested by the author and his collaborators over three decades of research and experience. All the material in the text is class-tested and incorporates the comments and input of both graduate and undergraduate students.

The book provides a unique integration of structural dynamics, structural vibrations, and structural mechanics principles, coupled with design and analysis. It integrates in a very logical order the static, dynamic, and instability responses of infrastructural components, combined with very powerful unique and existing methods of analysis and design. It emphasizes both theory and practical applications, and provides an extensive treatment regarding the elastic, inelastic, and nonlinear responses of infrastructural components of both uniform and variable stiffness.

The topics in the text are carefully selected in order to formulate a text that will help prepare the student and the practicing engineer for the challenges of the real engineering world, and will also prepare them for more advanced work on this and related subjects. It represents the long academic and industrial experiences of the author at Wayne State University, University of Iowa, University of Akron, NASA, Boeing, Michigan and Ohio Departments of Transportation, Lockheed, General Motors, Department of the Navy, and so on. The subject matter is treated in a way that progresses gradually from the simpler to the most challenging material for complete understanding of the topics included in the text. This helps the instructor to define the level of challenge for his/her class. A large number of solved example problems of varying difficulty are provided throughout the text, and a wide selection of problems at the end of each chapter is provided for the student to practice. Appendixes are also provided at the end of the text, which incorporate helpful information and computer programs that help the student and the practicing engineer to comprehend the material presented in the main body of the text.

On a chapter-by-chapter basis, Chapter 1 deals with basic principles of statics, the dynamic characteristics of highway bridges, and an introduction to earthquake engineering and design.

Chapter 2 deals with the fundamentals of structural mechanics, which serve as preparatory material for better comprehension of the material that follows.

Chapter 3 is devoted to the dynamics of infrastructural components such as beams, frames, buildings, bridges, and so on, that are subjected to various dynamic excitations and vibratory motions. Closed-form as well as numerical solutions are used to solve the structural dynamics problems. Practical design for earthquakes using available building codes is also covered.

Chapter 4 deals with the more advanced topics for the static and dynamic design and analysis of structural systems that are composed of members of any stiffness variation along their length. The author's unique method of the equivalent systems simplifies a great deal the solution of such complex problems and yields either exact results, if the exact method of the equivalent systems is used, or very accurate results if the approximate method of the equivalent systems is used. Helpful methods for the computation of deflections and rotations of structural members are also included in this chapter. The advanced material in this chapter is written in a way that helps both the advanced undergraduate and graduate student to comprehend the subject matter.

Chapter 5 also deals with more challenging material that builds up the interest of both the student and the practicing engineer for the inelastic and nonlinear analysis and design of nonlinear systems. The unique nonlinear method of the equivalent systems provides an excellent way to obtain a very accurate solution for the inelastic behavior of structural members and the computation of the large deformation of nonlinear systems. Dynamic and static instabilities of axially loaded members, fundamentals of suspension bridges, as well as discussion on suspension bridge failures and what we learn from it, are also included in this chapter. Both design and analysis characteristics are discussed.

Chapter 6 deals with the design and analysis of structures that are subjected to earthquake excitations and blast loadings. Both elastic and elastoplastic analyses are included. Lagrange's equation and the method of modal analysis for the computation of the dynamic response of structures subjected to earthquake accelerations are also discussed in this chapter.

Chapter 7 provides fundamental knowledge regarding energy concepts and methods and their application for the solution of structural problems. It also provides an introduction to the finite element and the finite difference methods with application to structures. The finite difference method is also applied for the solution of rectangular plates of both uniform and variable thickness. This chapter prepares the student for the more advanced work on these subjects that is essential to his/her work as a practicing engineer.

I wish to thank my graduate and undergraduate students for their help in locating errors and for their valuable suggestions regarding the material included in the text. My special thanks and gratitude go to my wife, Anna, a business consultant, for her constant encouragement, valuable suggestions, and endless discussions during the writing of this text, and for the typing of the manuscript. Special thanks are also due to Dr. J. Padovan, Distinguished Professor of Mechanical Engineering at the University of Akron, for his valuable suggestions and discussions on the various parts of the manuscript. Last, but not least, I wish to thank John Wiley & Sons for making my work available to the academic and professional audience.

DEMETER G. FERTIS

Akron, Ohio

1 Infrastructural Fundamentals

1.1 INTRODUCTION

This chapter provides a review of basic concepts and principles of mechanics that are essential in the design and analysis of structural and mechanical systems. It also establishes their correlation and importance in relation to the problems associated with the general infrastructure system and its components. The first section of the chapter briefly discusses the author's understanding of what infrastructure means to an engineer. The technical problems associated with the various types of structural elements and loading, are introduced next, along with general problems associated with these elements. Basic concepts of mechanics are considered, as are examples related to the equilibrium of particles and rigid bodies. Basic concepts related to moment, shear, and axial force are also covered, as are shear force and bending moment diagrams. The chapter finishes with a discussion of the dynamic aspects of component elements of infrastructure systems, such as bridges and buildings, by also examining earthquakes and their effects on structures. This chapter sets the tone and direction for the more detailed discussions in following chapters.

1.2 INFRASTRUCTURE SYSTEMS

An infrastructure system is usually established to accommodate definite needs of people, and to perform a definite function based on such needs. One convenient way of thinking about infrastructure systems is to classify them into major infrastructure systems (MISs) and component infrastructure systems (CISs). An MIS is composed of a definite number of interrelated CISs. For example, a national highway system may be thought of as an MIS that is composed of interdependent CISs such as: (a) arterial highways, that is, expressways and turnpikes; (b) local and regional roadways feeding the arterial highways; and (c) highway bridges of various types. The purpose of such MISs is to accommodate the movement of people and goods in every part of the country. It has to be safe, functional, economical, and it requires the construction of numerous interstate highways and thousands of highway bridges.

The creation and maintenance of such an MIS requires the development of new technical concepts and methodologies that take into consideration the ever-changing needs of our modern society. For example, as the quantity of goods and the number of people to be transported in the various parts of the country become progressively larger, the MIS could become obsolete even before its construction is completed. To maintain a balanced condition, the system must be flexible enough to incorporate such evolutionary changes, which usually requires the development of new technologies to be able to cope with such demands. Infrastructure mechanics should be developed in a way that provides the design directions for the development of advanced technologies that will fulfill established design requirements.

For example, the development of new materials may be required, with materials properties that fulfill special criteria in the construction of highway bridges and pavements. Material that can be used to repair cracked pavements and bridge decks may be needed, or material that will provide higher strength with good ductility. The purpose of infrastructure mechanics is to develop and analyze such new materials, and to provide the design engineer with the information required to come up with a functional and safe design.

Another trend today is the construction of highway bridges with much longer spans. Highway bridges with spans of 200 and 250 ft are a common occurrence today. Long spans make the highway bridge very flexible, and flexible bridges are associated with low fundamental natural frequencies of vibration. Such conditions can create various dynamical problems if the bridges are not analyzed and designed appropriately. Long-span bridges are difficult to design, and there are limits to allowable span lengths for given bridge design configurations. The fields of structural mechanics and structural dynamics, combined, should provide the required technology for the solution of such problems.

The following statistics serve to emphasize the importance of safety and maintenance of highway bridges. In the later part of the 1980–1990 decade, it was estimated that roadways in the United States involved 576,000 highway bridges. Federal Highway Administration (FHWA) inspection of these bridges [1, 2] showed that about 236,000 of those bridges, about 41 percent of all existing highway bridges, were deficient by today's standard. Inadequate and dangerous bridges are found in every part of the country. The country spends about $5 billion per year involving about 8,000 bridge improvement projects per year. Every two days, on average, a highway bridge somewhere will buckle, sag, or collapse, and a thousand Americans lose their lives every year because of poor bridge approaches and inadequate signs. Many catastrophic failures of bridges caused by floods have occurred during the past decades. For example, in 1985, 73 bridges were destroyed by flooding, including scour, in Virginia, West Virginia, and Pennsylvania. This illustrates the need for upgrading the system and for providing technology to construct safer and more economical highway bridges.

An air transportation system may also be thought of loosely as an MIS. The interdependent component CISs could be: (a) the commercial national interstate system, (b) commercial international system, (c) local transportation system, (d) space exploration, and (e) possibly the military air transport system.

The main purpose of these systems is to transport people and goods in every part of the globe. A second purpose is probably to perform a specific military function. Here, again, if it is necessary to design a new jet engine meeting specific requirements, it will usually take about fifteen years of extensive research, testing, and development before we are in a position to mount the new engine on an aircraft. Complete understanding of the mechanics and functional characteristics of such engines is a must for a successful design.

The demand for airfoil designs that provide a greater amount of lift without increasing drag, and that are also able to function at high angles of attack without stalling, is more pressing today than ever before. Such demands will become even more pressing as we enter the twenty-first century. Space exploration and travel will be top priorities during this century, and new technologies must be developed in order to cope with the problem. Space exploration requires the construction of space stations, such as the space station Freedom, which is under development today. Such space stations are subjected to various dynamic loads during shuttle docking, solar tracking, altitude adjustment, and other functional maneuvers. Accurate predictions regarding the natural frequencies and mode shapes of the space station components, including the solar arrays, will be important requirements in determining the structural adequacy of the components and in designing an appropriate dynamic control system [3, 4]. Understanding the mechanics and physical behavior of such engineering systems forms the basis for a successful design.

Proceeding with the same line of reasoning, one may construct many MISs, each fulfilling a definite need and function of our complex modern society. Since the needs are many and largely interdependent, the various MISs would also be interdependent. The development of one MIS would largely depend upon how the development of other MISs progresses. This would require the formation of a comprehensive master plan (CMP) that formulates the interrelated activities of all MISs. This is a tremendous task to undertake, and optimization parameters would have to be introduced for an optimum performance.

It should be pointed out that the above discussion is not intended to provide the foundation for the development of a new infrastructure theory. It is, however, intended to create a sense of responsibility for what is lying ahead. Infrastructure mechanics is an essential element for the development of functional infrastructure systems that are able to cope with the present and future demands of our complex society.

1.3 STRUCTURAL ELEMENTS AND LOADINGS

When a structure, or a mechanical system, is subjected to a certain static or dynamic loading, every component element of the structure or mechanical system will experience the effects of such loading. The simplest structure,

4 INFRASTRUCTURAL FUNDAMENTALS

consisting of only one component element, is a beam. A beam can be in itself a structure performing a certain function, or it can be a component element of a multielement structure. Trusses, columns, frames, plates, arches, and shells are other types of structural elements that form the skeleton of a structural or mechanical system. Each one, however, can be in itself a single element structure that is designed to be used for some practical purpose. Loadings and functional characteristics of such elements are briefly discussed below.

1.3.1 Beams, Columns, and Beam Columns

Beams are one-dimensional structural elements that are constructed to perform a certain structural function. A beam can be the girder of a bridge that supports the bridge deck and the vehicular traffic on the bridge, or it can be the structural element placed between girders to prevent them from moving closer to or farther away from each other. It can also be a part of a framework, thus subjected to the effects transmitted to this member by every member of the entire framework. A beam may be subjected to various types of loading, as shown in Fig. 1.1, and the boundary conditions at its ends can vary.

The beam in Fig. 1.1a is a uniform simply supported beam loaded by a uniformly distributed load q over its entire span. End A is prevented from deflecting in the vertical and horizontal direction by use of a hinge. End B is prevented only from moving in the vertical direction by use of a roller, but it is free to move in the horizontal direction.

The doubly tapered cantilever beam in Fig. 1.1b, is loaded with a concentrated load P at its free end and a concentrated bending moment as shown. End A is free to move in any direction, and end B is fixed. In other words, any movement of end B of the cantilever beam is prevented.

The beam in Fig. 1.1c is a uniform three-span beam subjected to a uniformly distributed load q over its entire length. End A is prevented from moving in the horizontal and vertical directions by use of a hinge, while supports B, C, and D are prevented only from moving in the vertical direction by use of rollers. The beams in Figs. 1.1a and 1.1b are statically determinate, because all the reactions at the end supports can be determined by using the equilibrium equations from statics. The beam in Fig. 1.1c, however, is statically indeterminate because the number of reactions at supports A, B, C, and D is larger than the number of equilibrium equations from statics. Since the continuous beam in Fig. 1.1c is loaded by a vertical load q on the plane of the paper, the reactions at A are R_A and H_A, and we have reactions R_B, R_C, and R_D at supports B, C, and D, respectively, acting as shown in the figure. Thus, we have a total of five unknown reactions and only three equations of statics, namely:

$$\sum F_x = 0 \qquad (1.1)$$

$$\sum F_y = 0 \qquad (1.2)$$

$$\sum M_0 = 0 \qquad (1.3)$$

STRUCTURAL ELEMENTS AND LOADINGS   5

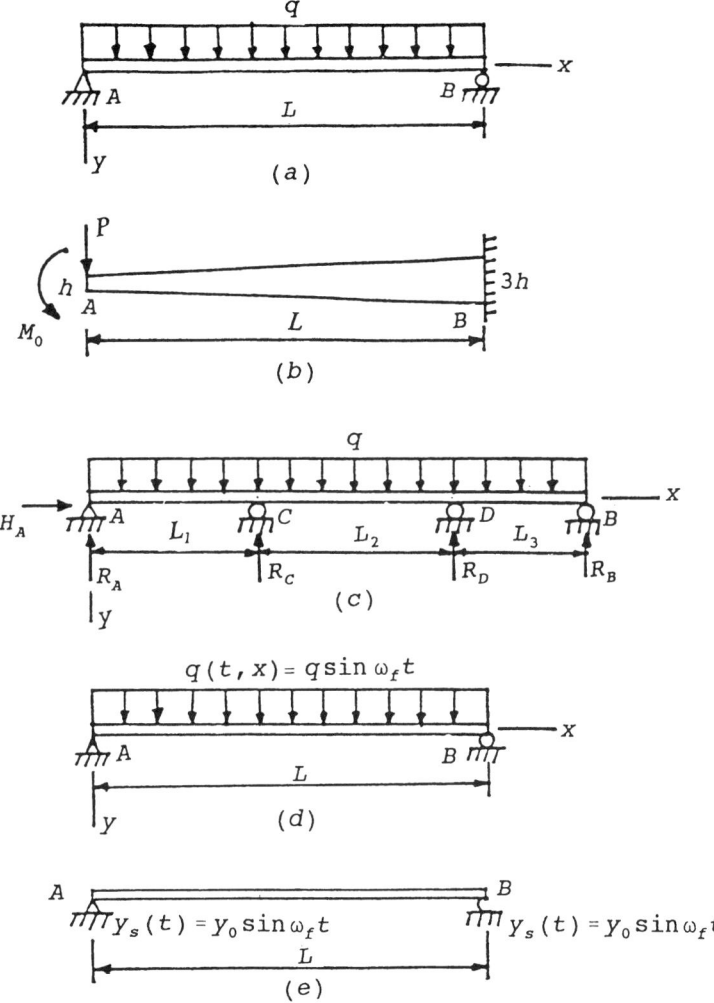

Figure 1.1. (a) Simply supported beam loaded with a uniformly distributed load q. (b) Doubly tapered cantilever beam loaded with a concentrated load P and bending moment M_0 at the free end. (c) Uniform three-span continuous beam loaded with a uniformly distributed load q. (d) Uniform simply supported beam subjected to a uniform external excitation $q(t, x)$. (e) Uniform simply supported beam subjected to a harmonic vertical support motion.

We need two additional boundary conditions in order to be able to solve for all five unknown reactions. Such types of problems are discussed in later chapters of this text.

The applied loading in Figs. 1.1a, 1.1b, and 1.1c is assumed to be a static load that is gradually applied to the member. Beams, however, can also be subject to loads that vary with time t. Such loads are called dynamic, and dynamic analysis must be used to secure the structural integrity of the member. For example, the simply supported beam in Fig. 1.1d is loaded with a dynamic load $q(t, x)$ that varies with time as follows:

$$q(t, x) = q \sin \omega_f t \qquad (1.4)$$

The dynamic loading in Eq. (1.4) is a harmonic load that is constant and equal to q along the x direction of the member, and its harmonic variation with time is represented by the function $\sin \omega_f t$, where ω_f is the frequency of the force. This is a very interesting load that requires particular attention. There are two distinct types of frequencies that are of particular interest to the design engineer. The first one is the forced frequency ω_f of the applied dynamic force $q(t, x)$, and the second type consists of the inherent free frequencies of vibration ω of the beam. For a beam with continuous mass and elasticity, such free frequencies of vibration are infinite in number, and they can be excited during the time of action of the dynamic force $q(t, x)$.

If the beam is not designed appropriately, it is possible for ω_f to coincide with one of the free frequencies ω of the beam ($\omega_f = \omega$), thus creating what is known as the phenomenon of resonance, or synchronism. This is a very dangerous condition, because when $\omega_f = \omega$, the vertical deflection of the member increases with time without any additional increase of the load $q(t, x)$, and it becomes infinite in value when time t reaches infinity. Since the beam can fail at some finite value of the vertical deflection, resonant phenomena should be prevented in order to maintain the structural integrity of the member. Free vibration analysis of the beam can provide the required information for a safe design. See, for example, the author's work in Reference [5].

The simply supported beam in Fig. 1.1e is subjected to a harmonic vertical support motion

$$y_s = y_0 \sin \omega_f t \qquad (1.5)$$

applied at its end supports A and B. The maximum vertical amplitude of the support motion is y_0, and it varies harmonically with time as $\sin \omega_f t$, where ω_f is the frequency of the support motion y_s. Again here, dynamic analysis must be used in order to be able to design the beam correctly, and vibration analysis must be used in order to make sure that the free frequencies of vibration of the beam do not coincide with the frequency ω_f of the support motion [5].

Columns are structural elements that are usually subjected to axial tensile or compressive loads, or acted upon by eccentrically applied loads. For example, the structural element in Fig. 1.2a is a column that is loaded axially by a compressive load P, as shown. Since the load is applied axially through the centroidal axis of the member, the column does not bend in any way,

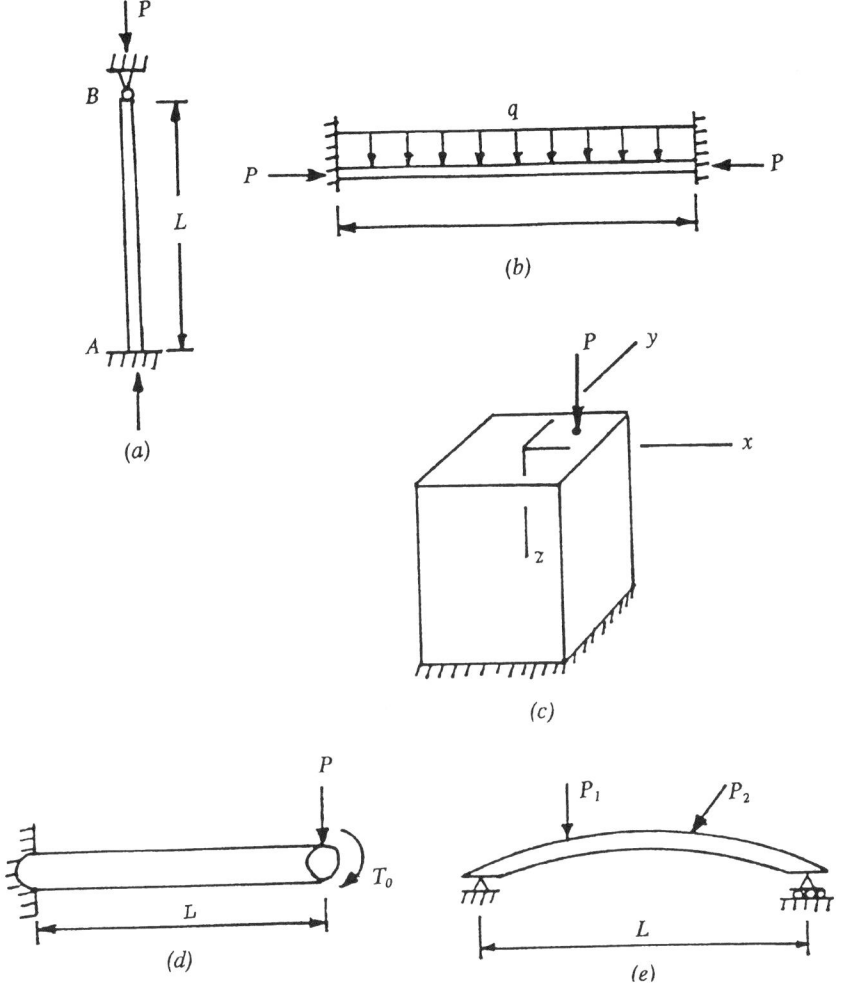

Figure 1.2. (*a*) Uniform column fixed at *A* and hinged at *B*. (*b*) Beam column fixed at both ends. (*c*) Eccentrically loaded column. (*d*) Cantilever beam of circular cross section subjected to a torsional moment T_0 and a vertical load *P* at its free end, as shown. (*e*) Curved beam.

provided that *P* does not exceed a certain value that would make the column bend in the lateral direction, which is a mode of failure known as buckling. Such critical axial compressive loads can be determined by studying the stability and instability characteristics of such columns. See, for example, Reference [6]. The column in Fig. 1.2c is loaded eccentrically by a force *P*, as shown. Since *P* is eccentrically applied, it will produce bending moments about both the *x* and *y* axes and, consequently, the column will also be subjected to bending moments about the x and y axes. The dead weight of a building is often carried out by columns as part of the framework of the building, which in turn transmit it to the foundation of the building and from there on to soil.

Members that are subjected to both bending and axial compression are known as beam columns. For example, the fixed-fixed member in Fig. 1.2b is a beam column, because it is loaded by the uniformly distributed load q that produces bending, and is also subject to an axial compressive force P. Since such members are usually part of the framework of a structure, the elastic restraints at their ends depend not only on the members framing directly to it, but also on the effects transmitted by every member of the entire framework. Thus, it becomes necessary to examine the stability, or instability, of the entire frame acting as a single unit.

Columns and beams may also be subjected to a combination of bending and twisting. For example, the circular cantilever beam in Fig. 1.2d is loaded by a vertical load P at its free end, which produces bending, and by a torque T_0 at the same end, which twists the member. If, hypothetically, an axial compressive force is also applied to the member, then we have an axially loaded member subjected to both bending and twisting.

Columns and beams may also fail by buckling when they are subjected to twisting moments, or to a combination of bending and twisting. For example, members that are made of thin-walled open sections may fail by torsional buckling and twisting. This can also happen to transversely loaded beams when the compression flange becomes unstable.

The beams shown in Fig. 1.1 are all assumed to be straight before the external load is applied. Any curvature in the member is assumed to be a result of the externally applied load. In some cases, as in the case of crane hooks, beams are curved before a bending moment is applied. Such members are known as curved beams. The member shown in Fig. 1.2e is a curved beam subjected to two concentrated loads P_1 and P_2, located as shown. If a member is sharply curved, the stress distribution at a cross section that is produced by bending is markedly different than the one produced by bending when the member is initially straight. For curved beams, such stress distribution becomes nonlinear, and the neutral surface must shift from the centroid of the section toward the axis of curvature of the member, and curved beam theory must be used for their analysis.

1.3.2 Trusses and Frames

Trusses: Trusses are used extensively in the construction of many engineering structures such as bridges, buildings, radar telescopes, transmission-line towers, space structures, cranes, and other related structures. Based on the types of loadings that can be applied to a truss structure, trusses are generally classified as plane trusses and space trusses. The plane truss may be defined as a system of bars all lying in one plane and joined together at their ends in a way that forms a rigid framework. The truss in Fig. 1.3a is a simple plane truss,

STRUCTURAL ELEMENTS AND LOADINGS   9

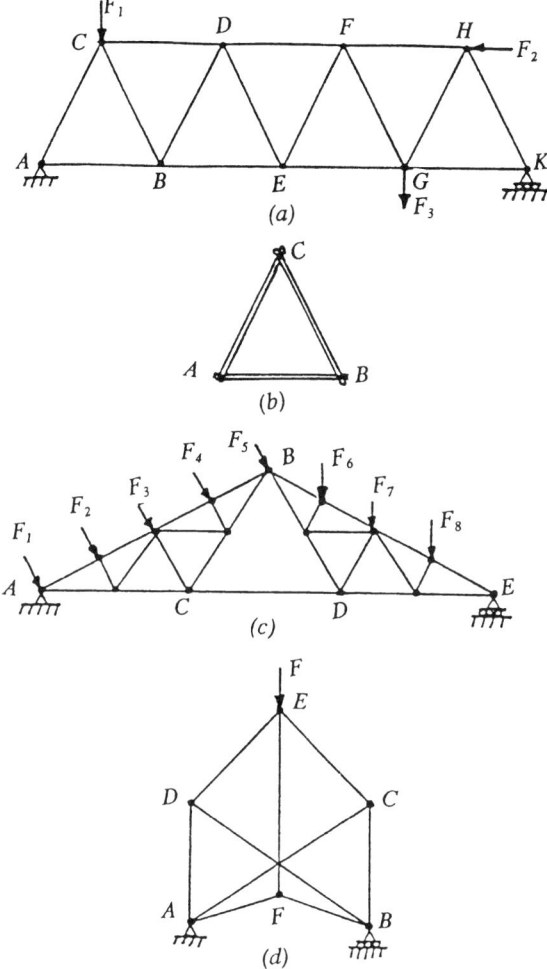

Figure 1.3. (*a*) Simple plane truss. (*b*) Triangle *ABC* of the truss formed by members *AB*, *AC*, and *BC*. (*c*) Compound truss. (*d*) Complex truss.

constructed by starting with the triangle *ABC*, which is formed by joining three bars together as shown in Fig. 1.3*b*. With triangle *ABC* as a rigid unit, joint *D* is formed by the addition of bars *DB* and *DC*, then joint *E* by the addition of bars *EB* and *ED*, and so on. In order to complete the constraint in its plane, the truss is attached to a foundation at point *A* by a hinge, and at point *K* by a roller.

The joints of the plane truss are assumed to be frictionless hinges, and forces F_1, F_2, and F_3 are applied at joints *C*, *H*, and *G*, respectively. All loads acting at the joints of a plane truss must lie on the same plane, which is the plane of the truss. Since all the loads of a plane truss must be applied at its joints, each member of a truss is acted upon by axial tensile or compressive forces only. Axial compressive forces in truss members can produce instabilities, such as buckling, if the truss is not appropriately designed. Many types of plane trusses have been developed by engineers and used in the construction of many engineering structures. Some bear their designers' names; others do not. Examples include Pratt, Howe, Fink, Baltimore, Warren, stadium, and K-trusses.

The axial forces in the members of a plane truss are commonly determined by the method of joints or the method of sections. Descriptions of both can be readily found in any book dealing with statics. The truss in Fig. 1.3c is a compound truss, formed by interconnecting the two simple trusses ABC and BDE by a hinge at B and a bar CD. Three bars appropriately located could also be used to interconnect these two simple trusses. The truss in Fig. 1.3d is a complex truss. We note that, as soon as the reactions at supports A and B are found by using static equilibrium, no further progress with the analysis can be made by using either the method of joints or the method of sections. Special methods of analysis, such as the Henneberg's method [7], must be used for the analysis of complex trusses.

Space trusses may be thought of as a system of bars in space that are jointed together at their ends to form a rigid space structure. Radar telescopes, cranes, and transmission-line towers utilize such types of trusses. A simple space truss could be constructed by attaching the first joint to a foundation by using three bars that do not lie in the same plane. Each additional joint is formed by using three more bars that do not lie in the same plane. The truss in Fig. 1.4a is a simple space truss that is constructed by attaching joint B to the foundation by using points A', B', and C', and bars $A'B$, $C'B$, and $B'B$ which do not lie on the same plane. Then we attach joint A to points C', A', and B, by using bars AC', AA', and AB, and so on.

The space truss in Fig. 1.4b is a compound space truss where the rigid square pyramid $ABCDE$ is attached to the foundation by seven bars that are arranged as shown in the figure. Load F_1 is applied at joint E, and a tensile force P is applied to bar DD' by using the turnbuckle F_2. Any self-contained simple space truss may be rigidly attached to a foundation by using six bars appropriately arranged. Such a structure constitutes a compound space truss. Extensive discussion on space trusses may be found in Reference [7].

A truss may be statically determinate or statically indeterminate. Based on the rules used to construct a truss, the following two equations may be used to relate the number of bars b and the number of joints j for plane trusses:

$$b = 2j \tag{1.6}$$

$$b = 2j - 3 \tag{1.7}$$

Equation (1.6) is used when the plane truss is attached to a wall at points A and K, as shown in Fig. 1.5a; the equation states that when the number of bars b is equal to twice the number of joints j, then the truss is statically determinate.

STRUCTURAL ELEMENTS AND LOADINGS   11

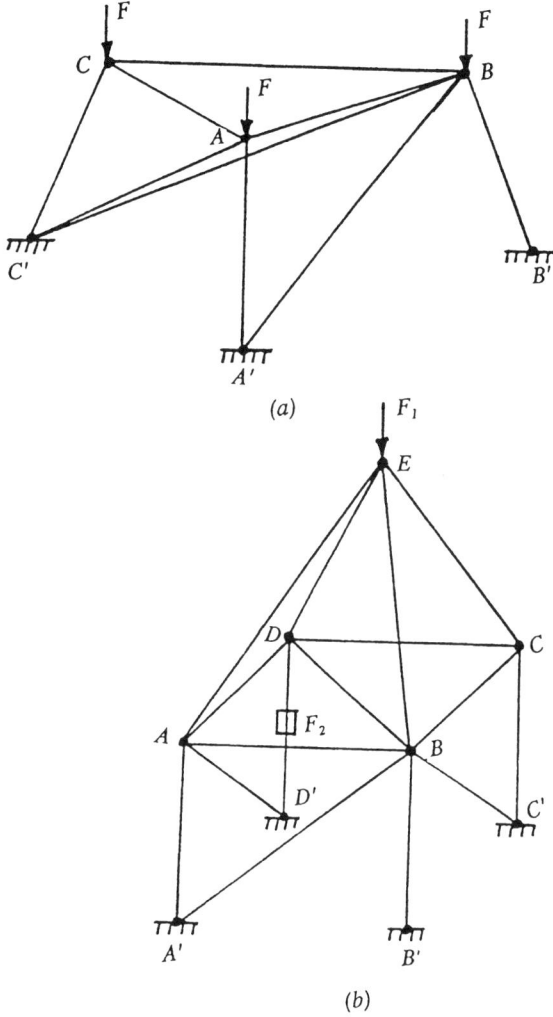

Figure 1.4. (a) Simple space truss. (b) Compound space truss.

The plane truss in Fig. 1.5a is statically indeterminate to the second degree, because it has 18 bars and 8 joints. Bars *KB* and *HD* are not required for static equilibrium.

Equation (1.7) applies to plane trusses that are supported to a foundation as shown in Fig. 1.5b. The number 3 in the right-hand side of this equation indicates that three additional supporting bars, or reactions, are necessary when the rigid plane truss is to be attached to a wall or foundation. The plane truss in Fig. 1.5b has 20 bars and 10 joints, which indicates that it is statically indeterminate to the third degree, because three of its bars are not required for static equilibrium. For example, bars *AI*, *DH*, and *EG* are not required. We

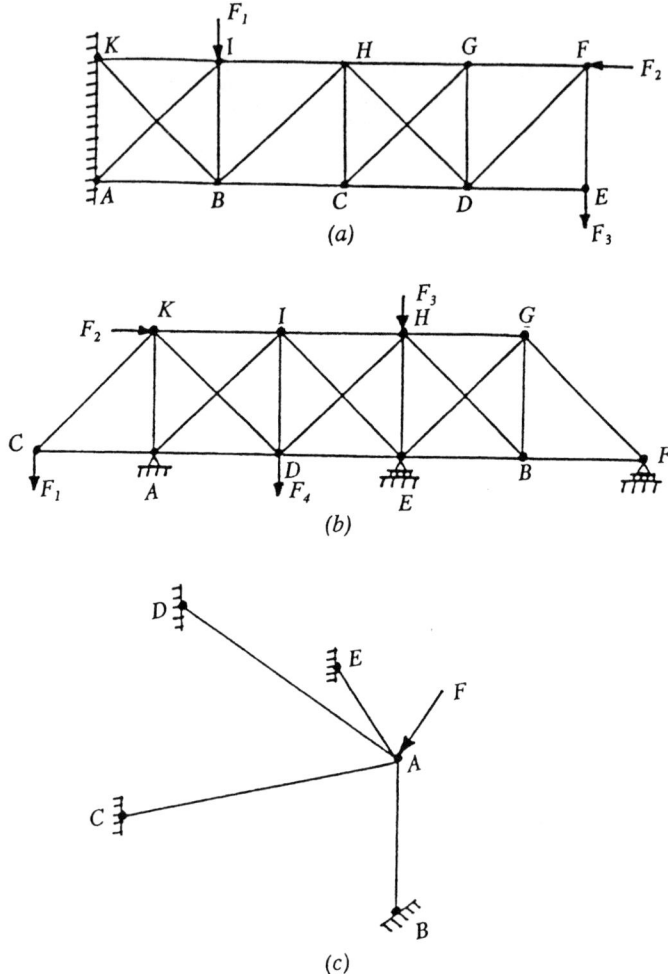

Figure 1.5. (*a*) Statically indeterminate plane truss attached to a wall. (*b*) Statically indeterminate plane truss. (*c*) Statically indeterminate space truss consisting of four bars.

also note that it is attached to the foundation by a hinge at point A, and by rollers at points E and F, yielding a total of four unknown reactions. Since we have only three static equilibrium equations available, the plane truss is also statically indeterminate to the first degree in terms of supporting reactions.

For space trusses, the analogous equations are

$$b = 3j \tag{1.8}$$

$$b = 3j - 6 \tag{1.9}$$

By using Eq. (1.8), we find that the space truss in Fig. 1.5c is statically indeterminate to the first degree because we have only one joint at A and four bars.

The number 6 in the right-hand side of Eq. (1.9) represents the number of external bars or supporting reactions that are required if the space truss is to be attached to a wall or a foundation. If more than six bars or supporting reactions are used, the truss is statically indeterminate externally to a degree equal to the number of supporting reactions that exceed six. The space truss shown in Fig. 1.6 is indeterminate to the second degree externally, because eight bars have been used. The same space truss is also statically indeterminate internally to the fourth degree because we have 34 bars, and in accordance with Eq. (1.9), we only need 30 bars to satisfy this equation. The bars shown by a dashed line in Fig. 1.6 are not needed for static equilibrium of the space truss.

Frames: A frame may be described as a structure composed of members designed to resist loads applied at any point along the length of its members. It may be either two-dimensional or three-dimensional. The joints of a frame may be either pinned, riveted, or welded, or the frame may be of continuous mass, such as a concrete frame. The members of a frame are designed to resist bending moments, shear, and axial forces. They are called multiforce members because three or more forces may be shown in their free-body diagrams. Some exceptions to the rule may exist for special cases where a frame may include cables or slender bars as part of its framework.

A frame loaded as shown in Fig. 1.7a is connected by a cable BC at B and C, by a pin at D, by a hinge at A, and it rests on a smooth surface of a wall at

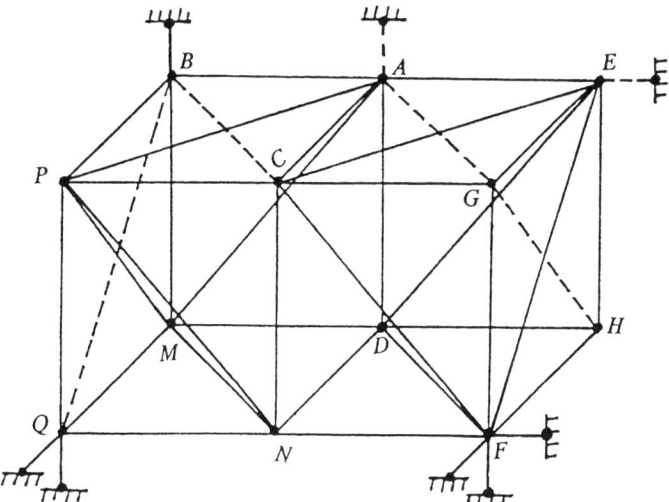

Figure 1.6. Statically indeterminate space truss.

14 INFRASTRUCTURAL FUNDAMENTALS

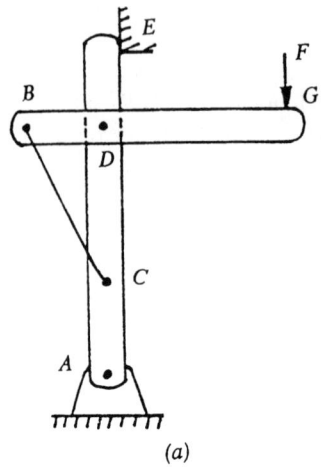

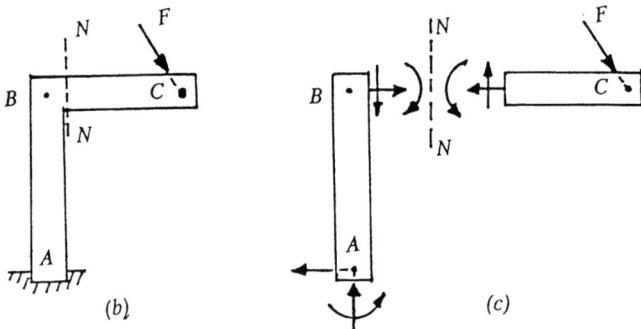

Figure 1.7. (a) Frame connected by a cable BC. (b) Monolithic frame. (c) Free-body diagrams of the two members of the monolithic frame.

E. A monolithic frame structure is shown in Fig. 1.7b. If we assume that the frame is cut at section $N-N$, the free-body diagrams of its component parts are shown in Fig. 1.7c, and they are multiforce members.

The type and number of external supports required for a frame depends mostly upon the purpose it serves and the forces acting on the frame. Some common types of frames are shown in Fig. 1.8. Component members of a frame may be curved, straight, or of variable thickness, as shown in the same figure. Special names, such as parabolic, arched, continuous parabolic, rectangular, two-hinge, and so on, are given to frames.

A three-dimensional frame is loaded as shown in Fig. 1.9a, and its free-body diagram is shown in Fig. 1.9b. At support A we have three unknown reactions, V_x, V_y, and V_z, and three unknown reactive moments, M_x, M_y, and M_z, as

STRUCTURAL ELEMENTS AND LOADINGS   15

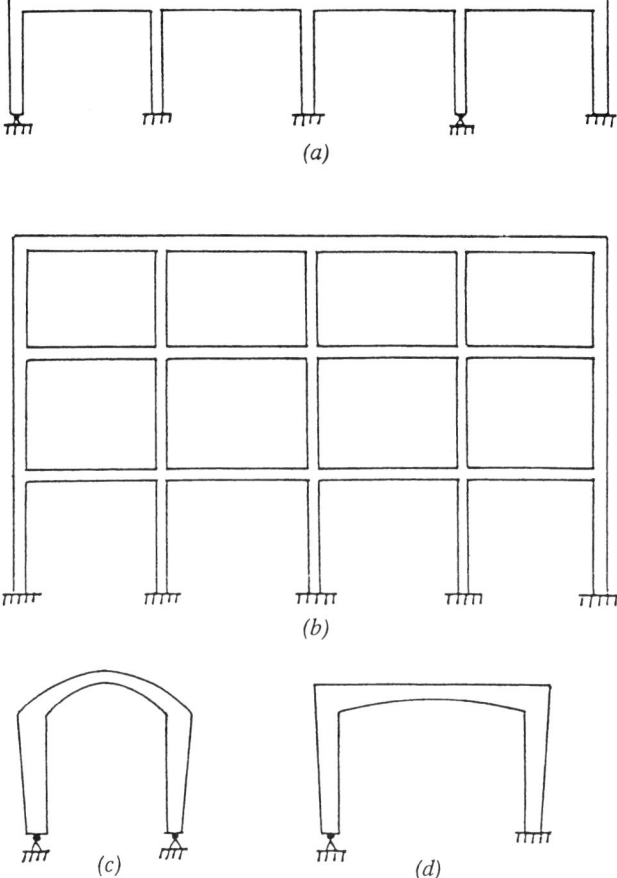

Figure 1.8. (*a*) Multiple-span continuous rectangular frame. (*b*) Multiple-story and multiple-span frame. (*c*) Arch frame with members of variable cross section. (*d*) Frame with members of variable cross section.

shown in Fig. 1.9*b*. They can be determined by applying the six static equations of equilibrium, namely,

$$\sum F_x = 0 \tag{1.10}$$

$$\sum F_y = 0 \tag{1.11}$$

$$\sum F_z = 0 \tag{1.12}$$

$$\sum M_x = 0 \tag{1.13}$$

$$\sum M_y = 0 \tag{1.14}$$

$$\sum M_z = 0 \tag{1.15}$$

16 INFRASTRUCTURAL FUNDAMENTALS

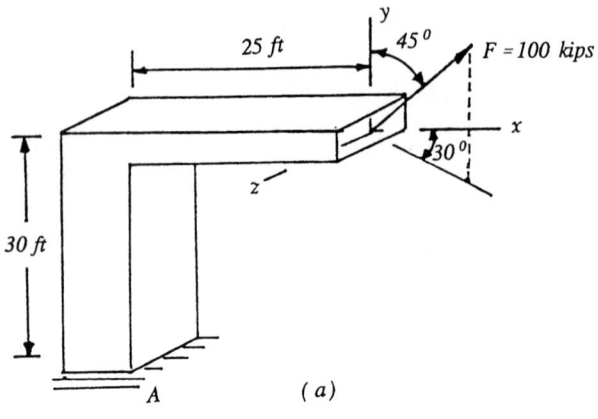

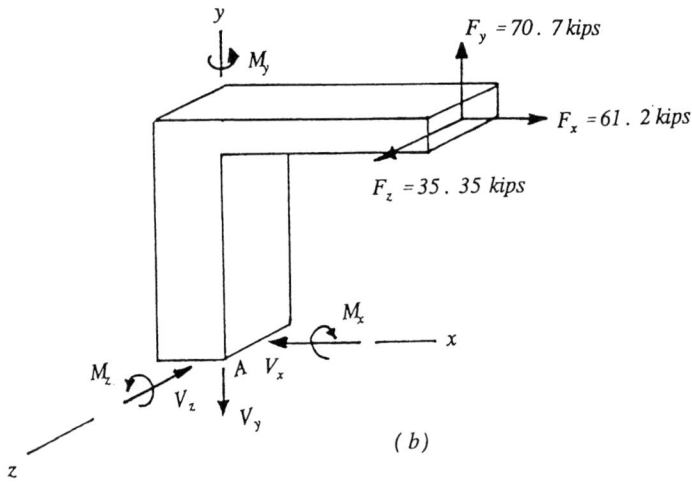

Figure 1.9. (*a*) Three-dimensional frame. (*b*) Free-body diagram of the three-dimensional frame.

1.3.3 Arches

An arch may be defined as a curved structure supported in a manner that permits it to develop horizontal supporting reactions acting toward the center of the arch span, as well as vertical reactions when vertical loads are applied to the arch. On this basis, the arch develops internal axial compressive and shearing forces, as well as moment to provide resistance to external loading. If the arch does not have the ability to create horizontal reactions at its end

STRUCTURAL ELEMENTS AND LOADINGS   17

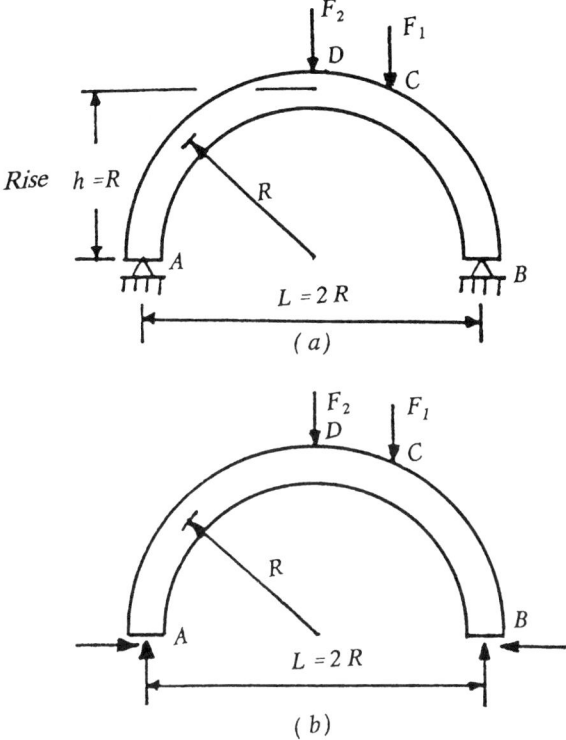

Figure 1.10. (a) Two-hinge arch. (b) Free-body diagram of the two-hinge arch.

supports when external vertical loading is applied, then the arch reduces to a curved beam. The two-hinge circular arch shown in Fig. 1.10a is acted upon by vertical forces F_1 and F_2, and develops the horizontal and vertical supporting reactions shown in Fig. 1.10b.

The applied loads on an arch may be concentrated, distributed, or a combination of both. The ideal shape for an arch is the one that has a rib axis exactly fitting the lines of action of the internal thrust and resultant supporting reactions. This ideal condition is rare in our modern arch structures, which are designed to resist heavy live loading. Thus, since the thrust line will often fall well outside the cross section of the arch, its cross-sectional dimensions should be designed to resist both bending and thrust.

Arch-type structures have been used extensively throughout the centuries. They provide a simple, practical, and artistic shape for the solution of many engineering problems. Common types of arches are shown in Fig. 1.11. The tied arch in Fig. 1.11d is primarily used where there is not sufficient foundation available to take the horizontal reaction component.

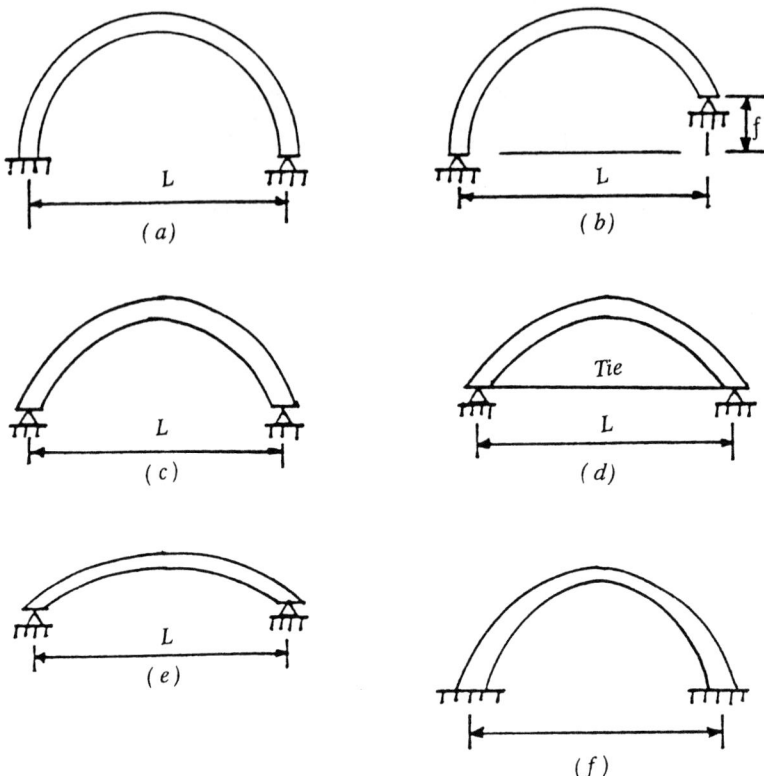

Figure 1.11. (*a*) Circular arch fixed at the one end and hinged at the other end. (*b*) Two-hinge arch with supports at different elevation. (*c*) Two-hinge parabolic arch. (*d*) Two-hinge arch with horizontal tie. (*e*) Two-hinge hollow arch. (*f*) Fixed-arch of variable cross section.

1.3.4 Plates and Shells

Plates and shells are used extensively in many engineering structures. Plates are used as concrete and reinforced concrete slabs subjected to lateral loading, and steel plates are used in ship hulls submitted to the action of water pressure, to name just a couple of examples. Domes, thin-walled tanks, and containers of various shapes subjected to external or internal pressure are often utilized. The design of boilers, locomotive engines, and steam turbines involves problems related to the bending of circular plates, or problems associated with spherical and conical shells.

A plate or shell can be subjected to bending, torsion, or axial tensile or compressive forces, and it can also fail by buckling. However, behavior of plates is very different from that of columns, and buckling behavior of shells is radically different from what is observed for columns and plates. When a column is loaded axially by a compressive load, its ability to resist axial load terminates as soon as buckling occurs. Thus, the critical load of a column is its failure load. The same would not be true for plates, because the buckling of a plate involves bending in two planes. When the critical load for a plate is reached, the plate will continue to resist increasing axial load and it will fail at a considerably larger load value, indicating that the critical load is not its failure load.

The main difference between a plate and a shell is that the plate is assumed to be initially flat, while the shell has a curvature in the unstressed stage. Therefore, the buckling behavior of axially compressed cylindrical shells, for example, would be radically different from the behavior exhibited by columns and plates. Imperfect cylindrical shells, to mention another example, will buckle at a stress that is considerably lower than the one obtained by linear theory.

Thin shells, such as domes, cylinders, or barrels, are extensively used in modern structures. Indoor stadiums, churches, and many public buildings are covered by domes. The design of these structures is very conservative, and they are much stronger even by the standards assumed by the most expert designer. Thin shells are three-dimensional structures requiring extremely complex rigorous mathematical analysis. Because of this complexity, simplifications are introduced in the analysis, or numerical methods, such as the finite element method, are used, which lead to a very conservative design.

The mathematical solution for thin plates started with Navier (1785–1836) in 1820, when he presented a memoir on the bending of plates to the Academy of Sciences. In 1821 his famous paper provided the equations of the mathematical theory of elasticity. Poisson (1781–1840), in his two memoirs in 1829 and 1831, obtained the two-dimensional solution for the lateral deflection of a loaded plate. In 1850, G. R. Kirchhoff (1824–1887) published his important paper on the theory of plates [8], in which we find the first satisfactory theory of bending of plates. He established the correct expression for the potential energy of a bend plate, and also applied his equations to the theory of vibration of circular plates. Sir Horace Lamb (1849–1934) took up the theory of plates and shells and investigated extentional vibrations of cylindrical and spherical shells, which were not considered in Rayleigh's [9, 10] approximate theory. The problem of shell vibrations, however, was first treated by Sophie Germaine (1776–1831) in about 1821 [11], where she assumed that the in-plane deflection of the neutral surface of a cylindrical shell was negligible. Some errors were contained in her results.

The above short history on plates and shells is only intended to provide some idea as to how the theory of plates and shells has been initiated, and the effort it took to solve the simplest problem on the subject.

1.3.5 Foundations and Footings

A foundation is the part of a structure that transmits the weight of the structure onto the natural ground. If we have a stratum of soil at a relatively shallow depth that is capable of sustaining the structure, then the structure is built directly on this soil and it is supported by a spread foundation. If the upper strata are weak, then the loads are transmitted to more suitable material at greater depth by means of piles or piers.

A single slab, known as a mat or raft foundation, may be used to cover the supporting stratum under the entire area of the superstructure. Alternately, spread footings may be used when various parts of the structure are supported individually. Individual footings may be used to support a group of columns, and a continuous footing may be used to support a wall.

The vertical distance between the base of the footing and the ground surface is known as the depth of the foundation. For building foundations, under normal circumstances, the minimum depth is usually governed by the requirement that the base of every part of the foundation should be located below the depth where the soil is subjected to seasonal volume changes caused by wetting and drying, and usually does not exceed four feet. For bridge foundations, the minimum depth for the foundation of a pier is determined by the level to which the river may scour during high water. When the water level of a river rises, the soil of the river bed moves throughout the greater part of the length and width of the river, and the bottom of the river moves downward, thus creating a condition known as scour. Scour does not always receive appropriate attention, and failure of bridge piers caused by scour is not uncommon, as briefly discussed in Section 1.2.

Since soil is compressible, every foundation resting on soil settles. The distribution of the settlement over the base area of the structure is extremely important, and it depends on the physical properties of the soil, the size of the area, the depth of the foundation, and the level of the water table. If the base becomes warped during the settlement, the structure can be damaged, but if it remains plane during settlement the structure will remain intact. Very conservative rules are usually used in foundation designs.

Rigid foundations are also often used to support mechanical systems directly on soil. Such mechanical systems usually involve rotating parts that produce critical dynamic excitations that are transmitted to the immediate or nearby environment. These excitations also produce effects that may be very damaging to the mechanical system itself if they are not controlled appropriately. A considerable amount of work on this subject was performed by the author and is discussed in Reference [5].

The material behavior of soil and soil-structure interaction are greatly affected when the forces entering the soil are dynamic. For example, if soils and structures are subjected to some earthquake activity, then some soils may compact, pore water pressure may increase, thus causing a loss of soil shear strength. This phenomenon is generally referred to as liquefaction. Gravel or clay soils are not susceptible to liquefaction. Dense sands are less likely to liquefy than loose sands, but hydraulically deposited sands are very vulnerable because of their uniformity. If it is determined that liquefication is likely to be a hazard, deep foundations or piling may be required in order to avoid unacceptable settlement or foundation failure during an earthquake.

The relationship between the periods of vibration of structures and the period of the supporting soil is very important when the seismic response of the structure is considered. During the 1970 earthquake at Gediz, Turkey, part of a factory was demolished in a town 135 kilometers from the epicenter, while no other buildings in the town were damaged. Studies revealed that the fundamental period of vibration of the factory was approximately equal to that of the underlying soil [12]. Further evidence of the importance of periods of vibration was obtained from the medium sized earthquake of Caracas in 1967, which completely destroyed four buildings and caused extensive damage to many others.

1.4 EQUILIBRIUM OF A PARTICLE

A particle in space may be acted upon by a system of two-dimensional coplanar concurrent forces, or by a system of three-dimensional concurrent forces. With respect to an x, y rectangular system of coordinate axes, the particle P in Fig. 1.12a will be subjected to a coplanar force system if forces F_1, F_2, and F_3 are all lying on the same x, y plane. The particle P will then be in equilibrium if the algebraic or vector summations of the components of all F_1, F_2, and F_3 forces in both x and y directions are in each case equal to zero. On this basis, we can write the following two general equilibrium equations for the particle:

$$\sum F_x = 0 \qquad (1.16)$$

$$\sum F_y = 0 \qquad (1.17)$$

With respect to an x, y, z system of rectangular coordinate axes, the particle P in Fig. 1.12b is subjected to a three-dimensional system of concurrent forces consisting of forces F_1, F_2, and F_3. One or more of these three forces must have components in all three axes of the coordinate system. In this case, equilibrium of the particle P can be maintained if the algebraic or vector sum of the components of F_1, F_2, and F_3, in the x, y, and z directions, is equal to zero in each case. Mathematically, we can write the following three general equations of equilibrium for the particle P:

$$\sum F_x = 0 \qquad (1.18)$$

$$\sum F_y = 0 \qquad (1.19)$$

$$\sum F_z = 0 \qquad (1.20)$$

Equations (1.16) and (1.17) are used extensively in the analysis of two-dimensional truss structures, particularly for the computation of the internal forces acting in the members of such trusses. The development of the well-known method of joints is based on the application of these two equation as

22 INFRASTRUCTURAL FUNDAMENTALS

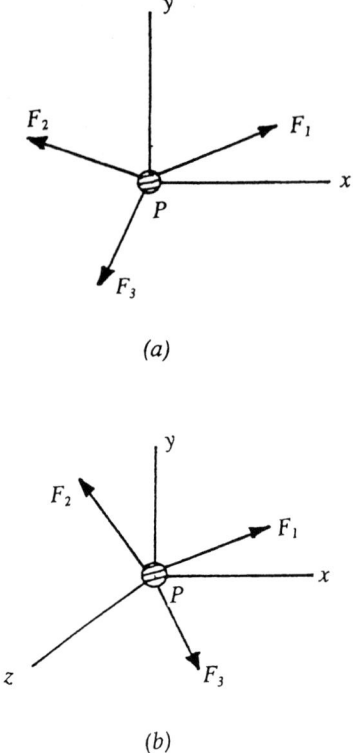

Figure 1.12. (*a*) Particle *P* subjected to a coplanar and concurrent system of forces. (*b*) Particle *P* subjected to a three-dimensional concurrent force system.

illustrated later in this section. Equations (1.18), (1.19), and (1.20) are extensively used for space trusses.

The following examples illustrate the application of Eqs. (1.16) through (1.20) to plane and space trusses by using the well-known method of joints.

Example 1.1: The cantilever plane truss in Fig. 1.13*a* is loaded at joint *C* by a vertical force of 20,000 lb, as shown. Determine the internal forces acting in each member of the truss by using the method of joints and Eqs. (1.16) and (1.17).

SOLUTION: The free-body diagrams of joints *C*, *D*, and *B* of the truss are shown in Figs. 1.13*b*, 1.13*c*, and 1.13*d*, respectively. For joint *C*, Eqs. (1.16) and (1.17) yield

$$F_{CD} - F_{CB} \cos 45° = 0 \qquad (1.21)$$

$$F_{CB} \cos 45° - 20{,}000 = 0 \qquad (1.22)$$

EQUILIBRIUM OF A PARTICLE   23

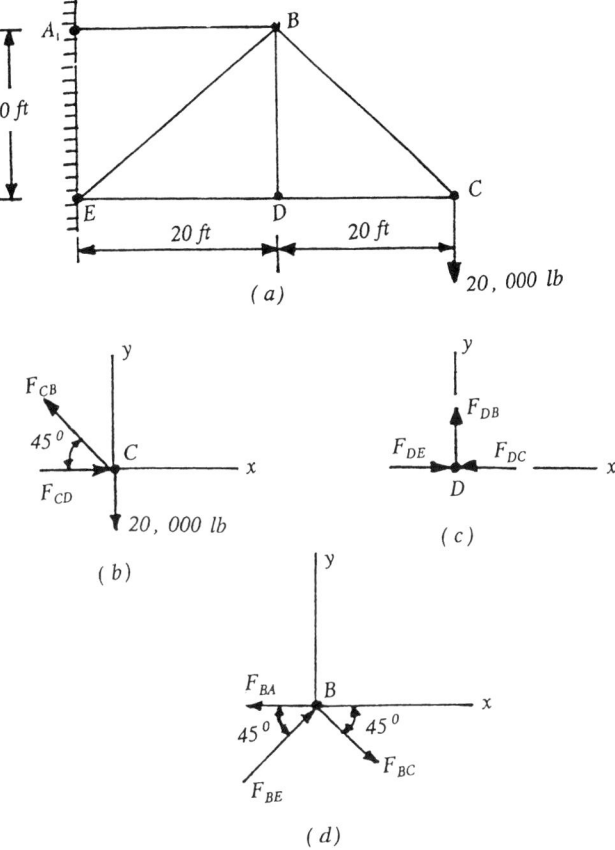

Figure 1.13. (*a*) Cantilever plane truss loaded as shown. (*b*) Free-body diagram of joint C. (*c*) Free-body diagram of joint D. (*d*) Free-body diagram of joint B.

Simultaneous solution of Eqs. (1.21) and (1.22) yields

$$F_{CB} = 28{,}300 \text{ lb} \quad \text{(tension)}$$
$$F_{CD} = 20{,}000 \text{ lb} \quad \text{(compression)}$$

In a similar manner, by using joints D and B, we obtain

$$F_{DB} = 0$$
$$F_{DE} = 20{,}000 \text{ lb} \quad \text{(compression)}$$
$$F_{BE} = 28{,}300 \text{ lb} \quad \text{(compression)}$$
$$F_{BA} = 40{,}000 \text{ lb} \quad \text{(tension)}$$

24 INFRASTRUCTURAL FUNDAMENTALS

It should be noted that the directions of the unknown forces in the free-body diagrams in Figs. 1.13b, 1.13c, and 1.13d, are chosen arbitrarily. When we apply Eqs. (1.16) and (1.17), the values of the forces that are chosen with the incorrect direction will be negative, indicating that their sense must be reversed in order to have equilibrium.

Example 1.2: For the three-dimensional truss loaded as shown in Fig. 1.14a, determine the internal forces in its members by using the method of joints and Eqs. (1.18) through (1.20). The plan view of the truss is shown in Fig. 1.14b.

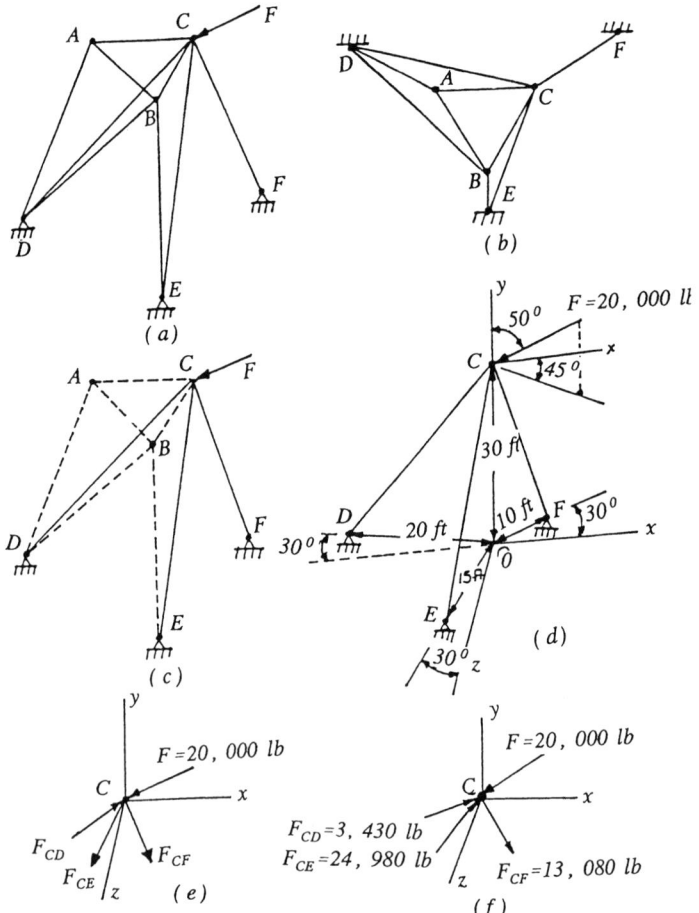

Figure 1.14. (a) Space truss loaded as shown. (b) Plan view of the space truss. (c) Space truss showing zero-force members. (d) Space truss showing the non-zero-force members only. (e) Assumed free-body diagram of joint C. (f) Free-body diagram of joint C showing correct direction and sense of force F_{CE}.

SOLUTION: By considering joint A of the space truss, we note that the internal forces in members AB, AC, and AD must be zero, because three concurrent forces can only be in equilibrium if they are coplanar. Since there is no external force acting at joint A and since AB, AC, and AD do not lie on the same plane, we have

$$F_{AB} = F_{AC} = F_{AD} = 0$$

The same reasoning may be used for joint B, because the internal forces F_{BC}, F_{BD}, and F_{BE} of members BC, BD, and BE, respectively, do not lie on the same plane. Also, the internal force F_{BA} in member BA is zero. Thus, the only members with active internal forces are members CD, CE, and CF, with forces designated as F_{CD}, F_{CE}, and F_{CF}, respectively. In Fig. 1.14c the zero-force members are represented by dashed lines, and a solid line is used to represent each non-zero-force member. The magnitude of the applied force F and the required dimensions of the truss are shown in Fig. 1.14d. The zero-force members are omitted in this figure for simplicity.

The free-body diagram of joint C is shown in Fig. 1.14e. The x, y, z components of F and the x, y, z components of members CD, CE, and CF are as follows:

$$F_x = F \cos 40° \cos 45° = 0.541 F$$

$$F_y = F \cos 50° = 0.643 F$$

$$F_z = F \cos 40° \cos 45° = 0.541 F$$

$$(CD)_x = 20 \cos 30° = 17.32 \text{ ft}$$

$$(CD)_y = 30 \text{ ft}$$

$$(CD)_z = 20 \cos 60° = 10 \text{ ft}$$

$$(CE)_x = 15 \cos 60° = 7.5 \text{ ft}$$

$$(CE)_y = 30 \text{ ft}$$

$$(CE)_z = 15 \cos 30° = 13 \text{ ft}$$

$$(CF)_x = 10 \cos 30° = 8.66 \text{ ft}$$

$$(CF)_y = 30 \text{ ft}$$

$$(CF)_z = 15 \cos 60° = 5 \text{ ft}$$

26 INFRASTRUCTURAL FUNDAMENTALS

By applying Eqs. (1.18) through (1.20), we obtain

$$\sum F_x = F_{CF}\frac{(CF)_x}{(CF)} - F_{CE}\frac{(CE)_x}{(CE)} + F_{CD}\frac{(CD)_x}{(CD)} - F_x = 0 \quad (1.23)$$

$$\sum F_y = -F_{CF}\frac{(CF)_y}{(CF)} - F_{CE}\frac{(CE)_y}{(CE)} + F_{CD}\frac{(CD)_y}{(CD)} - F_y = 0 \quad (1.24)$$

$$\sum F_z = -F_{CF}\frac{(CF)_z}{(CF)} + F_{CE}\frac{(CE)_z}{(CE)} + F_{CD}\frac{(CD)_z}{(CD)} + F_z = 0 \quad (1.25)$$

By substituting the appropriate values in Eqs. (1.23) through (1.25), we find

$$F_{CF} - 0.818F_{CE} + 1.755F_{CD} = 39{,}500 \quad (1.26)$$
$$-F_{CF} - 0.943F_{CE} + 0.875F_{CD} = 13{,}530 \quad (1.27)$$
$$-F_{CF} + 2.450F_{CE} + 1.753F_{CD} = -68{,}400 \quad (1.28)$$

By solving Eqs. (1.26) through (1.28) simultaneously, we obtain

$$F_{CE} = -24{,}980 \text{ lb}$$
$$F_{CD} = 3{,}430 \text{ lb}$$
$$F_{CF} = 13{,}080 \text{ lb}$$

The minus sign in F_{CE} indicates that its assumed direction in the free-body diagram in Fig. 1.14e is not correct, and it should act in the direction and sense shown in Fig. 1.14f. On this basis, the results are as follows:

$$F_{CE} = 24{,}980 \text{ lb} \quad \text{(compression)}$$
$$F_{CD} = 3{,}430 \text{ lb} \quad \text{(compression)}$$
$$F_{CF} = 13{,}080 \text{ lb} \quad \text{(tension)}$$

1.5 EQUILIBRIUM OF A RIGID BODY OR STRUCTURE

The equilibrium conditions of two-dimensional and three-dimensional rigid bodies or structures are briefly discussed here. When a two-dimensional structure or rigid body is subjected to coplanar force systems lying on the x, y plane of the rigid body or structure, equilibrium of the rigid body or structure is secured if the following static equilibrium equations are satisfied:

$$\sum F_x = 0 \quad (1.29)$$
$$\sum F_y = 0 \quad (1.30)$$
$$\sum M_0 = 0 \quad (1.31)$$

All applied and supporting resisting forces and moments must be considered in the application of Eqs. (1.29) through (1.31).

Alternate sets of equations that can establish equilibrium of the rigid body or structure are as follows:

$$\sum F_x = 0 \quad \sum M_A = 0 \quad \sum M_B = 0 \quad (1.32)$$

$$\sum M_A = 0 \quad \sum M_B = 0 \quad \sum M_C = 0 \quad (1.33)$$

The first two conditions in Eq. (1.32) suggest that the resultant of all resisting and applied forces must pass through point A and it must be zero, because the third condition in the same equation requires that its moment about point B should be zero. In this manner, equilibrium of the rigid body or structure is maintained.

In Eq. (1.33), the first condition requires the resultant of all applied and resisting forces to pass through point A, the second one requires this resultant to pass through point B, and the third one suggests that it should pass through point C. If points A, B, and C are not located in a straight line, then the resultant must be zero and equilibrium of the rigid body or structure is maintained. The requirement therefore is that points A, B, and C must not be on the same straight line in order to secure equilibrium of the rigid body or structure.

With reference to an x, y, z system of coordinate axes, a rigid body or structure supported in space will be in equilibrium if the noncoplanar system of the external and resisting forces and moments does not produce motion along the x, y, or z directional system of axes or rotation about any of these three axes. On this basis, the following six equations of equilibrium must be satisfied:

$$\sum F_x = 0 \quad (1.34)$$

$$\sum F_y = 0 \quad (1.35)$$

$$\sum F_z = 0 \quad (1.36)$$

$$\sum M_x = 0 \quad (1.37)$$

$$\sum M_y = 0 \quad (1.38)$$

$$\sum M_z = 0 \quad (1.39)$$

When an applied noncoplanar force system is acting on a rigid body or structure, a maximum of six unknown reactive or supporting forces and moments may be determined by applying Eqs. (1.34) through (1.39).

The following examples illustrate the application of the above theory and equilibrium equations.

28 INFRASTRUCTURAL FUNDAMENTALS

Example 1.3: The three-hinge arch shown in Fig. 1.15a is loaded by two vertical forces as shown. Determine all reactions at supports A and B by applying appropriate equations of equilibrium.

SOLUTION: The free-body diagram of the three-hinge arch is shown in Fig. 1.15b. By applying Eqs. (1.29) through (1.31) we find

$$\sum F_x = H_A - H_B = 0 \qquad (1.40)$$

$$\sum F_y = V_A - 5 - 20 + V_B = 0$$

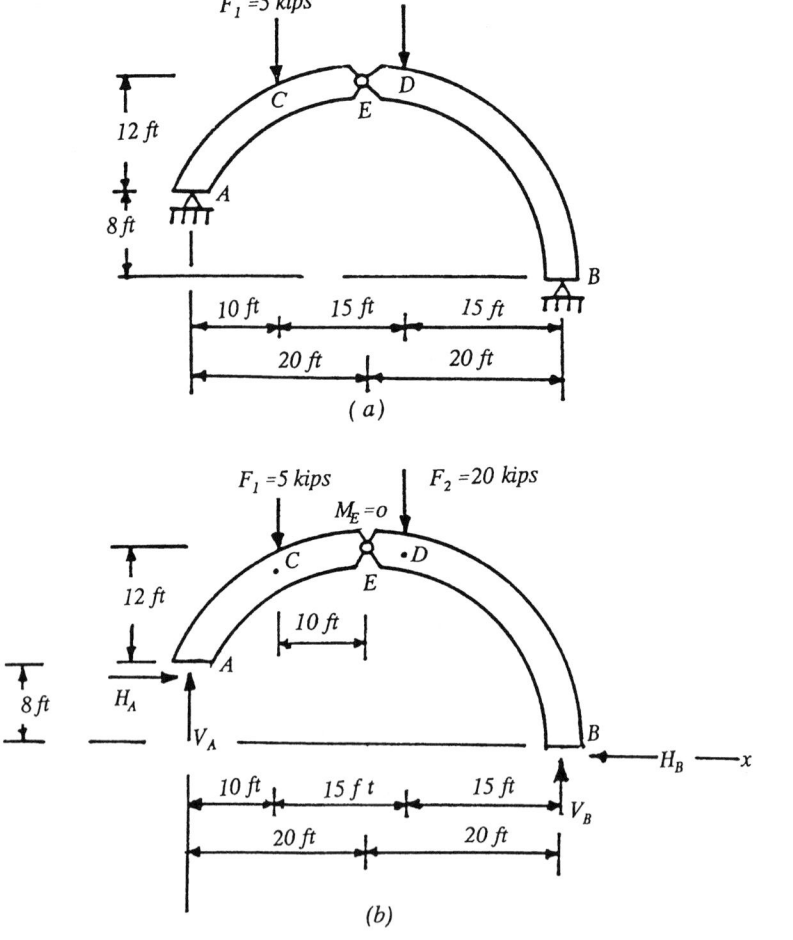

Figure 1.15. (a) Three-hinge arch loaded as shown. (b) Free-body diagram of the three-hinge arch

or

$$V_A + V_B = 25 \tag{1.41}$$

$$\sum M_B = -V_A(40) - H_A(8) + (5)(30) + (20)(15) = 0$$

or

$$40V_A + 8H_A = 450 \tag{1.42}$$

We note here that we have only three equations of statics and they are not sufficient to solve for the four unknown supporting reactions V_A, V_B, H_A, and H_B. An additional independent equation, however, may be obtained by realizing that the moment at the internal hinge E of the arch should be zero, because an internal hinge cannot resist bending moment. On this basis, the sum of the moments about point E of all the forces to the left of E, or the sum of the moments about E of all the forces to the right of E, must be zero. By considering the sum $\sum M_E$ of all the forces to the left of E, we have

$$\sum M_E = H_A(12) - V_A(20) + (5)(10) = 0$$

or

$$12H_A - 20V_A = -50 \tag{1.43}$$

By solving Eqs. (1.40) through (1.43) simultaneously, we find

$$H_A = 10.94 \text{ kips} \rightarrow$$
$$V_A = 9.06 \text{ kips} \uparrow$$
$$H_B = 10.94 \text{ kips} \leftarrow$$
$$V_B = 15.94 \text{ kips} \uparrow$$

Example 1.4: A rigid slab is supported by six bars, as shown in Fig. 1.16a, and is loaded by a vertical force $F_1 = 20$ kips and by a horizontal force $F_2 = 10$ kips, located as shown in the same figure. Determine the resisting forces acting along the centroidal axes of the six supporting bars.

SOLUTION: The free-body diagram of the slab is shown in Fig. 1.16b. The sense of the internal force in each of the six bars is arbitrarily chosen. By applying Eqs. (1.34) through (1.39), we find

$$\sum F_x = H_C + H_E = 0 \tag{1.44}$$

$$\sum F_y = V_A - V_C - V_G - 20 = 0 \tag{1.45}$$

$$\sum F_z = H_B = 0 \tag{1.46}$$

$$\sum M_x = (20)(7) - V_A(10) - H_B(0.50) - V_C(10) + V_G(10) = 0$$

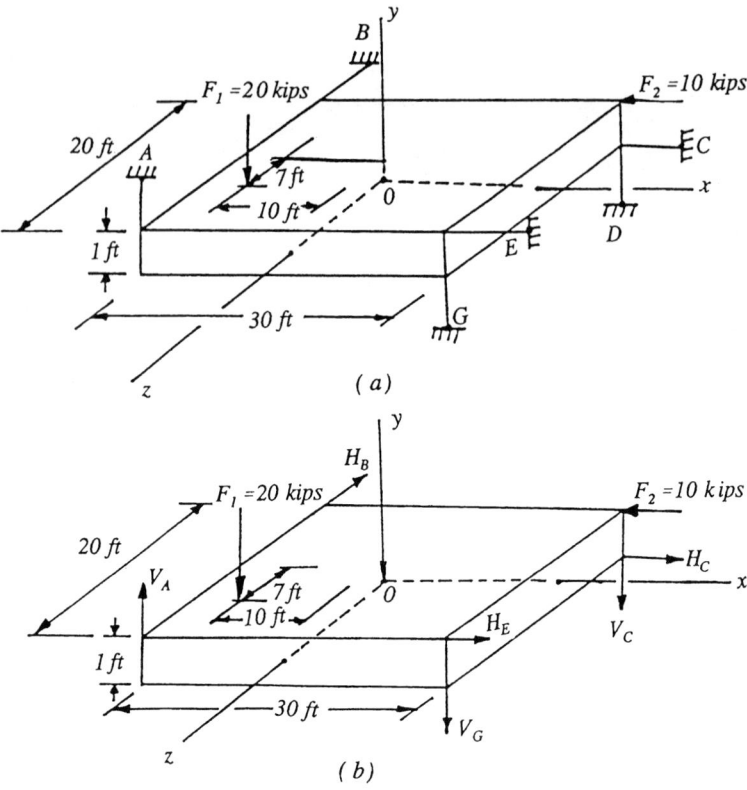

Figure 1.16. (a) Rigid slab supported and loaded as shown. (b) Free-body diagram of the slab.

or

$$-10V_A - 0.50H_B - 10V_C + 10V_G = -140 \qquad (1.47)$$

$$\sum M_y = -H_B(15) + (10)(10) - H_C(10) + H_E(10) = 0$$

or

$$-15H_B - 10H_C + 10H_E = -100 \qquad (1.48)$$

$$\sum M_z = (20)(10) - V_A(15) + (10)(0.50) + H_C(0.50) - V_C(15)$$
$$- H_E(0.50) - V_G(15) = 0$$

or

$$-15V_A + 0.50H_C - 15V_C - 0.50H_E - 15V_G = -205 \qquad (1.49)$$

Simultaneous solution of Eqs. (1.44) through (1.49) yields

$$V_A = 17 \text{ kips}$$
$$V_C = -3 \text{ kips}$$
$$V_G = 0$$
$$H_B = 0$$
$$H_C = 10 \text{ kips}$$
$$H_E = 0$$

The negative sign in V_C indicates that its assumed sense in the free-body diagram in Fig. 1.16b is not correct and it must be reversed. Thus, we have

$$V_A = 17 \text{ kips} \uparrow$$
$$V_C = 3 \text{ kips} \uparrow$$
$$V_G = 0$$
$$H_B = 0$$
$$H_C = 10 \text{ kips} \rightarrow$$
$$H_E = 0$$

1.6 MOMENT, SHEAR, AND AXIAL FORCE EQUATIONS AND DIAGRAMS

Internal forces usually vary along the length of a member, and it is often convenient to express such variations in equation forms that cover the entire length of the member. Since the applied loading may not be continuous over the whole length of the member and concentrated loads and moments may also be present, it is often required to use more than one equation in order to be able to cover the whole length of the member or span. For a beam or beam span that is loaded with a coplanar force system lying in its vertical x, y plane passing through the centroidal axis of the member, the internal forces are usually an axial force, a shear, and a bending moment. If their variations along the length of the member are plotted, we have what are known as axial force, shear force, and bending moment diagrams. Such diagrams are extremely useful to the practicing design engineer, because they readily help to establish maximum conditions, thus letting the engineer design the member so it conforms with allowable levels of stress.

The following examples illustrate how equations for internal forces can be derived, and also provide a brief illustration as to how shear force, axial force, and bending moment diagrams may be developed and plotted.

Example 1.5: Determine the bending moment and shear force equations for the beam shown in Fig. 1.17a, which is loaded as shown in the same figure. Also plot the shear force and bending moment diagrams for the beam.

SOLUTION: The free-body diagram of the beam is shown in Fig. 1.17b. By

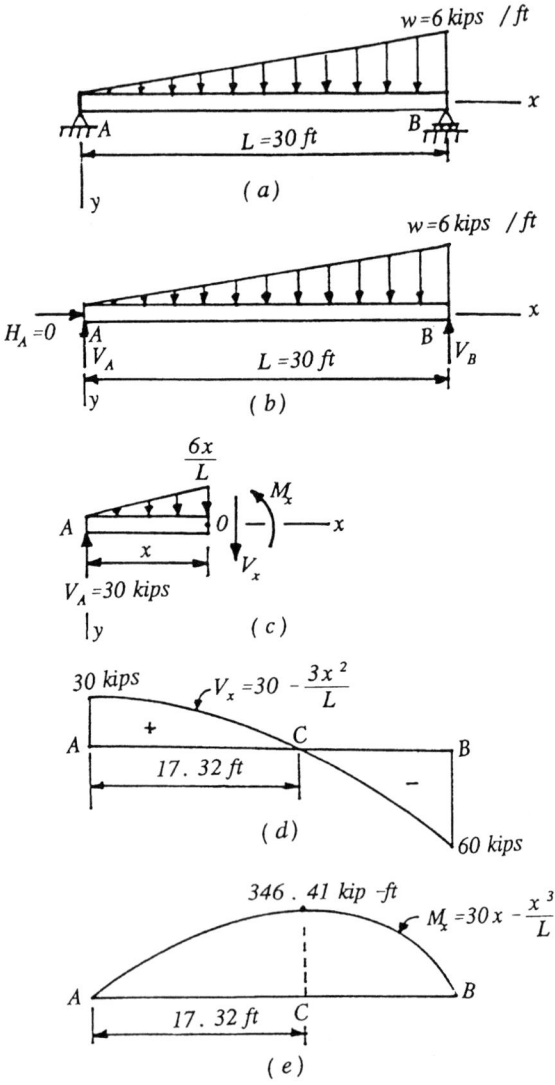

Figure 1.17. (a) Simply supported beam loaded as shown. (b) Free-body diagram of the simply supported beam. (c) Free-body diagram of a portion x of the beam. (d) Shear force diagram. (e) Bending moment diagram.

applying the static equilibrium equations given by Eqs. (1.29) through (1.31), we find

$$V_A = 30 \text{ kips}$$
$$V_B = 60 \text{ kips}$$

The shear force and bending moment equations at any distance x from support A may be obtained by using the free-body diagram in Fig. 1.17c. By using this free-body diagram and taking moments about point 0, we find

$$\sum M_0 = -(30)x + \left(\frac{6x}{L}\right)\left(\frac{x}{2}\right)\left(\frac{x}{3}\right) + M_x = 0$$

or

$$M_x = 30x - \frac{x^3}{L} \qquad (1.50)$$

Also, equilibrium in the vertical direction yields

$$\sum F_y = -30 + \left(\frac{6x}{L}\right)\left(\frac{x}{2}\right) + V_x = 0$$

or

$$V_x = 30 - \frac{3x^2}{L} \qquad (1.51)$$

Equations (1.50) and (1.51) are the bending moment and shear force expressions, respectively, of the beam. The shear force and bending moment diagrams may be plotted from these two equations for various values of $0 \leqslant x \leqslant 30$ ft, and they are shown in Figs. 1.17d and 1.17e, respectively.

The maximum bending moment occurs at the value of x that makes the shear force V_x equal to zero. Thus, from Eq. (1.51), we have

$$30 - \frac{3x^2}{30} = 0,$$

which yields $x = 17.32$ ft. For this value of x, Eq. (1.50) yields

$$M_{\max} = (30)(17.32) - \frac{(17.32)^3}{30}$$

$$= 346.41 \text{ kip-ft}$$

34 INFRASTRUCTURAL FUNDAMENTALS

Example 1.6: For the three hinge rigid frame loaded as shown in Fig. 1.18a, determine the shear force and bending moment equations of its members. Plot the shear force and bending moment diagrams of the frame.

SOLUTION: The free-body diagram of the three hinge frame is shown in Fig. 1.18b. The four reactions at the supports A and B may be determined by using the three equilibrium equations from statics and the additional independent condition that the bending moment at the hinge C must be zero. On this basis, we have

$$\sum M_B = -(40)V_A - (20)(35) + (3)(40)(20) + (10)(15) = 0$$

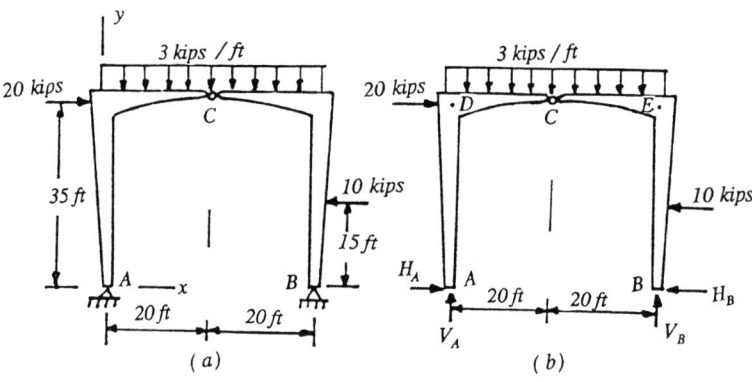

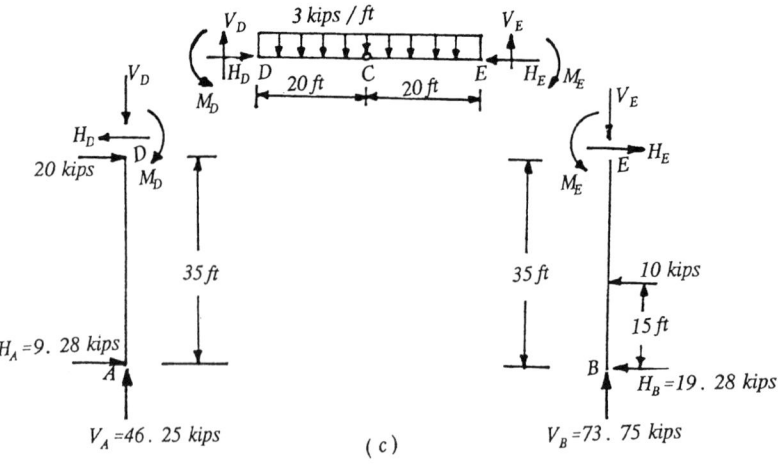

Figure 1.18. (a) Three-hinge frame loaded as shown. (b) Free-body diagram of the three-hinge frame. (c) Free-body diagrams of the members of the frame.

or

$$V_A = 46.25 \text{ kips} \uparrow$$

By considering only the forces to the left of C, we have

$$\sum M_C = -(20)V_A + (35)H_A + (3)(20)(10) = 0$$

or

$$H_A = 9.28 \text{ kips} \rightarrow$$

We also have

$$\sum F_x = 20 + H_A - 10 - H_B = 0$$

or

$$H_B = 18.28 \text{ kips} \leftarrow$$
$$\sum F_y = (3)(40) - V_A - V_B = 0$$

or

$$V_B = 73.75 \text{ kips} \uparrow$$

The free-body diagrams of the individual members of the frame are shown in Fig. 1.18c. The centroidal axes of these members are assumed to be straight, because any deviations from this assumption caused by the fact that the members are of variable cross section are small compared to their longitudinal dimensions. This is a commonly used assumption in practice.

By using the free-body diagram of member AD and applying the three static equations of equilibrium, we find that the internal forces V_D, H_D, and bending moment M_D at D, are as follows:

$$V_D = 36.25 \text{ kips} \downarrow$$
$$H_D = 29.28 \text{ kips} \leftarrow$$
$$M_D = 324.50 \text{ kip-ft} \quad \text{(clockwise)}$$

In a similar manner, by considering member BE, the internal forces V_E, H_E, and bending moment M_E are determined. They are

$$V_E = 73.75 \text{ kips} \downarrow$$
$$H_E = 29.28 \text{ kips} \rightarrow$$
$$M_E = 874.00 \text{ kip-ft} \quad \text{(counterclockwise)}$$

The internal forces and bending moments at points D and E of member DE are, respectively, equal and opposite to the ones determined at D and E for members AD and BE. Since all internal forces at the ends of each member in Fig. 1.18c are determined, the equations for the internal shear force and bending moment of each component member of the frame may be determined in a manner similar to the one used in Example 1.5.

By considering member AD in Fig. 1.19a, a free-body diagram for a portion of the member may be drawn as shown in Fig. 1.19b. By applying to this portion the static equations of equilibrium, we find

$$V_y = 46.25 \text{ kips} \quad \text{(constant)}$$
$$H_y = 9.28 \text{ kips} \quad \text{(constant)}$$
$$M_y = 9.28y$$

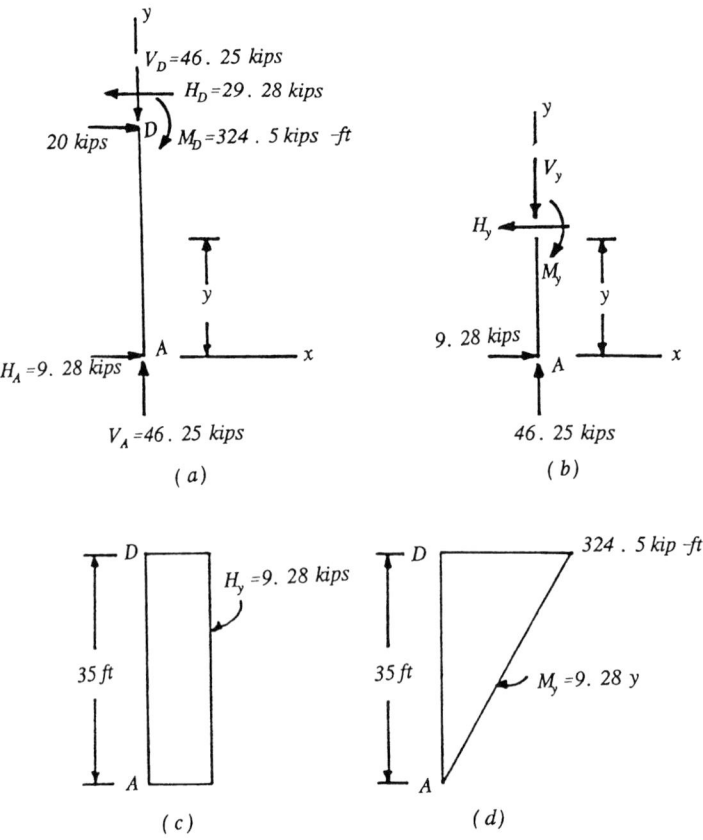

Figure 1.19. (a) Free-body diagram of member AD. (b) Free-body diagram of a portion of member AD. (c) Shear force diagram of member AD. (d) Bending moment diagram of member AD.

MOMENT, SHEAR, AND AXIAL FORCE EQUATIONS AND DIAGRAMS 37

The shear force and bending moment diagrams of member AD are shown in Figs. 1.19c and 1.19d, respectively.

A similar procedure may be followed for member BE of the frame, which is shown in Fig. 1.20a. In this case, however, separate expressions for the portions BG and GE of the member are required. By using the free-body diagram of the

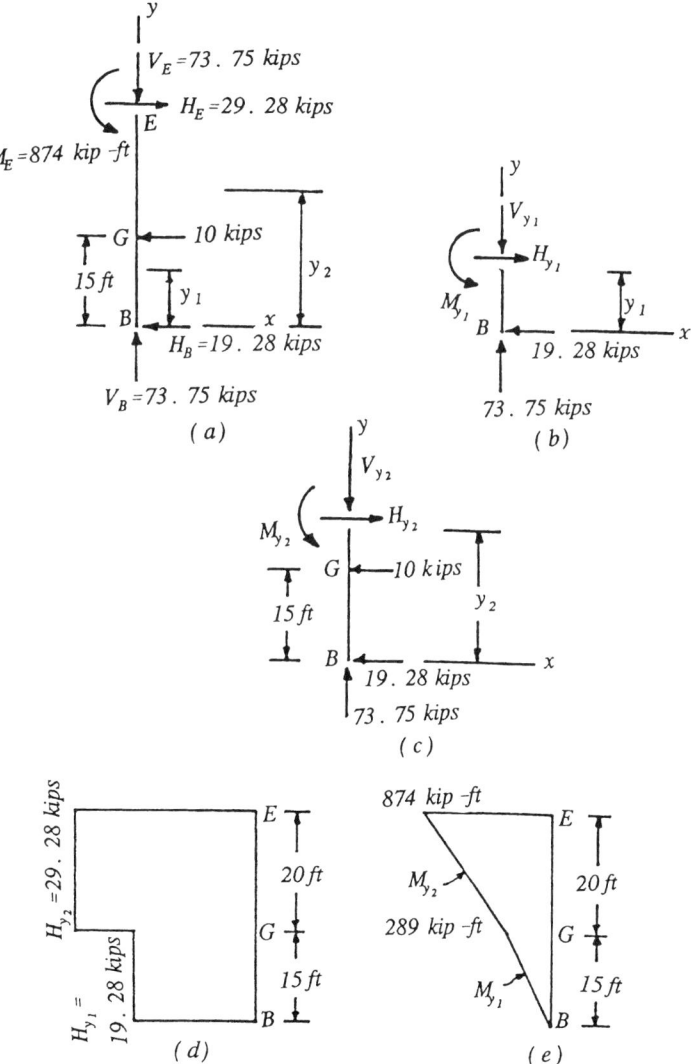

Figure 1.20. (a) Free-body diagram of member BE of the frame. (b) Free-body diagram of a portion for part BG of the member. (c) Free-body diagram of a portion for part GE of the member. (d) Shear force diagram for member BE. (e) Bending moment diagram for member BE.

38 INFRASTRUCTURAL FUNDAMENTALS

portion in Fig. 1.20b and applying the three equations of equilibrium, we find the following expressions for portion BG:

$$V_{y_1} = 73.75 \text{ kips} \quad \text{(constant)} \quad 0 \leqslant y_1 \leqslant 15 \text{ ft}$$

$$H_{y_1} = 19.28 \text{ kips} \quad \text{(constant)} \quad 0 \leqslant y_1 \leqslant 15 \text{ ft}$$

$$M_{y_1} = 19.28 y_1 \quad 0 \leqslant y_1 \leqslant 15 \text{ ft}$$

In a similar manner for portion GE, by using the free-body diagram in Fig. 1.20c we find

$$V_{y_2} = 73.75 \text{ kips} \quad \text{(constant)} \quad 15 \leqslant y_2 \leqslant 35 \text{ ft}$$

$$H_{y_2} = 29.28 \text{ kips} \quad \text{(constant)} \quad 15 \leqslant y_2 \leqslant 35 \text{ ft}$$

$$M_{y_2} = 19.28 y_2 + 10(y_2 - 15) \quad 15 \leqslant y_2 \leqslant 35 \text{ ft}$$

The shear force and bending moment diagrams of member BE are shown plotted in Figs. 1.20d and 1.20e, respectively.

For member DE in Fig. 1.21a, the expressions for the internal shear force V_x and internal bending moment M_x may be determined by using the free-body diagram in Fig. 1.21b and applying the static equations of equilibrium as in the preceding member cases. The expression for the internal forces V_x, H_x, and internal moment M_x are as follows:

$$V_x = 46.25 - 3x \quad 0 \leqslant x \leqslant 40 \text{ ft}$$

$$M_x = \frac{3x^2}{2} - 46.25x + 324.50 \quad 0 \leqslant x \leqslant 40 \text{ ft}$$

$$H_x = 29.28 \text{ kips} \quad \text{(constant)}$$

The shear force and bending moment diagrams for member DE are shown plotted in Figs. 1.21c and 1.21d, respectively.

The shear force and bending moment diagrams of the whole frame are shown plotted in Figs. 1.22a and 1.22b, respectively.

1.7 RELATIONS BETWEEN LOAD, SHEAR, AND BENDING MOMENT

The construction of shear and bending moment diagrams may be facilitated by establishing relations between applied load and shear force, and between shear force and bending moment. Consider, for example, the arbitrarily loaded simply supported beam in Fig. 1.23a. The free-body diagram of a portion of the beam of length dx is shown in Fig. 1.23b. By applying the static equilibrium

RELATIONS BETWEEN LOAD, SHEAR, AND BENDING MOMENT   39

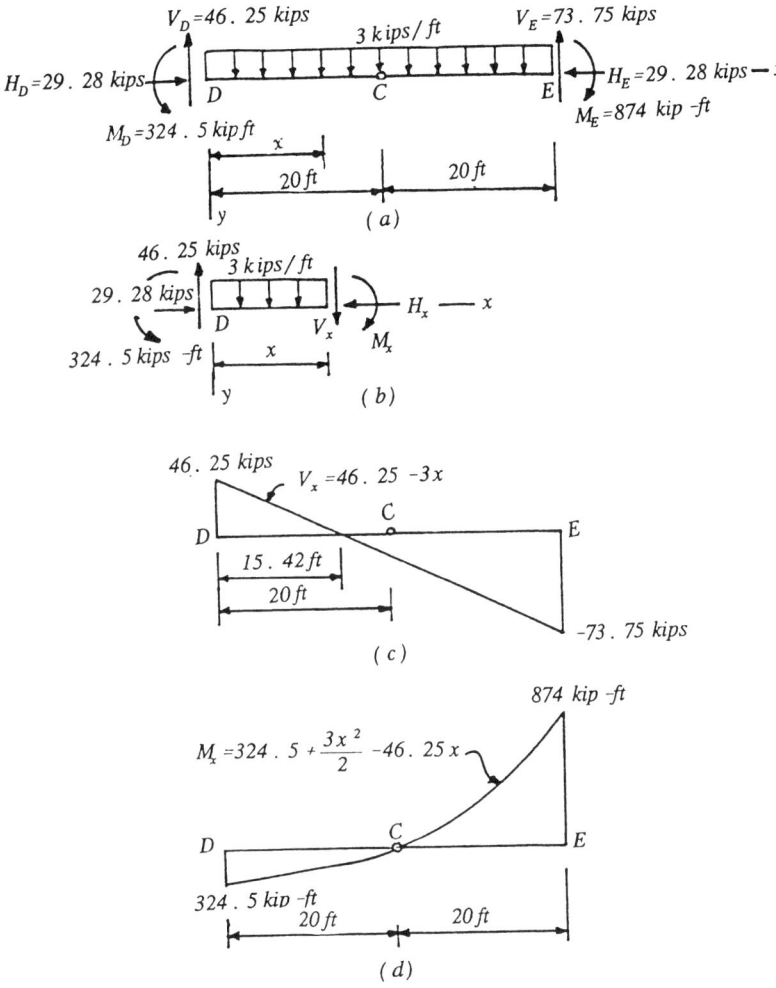

Figure 1.21. (a) Free-body diagram of member DE of the frame. (b) Free-body diagram of a portion of member DE. (c) Shear force diagram of member DE. (d) Bending moment diagram of member DE.

equation in the vertical direction, we have

$$\sum F_y = -V + w\,dx + (V + dV) = 0$$

or

$$w = -\frac{dV}{dx} \tag{1.52}$$

40 INFRASTRUCTURAL FUNDAMENTALS

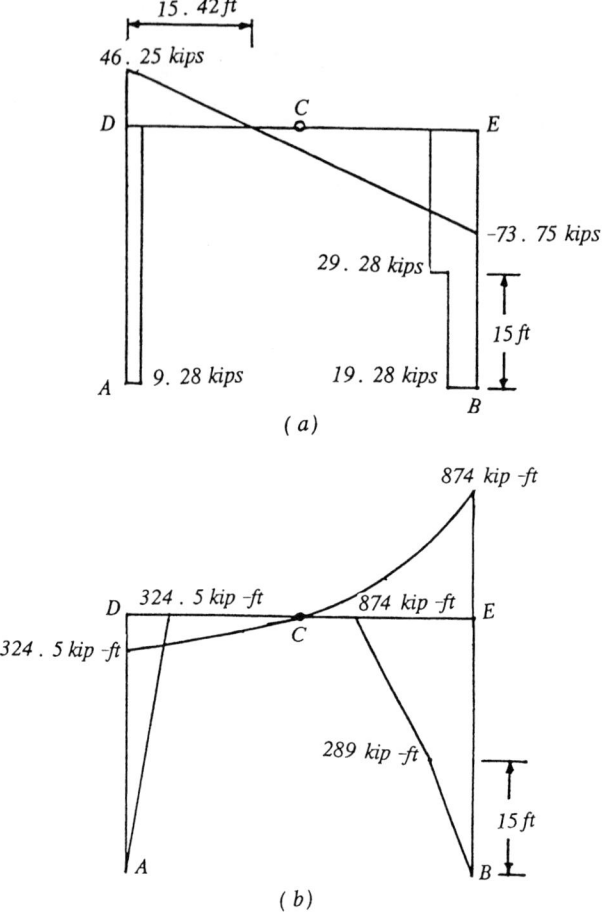

Figure 1.22. (a) Shear force diagram of the frame. (b) Bending moment diagram of the frame.

Equation (1.52) states that the negative rate of change of the shear force is equal to the external loading acting on the member. Thus, Eq. (1.52) establishes a relationship between shear force and applied loading.

We rewrite Eq. (1.52) as follows:

$$-dV = w\, dx \tag{1.53}$$

If Eq. (1.53) is integrated between sections x_1 and x_2 of the member in Fig. 1.23a, we have

$$-\int_{V_1}^{V_2} dV = \int_{x_1}^{x_2} w\, dx$$

RELATIONS BETWEEN LOAD, SHEAR, AND BENDING MOMENT   41

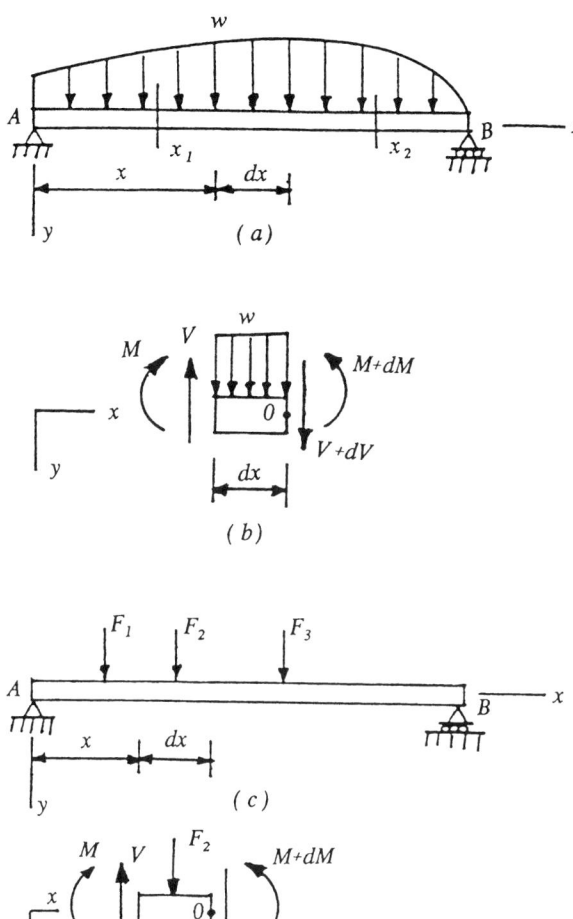

Figure 1.23. (a) Simply supported beam loaded by an arbitrarily distributed load w. (b) Free-body diagram of a portion of the member of length dx. (c) Simply supported beam loaded with vertical concentrated loads. (d) Free-body diagram containing force F_2.

or

$$V_2 - V_1 = -\int_{x_1}^{x_2} w\, dx \qquad (1.54)$$

Equation (1.54) indicates that the change $(V_2 - V_1)$ in shear between sections at x_1 and x_2 is equal to the negative area under the load diagram between the two sections.

Returning to the free-body diagram in Fig. 1.23b and taking moments about point 0, we have

$$\sum M_0 = -M - V\,dx + w\,dx\left(\frac{dx}{2}\right) + (M + dM) = 0 \qquad (1.55)$$

By neglecting the higher order term $w(dx)^2/2$ because it is very small compared to the other terms, the following result may be obtained from Eq. (1.55):

$$V = \frac{dM}{dx} \qquad (1.56)$$

Equation (1.56) shows that the shear force is equal to the rate of change (or slope) of the moment. Thus, at sections where the bending moment is maximum, the shear force would be zero.

We rewrite Eq. (1.56) as follows:

$$dM = V\,dx \qquad (1.57)$$

By integrating Eq. (1.57) between the limits defined by sections at x_1 and x_2 of the member, we find

$$\int_{M_1}^{M_2} dM = \int_{x_1}^{x_2} V\,dx$$

or

$$M_2 - M_1 = \int_{x_1}^{x_2} V\,dx \qquad (1.58)$$

Equation (1.58) shows that the change in bending moment between sections at x_1 and x_2 is equal to the area under the shear force diagram between the two sections.

Equations (1.54) and (1.58) are extensively used for the construction of shear force and bending moment diagrams, as illustrated in the examples of this section.

Consider now the case where a beam is loaded by concentrated loads, as shown in Fig. 1.23c. In Fig. 1.23d, the free-body diagram of an element of length dx that contains the concentrated force F_2 is shown. Summing up all the forces in the vertical direction and applying equilibrium, we have

$$V - F_2 - V_1 = 0$$

RELATIONS BETWEEN LOAD, SHEAR, AND BENDING MOMENT   43

or
$$V_1 = V - F_2 \tag{1.59}$$

Equation (1.59) indicates that there is an abrupt change in the shear force over the length dx if a concentrated force F_2 is acting as shown in the free-body diagram in Fig. 1.23d. The following examples illustrate the application of the above theory.

Example 1.7: The bending moment diagram in a portion of a member is as shown in Fig. 1.24a. If the shear force at A is -40 kips, draw the shear force and load diagrams between A and C. The variation of the bending moment between A and B is linear, and between B and C it is parabolic.

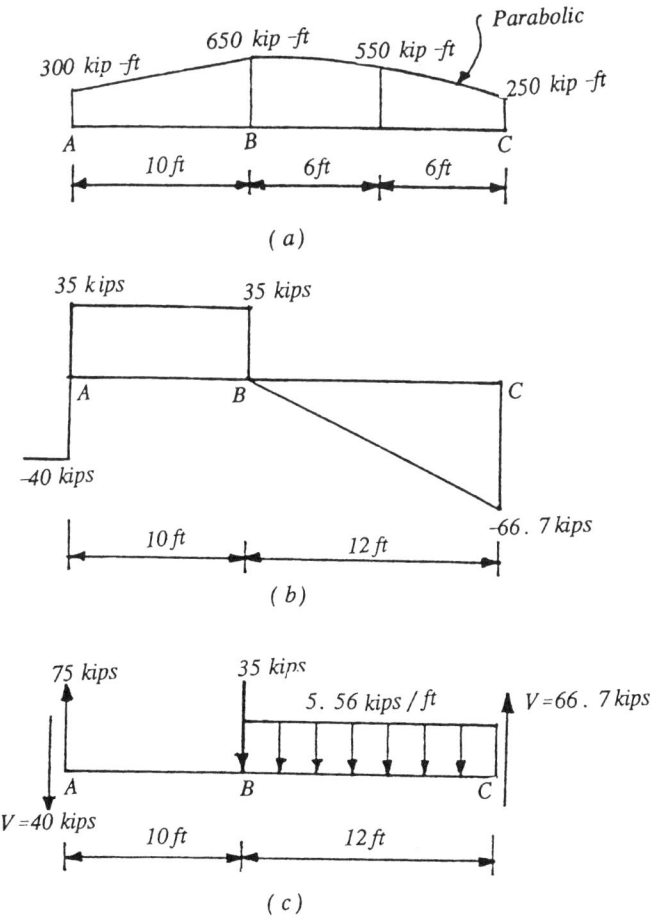

Figure 1.24. (a) Bending moment diagram. (b) Shear force diagram. (c) Loading diagram.

SOLUTION: The parabolic variation of the bending moment M_x between B and C may be expressed as

$$M_x = m_1 + m_2 x + m_3 x^2 \qquad (1.60)$$

where m_1, m_2, and m_3 are constants that can be determined by applying appropriate boundary conditions.

By considering the portion between points A and B, we note that the slope of the moment diagram in this region is constant. Consequently, the shear force is constant within this region. Thus, we have

$$\begin{aligned} V_{AB} &= \frac{\Delta M}{\Delta x} \\ &= \frac{650 - 300}{10} = 35 \text{ kips} \end{aligned} \qquad (1.61)$$

Since the boundary condition at A requires that the shear force should be equal to -40 kips, it is evident that an abrupt change in shear, or discontinuity, exists at point A. An upward load of 75 kips should act at A, so that the shear V at a section a little to the right of A would have the value

$$V = 75 - 40 = 35 \text{ kips}$$

in order to be able to satisfy Eq. (1.61).

We consider now portion BC with B as the origin of the coordinate system of axes. At $x = 0$ the moment $M(0) = M_B = 650$ kip-ft. Thus, for $x = 0$, Eq. (1.60) yields $m_1 = 650$ kip-ft. At $x = 6$ ft and $x = 12$ ft, the bending moments are 550 kip-ft and 250 kip-ft, respectively. On this basis, Eq. (1.60) yields

$$550 = 650 + 6m_2 + 36m_3 \qquad (1.62)$$

$$250 = 650 + 12m_2 + 144m_3 \qquad (1.63)$$

Simultaneous solution of Eqs. (1.62) and (1.63) yields

$$m_2 = 0 \qquad (1.64)$$

$$m_3 = -2.78 \qquad (1.65)$$

By substituting the values of m_1, m_2, and m_3 into Eq. (1.60), we find

$$M_x = 650 - 2.78 x^2 \qquad (1.66)$$

Equation (1.66) expresses the variation of the bending moment between points B and C.

RELATIONS BETWEEN LOAD, SHEAR, AND BENDING MOMENT   45

The shear force variation V_x between points B and C may be determined by using Eq. (1.56); that is

$$V_x = \frac{dM}{dx} = -5.56x \qquad (1.67)$$

The section at B that would satisfy the shear force expression given by Eq. (1.67) should be taken at a small distance ahead of B, because just to the left of B the value of the shear force is 35 kips. Thus, a discontinuity exists at B, introduced by the application of a downward concentrated load of 35 kips acting at point B.

Since the shear force is constant between points A and B, Eq. (1.52) suggests that the load w is zero within this region. Between points B and C, the expression for the load w may be determined by using Eqs. (1.52) and (1.67). That is,

$$w = -\frac{dV_x}{dx} = 5.56 \text{ kips/ft} \quad \text{(constant)} \qquad (1.68)$$

The shear force and load diagrams are shown plotted in Figs. 1.24b and 1.24c, respectively.

Example 1.8: For the articulated beam loaded as shown in Fig. 1.25a, determine the shear force and bending moment diagrams by using Eqs. (1.54) and (1.58), respectively.

SOLUTION: The free-body diagram of the member is shown in Fig. 1.25b. The supporting reactions and supporting moment are H_A, V_A, M_A, V_B, and V_C. These five unknown resistances may be determined by using the three static equations of equilibrium, and the two additional independent conditions, which suggest that the bending moments at the internal hinges located at points D and E of the member must be zero. The results are

$$H_A = 0$$
$$V_A = 5 \text{ kips} \downarrow$$
$$M_A = 300 \text{ kip-ft} \quad \text{(clockwise)}$$
$$V_B = 70 \text{ kips} \uparrow$$
$$V_C = 15 \text{ kips} \downarrow$$

According to Eq. (1.54), the change in shear between sections located at x_1 and x_2 is equal to the negative area under the load diagram between the two sections. With this in mind and knowing that $V_A = -5$ kips, the ordinate V_D of

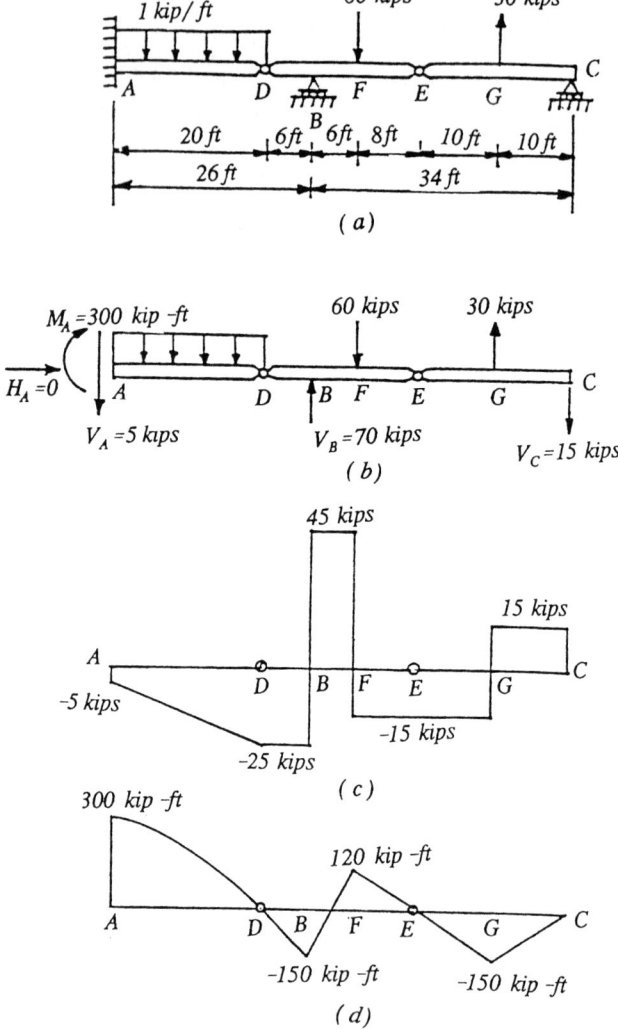

Figure 1.25. (*a*) Articulated beam loaded as shown. (*b*) Free-body diagram of the beam. (*c*) Shear force diagram of the beam. (*d*) Bending moment diagram of the beam.

the shear force diagram at D is

$$V_D = -5 - \int_{x_A}^{x_D} w\, dx$$

$$= -5 - \int_0^{20'} dx = -25 \text{ kips}$$

Between points D and B the shear force is constant and equal to -25 kips, because there is no load applied to this portion of the member. At B there is a discontinuity caused by the reaction $V_B = 70$ kips, and the ordinate of the shear force just to the right of B is

$$-25 + 70 = 45 \text{ kips}$$

From B to F the shear force remains constant and equal to 45 kips. Just to the right of F, because of the concentrated force of 60 kips, the ordinate of the shear force diagram is

$$45 - 60 = -15 \text{ kips}$$

and it remains constant until we reach point G. From G to C is again constant and equal to $(-15 + 30) = 15$ kips. The complete shear force diagram of the member is shown plotted in Fig. 1.25c.

The bending moment diagram of the member may be plotted by using the shear force diagram in Fig. 1.25c and Eq. (1.58), which states that the change in bending moment between sections at x_1 and x_2 is equal to the area under the shear force diagram between the two sections. We note that $M_A = 300$ kip-ft. Thus, the ordinate M_D of the bending moment at D is

$$M_D = 300 - \int_{x_A}^{x_D} V \, dx$$

$$= 300 - 300 = 0$$

The variation of the bending moment between A and D is represented by a second-degree parabola, because the shear force variation within the same region is linear.

The bending moment at B is equal to the bending moment at D, plus the algebraic sum of the area of the shear force diagram between points D and B. The value at B is -150 kip-ft, and the variation between D and B is linear, because the shear force is constant within this interval. Proceeding in the same manner, we find that the bending moment at F is 120 kip-ft, it is zero at E, -150 kip-ft at G, and zero at C. The complete bending moment diagram of the member is shown plotted in Fig. 1.25d.

1.8 DYNAMICAL ASPECTS OF COMPONENT ELEMENTS OF AN INFRASTRUCTURE SYSTEM

In an infrastructure system, component elements, such as bridges, buildings, mechanical systems, airport and space structures, and so on, are subjected not only to static forces, but also to various types of dynamic loading conditions that can produce catastrophic effects if they are not properly accounted for. A brief discussion regarding some of these problems is provided in this section, by considering the dynamic characteristics of highway bridges.

1.8.1 Dynamic Characteristics of Highway Bridges

The study of the dynamics of highway bridges [13, 14, 15] has been characterized through the years by a continuous refinement in both the problem formulation and the methods used for its solution. Initially, the studies were restricted to very specialized cases of bridge dynamics that involved major simplifying assumptions. With developments in technology, more aspects of the problem were considered, which provided more reliable solutions to the bridge dynamics problem. For example, the bridge-vehicle system was initially assumed to consist of a massless beam traversed by a constant moving force, and exact solutions to this problem were obtained. When the effects of the mass of the moving load and the mass of the bridge were incorporated in the solution, the complexity of the problem increased many times over, and it required the utilization of approximate methods of analysis in order to be able to obtain a solution.

Historically, a series of tests were devised in about the middle of the nineteenth century in order to study the dynamic characteristics of railway bridges [13]. However, there was little agreement among engineers over the effect of a moving load on a bridge and two schools of thought developed. The one group believed that a load that moves at high speed resembles a suddenly applied force, and produces responses that are much higher than those produced by the static application of the load. The second group believed that a load moving at a high speed will not be able to follow the contour of the deflected beam and, consequently, will have little or no effect.

Tests conducted on a specially constructed scaffold showed that dynamic deflections produced by a moving carriage were two to three times larger than the static ones. Tests on actual bridges, however, produced much smaller dynamic deflections. This inconsistency was resolved by R. Willis and later by G. G. Stokes [13], by the preparation of simplified analytical models of specialized cases, models capable of producing reliable approximate and/or exact mathematical solutions. Their conclusion was that the dynamic effects in an actual bridge are relatively small.

In 1934, the single most important analysis was completed by C. E. Inglis [16], and published under the title "A Mathematical Treatise on Vibration in Railway Bridges." It provided the foundation for the research on railway bridge vibrations that followed. Certain principles of this research were adapted later on in the dynamic analysis of highway bridges. The earliest major contributions were made by Hillerborg [17], Biggs et al. [18], Tung et al. [19], and Jacobsen and Ayre [20].

The most comprehensive studies on the dynamics of bridges were performed in about 1950 at the University of Illinois, with the participation of various investigators such as Newmark, Looney, Wen, Veletsos, Huang, and others. The AASHO Road Test [21] involved the first comprehensive dynamic testing program on highway bridges, and incorporated the parameters affecting the dynamic behavior of bridge-vehicle systems. The study by Veletsos and Huang [22] represents, perhaps, the most general form of analysis, and provided results that were in good agreement with certain experimental results.

A great deal of analytical and experimental work [23] was also conducted in the Soviet Union and several other eastern European countries such as Czechoslovakia, work which involved the more complex aspects of railway bridge dynamics. A ten-year study was also performed jointly by the Association of American Railroads and the U.S. and Canadian governments.

In the analytical investigations of bridges, the development of any mathematical model, regardless of its degree of sophistication, should incorporate three fundamental assumptions regarding the nature of: (a) the moving vehicular load; (b) the bridge; and (c) the dynamic deflection. The model used by Willis [24], one of the first of its kind, includes an unsprung mass moving across a massless beam. He assumed that the dynamic effect was small and the deflected shape was approximated by the static deflection produced by the gravitational force of the moving mass. Some years later, Stokes [25] obtained an exact solution to the case originally considered by Willis. He also obtained an approximate solution for a constant force moving across a beam by considering the mass of the beam and neglecting the mass of the force. Only the fundamental mode of vibration was considered.

Krylov [26] incorporated into Stoke's work an infinite degree of freedom system by assuming a beam with uniformly distributed mass. However, he neglected the mass of the moving load in order to be able to express its deflected shape as an infinite series involving the normal modes of vibration, which simplifies a great deal the solution of the problem. This case was further investigated by Timoshenko [27], who also considered the case of a pulsating force. He assumed that the deflected shape is represented by an infinite series with terms that are products of a normal mode deflection and a time function. He also examined the possibility of resonance associated with a moving load. A wide variety of cases were investigated by Inglis [16], who applied his method of harmonic analysis in each case. It should be noted, however, that all cases cited to this point involved only an unsprung mass, and many of the essential features of an actual vehicle were neglected.

Many investigators, such as Hillerborg [17], Tung et al. [19], Biggs et al.[18], and Scheffey [28], tried to prepare a more reasonable vehicle model by including the effects of the suspension system and tires, thus incorporating a system composed of a sprung mass, or a combination of a sprung mass and an unsprung mass. Hillerborg dealt with a sprung mass, while the others investigated the strung-unsprung system. Hillerborg assumed that the dynamic deflection shape of the bridge is proportional to its instantaneous static deflection curve produced by the moving load. That is, the dynamic deflection curve is proportional to the deflection curve that is produced if the load at that instant is statically applied. All these vehicle idealizations may be properly categorized as single-axle systems.

The development of multiple-axle systems may be attributed to the models developed by Wen (in Wen and Veletsos [29], and Wen [30], Veletsos and Huang [22], and Fryba [23]. These systems, three in all, represent a class of their own. The principle of superposition cannot be used because the deflected shape is a function of time. None of these models include lateral effects, and the bridge is treated as a single beam.

The model by Wen considered the single-axle analysis of Tung, based on Hillerborg's assumption, and adapted it to a two-axle system. He modified Hillerborg's assumption slightly by including the dead weight of the simple-span bridge in the calculation of the instantaneous static deflection. He also incorporated the rotary inertia effects of the sprung masses and the surface irregularities of the bridge and its approach. He neglected bridge and vehicle damping and the oscillation of the axles, known as "tire hop," and he assumed that the wheels of the vehicle are in contact with the roadway surface at all times. Wen's model provided results that were in good agreement with certain experimental results obtained from the AASHO Road Test.

Fryba's model is rather similar to Wen's, but it includes damping in both bridge and vehicle, and the oscillatory effects of the axles on their respective tires by introducing linear springs to represent the vehicle tires. He examined primarily the dynamics of railway bridges, but he also suggested the application of his methodology and model to highway systems. His model is designed to be used for simple span bridges and incorporates initial vehicle or bridge oscillations and road unevenness.

The third and most sophisticated model was prepared by Veletsos, where the vehicle is represented by a three-axle sprung unit load. Other loads can also be considered if they are treated as special cases of the more general case. The most unique feature of this vehicle idealization is the use of a bilinear hysteresis-type friction mechanism to represent an interleaf suspension system. The bridge can be a girder bridge of any number of spans, and in each span the mass is assumed to be lumped in a series of concentrated point masses, but the flexibility of the beam is maintained to be distributed throughout its length. The number of degrees of freedom of the bridge is characterized by the number of its point masses. Dynamic influence coefficients are used to carry out the analysis, and the model incorporates bridge surface irregularities, initial oscillations of both bridge and vehicle, and damping of both bridge and vehicle. It can also handle inelastic yielding if it is appropriately modified. It neglects, however, "tire hop."

Resonant vibration is also a very important consideration in the dynamic analysis of highway bridges, because it could occur between elements of the bridge-vehicle system. In the very simplified cases, such as the case of a single pulsating force moving across a simple span with the mass of the moving load neglected, the condition of resonance could be more easily analyzed. In such cases, Fryba [23] has shown that resonance is a very important factor in bridge dynamics. For the more realistic vehicle idealization, the complex interaction between the various components of the system makes it extremely difficult to predict the effects of resonance. A combination of experimental and theoretical investigations performed by various researchers was aimed at producing evidence regarding the importance of such effects.

The experimental results by Biggs et al. [18] related the amplitude of vibration to the ratio of the frequencies of the bridge and vehicle. The studies by Hayes and Sbarounis [31], and Foster and Oehler [32] demonstrated the existence of a type of resonance caused by the successive application of the axle loads at or near a bridge frequency. Wen used the term "quasi resonance" for this effect, and he showed that it is clearly a function of both vehicle velocity and axle spacing. Resonant effects that were noted in railway bridges are discussed by Kolousek [33] and Fryba [23].

The question, however, regarding the existence and nature of resonance is not yet resolved and remains open to debate.

1.8.2 Dynamic Bridge Studies by Fertis and Bender

Fertis [14] and Bender [15] performed analytical studies regarding the dynamic response of highway bridges subjected to heavy vehicle loads, with emphasis on long-span and low fundamental frequency bridges. The particular problems considered are: (a) safety aspects of low frequency long-span highway bridges in the range of 2 to 7.75 Hz; (b) the effects of damping; (c) the effects of vehicle speed in the range of 15 to 55 mph; (d) the effects of axle spacing; and (e) the effects of resonant frequencies on deflection buildup. The system of the governing differential equations of motion was solved for the first time by using the powerful numerical method known as the "acceleration impulse extrapolation method," and the results, when applicable, were compared with existing analytical and experimental data.

In particular, the purpose of the research was to study the dynamic response of a single-span highway bridge subjected to a two-axle moving load, by taking into consideration the effects of damping, vehicle velocity, axle spacing, and resonance, as well as low frequency and long-span aspects of the highway bridge. The bridge surface profile and the initial oscillations of the bridge and vehicle were not taken into consideration. The mathematical analysis utilizes an idealized bridge-vehicle model in a way similar to those previously introduced by Wen [30] and Fryba [23].

The model itself consists of a two-axle vehicle load moving across a single-span bridge, and assumes that the static and dynamic deflections are related to one another by a magnification factor that is a function of time. The application of the acceleration impulse extrapolation method leads to a single matrix equation that is readily solved by standard computer techniques.

The model and general method of analysis just described constitute a significant improvement on those of both Wen and Fryba. The model includes both bridge and vehicle damping, and also the weight of the vehicle axles and suspension system, none of which were considered by Wen. In addition, the basic assumptions regarding the dynamic deflection are more realistic than those used by Fryba. Finally, the simplicity and high accuracy of the numerical method makes it advantageous to use in place of trial-and-error methods.

The idealized bridge-vehicle system that is used in the research work is shown in Fig. 1.26, where

m_1 = mass of front tire and axle
m_2 = mass of rear tire and axle
m_3 = mass of chassis and payload
c_{v1} = equivalent viscous damping coefficient of the damping mechanism in the front suspension
c_{v2} = equivalent viscous damping coefficient of the damping mechanism in the rear suspension
k_{s1} = equivalent spring constant for the front suspension system
k_{s2} = equivalent spring constant for the rear suspension system

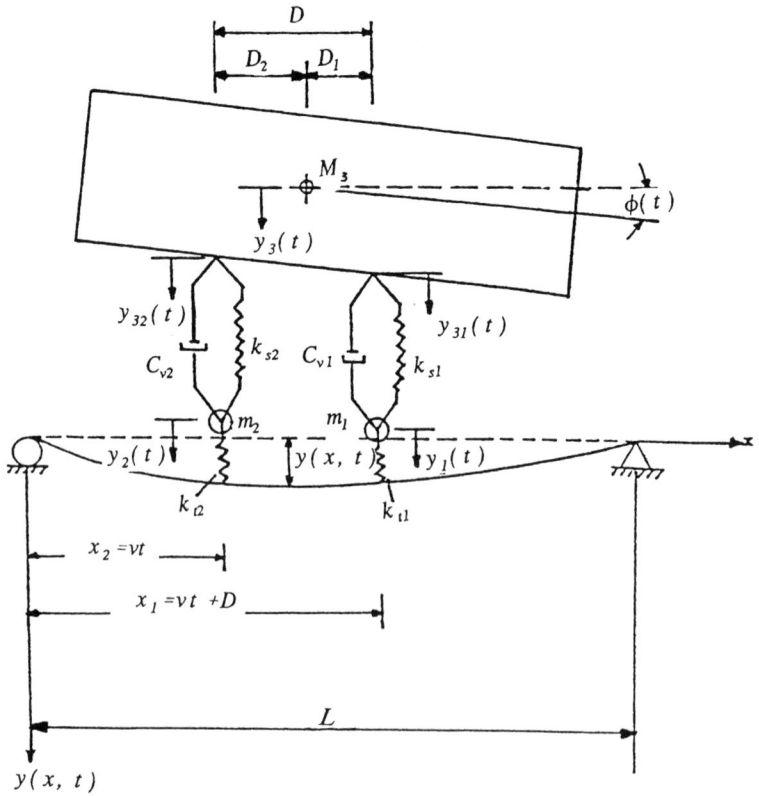

Figure 1.26. Two-axle bridge-vehicle model. Reference [14].

K_{t1} = equivalent spring constant for the front tire
K_{t2} = equivalent spring constant for the rear tire
D_1 = horizontal distance from the centroid of the sprung mass to the front axle
D_2 = horizontal distance from the centroid of the sprung mass to the rear axle
D = axle spacing $(D_1 + D_2)$
x_1 = location of the front axle at time t, measured from the left support
x_2 = location of the rear axle at time t, measured from the left support
$\phi(t)$ = angular rotation of the sprung mass about its centroid, measured clockwise from the horizontal
$y_{31}(t)$ = vertical displacement of the sprung mass at the front axle, measured from its equilibrium position when the upper springs are deformed by the weight $m_3 g$
$y_{32}(t)$ = vertical displacement of the sprung mass at the rear axle, measured from its equilibrium position
$y_1(t)$ = vertical displacement of mass m_1, measured from its equilibrium position when the lower springs are underformed
$y_2(t)$ = vertical displacement of mass m_2, measured from its equilibrium position
$y(x, t)$ = vertical displacement of the neutral axis of the beam, measured from the static equilibrium position

The important features of the bridge-vehicle model in Fig. 1.26 may be listed briefly as follows:

1. The vehicle consists of rigid sprung mass connected to two unsprung masses by means of a linearly elastic spring.
2. A dashpot is placed between each unsprung mass and the sprung mass to provide for damping in the system.
3. The unsprung masses are connected to the bridge surface by a second set of linearly elastic springs.
4. The upper and lower linear springs simulate the flexibility of the suspension system and tires, respectively.
5. Damping of the suspension system is accounted for by means of the dashpots while damping of the tires is not considered.
6. The bridge is simply supported and free of all surface irregularities.
7. The five degrees of freedom associated with this system are the vertical displacement and rotation of the sprung mass, the vertical displacements of the unsprung masses, and the vertical displacement of the bridge surface.

From the results of the study it may be concluded that: (a) the acceleration impulse extrapolation method is especially well-suited for dynamic analysis of highway bridges; (b) the method of analysis and bridge-vehicle system used in this study can duplicate earlier theoretical studies, it compares very well with experimental results from AASHO Road Tests, and it can incorporate dynamic characteristics that are very essential in understanding the dynamic behavior of highway bridges; (c) a low bridge natural frequency was shown to have critical effects on the static response of the bridge, that is, the magnitude of the static deflections increased by 200 percent for only a 20 percent decrease in the bridge natural frequency; (d) bridge damping cannot be counted on to prevent the build-up of dynamic oscillations; (e) the synchronization of axle load applications with the bridge frequency may result in increased dynamic response; (f) data representing more than 500 vehicle passages over the bridge span indicate that certain peak responses are present at particular bridge frequencies; (g) bridge amplitudes at resonance, or near resonance, may be a function of both bridge natural frequency and vehicle velocities; and (h) the amplitude buildup of bridge oscillations becomes more critical for lower vehicle velocities and longer span bridges where a substantial number of bridge oscillations are involved.

Complete details regarding the above conclusions, and more, may be found in the studies by Fertis [14] and Bender [15] cited earlier in the section.

1.9 THE EARTHQUAKE PROBLEM

Modern theory indicates that the earth consists of a central core of molten nickel-iron 4,350 miles in diameter. The core is enveloped first by three concentric layers of dense material totaling 1,050 miles, then by a mantle of highly elastic substance 500 miles thick, and finally by the outer part of the earth, which is a shell of rock 50 miles thick, known as the crust.

The crust is divided by planes of cleavage known as faults into immense irregular blocks called fault blocks. The surface trace of a fault plane is known as a rift. Normally, the fault blocks are in equilibrium, but sometimes adjacent blocks move with respect to one another due to forces acting under certain definite but unknown physical laws. This relative motion is called slip, and it may be horizontal, vertical, or a combination of both.

A slip may be gradual or abrupt. If it takes place gradually, there is a similar adjustment in the continuous strata, which produces no immediately observable consequence. If the slip is abrupt, the surrounding strata are fractured, and a series of local movements begins. The elastic earth waves, and the surface vibration caused by these waves in passing, is known as an earthquake. Earthquakes caused by slip are classified as tectonic, and earthquakes caused by volcanic explosions are called volcanic.

The epicenter of an earthquake is the point of the earth's surface directly over the origin or focus of the earthquake, and the epicentral zone is the area about the epicenter where the earthquake is most destructive.

If the period of the periodic force approaches the value of a free period of a structure, the phenomenon is known as resonance or synchronism. The consensus of scientific opinion seems to be that destructive earthquakes have associated with them a definite bracket of periods ranging between 1 and 1.5 sec. This is known as the dangerous bracket of periods. It is also indicated experimentally that a particle located at some depth is displaced less than one near the surface.

Earthquakes are possible in all localities. A report published in 1931 by Heck and Bodle of the U.S. Coast and Geodetic Survey refers to the earthquake activity of a specific year, 1929, typical of any year. It shows that there is probably no region within the United States and its possessions that is immune from the possibility of earthquake damage. This report shows that in 1929 24 states had experienced the effects of minor, moderate, and severe earthquake shocks. Major and severe earthquakes in that year were observed in California, Ohio, Oklahoma, South Dakota, Alaska, and the Hawaiian Islands [34, 37].

The design of a structure to resist the effects of an earthquake is a very difficult task, primarily because it is almost impossible to predict the character and intensity of the earthquake to which the structure may be subjected during its lifetime. Some indication may be obtained from earthquakes that have been recorded as they occurred during the past decades. The intensities of such earthquakes are usually obtained by measuring their respective accelerations as a function of time in terms of both vertical and horizontal components. With known accelerations, the ground velocities and displacements as a function of time can be approximately determined by integration. A sample of what is considered to be a strong earthquake is shown in Fig. 1.27. Such records provide a useful history of past earthquake activity, which can be used to establish some reasonable criteria for the type and intensity of an earthquake that the structure should be designed to resist.

The ground accelerations that are produced by an earthquake are essentially random, as you can see in Fig. 1.27a. The positive and negative peaks occur at irregular times and have different amplitudes. In practice, four types of approaches are commonly used in the design of structures to resist the earthquake catastrophic effects. All approaches have advantages and disadvantages, but none has been developed to a desirable extent.

The first one utilizes a standard earthquake of certain amplitude of acceleration and time variation, and the analysis for earthquake response is carried out by using this standard earthquake. The disadvantages regarding this approach are that actual earthquakes vary in intensity and time variation, while the response of a structure is very much dependent on these properties.

The second approach is considered by many engineers to be the most practical and involves the utilization of earthquake response spectra [35]. This idea was first introduced by M. A. Biot, E. C. Robinson, G. Housner, and others, and response spectra from real earthquake records have been determined and used directly for earthquake analysis and design. This is a rather reasonable simplification of the problem, and by many engineers is considered to be one of the most significant contributions to earthquake engineering.

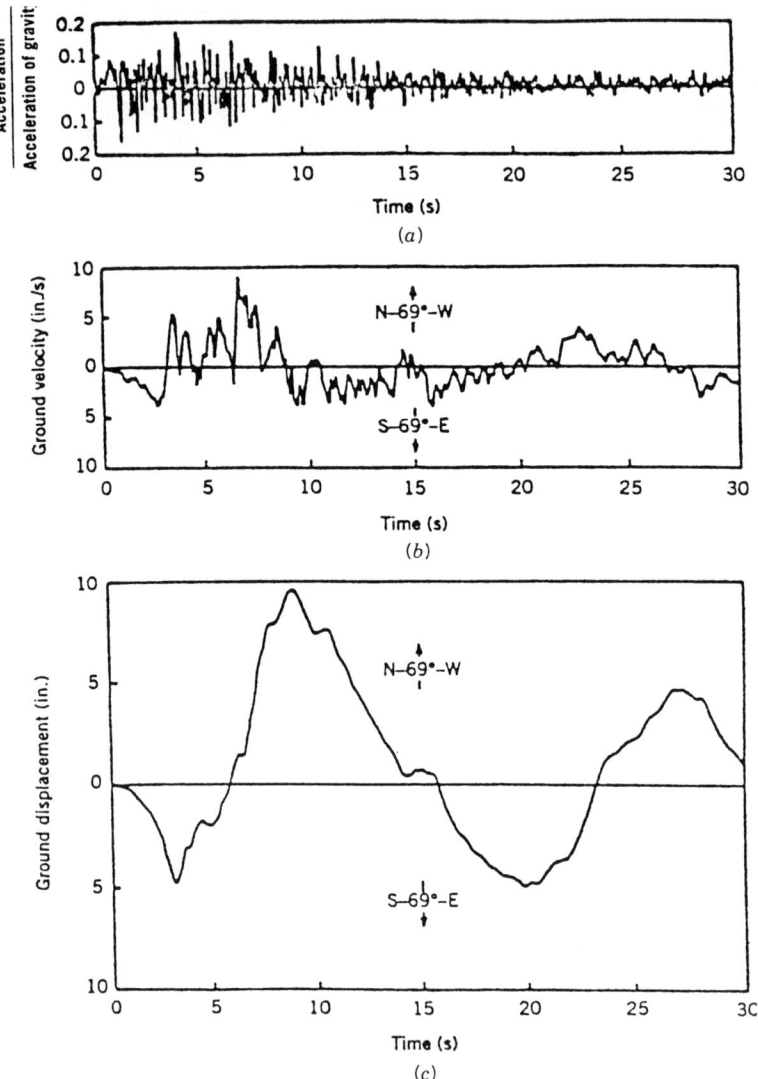

Figure 1.27. (*a*) Ground accelerations at a distance of 35 miles for the Kern County, California, earthquake. (*b*) Corresponding velocities. (*c*) Corresponding displacements. (Acceleration record courtesy of the United States Coast and Geodetic Survey.)

The third approach is to treat earthquake ground motions as random variables, which is a more logical way of thinking since the input is nondeterministic (random). This type of analysis involves the element of prediction, and stochastic methods of analysis should be used. On this basis, if a structure is to be built in a region of known earthquake history, the object would be to design the structure for the least favorable among all foreseeable earthquake excitations. With this in mind, a structure could be designed so that the probability that its maximum stresses and maximum relative displacements to which it would be subjected would not exceed specified values. The disadvantage of this methodology results from the hard fact that we do not have sufficient record of earthquake activity in every part of the globe to be able to carry out a reasonable analysis.

In the fourth methodology, the design of many structures to resist the catastrophic effects of an earthquake is based on the utilization of engineering design codes. The development of such codes is based on experience gained from previous observations of earthquake damage, as well as on current developments in the field of earthquake design. The codes are prepared to provide a simple but adequate design for many types of structures, but more elaborate analysis is required for the more specialized cases.

The main objectives in a design code are to provide construction and design requirements that should be followed by the practicing engineer in order to protect the structure from undesirable damage, and prevent loss of life and injury from earthquake activity. The requirements may vary among the various codes, because the geological formations, earthquake intensities, and engineering technology are not the same from one locality to another. Some of the widely used codes are the ones prepared by the Structural Engineers Association of California (SEAOC), the Uniform Building Code (UBC), the National Japanese Building Code (JBC), and the Joint Committee Code (JCC). In these codes, the maximum conditions of dynamic response are usually converted into a set of equivalent static forces, and the actual design is carried out by using static analysis.

1.9.1 Practical Design Considerations

The following design considerations may be used in order to design an earthquake-resistant structure:

1. In investigating the site of an important structure we should determine: (a) the proximity of rifts and faults; (b) the dominant periods of vibration; (c) the soil characteristics; and (d) the character of substrata.

2. It is not right, even in the absence of all signs of seismic activity, to locate a building over a known fault or rift, for it is highly probable that in the event of a slip of any magnitude the building will be destroyed.

3. The closer the periods approach those in the dangerous bracket of periods, or those of the proposed building, the larger the amplitudes assumed.

4. The nature of the materials constituting the site, their elasticity, moisture content, and uniformity determine the damping value of the site, its probable amplitude during a destructive earthquake, and the character of foundation required. For example, if the material is bedrock whose

elastic value is high, the damping value is poor and the amplitude assigned is low. If the site is composed of dry, compact, uniform alluvial or diluvial deposits whose elastic value is less than that of bedrock, the damping value is better and the amplitude assigned is larger. Where the materials contain an appreciable amount of moisture, the elastic value becomes low, the damping is excellent, and the amplitude assumed is high, its value depending on the cohesive strength of the soil and on how closely it approaches the condition of semifluidity. Finally, if the soil is not uniform but consists of large adjoining deposits of varying physical properties, the damping value is good and the amplitude assumed is an average value depending on the nature of the component materials.

5. During an earthquake, the soil forces the foundation of a building through a definite amplitude. An effective way to decrease that amplitude is to insulate the foundation from the native soil by eliminating any condition of fixity between the foundation and the soil. That means by creating discontinuity. A three-foot layer of gravel or broken stone introduced between the foundation and the native soil is deep enough to create discontinuity. By experiment a three-foot bed of gravel or broken stone will decrease the amplitude transmitted to the foundation by one half.

6. Locating foundation at a depth greater than is required ordinarily does not materially decrease the amplitude transmitted. An aseismic foundation is a beam-and-slab mat. It cancels the adverse effect of differential soil movements. In addition, it doesn't have footings digging themselves in and destroying the desired state of discontinuity. In this manner we have a rigid and monolithic design.

7. If individual footings are used, the elevations of the bottoms of the footings should all be identical and the footings should be interconnected by beams able to maintain them in that same relative position.

8. Owing to the oscillation of the center of gravity of a building during vibration, the exterior parts of the foundation should be designed to withstand soil pressures as high as four times the static soil pressure.

9. If the soil is not uniform, yields excessively under pressure, or contains considerable moisture, a beam-and-slab foundation is indicated.

10. Where the site consists of wet, loose fill, or approaches a state of semifluidity, a beam-and-slab foundation on piles is required.

11. Where soil values permit, short piles, which allow the building to rotate in a vertical plane, should be used rather than long piles driven to hardpan.

12. The fundamentals of satisfactory earthquake-proof design are: symmetry of plan, mass, and rigidity. The center of mass should coincide with the center of rigidity. Both mass and rigidity should be as nearly

uniform throughout the building as possible, so that the building cannot rotate about a vertical axis owing to one part being heavier or stiffer than another.

13. Rigid bents should be symmetrical with respect to the floor plan.
14. All parts of the building should be so connected to the frame that the building will vibrate as a unit, the connections being strong enough to overcome the inertia of the separate parts.
15. In tied columns, the ties should be spaced closely for some distance above and below the floors. Diagonal bands of steel should extend back from the corner column into the floor slab. Provide additional end restraint for columns by means of extra bars in the floor slab and across the columns.
16. The floors act as horizontal girders and should be designed as such.
17. The floor system should be so braced diagonally that it will deflect as a unit and will be strong enough to transmit the lateral forces to all the bents.
18. The full depth of the wall spandrels should be designed as beams, so as to attain the maximum possible floor rigidity.
19. Floor slabs should be of rock concrete, rigid and monolithic.
20. Horizontal diagonal bars should be used between floor beams connecting interior columns.
21. Exterior walls should be of reinforced concrete, of uniform thickness, and designed as vertical beams. Use ties and diagonal bars.
22. Diagonal bars should be used across all wall openings.
23. All construction joints should be dowelled, as this increases their reliability.
24. Wall bents should be continuous for the full height of the building.
25. Temperature steel should be placed diagonally.
26. The end conditions at the top of a building should be considered to be free.
27. The end condition at the bottom of a building should be determined by the nature of the soil, the type of foundation, and the state of discontinuity. For example:
    a. If the footings are fixed into bedrock, or into dry, compact, alluvial or diluvial deposits, the end condition should be assumed to be *fixed*.
    b. Where a layer of gravel separates the foundation from the native soil, and the construction is designed to permit the foundation to deflect laterally, the end condition should be considered to be *sliding*.
    c. If the foundation is fixed into wet, compressible deposit of such nature that rotation in a vertical plane can occur, the end condition should be assumed to be *hinged*

d. Where a layer of gravel separates the foundation from the soil, and the building is free to move in a horizontal as well as in a vertical plane, the end condition should be considered to be approximately *free*.

28. After all the conditions involved are evaluated, it becomes a matter of engineering judgment to decide how closely any of the arbitrary standards are approached, and decide on the methodology to be used for the computation of the fundamental period of vibration of the structure. Practical formulas and methods of analysis may be used for this purpose [5, 36, 37].

29. The gravest stresses will occur when the period of the earthquake synchronizes with that of the structure. The worst case that needs to be considered in a practical situation may be covered if we assume that the ratio of the period of the structure and earthquake is 0.9 or 1.1, because complete synchronism, or resonance, is highly improbable. If, however resonance does occur, the inherent factor of safety of the structure will be able to cope with the temporary increased stresses, provided that the structure is designed for 0.9 or 1.1.

30. A complete investigation of the earthquake-resistant design of a building usually includes the following steps:
    a. A seismological study of the site and locality with particular reference to its earthquake history, dominant periods of vibration, and proximity to faults and rifts.
    b. A geological survey of the site to determine the nature of the soil, the depth of the strata, and the elevation of the groundwater level.
    c. Selection of the type of foundation to be used.
    d. Decision as to the probable end conditions of the building and computation of its required period of vibration.
    e. Determination of the foundation amplitude.
    f. Design of the structural frame for the statical loads, dead, live and wind loads, making the floor systems as deep and rigid as possible.
    g. Symmetrical distribution of the vertical reinforced concrete (R/C) beams.
    h. Calculation of the transverse fundamental periods of the building about the X–X and Y–Y axes.
    i. Determination of the ratio of the period of vibration of the building with that of the earthquake. If it falls in the dangerous bracket, assume a value of 0.9 or 1.1.
    j. Determination of maximum deflections, moments, and shears, using a suitable methodology.
    k. For very important structures, confirmation of these values by tests on models is advisable.
    l. Distribution of seismic shears and moments among the vertical beams and confirmation of the stresses in the vertical beams.

m. Design, detailing, and construction of the building in conformity with the principles outlined previously.

n. Confirmation that the column sections are adequate for the combined stresses due to the statical and seismic loadings.

o. Measurement of the periods of the completed structure and confirmation of the assumed end conditions, computed periods, period ratios, deflections, moments, and shears.

1.10 THE LOS ANGELES EARTHQUAKE

In order to obtain some fair idea regarding the catastrophic effects of an earthquake, a brief discussion of the 1971 Los Angeles earthquake is provided in this section. It represents a summary of a preliminary report from the Prestressed Concrete Institute, prepared in 1971. Although, comparatively speaking, this earthquake is not considered to be very strong, it still produced considerable damage in the Los Angeles region because of its inherent peculiarities, peculiarities deserving attention.

This earthquake occurred at 6:01 a.m., Tuesday, February 9, 1971, in southern California. Its epicenter was about 26 miles north of Los Angeles in the San Gabriel Mountains. The magnitude, as reported by the news media, ranged between 6.5 and 6.8 on the Richter scale. This scale provide a measure of the energy released by an earthquake, but it does not necessarily provide a true measure of the damage that could occur at any given site. The extent of an earthquake's damage depends not only on the size or intensity of the earthquake, but also on the dynamic characteristics of the infrastructural element of the earthquake site.

The major damage in this instance was centered between the Sylmar area and the Van Norman Dam reservoir. The buildings of the San Fernando Veterans Administration Hospital collapsed, and extensive failures were reported on the Olive View Tuberculosis Sanitarium in Sylmar, California. The Veterans Administration Hospital had been built in the 1920s and the buildings were constructed using cast-in-place reinforced concrete. The Tuberculosis Sanitarium was completed not long before the earthquake occurred, and was a cast-in-place moment-resisting reinforced concrete framed structure. The exterior of the sanitarium building was made of non-load-bearing precast concrete curtain wall panels.

The Van Norman Lake Dam, an earth-filled structure, suffered damage so severe that the housing areas below the dam were evacuated. Extensive damage also occurred to homes and to highway structures in these two areas. At some of the collapsed bridges there was evidence of severe vertical displacement of the ground of up to 3 ft, and horizontal ground displacement of up to 6 ft. Preliminary reports have also indicated that superimposed loads on structures may have reached about 25 percent of gravity. In the Los Angeles city area,

62 INFRASTRUCTURAL FUNDAMENTALS

precast concrete reacted well to the earthquake in both tall buildings and smaller structures.

The Formigli Corporation reported that completed buildings performed well. Only one job of several that were under construction sustained some damage. On a shopping center project near Sylma, 17 out of the 400 cored floor slabs placed on flanges of steel beams fell down. Freyssinet Company, Inc., reported that the Palmdale interchange upper-level bridge, of about 200-feet elevation above grade, moved laterally off the hinge points and collapsed onto the lower-level bridges, which then collapsed. A two-span continuous bridge also failed when the center pier was completely knocked out because of the severe lateral ground movement.

The earthquake occurred over a long duration of 60 seconds. Within a radius of 2.5 miles of the heavily damaged Veterans Hospital in Sylmar and the Van Norman Dam lies a heavy network of freeways. Many of the more than 100 bridges in this area were damaged by the earthquake and several of them collapsed. The collapsed bridges included conventionally reinforced box girders, structural steel girders, prestressed, post-tensioned box girders, and precast concrete girders. Three post-tensioned bridges collapsed, all of which appeared to be the result of substructure failure. No evidence, however, of any superstructure failure or tendon failure was found.

The extensive horizontal and up and down movements of the ground produced by the earthquake are considered to be the main forces causing the indicated destruction. The main question that triggers in the minds of our engineers is how can we design an infrastructure system that can be immune to such forces of nature. This problem is not yet solved, and a great deal of research is required for a reasonably happy ending.

PROBLEMS

1.1 An unsymmetrical triangular roof truss is loaded as shown in Fig. P1.1. Determine the reactions at the supports A and B, and the internal axial forces in the members of the truss.

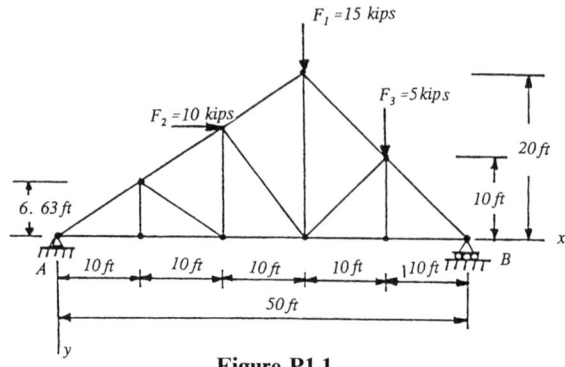

Figure P1.1.

1.2 A pratt truss is loaded as shown in Fig. P1.2. By using the method of sections and a section through members *CD*, *CF*, and *EF*, determine the internal forces in members *CD*, *CF*, and *EF*.

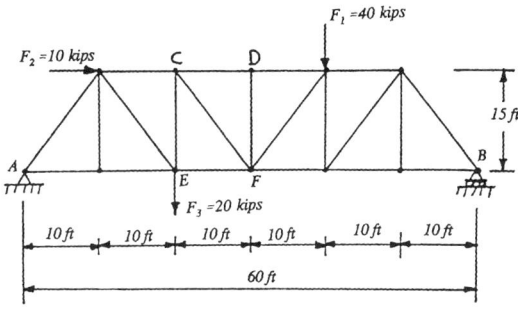

Figure P1.2.

1.3 By utilizing the method of joints, the method of sections, or a combination of both, determine the internal forces in the members of the plane trusses loaded as shown in Fig. P1.3. Also determine all reactions at the supports.

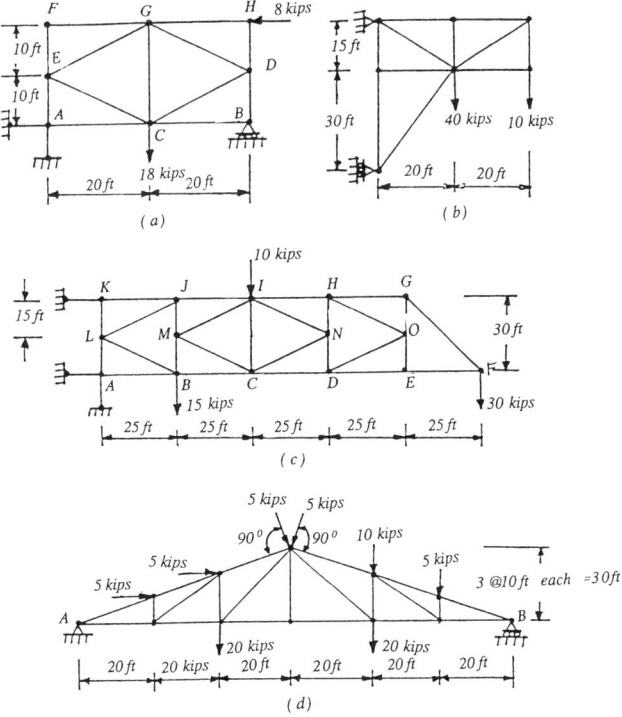

Figure P1.3.

1.4 By inspection, determine the zero-force members in the trusses loaded as shown in Fig. P1.3.

1.5 For the two-dimensional structural elements shown in Fig. P1.5, determine all reactions at their supports.

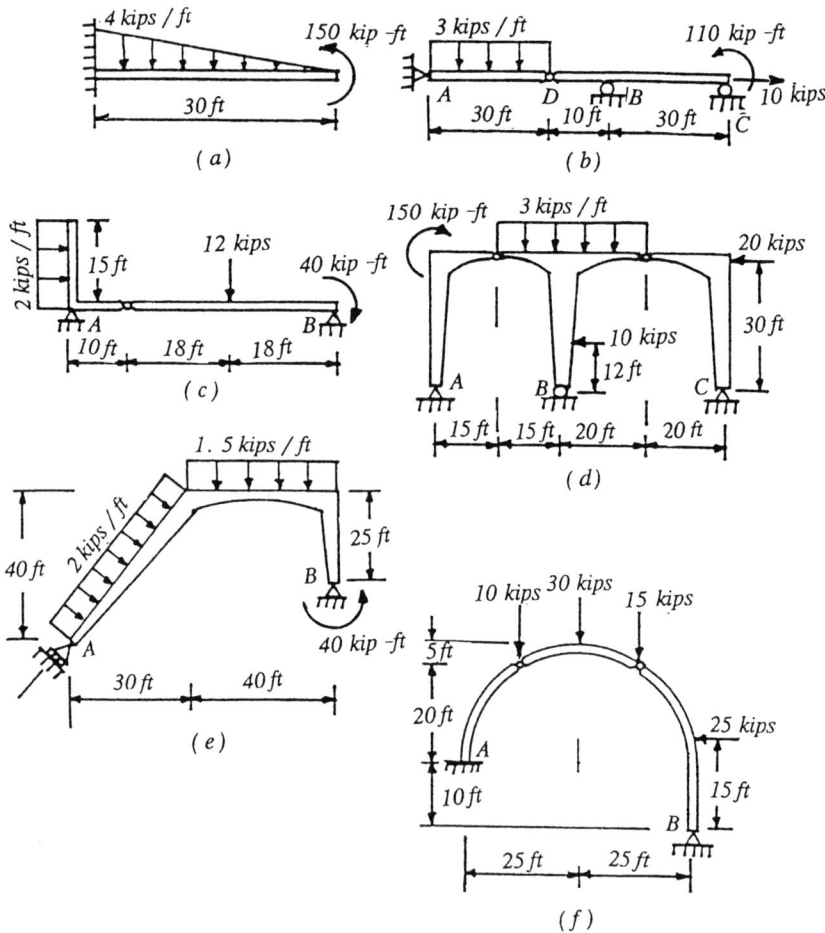

Figure P1.5.

1.6 By utilizing the method of joints, or the method of sections, determine the internal forces in the members of each of the space trusses shown in Fig. P1.6. Also determine the external supporting reactions of the space trusses in Figs. P1.6c and P1.6d.

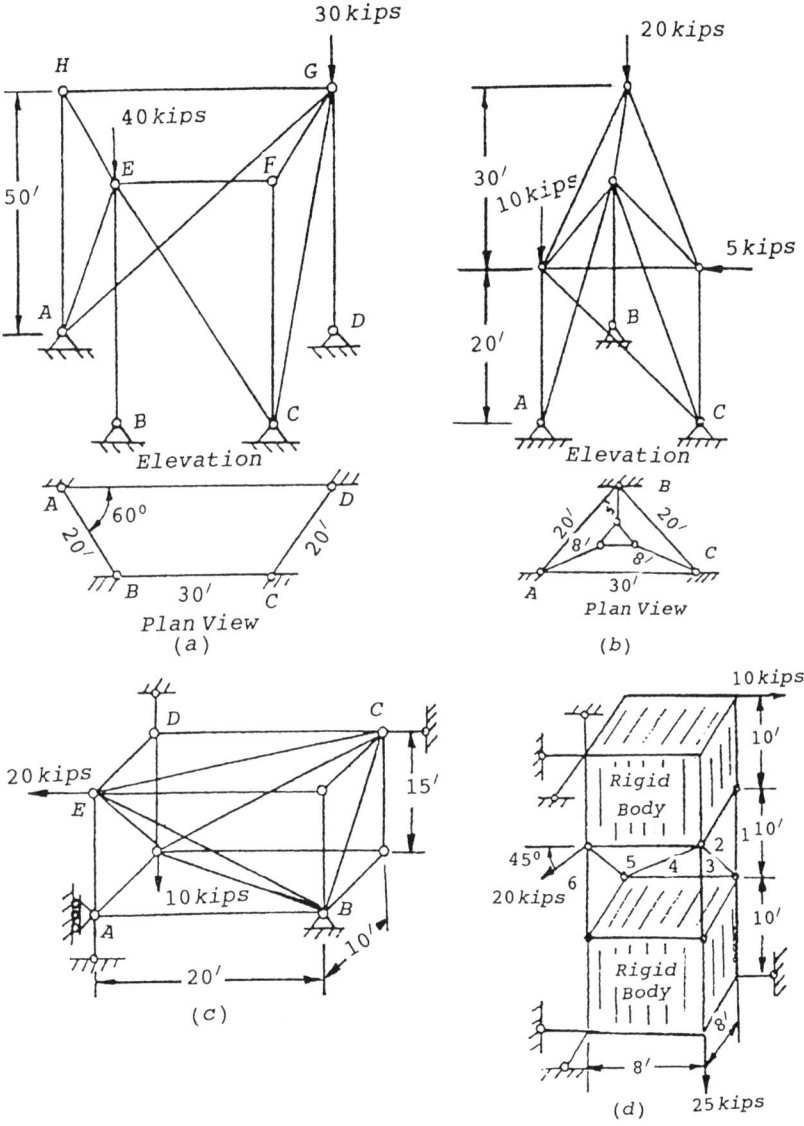

Figure P1.6.

1.7 Examine the three-dimensional truss loaded as shown in Fig. P1.7 and identify by inspection the inactive members (zero-force members). Draw these members with dashed lines on a carefully prepared sketch of the truss structure.

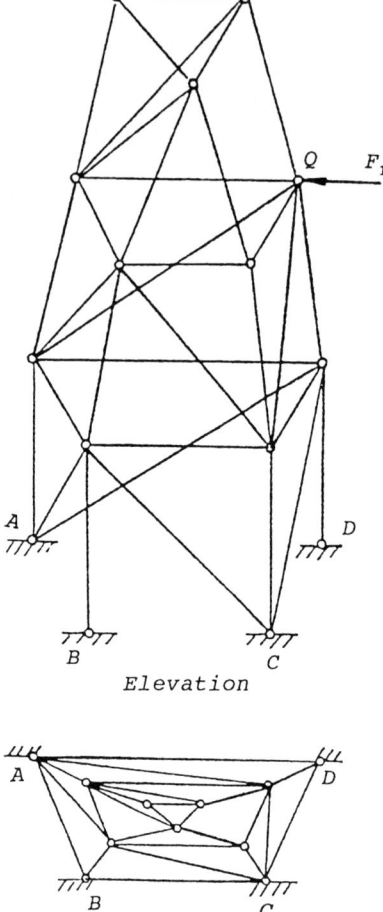

Elevation

Plan View

Figure P1.7.

1.8 Determine the reactions at the supports of each of the structures loaded as shown in Fig. P1.8.

1.9 For each of the structures loaded as shown in Fig. P1.9, state which law, or laws, of static equilibrium is or are violated, and briefly discuss its effects on the structure.

1.10 In each of the structures shown in Fig. P1.9, correct for the structural discrepancy so that all structures satisfy the corresponding necessary and sufficient conditions for static equilibrium.

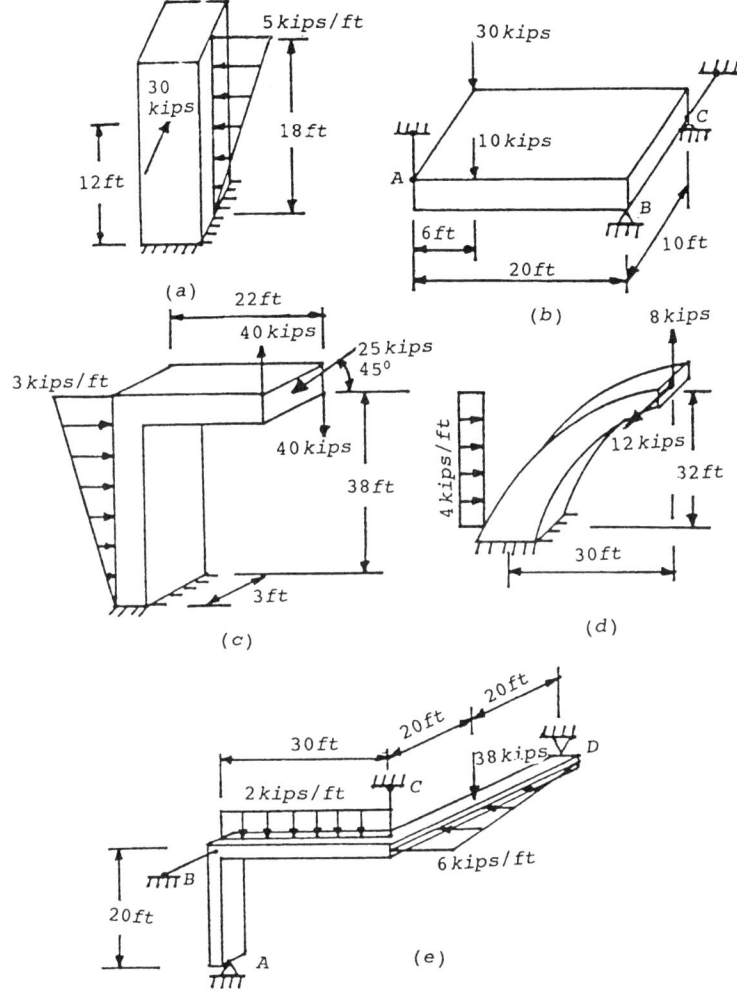

Figure P1.8.

1.11 For the semicircular three-hinge arch loaded as shown in Fig. P1.11, determine the internal forces at section $N-N$ located as shown in the same figure.

1.12 For each of the structures shown in Fig. P1.12, determine the internal forces at the indicated $N-N$ sections.

1.13 For each of the structures shown in Fig. P1.12, derive the axial force, shear force, and bending moment equations.

68 INFRASTRUCTURAL FUNDAMENTALS

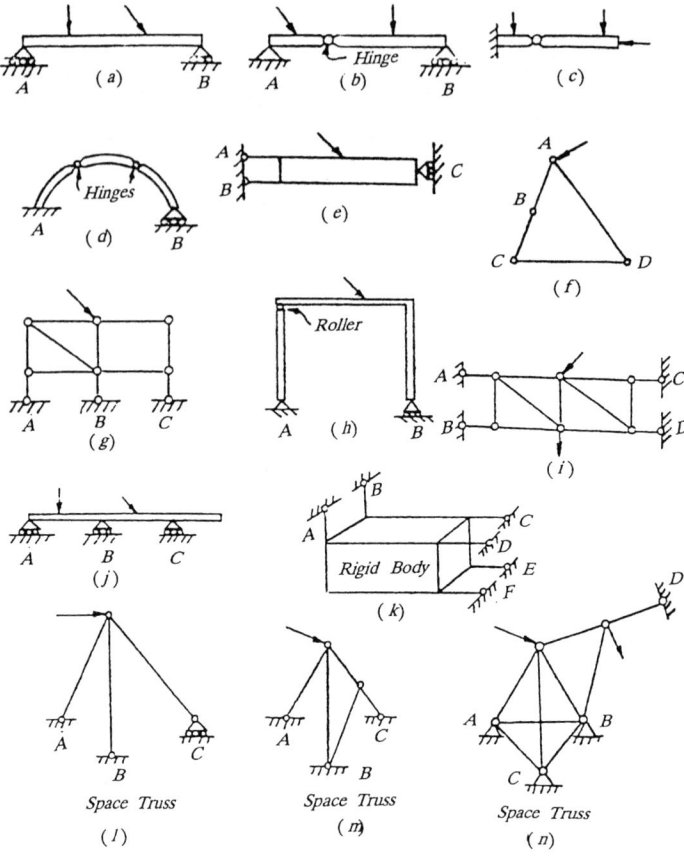

Figure P1.9.

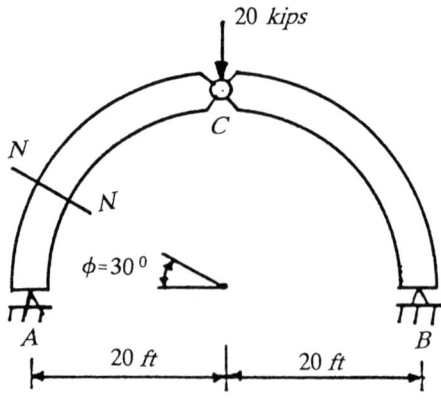

Figure P1.11.

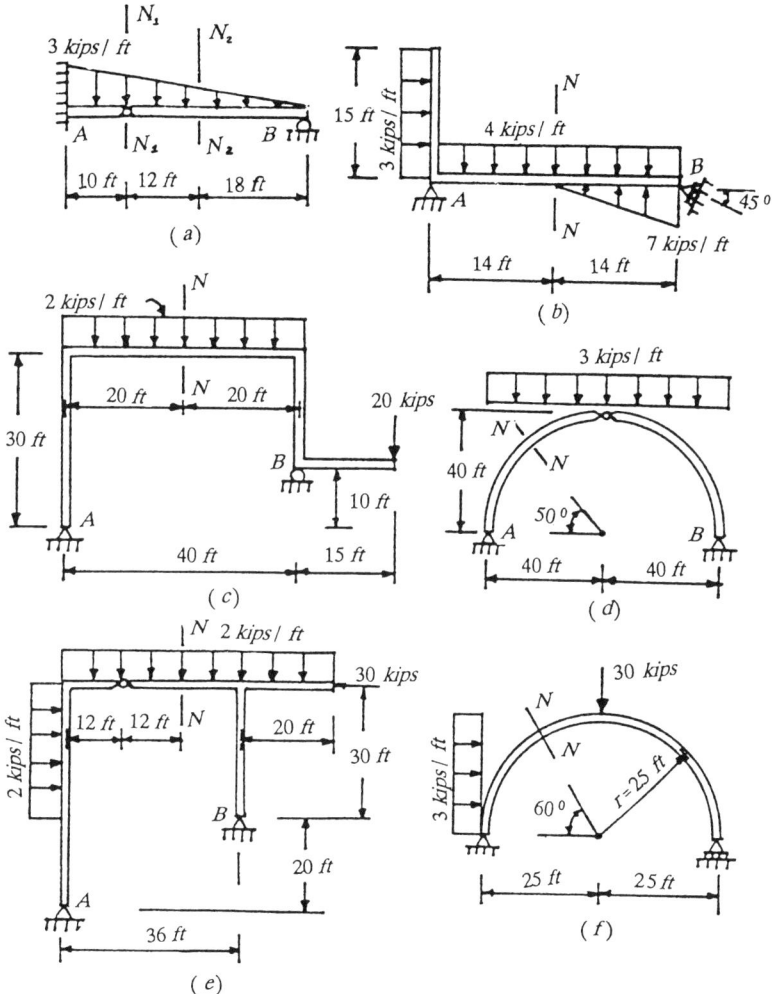

Figure P1.12.

1.14 For each of the structures shown in Fig. P1.12, draw the axial force, shear force, and bending moment diagrams by using the expressions derived in Problem 1.13.

1.15 By using Eqs. (1.54) and (1.58), construct the shear force and bending moment diagrams for each of the structures loaded as shown in Fig. P1.12.

1.16 The shear force diagram in a segment of a beam is as shown in Fig. P1.16. If the bending moment at A is 165 kip-ft, determine the bending moments at B and C.

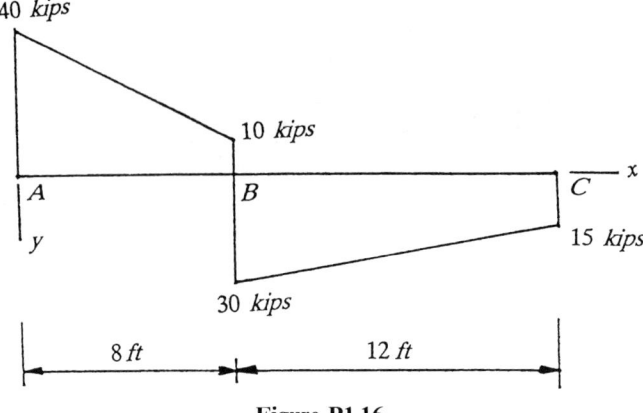

Figure P1.16.

1.17 A two-span continuous beam is loaded as shown in Fig. P1.17. If the internal bending moment M_B at support B is -162.5 kip-ft, plot the shear force and bending moment diagrams of the beam. Show all maximum and minimum ordinates for both shear and bending moment, and give a brief discussion in regards to their relationship.

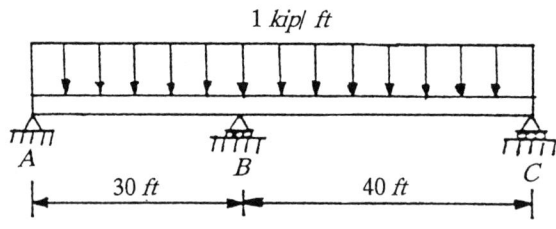

Figure P1.17.

2 Stress and Deformation

2.1 INTRODUCTION

Structural and mechanical systems are essential elements of an infrastructure system, and their structural integrity must be maintained when they are under the action of various types of force systems and the forces of nature. They must be designed to perform successfully under well-defined, or random, kinematic conditions, should satisfy appropriate equilibrium conditions, and should also conform with design requirements in terms of stress and deformation. Stress and deformation characteristics in a structural or a mechanical system will depend on the magnitude and type of the applied loading, its geometric characteristics, and the mechanical properties of the material used to construct such systems.

The subjects of stress and deformation are vital in the analysis of infrastructural components, and thorough understanding of such responses leads to successful and economical designs. For example, frames of buildings cannot deflect excessively, because such deformations may develop undesirable cracks in plastered walls and ceilings. Stresses and deformations beyond established criteria may cause new types of failure, develop additional forces that were not accounted for in the analysis, or they may affect the intended artistic appearance.

The purpose in this chapter is to study the stress and deformation characteristics of structural elements that are subjected to various types of loadings, and to provide methodologies for their evaluation. Such information constitutes an important part of the design and analysis of efficient infrastructure systems.

2.2 FUNDAMENTALS OF STRESS, STRAIN, AND DEFORMATION

The applied forces and moments on a member may include axial forces acting through the centroidal axis of the member, twisting moments that twist the member along its centroidal axis, and bending moments that produce bending of the member. Combinations of such loadings may also act on a member. The applied forces and moments are balanced by appropriate reactive forces and moments developed at the supports of the member, so that static equilibrium of the member is satisfied.

When a member is subjected to externally applied and resisting forces and moments, it deforms, and the position of each particle of the member may change, depending upon the boundary conditions of the member. Such deformations are associated with internal strains and stresses developed inside the member that hold together the various particles of the member. The deformation configuration and the type of stresses and strains developed inside the member are characterized by the external loading type and by the member's boundary conditions at the supports.

The following discussion concentrates on the static response of infrastructural elements subjected to various types of loading.

2.2.1 Axially Loaded Members

Consider the doubly tapered member illustrated in Fig. 2.1a, which is loaded by an axial tensile force F as shown. Since the ends of the member are free of any boundary restraints and the force F is axial, the length L of the member will increase in the longitudinal direction by an amount equal to δ. If A is considered to be the origin of the x, y coordinate system, an expression for the deflection δ_x at any distance x along the length of the member may be derived as follows:

At a distance x from the origin, the deformation of an element of the member of length dx is shown by the dashed line of its free-body diagram in Fig. 2.1b. The undeformed length of the element is dx and $dx + \varepsilon_x dx$ is its deformed length. We assume here that the member is elastic, and ε_x is the change per unit of length in the longitudinal direction of the member, known

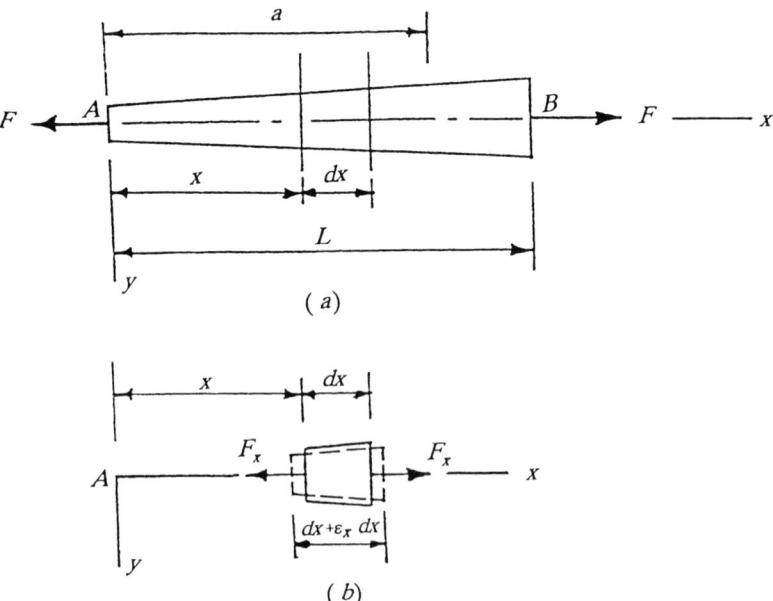

Figure 2.1. (a) Doubly tapered beam subjected to an axial force F. (b) Free-body diagram of an element of the beam of length dx.

FUNDAMENTALS OF STRESS, STRAIN, AND DEFORMATION   73

as longitudinal strain. On this basis, the infinitesimal change $d\delta$ of the length dx of the element is

$$d\delta = \varepsilon_x \, dx \qquad (2.1)$$

If we assume that the longitudinal stress σ_x is proportional to the longitudinal strain ε_x, which means that Hooke's law applies, we can write the following equation relating the stress σ_x with the strain ε_x:

$$\sigma_x = E\varepsilon_x \qquad (2.2)$$

where E is the constant of proportionality known as the modulus of elasticity. In practice, the stress σ_x is usually assumed to be constant over the cross-sectional area A_x of the member. On this basis, we can write the following expressions for σ_x and ε_x:

$$\sigma_x = \frac{F_x}{A_x} \qquad (2.3)$$

$$\varepsilon_x = \frac{\sigma_x}{E} = \frac{F_x}{A_x E} \qquad (2.4)$$

The internal axial force F_x at any coordinate x may be constant or variable, depending upon the variation of the applied axial force along the length of the member.

By substituting Eq. (2.4) into Eq. (2.1), we find

$$d\delta = \frac{F_x}{A_x E} \, dx \qquad (2.5)$$

The elongation δ_x at any distance x along the length of the member may be obtained by integrating Eq. (2.5); that is,

$$\delta_x = \int_0^x \frac{F_x \, dx}{A_x E} \qquad (2.6)$$

The modulus of elasticity E in Eq. (2.6) is assumed to be constant.

Equation (2.6) may be used for any portion of a member, provided that the continuous functions for F_x and A_x are known. If an expression for F_x and A_x may be written for the whole length L of the member, the total elongation δ of the member may be determined from the equation

$$\delta = \int_0^L \frac{F_x \, dx}{A_x E} \qquad (2.7)$$

74 STRESS AND DEFORMATION

If the axial force F_x and the cross-sectional area A_x are constant throughout the length of the member and equal to F and A, respectively, Eq. (2.7) yields

$$\delta = \frac{FL}{AE} \tag{2.8}$$

Equation (2.8) may also be used to determine the total elongation of a portion of a member of length L_i, when F_i, A_i, and E are constant over the whole length of the portion. Thus, for a member consisting of n such portions, its total deformation δ may be obtained from the equation

$$\delta = \sum_{n=1}^{n} \frac{F_i L_i}{A_i E} \tag{2.9}$$

The following examples illustrate the application of the above theory.

Example 2.1: The two-member truss in Fig. 2.2a is loaded by a vertical force at joint A as shown. The cross-sectional area of each member is $3\,cm^2$. Determine the maximum value of P so that the tensile or compressive stress in either of the two members does not exceed 140 MPa.

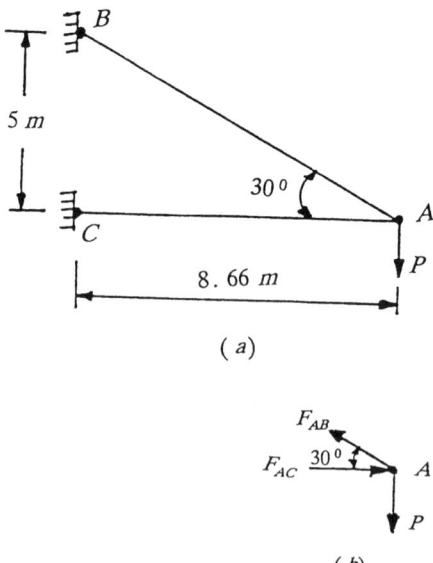

Figure 2.2. (*a*) Two-member truss. (*b*) Free-body diagram of joint A of the truss.

SOLUTION: The free-body diagram of joint A is shown in Fig. 2.2b. The static equilibrium equation in the vertical direction yields

$$-P + F_{AB} \cos 60° = 0$$

or

$$F_{AB} = 2P$$

The static equilibrium equation in the horizontal direction yields

$$F_{AC} - 2P \cos 30° = 0$$

or

$$F_{AC} = 1.7321P$$

For member AB, by using Eq. (2.3), we have

$$140\,\text{MPa} = \frac{2P}{(3)(1/100\ \text{m})^2}$$

or, solving for P, we find

$$P = 21{,}000\ \text{N}$$

In a similar manner, for member AC, we find

$$140\,\text{MPa} = \frac{1.7321P}{(3)(1/100\ \text{m})^2}$$

or

$$P = 24{,}248\ \text{N}$$

Thus, the maximum load P that can be applied at joint A is 21,000 N. Member AB governs the design.

Example 2.2: The member in Fig. 2.3a is loaded by a linearly varying axial force $f(x)$ and by an axial force F applied at the free end, as shown. The linear variation of $f(x)$ is shown in Fig. 2.3b. Determine the expression for the elongation δ_x at any distance x from the fixed support A, and the total elongation of the member. The modulus of elasticity E and cross-sectional area A are constant along the length of the member.

76   STRESS AND DEFORMATION

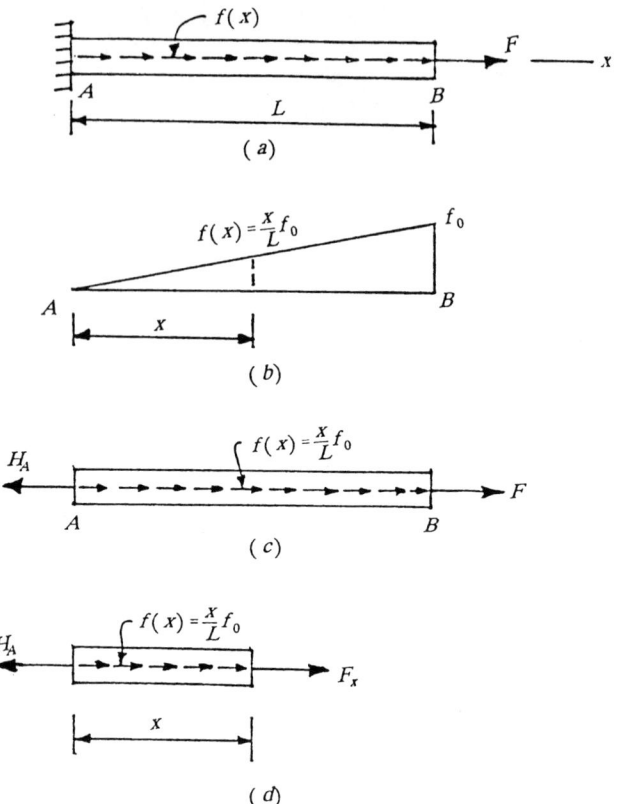

Figure 2.3. (*a*) Axially loaded member. (*b*) Variation of the axial force $f(x)$. (*c*) Free-body diagram of the member. (*d*) Free-body diagram of a portion x of the member.

SOLUTION: By applying static equilibrium to the member in Fig. 2.3c, we find

$$-H_A + \tfrac{1}{2}f_0 L + F = 0$$

or

$$H_A = F + \frac{f_0 L}{2} \tag{2.10}$$

By using the free-body diagram in Fig. 2.3d and applying static equilibrium in the x direction, we find

$$-H_A + \frac{1}{2}\frac{x}{L}f_0 x + F_x = 0 \tag{2.11}$$

FUNDAMENTALS OF STRESS, STRAIN, AND DEFORMATION 77

By using Eqs. (2.10) and (2.11) and solving for F_x, we obtain

$$F_x = F + \frac{f_0 L}{2} - \frac{f_0}{2L} x^2 \qquad (2.12)$$

Equation (2.12) provides the variation of the axial force F_x along the length of the member.

The elongation δ_x at any distance x from the fixed support A may be obtained by using Eq. (2.6). With E and A as constants, this equation yields

$$\begin{aligned}
\delta_x &= \int_0^x \frac{F_x \, dx}{EA} \\
&= \frac{1}{AE} \int_0^x \left[F + \frac{f_0 L}{2} - \frac{f_0}{2L} x^2 \right] dx \\
&= \frac{1}{AE} \left[Fx + \frac{f_0 L}{2} x - \frac{f_0}{6L} x^3 \right]_0^x \qquad (2.13) \\
&= \frac{1}{AE} \left[Fx + \frac{f_0 L}{2} x - \frac{f_0}{6L} x^3 \right]
\end{aligned}$$

At $x = 0$, Eq. (2.13) yields

$$\delta_{x=0} = 0 \quad \text{(as expected)} \qquad (2.14)$$

At $x = L$, Eq. (2.13) gives

$$\begin{aligned}
\delta_{x=L} &= \frac{1}{AE} \left[FL + \frac{f_0 L^2}{2} - \frac{f_0 L^3}{6L} \right] \\
&= \frac{L}{AE} \left[F + \frac{f_0 L}{3} \right] \qquad (2.15)
\end{aligned}$$

Equation (2.15) provides the total elongation of the member produced by the axial loading. Note that the first term of Eq. (2.15) shows the total elongation produced in the member by the force F, and the second one gives the total elongation produced by the linearly varying force $f(x)$.

2.2.2 Members Subjected to Torsional Moments

Consider the case where a member of solid circular cross section is subjected to pure torsion, as shown in Fig. 2.4a. An element of length dx is taken apart from the member, and its free-body diagram is shown in Fig. 2.4b. We assume here that plane sections remain plane after the application of the torsional

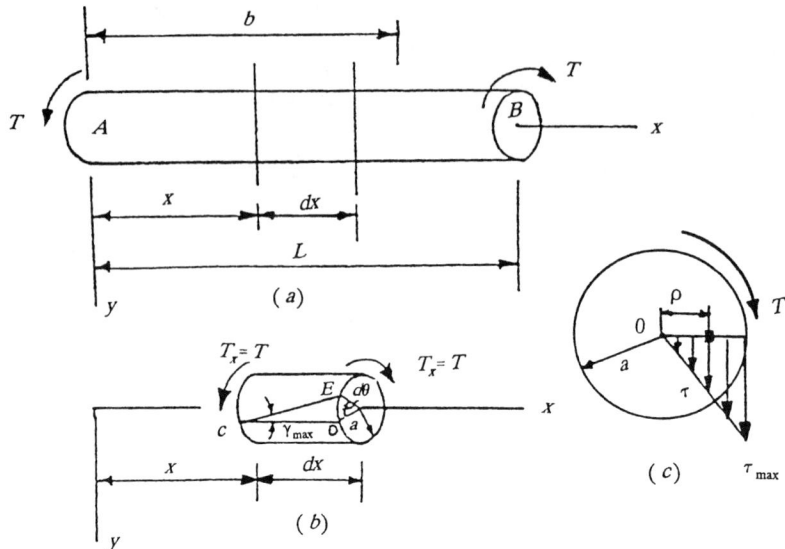

Figure 2.4. (*a*) Circular member subjected to pure torsion. (*b*) Free-body diagram of an element of the member. (*c*) Shear stress distribution at a cross section of the circular member.

moment T, and that the deformation of the element is as shown in Fig. 2.4*b*. Line *CD* will move to position *CE* and will remain straight, and the shear strain γ varies linearly from the central axis x to the outer surface of the element. The maximum shear strain γ_{max} is given by the angle *DCE* in Fig. 2.4*b*.

From the geometry of the deformation of the element in Fig. 2.4*b*, we have

$$\text{Arc } DE = \gamma_{max} \, dx \tag{2.16}$$

$$\text{Arc } DE = a \, d\theta \tag{2.17}$$

where a in Eq. (2.17) is the radius of the circular cross section and $d\theta$ is the angular twist produced on the element by the torsional moment $T_x = T$.

Equations (2.16) and (2.17) yield

$$\gamma_{max} \, dx = a \, d\theta \tag{2.18}$$

or

$$\gamma_{max} = a \frac{d\theta}{dx} = a\varphi \tag{2.19}$$

where

$$\varphi = \frac{d\theta}{dx} \tag{2.20}$$

FUNDAMENTALS OF STRESS, STRAIN, AND DEFORMATION   79

is the angular twist per unit of length of the member, or, simply, the rate of change of the angular twist θ with respect to x. At any distance of radius ρ from the center of the circular cross section, the shear strain γ is

$$\gamma = \rho\varphi \qquad (2.21)$$

which is obtained from Eq. (2.19) by replacing a with ρ.

If the material of the member is elastic and Hooke's law applies, the shear stress τ at a circular cross section would be proportional to the shear strain γ, and at any radius ρ from the center of the circular cross section, we have

$$\tau = G\gamma = G\rho\varphi \qquad (2.22)$$

where G is the constant of proportionality known as the shear modulus. Since the variation of γ is linear, the variation of τ would also be linear, and it would have the variation shown in Fig. 2.4c. It would be zero at the origin 0, and maximum when $\rho = a$. The shear stress τ would be a vector that is tangent to the periphery of any circle of radius ρ and of constant magnitude along the periphery.

From basic mechanics [38], it can be shown that τ is related to the twisting moment T by the expression

$$\tau = \frac{T\rho}{J} \qquad (2.23)$$

where J is the polar moment of inertia about the centroidal x axis. The maximum shear stress τ_{max} will occur when $\rho = a$, and, consequently,

$$\tau_{max} = \frac{Ta}{J} \qquad (2.24)$$

The maximum shear strain γ_{max} is

$$\gamma_{max} = \frac{\tau_{max}}{G} \qquad (2.25)$$

By using Eqs. (2.18), (2.24), and (2.25), we find

$$d\theta = \frac{T\,dx}{JG} \qquad (2.26)$$

which provides the relative angular twist of two adjoining sections of the member of distance dx apart. If T and J are variable along the length of the

80 STRESS AND DEFORMATION

member, Eq. (2.26) can be written as

$$d\theta = \frac{T_x \, dx}{J_x G} \tag{2.27}$$

where T_x and J_x are, respectively, the twisting moment and polar moment of inertia of the cross section.

The total angular twist θ between sections at points A and B of the member may be obtained by integrating Eq. (2.27). That is,

$$\theta = \int_A^B \frac{T_x \, dx}{J_x G} \tag{2.28}$$

or, for a length L of the member,

$$\theta = \int_0^L \frac{T_x \, dx}{J_x G} \tag{2.29}$$

Equations (2.28) and (2.29) are general to the extent that they may be used for members of circular or tubular cross sections, and where T_x and J_x are either constant or variable along the length of the member. If they are constant, Eq. (2.29) yields

$$\theta = \frac{TL}{JG} \tag{2.30}$$

The length L in Eq. (2.30) may be the length of the member, or the length b of a portion of a member where T and J are constant, as shown in Fig. 2.4a.

In the case where the cross section of a member is rectangular, a complete solution to this torsion problem is given by Saint Venant in his "Memoirs des Savants Strengers," published in 1855. See also references [6, 49]. Investigation of this problem reveals that the maximum shear stress τ_{max} occurs in the middle of the longer side of the rectangular cross section, and it is given by the equation

$$\tau_{max} = \frac{T}{\alpha b h^2} \tag{2.31}$$

The angular twist φ per unit of length of the member may be obtained from the equation

$$\varphi = \frac{T}{\beta b h^3 G} \tag{2.32}$$

FUNDAMENTALS OF STRESS, STRAIN, AND DEFORMATION 81

TABLE 2.1. Numerical Values of α and β for Various Ratios of b/h

b/h	α	β
1.00	0.208	0.141
1.50	0.231	0.196
1.75	0.239	0.214
2.00	0.246	0.229
2.50	0.258	0.249
3.00	0.267	0.263
4.00	0.282	0.281
6.00	0.299	0.299
8.00	0.307	0.307
10.00	0.312	0.312
∞	0.333	0.333

In Eqs. (2.31) and (2.32), T is the torsional moment, b is the longer side of the rectangular cross section, h is the shorter side of the cross section, α and β are numerical factors, and G is the shear modulus. Values of the numerical factors α and β for various ratios of b over h are provided in Table 2.1. From this table, we can conclude that for very narrow rectangular cross sections, Eqs. (2.31) and (2.32) reduce to the following form:

$$\tau_{max} = \frac{3T}{bh^2} \qquad (2.33)$$

$$\varphi = \frac{3T}{bh^3 G} \qquad (2.34)$$

Equations (2.33) and (2.34) provide a reasonable approximate solution to problems where the width of the cross section is small compared to its depth. Thin cylindrical shells, L-shaped cross sections, and so on, are examples of such categories. For such cases, the quantity b in Eqs. (2.33) and (2.34) may be considered to be the developed length of its center line.

The following examples illustrate the application of the above torsion theory.

Example 2.3: A stepped cantilever shaft of circular cross section is subjected to three concentrated twisting moments located as shown in Fig. 2.5a. Portion AB of the shaft is hollow, as shown in the same figure. Determine the magnitude and location of the maximum shear stress in the shaft, its total angular twist, and plot the variation of the angular twist θ along the length of the shaft. The shear modulus $G = 6 \times 10^6$ psi.

82   STRESS AND DEFORMATION

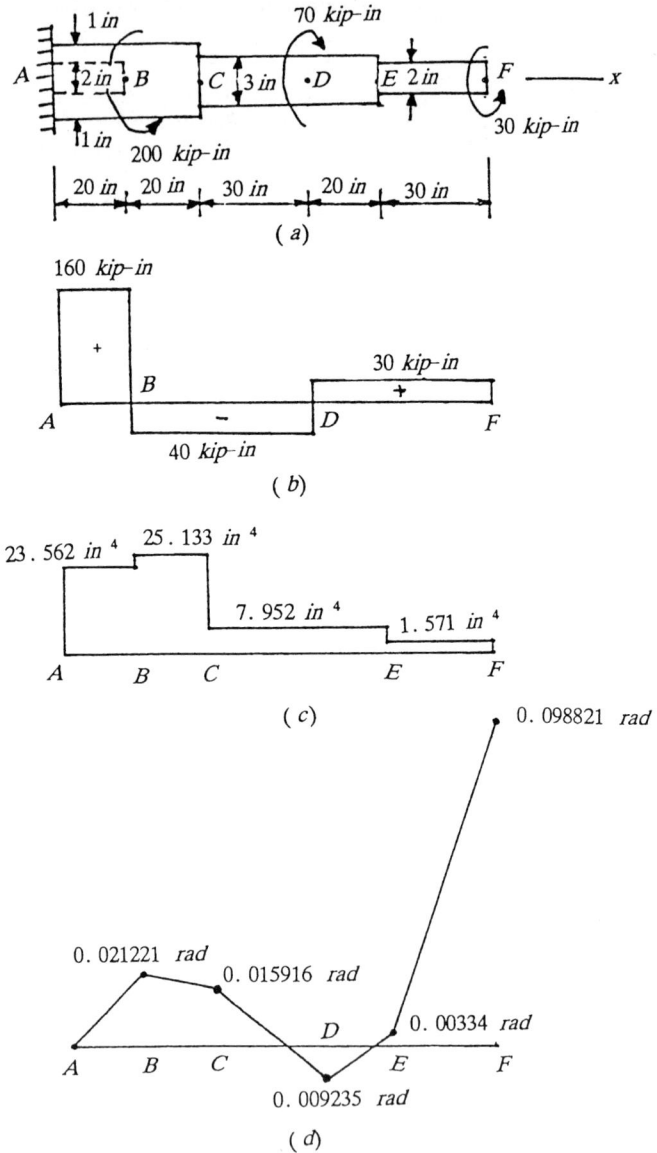

Figure 2.5. (*a*) Stepped shaft subjected to three twisting moments located as shown. (*b*) Twisting moment diagram. (*c*) Variation of *J* along the length of the shaft. (*d*) Angular twist diagram.

FUNDAMENTALS OF STRESS, STRAIN, AND DEFORMATION   83

SOLUTION: The variation of the internal twisting moment along the length of the shaft may be obtained by using appropriate free-body diagrams, and applying static equilibrium equations about the centroidal x axis. The twisting moment diagram is shown plotted in Fig. 2.5b. Counterclockwise twisting moments are considered positive.

The polar moments of inertia J of the various portions of the shaft may be determined by using one of the following two well-known equations:

$$J = \frac{\pi a^4}{2} \qquad (2.35)$$

$$J = \frac{\pi}{2}(b^4 - a^4) \qquad (2.36)$$

Equation (2.35) applies to solid circular sections of radius a, and Eq. (2.36) applies to hollow circular cross sections of external radius b and internal radius a.

The polar moments of inertia for the shaft in Fig. 2.5a are determined by using Eq. 2.36 for portion AB, and Eq. 2.35 for the remaining length of the shaft. The variation of J along the length of the shaft is shown plotted in Fig. 2.5c.

The maximum shear stress τ_{max} occurs at the surface of the shaft, and it can be determined by using Eq. 2.24. For the hollow part of the shaft, a in this equation should be replaced by the external radius b of the hollow section. On this basis, we have:

Portion AB:

$$\tau_{max} = \frac{Tb}{J} = \frac{(160)(2)}{23.562}$$

$$= 13.581 \text{ ksi}$$

Portion BC:

$$\tau_{max} = \frac{Ta}{J} = \frac{(-40)(2)}{25.133}$$

$$= -3.183 \text{ ksi}$$

Portion CD:

$$\tau_{max} = \frac{Ta}{J} = \frac{(-40)(1.5)}{7.952}$$

$$= -7.545 \text{ ksi}$$

Portion DE:

$$\tau_{max} = \frac{Ta}{J} = \frac{(30)(1.5)}{7.952}$$

$$= 5.659 \text{ ksi}$$

Portion EF:

$$\tau_{max} = \frac{Ta}{J} = \frac{(30)(1)}{1.571}$$

$$= 19.096 \text{ ksi}$$

The above results show that the maximum numerical value of the maximum shear stress τ_{max} occurs along section EF of the shaft, and it is equal to 19.096 ksi.

To determine the total angular twist of the shaft, we can start from the fixed end A where θ is zero. The total angular twist of the shaft may be determined by using Eq. (2.30) for the various portions of the shaft where T and J are constant. If we start from A and move along the length of the shaft, the angular twist θ_F at the free end of the shaft would be the total angular twist of the shaft. Thus, we have

$$\theta_F = \frac{T_{AB}L_{AB}}{J_{AB}G} + \frac{T_{BC}L_{BC}}{J_{BC}G} + \frac{T_{CD}L_{CD}}{J_{CD}G} + \frac{T_{DE}L_{DE}}{J_{DE}G} + \frac{T_{EF}L_{EF}}{J_{EF}G}$$

$$= \frac{(150)(20)}{(23.562)G} + \frac{(-40)(20)}{(25.133)G} + \frac{(-40)(30)}{(7.952)G} + \frac{(30)(20)}{(7.952)G} + \frac{(30)(30)}{(1.571)G}$$

$$= \frac{1}{G}[127.3237 - 31.8307 - 150.9054 + 75.4527 + 572.8835] \quad (2.37)$$

$$= 0.021221 - 0.005305 - 0.025151 + 0.012575 + 0.095481$$

or

$$\theta_F = 0.098821 \text{ rad}$$
$$= 5.662° \quad \text{(counterclockwise)} \quad (2.38)$$

To plot the variation of θ, we start at A where θ is zero. At point B, the angular twist θ_B would be equal to 0.021221 rad, which is the first term of Eq. (2.37). At point C, the angular twist θ_C is

$$\theta_C = 0.021221 - 0.005305 = 0.015916 \text{ rad}$$

FUNDAMENTALS OF STRESS, STRAIN, AND DEFORMATION 85

which is the algebraic sum of the first two terms of Eq. (2.37). In a similar manner,

$$\theta_D = \theta_C - 0.025151 = -0.009235 \text{ rad}$$

$$\theta_E = \theta_D + 0.012575 = 0.00334 \text{ rad}$$

and

$$\theta_F = \theta_E + 0.095481 = 0.098821 \text{ rad}$$

The results are plotted in Fig. 2.5d. Note that the variation of θ within each portion is linear because T, J, and G are constant within each portion. The variation of θ within each portion of a member can be easily obtained by examining Eqs. (2.28) and (2.29).

Example 2.4: For the rectangular beam loaded as shown in Fig. 2.6a, determine the maximum shear stress in the beam and the angular twist at its free end. If the rectangular cross section is replaced by a circular one, Fig. 2.6b, of the same area, determine its maximum shear stress and the angular twist at the free end and compare the results. The shear modulus $G = 10 \times 10^6$ psi.

SOLUTION: For the rectangular cross section, the maximum shear stress occurs in the middle of its longer side, and it is given by Eq. (2.31). Thus,

$$\tau_{max} = \frac{(60{,}000)}{\alpha(5)(2)^2} = \frac{3{,}000}{\alpha}$$

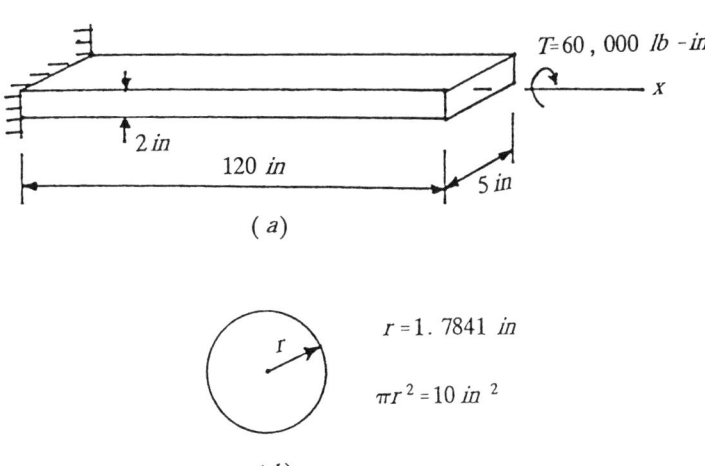

Figure 2.6. (a) Rectangular beam loaded as shown. (b) Circular cross section.

86 STRESS AND DEFORMATION

From Table 2.1 we find that $\alpha = 0.258$. Therefore,

$$\tau_{max} = \frac{3{,}000}{0.258} = 11{,}628 \text{ psi}$$

At the free end, the angular twist θ is

$$\theta = \varphi L = \frac{TL}{\beta bh^3 G}$$

$$= \frac{(60{,}000)(120)}{\beta(5)(2)^3(10)(10)^6}$$

$$= \frac{18{,}000}{\beta(10)^6}$$

From Table 2.1 we find that $\beta = 0.249$. Thus,

$$\theta = \frac{18{,}000}{(0.249)(10)^6} = 0.072289 \text{ rad}$$

$$= 4.14° \quad \text{(clockwise)}$$

If the member is circular, then

$$\pi r^2 = 10$$

$$r = 1.7841 \text{ in.}$$

By using Eq. (2.24), we find

$$\tau_{max} = \frac{Tr}{J} = \frac{(60{,}000)(1.7841)}{(\pi/2)(1.7841)^4}$$

$$= 6{,}726 \text{ psi}$$

From Eq. (2.30), the angular twist at the free end is

$$\theta = \frac{(60{,}000)(120)}{(\pi/2)(1.7841)^4(10)(10)^6}$$

$$= 0.045241 \text{ rad}$$

$$= 2.59° \quad \text{(clockwise)}$$

The results show that the circular cross section is much better than the rectangular one, because both maximum shear stress and angular twist are much lower. In practice, circular members are usually used when their primary loading is torsion.

Example 2.5: A thin-walled tube is acted upon by a torsional moment $T = 20,000$ in.-lb. Its cross section is split as shown in Fig. 2.7. By using Eq. (2.34), determine the angular twist of the tube when its length is 16 in. Also, determine the angular twist when the tube is seamless and compare the results. The shear modulus $G = 10 \times 10^6$ psi. Assume that the cross section is free to warp.

SOLUTION: By using Eq. (2.34), the angular twist θ_1 of the split tube is

$$\theta_1 = \varphi L = \frac{3TL}{bh^3G} \qquad (2.39)$$

In Eq. (2.39), the developed length of the mean circle of the cross-sectional area of the tube will be substituted for b; that is,

$$b = 2\pi r = 2\pi(3) = 6\pi \qquad (2.40)$$

where $r = 3$ in. is the mean radius of the cross section and $h = 0.2$ in. is the thickness of the tube.

By substituting into Eq. (2.39), we find

$$\theta_1 = \varphi L = \frac{3TL}{6\pi h^3 G}$$

$$= \frac{(3)(20,000)(16)}{6\pi(0.2)^3(10)^7} = \frac{2}{\pi}$$

If the tube is seamless, its angular twist θ_2 is

$$\theta_2 = \varphi L = \frac{TL}{2\pi r^3 hG}$$

$$= \frac{(20,000)(16)}{2\pi(3)^3(0.2)(10)^7} = \frac{8}{2,700\pi}$$

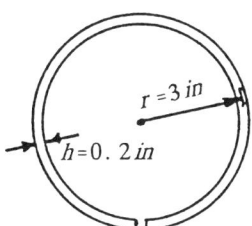

Figure 2.7. Thin-walled tube subjected to torsion.

88 STRESS AND DEFORMATION

The ratio θ_1/θ_2 is

$$\frac{\theta_1}{\theta_2} = \frac{2{,}700}{4} = 675 \tag{2.41}$$

Equation (2.41) shows that the angular twist of the split tube is 675 times that of the seamless tube. Consequently, a split tube is a very weak structural element to be used for torsional resistance.

2.3 MEMBERS SUBJECTED TO BENDING

We consider now the member in Fig. 2.8a, which is subjected to bending by the application of the concentrated loads F_1, F_2, F_3, distributed load w, and bending moment M_B, located as shown. The undeformed shape of an element of length dx of the member, at a distance x from support A, is shown in Fig. 2.8b. In the same figure, AB is the centroidal (or neutral) axis of the member, and GH is the centroidal axis of the element.

When the loads w, F_1, F_2, F_3, and moment M_B are applied, the member will bend, and the element $CDEF$ will deform as shown in Fig. 2.8c. Points C, D, E, and F will move to positions C', D', E', and F', respectively, after deformation, while length GH remains the same, because we assume that no normal stress exists at the neutral axis AB. We also assume that plane sections remain plane after deformation, and, consequently, lines $C'E'$ and $D'F'$ will remain straight. On the other hand, lines $C'D'$ and $E'F'$ become curved as a result of the action of the internal bending moments developed at the cross sections of the member. The deflected configuration of the centroidal axis of the member is known as the elastic line of the member produced by bending, and it needs to be determined. The deflections caused by the shear forces at cross sections along the length of the member are neglected as being small compared to ones produced by bending.

The solid line in Fig. 2.9a shows the enlarged deformed configuration of the element, while the dashed line is used for the undeformed configuration. The deformation of the member is assumed to be small compared to its other dimensions, such as length, and consequently, the very small longitudinal movements of points G and H are neglected. The angle $d\theta$ between the tangents at G and H and the one between the lengths GO and HO are equal in magnitude since their sides are mutually perpendicular.

By considering the deformed configuration of the element in Fig. 2.9b and the fiber KL of distance y from GH, and drawing the line HL' parallel to line $C'E'$, we note that sectors $HL'L$ and OGH are similar. Thus, from geometry,

$$\frac{L'L}{GH} = \frac{HL'}{OG} \tag{2.42}$$

MEMBERS SUBJECTED TO BENDING   89

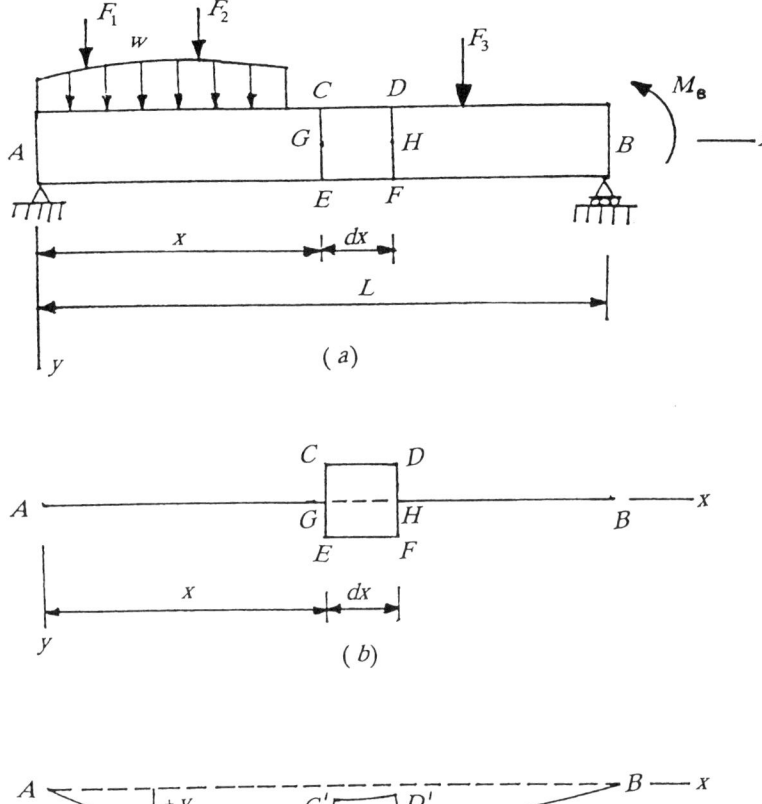

Figure 2.8. (a) Member subjected to bending. (b) Undeformed configuration of an element of the member of length dx. (c) Deformed configuration of the element and centroidal axis.

In Eq. (2.42), we note that OG is the radius of curvature ρ shown in Fig. 2.9a, HL' is y, and by definition, $L'L/GH$ is the longitudinal strain ε, since $GH = KL'$. Thus, Eq. (2.42) yields

$$\varepsilon = \frac{y}{\rho} \qquad (2.43)$$

90 STRESS AND DEFORMATION

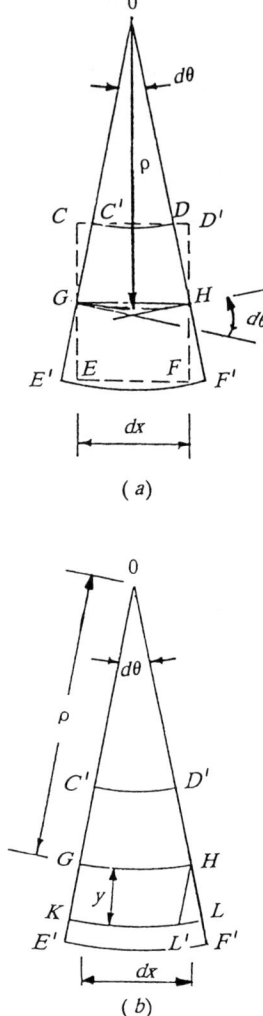

Figure 2.9. (*a*) Deformed and undeformed configurations of a beam element. (*b*) Deformed configuration of the beam element.

which shows that the longitudinal strain ε is proportional to the distance y from the neutral (centroidal) axis.

For linearly elastic systems where Hooke's law applies, the associated normal stress σ produced by bending is proportional to the strain ε. Thus, we have

$$\sigma = E\varepsilon = \frac{Ey}{\rho} \tag{2.44}$$

where E is the constant of proportionality known as Young's modulus of elasticity.

From mechanics of solids, we know that the bending stress σ at any distance y from the neutral axis is given by the equation

$$\sigma = \frac{My}{I} \tag{2.45}$$

where M is the bending moment at the cross section, and I is the moment of inertia of the cross-sectional area about its centroidal axis. By substituting Eq. (2.45) into Eq. (2.44) and solving for the curvature $1/\rho$, we find

$$\frac{1}{\rho} = \frac{M}{EI} \tag{2.46}$$

The product EI in Eq. (2.46) is known as the bending stiffness of the member.

Equation (2.46) relates the bending moment at cross sections along the length of the member with the curvature $1/\rho$ of the elastic line. The moment M, as well as the moment of inertia, may vary along the length of the member, or they may be constant. If both M and EI are constant, then the curvature $1/\rho$ is constant along the length of the member, and, consequently, the elastic line of the member would be an arc of a circle of radius ρ.

From differential calculus and with no regard to sign, the curvature $1/\rho$ is given by the expression

$$\frac{1}{\rho} = \left| \frac{y''}{[1 + (y')^2]^{3/2}} \right| \tag{2.47}$$

where

$$y' = \frac{dy}{dx} \tag{2.48}$$

$$y'' = \frac{d^2y}{dx^2} \tag{2.49}$$

For small deformations, the term $(y')^2$ in the denominator of Eq. (2.47) would be small compared to one, and it could be neglected. On this basis, Eq. (2.47) yields

$$\frac{1}{\rho} = |y''| \tag{2.50}$$

By substituting Eq. (2.50) into Eq. (2.46), we find

$$\frac{d^2y}{dx^2} = \pm \frac{M}{EI} \tag{2.51}$$

TABLE 2.2. First Quadrant and Fourth Quadrant Sign Conventions

Consistent Notation (All Signs Positive)	First Quadrant Coordinates	Fourth Quadrant Coordinates
y		
$\dfrac{dy}{dx} = \theta$		
$\dfrac{d^2y}{dx^2} = \dfrac{M}{EI}$		
$\dfrac{d^3y}{dx^3} = \dfrac{V}{EI}$		
$\dfrac{d^4y}{dx^4} = \dfrac{w}{EI}$		

Equation (2.51) is the differential equation of an elastic line, and it was formulated by the Swiss mathematician James Bernoulli in 1694.

If we decide to use the fourth quadrant coordinate system of axes in Fig. 2.8 with downward deflection y as positive, we have

$$\frac{d^2y}{dx^2} = -\frac{M}{EI} \tag{2.52}$$

With this sign convention in mind, we have the following relations between deflection y, slope θ, bending moment M, shear force V, and loading w:

$$
\begin{aligned}
&y &&\text{(positive downward)} \\
&\frac{dy}{dx} = \theta &&\text{(positive when } y \text{ increases with increasing } x\text{)} \\
&\frac{d^2y}{dx^2} = -\frac{M}{EI} &&\text{(positive when curvature is concave upwards)} \\
&\frac{d^3y}{dx^3} = -\frac{V}{EI} &&\text{(shear is positive when } \uparrow +V \downarrow\text{)} \\
&\frac{d^4y}{dx^4} = \frac{w}{EI} &&\text{(downward } w \text{ is positive)}
\end{aligned}
\tag{2.52a}
$$

It is often convenient to adapt a sign convention for M, V, and w so that all signs in the above expressions are positive. Two examples of such sign conversions are given in Table 2.2. The reader should become familiar with all three foregoing sign conventions and use whichever is most convenient in any given practical problem.

The following examples illustrate the application of the theory to practical problems.

Example 2.6: By utilizing the expressions in Table 2.2 and first quadrant sign convention, determine the equation of the elastic line of the simply supported beam in Fig. 2.10a, which is loaded by a distributed load w as shown in the figure. Also determine the maximum deflection caused by the bending of the member. The stiffness EI is constant throughout the length of the beam.

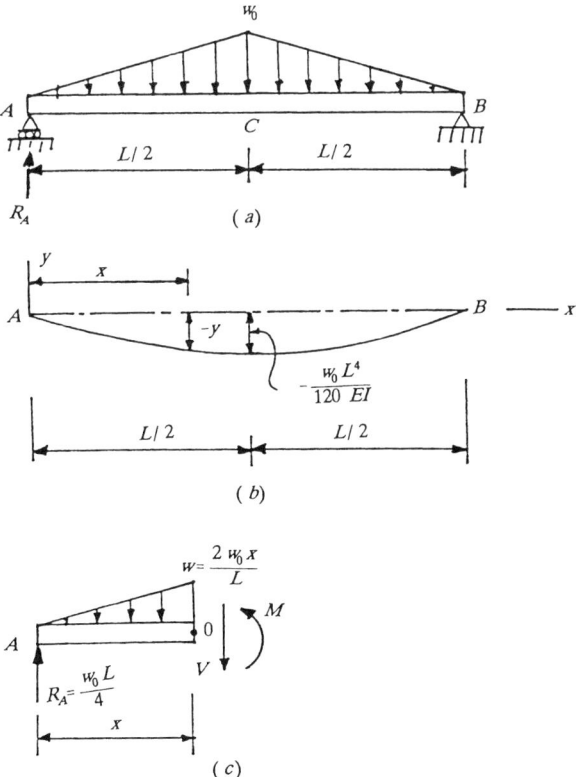

Figure 2.10. (a) Uniform simply supported beam loaded as shown. (b) Shape of the elastic line of the member. (c) Free-body diagram of a portion of the member of length x.

94 STRESS AND DEFORMATION

SOLUTION: The load distribution w of the beam is symmetrical with respect to its center C, and, consequently, the elastic line of the member would be symmetrical about this point. On this basis, the derivation of the equation of the elastic line for portion AC of the member would be sufficient, because the other half is identical. The shape of the elastic line is shown schematically in Fig. 2.10b, together with the first quadrant coordinate system of axes that needs to be used.

At any distance x from support A, the load w is given by the expression

$$w = -\frac{2w_0 x}{L} \qquad 0 \leqslant x \leqslant \frac{L}{2} \qquad (2.53)$$

and it is negative according to the first quadrant sign convention.

From Table 2.2, we have

$$\frac{d^4 y}{dx^4} = \frac{w}{EI} \qquad (2.54)$$

The expression for the elastic line y of the member may be determined from Eq. (2.54) by integrating four times. This is known as the method of integration. The four constants of integration may be determined by using appropriate boundary conditions, as shown later in this example.

By substituting Eq. (2.53) into Eq. (2.54), have

$$\frac{d^4 y}{dx^4} = -\frac{2w_0 x}{EIL} \qquad 0 \leqslant x \leqslant \frac{L}{2} \qquad (2.55)$$

Integration of Eq. (2.55) yields the expression for the shear force V divided by the stiffness EI of the member. This yields

$$\begin{aligned}\frac{V}{EI} &= \int -\frac{2w_0 x}{EIL} \, dx + C_1 \\ &= -\frac{w_0 x^2}{EIL} + C_1 \qquad 0 \leqslant x \leqslant \frac{L}{2}\end{aligned} \qquad (2.56)$$

Integration of Eq. (2.56) yields the expression for the bending moment of the member divided by the stiffness EI; that is,

$$\begin{aligned}\frac{M}{EI} &= \int -\frac{w_0 x^2}{EIL} \, dx + \int C_1 \, dx + C_2 \\ &= -\frac{w_0 x^3}{3EIL} + C_1 x + C_2 \qquad 0 \leqslant x \leqslant \frac{L}{2}\end{aligned} \qquad (2.57)$$

Integration of Eq. (2.57) yields the rotation (or slope) θ of the elastic line of the member. Thus, we have

$$\theta = \frac{dy}{dx} = \int -\frac{w_0 x^3}{3EIL} dx + \int C_1 x\, dx + \int C_2\, dx + C_3$$
$$= -\frac{w_0 x^4}{12EIL} + C_1 \frac{x^2}{2} + C_2 x + C_3 \qquad 0 \leqslant x \leqslant \frac{L}{2} \qquad (2.58)$$

Finally, integration of Eq. (2.58) yields the expression y for the elastic line of the member. We have

$$y = \int \frac{w_0 x^4}{3EIL} dx + \int C_1 \frac{x^2}{2} dx + \int C_2 x\, dx + \int C_3\, dx + C_4$$
$$= -\frac{w_0 x^5}{60EIL} + C_1 \frac{x^3}{6} + C_2 \frac{x^2}{2} + C_3 x + C_4 \qquad 0 \leqslant x \leqslant \frac{L}{2} \qquad (2.59)$$

The four boundary conditions that are needed for the computation of the four constants of integration, C_1, C_2, C_3, and C_4, may be taken as follows:

At $x = 0$:

$$y = 0 \qquad (2.60)$$
$$M = 0 \qquad (2.61)$$
$$V = -\frac{w_0 L}{4} \qquad (2.62)$$

At $x = L/2$:

$$\theta = 0 \quad \text{(symmetry)} \qquad (2.63)$$

By using Eq. (2.59) and applying the boundary condition given by Eq. (2.60), we find

$$C_4 = 0 \qquad (2.64)$$

Also, by using Eq. (2.57) and applying the boundary condition given by Eq. (2.61), we find

$$C_2 = 0 \qquad (2.65)$$

By using Eq. (2.56) and the boundary condition given by Eq. (2.62), we find

$$C_1 = \frac{w_0 L}{4EI} \qquad (2.66)$$

96 STRESS AND DEFORMATION

From Eq. (2.58) and the boundary condition given by Eq. (2.63), we find

$$-\frac{w_0 L^4}{192 EIL} + \frac{w_0 L^3}{32 EI} + C_3 = 0$$

or

$$C_3 = -\frac{5 w_0 L^4}{192 EIL} \tag{2.67}$$

By substituting the values of C_1, C_2, C_3, and C_4 given by Eqs. (2.66), (2.65), (2.67), and (2.64), respectively, into Eq. (2.59), we find

$$y = -\frac{w_0 x^5}{60 EIL} + \frac{w_0 L^2 x^3}{24 EIL} - \frac{5 w_0 L^4 x}{192 EIL} \qquad 0 \leqslant x \leqslant \frac{L}{2} \tag{2.68}$$

Equation (2.68) is the expression for the elastic line of the member with $0 \leqslant x \leqslant L/2$. The remaining half portion of the elastic line is symmetrical.

Since the loading w and the elastic line of the member are symmetrical with respect to the center C of the beam, the maximum deflection is at the center C. By using Eq. (2.68) and $x = L/2$, we find

$$y_{max} = y_{x=L/2} = -\frac{w_0 L^4}{120 EI} \tag{2.69}$$

The negative sign in Eq. (2.69) indicates that the deflection y_{max} is in the downward direction, which agrees with the first quadrant sign convention that was selected.

The slope θ of the elastic line y, at any $0 \leqslant x \leqslant L/2$, may be obtained from Eq. (2.58) since the constants C_1, C_2, and C_3 are known. By substituting the values of the constants, Eq. (2.58) yields

$$\theta = -\frac{w_0 x^3}{12 EI} + \frac{w_0 L x^2}{8 EI} - \frac{5 w_0 L^3}{192 EI} \tag{2.70}$$

At $x = 0$, which is support A in Fig. 2.10b, we have

$$\theta_{x=0} = -\frac{5 w_0 L^3}{192 EI} \tag{2.71}$$

The negative sign indicates that the rotation is clockwise.

Using the Double Integration Method: The equation of the elastic line y of the member in Fig. 2.10a may also be determined by using Eq. (2.52) and integrating twice. This is known as the double integration method. The two constants of integration may be determined by using the slope and deflection boundary conditions of the elastic line. The use of Eq. (2.52), however, requires the expression for the bending moment M along the length of the member, which can be determined by using appropriate free-body diagrams and the equilibrium equations from statics, as discussed in the preceding chapter.

By applying statics, the reaction R_A at support A of the member may be determined and it is equal to $w_0 L/4$. Thus, by using the free-body diagram in Fig. 2.10c and taking moments about point 0, the bending moment M, at any $0 \leqslant x \leqslant L/2$, is given by the following expression:

$$M = \frac{w_0 L x}{4} - \frac{w_0 x^3}{3L} \qquad 0 \leqslant x \leqslant \frac{L}{2} \qquad (2.72)$$

By substituting Eq. (2.72) into Eq. (2.52), we find

$$\frac{d^2 y}{dx^2} = -\frac{1}{EI}\left(\frac{w_0 L x}{4} - \frac{w_0 x^3}{3L}\right) \qquad 0 \leqslant x \leqslant \frac{L}{2} \qquad (2.73)$$

We apply now the method of double integration. By integrating Eq. (2.73) once, we obtain

$$\theta = \frac{dy}{dx} = -\frac{1}{EI}\left(\frac{w_0 L x^2}{8} - \frac{w_0 x^4}{12L}\right) + C_1 \qquad 0 \leqslant x \leqslant \frac{L}{2} \qquad (2.74)$$

Integration of Eq. (2.74) yields

$$y = -\frac{1}{EI}\left(\frac{w_0 L x^3}{24} - \frac{w_0 x^5}{60L}\right) + C_1 x + C_2 \qquad 0 \leqslant x \leqslant \frac{L}{2} \qquad (2.75)$$

The constants C_1 and C_2 may be determined by using the following boundary conditions:

$$\text{At } x = 0 \qquad y = 0 \qquad (2.76)$$
$$\text{At } x = L/2 \qquad \theta = 0 \qquad (2.77)$$

By using Eq. (2.75) and applying the boundary condition given by Eq. (2.76), we find $C_2 = 0$. By using Eq. (2.74) and applying the boundary condition given by Eq. (2.77), we find

$$-\frac{1}{EI}\left(\frac{w_0 L^3}{32} - \frac{w_0 L^4}{192L}\right) + C_1 = 0$$

98 STRESS AND DEFORMATION

or

$$C_1 = \frac{5w_0 L^3}{192EI} \qquad (2.78)$$

With C_1 and C_2 known, Eq. (2.75) yields

$$y = -\frac{1}{EI}\left(\frac{w_0 L x^3}{24} - \frac{w_0 x^5}{60L} - \frac{5w_0 L^3 x}{192}\right) \qquad 0 \leq x \leq \frac{L}{2} \qquad (2.79)$$

Equation (2.79) is identical to Eq. (2.68). The negative sign in Eq. (2.79) is associated with sign convention, because in the first quadrant sign convention upward deflection is assumed positive. In Eq. (2.52), downward deflection is assumed positive. At $x = L/2$, Eq. (2.79) yields

$$y_{x=L/2} = \frac{w_0 L^4}{120EI} \quad \text{(downward)} \qquad (2.80)$$

The double integration method is easier to use for this problem and less time consuming.

Example 2.7: The steel beam in Fig. 2.11a is loaded by a uniformly distributed load w throughout its length. The uniform cross section of the member has the shape shown in Fig. 2.11b. If the material of the beam is equally strong in both tension and compression, determine the maximum load w that can be applied to the member if the allowable bending stress $\sigma_{al} = 20,000$ psi. The modulus of elasticity $E = 30 \times 10^6$ psi.

SOLUTION: By applying statics, reactions R_A and R_B in Fig. 2.11a are found to be 18.25w and 31.75w, respectively. By using the free-body diagram in Fig. 2.11d and taking moments about point 0, we find

$$M = 18.25wx - \frac{wx^2}{2} \qquad (2.81)$$

Equation (2.81) represents the variation of the bending moment M for span AB of the member.

The location of the maximum value of M may be determined by setting equal to zero the derivative of M with respect to x. This yields

$$\frac{dM}{dx} = 18.25w - wx = 0$$

or

$$x = 18.25 \text{ ft}$$

MEMBERS SUBJECTED TO BENDING 99

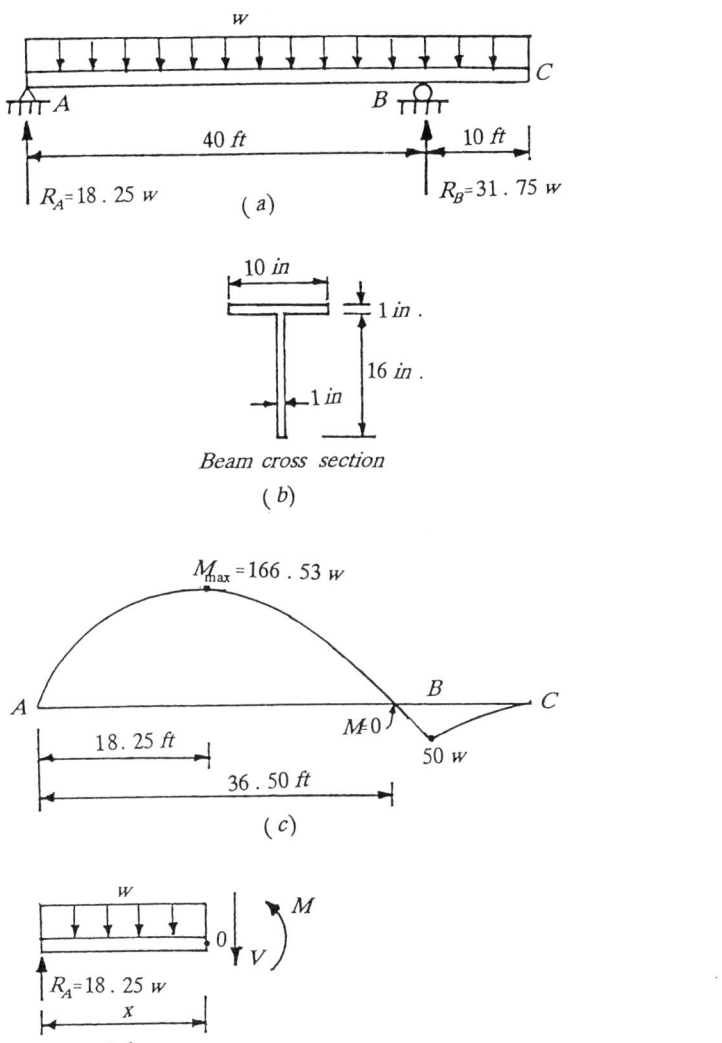

Figure 2.11. (a) Uniform beam loaded as shown. (b) Cross-sectional shape of the beam. (c) Bending moment diagram of the beam. (d) Free-body diagram.

By substituting $x = 18.25$ ft into Eq. (2.81), we find

$$M = M_{max} = 18.25w(18.25) - \frac{w(18.25)^2}{2}$$
$$= 166.53w$$
(2.82)

The complete bending moment diagrams of the member is shown in Fig. 2.11c.

In order to be able to locate the maximum bending stress σ_{max}, we must first determine the location of the neutral axis where the bending stress σ is zero. Since we have no axial load acting on the beam, the neutral axis coincides with the centroidal axis of the cross section of the member. By taking first moments about point Q in Fig. 2.12a, we find

$$[(10)(1) + (16)(1)]\bar{y} = (10)(1)(16.5) + (16)(1)(8)$$

or

$$\bar{y} = 11.27 \text{ in.} \tag{2.83}$$

By applying the parallel axis theorem, the moment of inertia $I_{N.A.}$ of the cross section about the neutral axis is

$$I_{N.A.} = \frac{(10)(1)^3}{12} + (10)(1)(16.50 - 11.27)^2 + \frac{(1)(16)^3}{12} + (16)(1)(11.27 - 8.00)^2$$

$$= 0.83 + 273.53 + 341.33 + 171.09 \tag{2.84}$$

$$= 786.78 \text{ in.}^4$$

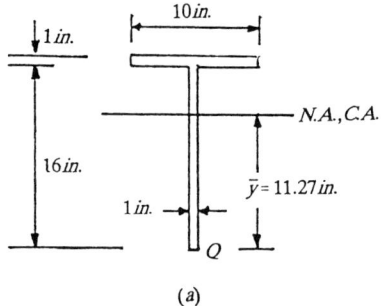

(a)

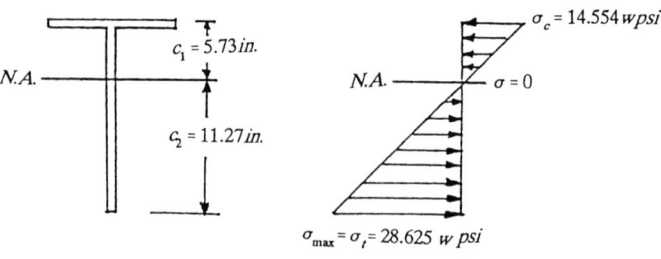

(b)

Figure 2.12. (a) Location of neutral (N.A.) and centroidal (C.A.) axes. (b) Bending stress distribution at the cross section of maximum bending moment.

MEMBERS SUBJECTED TO BENDING   101

We recall now Eq. (2.45), which we rewrite again below:

$$\sigma = \frac{My}{I} \tag{2.85}$$

Examining Eq. (2.85), we note that the bending stress σ is zero at the neutral axis, it is inversely proportional to the moment of inertia I about the neutral axis, and it is proportional to the bending moment M and the distance y from the neutral axis. Since I is constant throughout the length of the member, the maximum values of M and y will yield the maximum bending stress σ_{max}. From the bending moment diagram in Fig. 2.11c, we note that $M_{max} = 166.53w$, and from Fig. 2.12b, the maximum distance from the neutral axis is $c_2 = 11.27$ in. Thus, we have

$$\sigma_{max} = \frac{166.53w(11.27)}{786.78} \quad (12)$$
$$= 28.625w \text{ psi} \quad \text{(tension)} \tag{2.86}$$

The factor of 12 in the above equation is used to convert feet into inches for the units of the bending moment. The bending stress distribution along the cross section of the maximum bending moment is shown in Fig. 2.12b, where σ_c and σ_t indicate compressive and tensile bending stress, respectively.

The allowable bending stress, however, is 20,000 psi. On this basis, Eq. (2.86) yields

$$20,000 = 28.625w$$

or

$$w = 698.69 \text{ lb/in.}$$

This is the maximum load w that can be applied to the member in Fig. 2.11a in order to conform with the requirement of 20,000 psi allowable bending stress.

Example 2.8: The simply supported beam in Fig. 2.13a is loaded by a concentrated force F, as shown. By using the double integration method, determine the slope and deflection equations of the elastic line of the member. Also, determine the slope and deflection at point C when $F = 20$ kips, $a = 10$ ft, and $b = 30$ ft. The stiffness EI is constant throughout the length of the member.

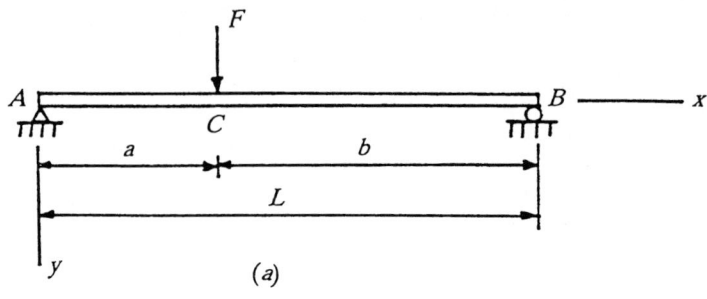

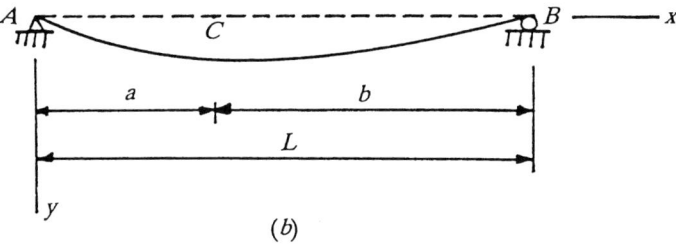

Figure 2.13. (a) Simply supported beam loaded by a force F as shown. (b) Elastic line of the member.

SOLUTION: The reactions R_A and R_B at supports A and B, respectively, are

$$R_A = \frac{Fb}{L} \tag{2.87}$$

$$R_B = \frac{Fa}{L} \tag{2.88}$$

By considering support A as the origin of the x, y coordinate system of axes, the expressions for the bending moment M_x for portions AC and CB of the member are derived by using statics, and they are as follows:

$$M_x = \frac{Fb}{L} x \qquad 0 \leqslant x \leqslant a \tag{2.89}$$

$$M_x = \frac{Fb}{L} x - F(x-a) \qquad a \leqslant x \leqslant L \tag{2.90}$$

By applying Eq. (2.52), the curvatures for portions AC and CB of the member are as follows:

$$\frac{d^2y}{dx^2} = -\frac{1}{EI}\left(\frac{Fb}{L}x\right) \qquad 0 \leqslant x \leqslant a \qquad (2.91)$$

$$\frac{d^2y}{dx^2} = -\frac{F}{EI}\left[\frac{b}{L}x - (x-a)\right] \qquad a \leqslant x \leqslant L \qquad (2.92)$$

Note that Eq. (2.52) is used twice, since the moment expressions for portions AC and CB are different.

By integrating Eq. (2.91) once and twice, we have

$$\theta = \frac{dy}{dx} = -\frac{1}{EI}\left(\frac{Fb}{2L}x^2\right) + C_1 \qquad 0 \leqslant x \leqslant a \qquad (2.93)$$

$$y = -\frac{1}{EI}\left(\frac{Fb}{6L}x^3\right) + C_1 x + C_2 \qquad 0 \leqslant x \leqslant a \qquad (2.94)$$

Integrating Eq. (2.92) once and twice, we find

$$\theta = \frac{dy}{dx} = -\frac{F}{EI}\left[\frac{b}{2L}x^2 - \frac{(x-a)^2}{2}\right] + C_3 \qquad a \leqslant x \leqslant L \qquad (2.95)$$

$$y = -\frac{F}{EI}\left[\frac{b}{6L}x^3 - \frac{(x-a)^3}{6}\right] + C_3 x + C_4 \qquad a \leqslant x \leqslant L \qquad (2.96)$$

The constants of integration, C_1, C_2, C_3, and C_4, may be determined by using the boundary conditions of the elastic line of the member. They are:

At $x = 0$	$y = 0$	(2.97)
At $x = L$	$y = 0$	(2.98)
At $x = a$	$(\theta_{AC})_{x=a} = (\theta_{CB})_{x=a}$	(2.99)
At $x = a$	$(y_{AC})_{x=a} = (y_{CB})_{x=a}$	(2.100)

The boundary conditions given by Eqs. (2.99) and (2.100) must be used, because they satisfy the continuity conditions for slope θ and deflection y, respectively, of the two portions at $x = a$.

By using Eq. (2.94) and the boundary condition given by Eq. (2.97), we find

$$C_2 = 0 \qquad (2.101)$$

By using Eqs. (2.93) and (2.95) and the boundary condition given by Eq. (2.99), we find

$$C_1 = C_3 \qquad (2.102)$$

From Eqs. (2.94) and (2.96) and the boundary condition given by Eq. (2.100), we obtain

$$C_4 = 0 \qquad (2.103)$$

From Eq. (2.96) and the boundary condition given in Eq. (2.98), we find

$$C_3 = \frac{Fb}{6EIL}(L^2 - b^2) \qquad (2.104)$$

and, consequently, from Eq. (2.102),

$$C_1 = \frac{Fb}{6EIL}(L^2 - b^2) \qquad (2.105)$$

With known C_1, C_2, C_3, and C_4, Eqs. (2.93) through (2.96) yield

$$\theta = \frac{Fb}{6EIL}(L^2 - b^2 - 3x^2) \qquad 0 \leqslant x \leqslant a \qquad (2.106)$$

$$\theta = \frac{Fb}{6EIL}\left[(L^2 - b^2) + \frac{3L}{b}(x-a)^2 - 3x^2\right] \qquad a \leqslant x \leqslant L \qquad (2.107)$$

$$y = \frac{Fbx}{6EIL}(L^2 - b^2 - x^2) \qquad 0 \leqslant x \leqslant a \qquad (2.108)$$

$$y = \frac{Fb}{6EIL}\left[(L^2 - b^2)x + \frac{L}{b}(x-a)^3 - x^3\right] \qquad a \leqslant x \leqslant L \qquad (2.109)$$

The slope at C may be evaluated from either Eq. (2.106) or Eq. (2.107) by substituting a for x. This yields

$$\theta_{x=a} = \frac{Fb}{6EIL}(L^2 - b^2 - 3a^2) \qquad (2.110)$$

By using either Eq. (2.108) or Eq. (2.109), the vertical deflection at C may be determined by substituting a for x, which yields

$$y_{x=a} = \frac{Fba}{6EIL}(L^2 - b^2 - a^2) \qquad (2.111)$$

Thus, when $F = 20$ kips, $a = 10$ ft, and $b = 30$ ft, Eqs. (2.110) and (2.111) yield

$$\theta_{x=a} = \frac{(20)(30)(12)^2}{(6)(40)EI}(1,600 - 900 - 300)$$

$$= \frac{(144)(10)^3}{EI}$$

$$y_{x=a} = \frac{(20)(30)(10)(12)^3}{(6)(40)EI}(1,600 - 900 - 100)$$

$$= \frac{25.92(10)^6}{EI}$$

The units of E should be kips/in.^{2}, and for I should be in.^{4}

The elastic line of the member is shown in Fig. 2.13b. Note that because the rotation θ is not zero at point C, the maximum deflection y does not occur under the load concentration point C. It is, however, fairly close to its maximum value under the load concentration point C.

2.4 COMBINED LOADING

In Sections (2.2) and (2.3), structural members were subjected to axial loadings, torsional moments, and loads that caused the member to bend. For each type of loading case, stress and deformation characteristics of the member were investigated. In this section, the structural member will be subjected to a combination of loading types and its stress characteristics will be examined. For example, a member may be subjected simultaneously to axial loadings and loads that produce bending. It could be also subjected to a combination of bending, torsion, and axial loadings. In practice, such loading combinations are a common occurrence, and special consideration must be provided by the design engineer in order to maintain their structural integrity.

2.4.1 Bending Combined with Axial Loading

When a member is subjected only to bending, the neutral axis and the centroidal axis at a cross section of the member coincide, and the cross-sectional bending stress distribution is given by Eq. (2.45). If an axial load is also acting on the member, then the neutral and centroidal axes will not coincide, and the stress distribution at a cross section of the member is given by the following equation:

$$\sigma_t = \pm \frac{My}{I} \pm \frac{P}{A} \qquad (2.112)$$

106 STRESS AND DEFORMATION

Equation (2.112) combines the normal stress contributions of both bending and axial loadings. The total normal stress σ_t is the algebraic sum of the one produced by bending, which is the first term on the right-hand side of Eq. (2.112), and the one produced by the axial force P, which is the second term on the right-hand side of Eq. (2.112). The letter y in the same equation denotes the distance from the centroidal axis of the cross section.

The neutral axis for the combined loading, that is, the axis where σ_t in Eq. (2.112) is zero, no longer coincides with the centroidal axis and its location must be determined. The following examples illustrate the application of the theory.

Example 2.9: A cantilever beam of uniform rectangular cross section is loaded by a uniformly distributed load $w = 5$ lb/in. and by an axial tensile force $P = 24{,}000$ lb, as shown in Fig. 2.14a. Determine the normal stress distribution at the cross section of maximum normal stress produced by the combined

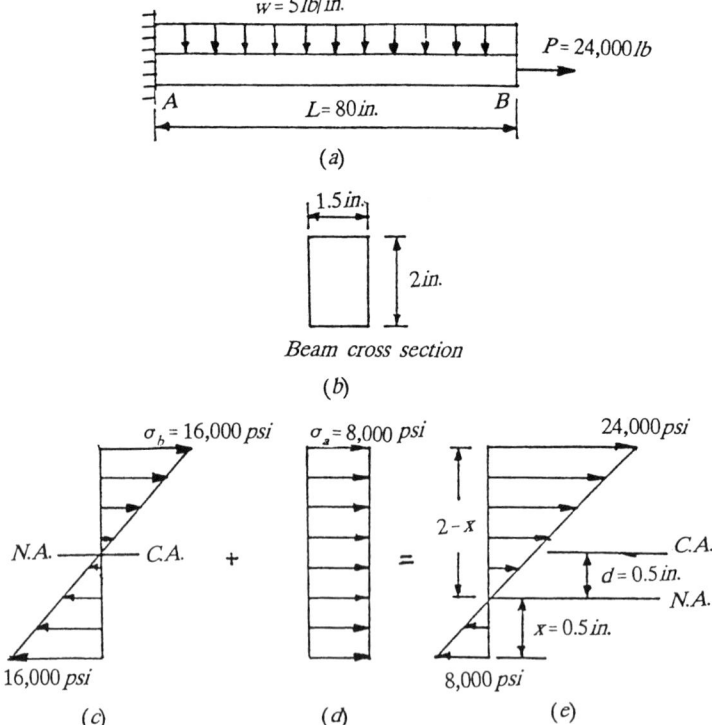

Figure 2.14. (a) Cantilever beam subjected to bending and axial loadings. (b) Rectangular cross section of the member. (c) Bending stress distribution at the fixed end. (d) Normal stress distribution at the fixed end produced by P. (e) Resultant normal stress distribution at the fixed end.

loading, and the location of the neutral axis. Also, plot the variation of the location of the neutral axis at cross sections along the length of the member.

SOLUTION: The rectangular cross section of the member is shown in Fig. 2.14b. Its moment of inertia I about the centroidal axis is

$$I = \frac{(1.5)(2)^3}{12} = 1 \text{ in.}^4$$

The cross section of maximum normal stress is located at the fixed end A of the member. At this section, the bending moment M_A is

$$M_A = \frac{(5)(80)^2}{2} = 16,000 \text{ in.-lb}$$

Bending of the member produces tensile normal stress at the top surface of the beam, and compressive stress at its bottom surface. If the plus sign indicates tension, the bending stress σ_b at the top and bottom surfaces at the fixed end is

$$\sigma_b = \pm \frac{M_A c}{I} = \pm \frac{(16,000)(1)}{1} = \pm 16,000 \text{ psi}$$

At the same cross section, the normal stress σ_a, produced by the application of the axial force P, is

$$\sigma_a = \frac{P}{A} = \frac{24,000}{(2)(1.5)} = 8,000 \text{ psi}$$

Figure 2.14c shows the normal stress distribution at the fixed end caused by bending, and Fig. 2.14d shows the normal stress distribution produced by the axial force P. Superposition of the normal stress distributions given in Figs. 2.14c and 2.14d yields the resultant normal stress distribution at the fixed end of the member.

The maximum normal tensile stress occurs at the top of the section and is equal to 24,000 psi, while the maximum numerical value of the compressive normal stress is at the bottom of the cross section and is equal to 8,000 psi. The neutral axis is located at a distance x from the bottom, as shown in Fig. 2.14e, and it can be determined by using similar triangles. That is,

$$\frac{8,000}{x} = \frac{24,000}{2-x}$$

or

$$x = 0.50 \text{ in.}$$

The distance d between the centroidal and neutral axes is

$$d = 1 - 0.5 = 0.50 \text{ in.}$$

The distance d varies as we consider cross sections along the length of the member, and it will approach infinity ($d \to \infty$) at the free end. For example, by following a similar procedure, the bending stress distribution at midspan is shown in Fig. 2.15a, the normal stress distribution due to the axial force P is

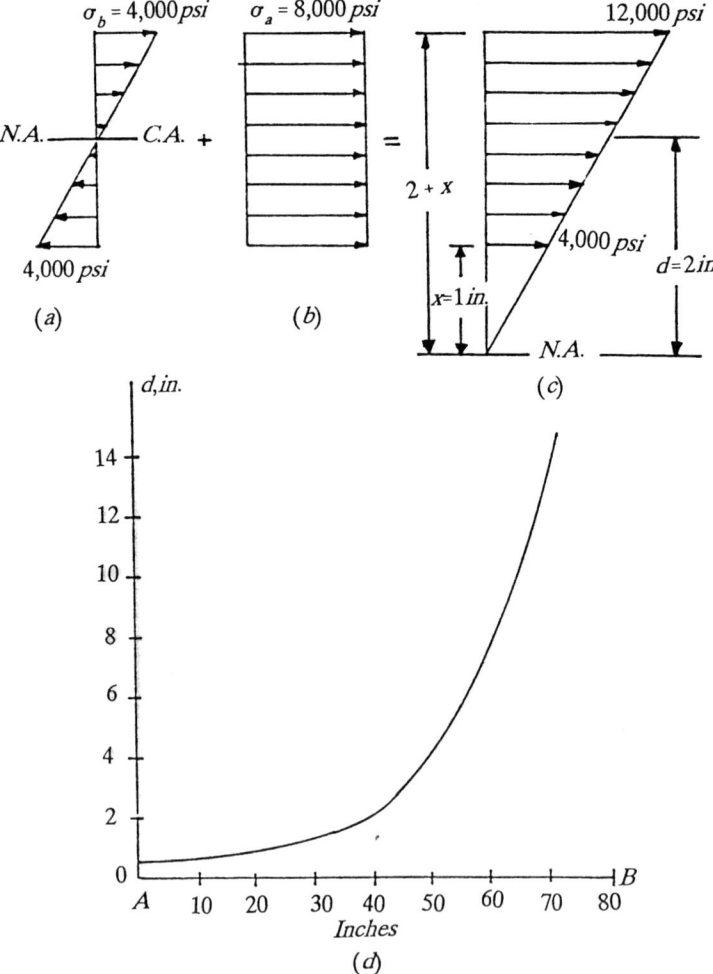

Figure 2.15. (a) Bending stress distribution at midspan. (b) Normal stress distribution at midspan produced by P. (c) Resultant normal stress distribution at midspan. (d) Variation of neutral axis location d at cross sections along the length of the member.

shown in Fig. 2.15b, and the resultant normal stress distribution is shown in Fig. 2.15c. In this case, $x = 1$ in. and $d = 2$ in., indicating that the neutral axis is located outside the cross section and, consequently, the whole cross section is in tension. Figure 2.15d shows the variation of d at cross sections along the length of the member.

Example 2.10: Determine the maximum load w that can be applied to the cantilever beam in Fig. 2.14a if the allowable normal stress $\sigma_{al} = 20,000$ psi. The axial force $P = 24,000$ lb.

SOLUTION: The maximum normal stress occurs at the fixed end A of the member. At A, the fixed-end moment M_A is

$$M_A = (w)(80)(40) = 3{,}200w \text{ in.-lb}$$

The maximum stress at the fixed end is

$$\sigma_{A\max} = \frac{3{,}200w(1)}{1} + \frac{24{,}000}{(1.5)(2)}$$

$$= 3{,}200w + 8{,}000$$

For an allowable stress of 20,000 psi, we have

$$20{,}000 = 3{,}200w + 8{,}000$$

or

$$w = 3.75 \text{ lb/in.}$$

Thus, the maximum load w that can be applied to the member is 3.75 lb/in.

2.4.2 Members Subjected to Combined Bending, Torsion, and Axial Force

We examine here a practical situation where a circular solid shaft is subjected to a loading combination that includes bending, torsion, and axial force. The following example illustrates the stress combinations that are developing at various points in the shaft by using infinitely small elements of sides $dx, dy,$ and dz.

Example 2.11: The circular shaft in Fig. 2.16a, of radius $r = 1$ in., is subjected to a uniformly distributed load $w = 5$ lb/in., a torsional moment $T = 8{,}000$ in.-lb, and an axial force $P = 12{,}000$ lb, as shown in the figure. Determine the stress condition of the shaft at points Q and O' by using infinitely small elements of sides $dx, dy,$ and dz. Discuss the results.

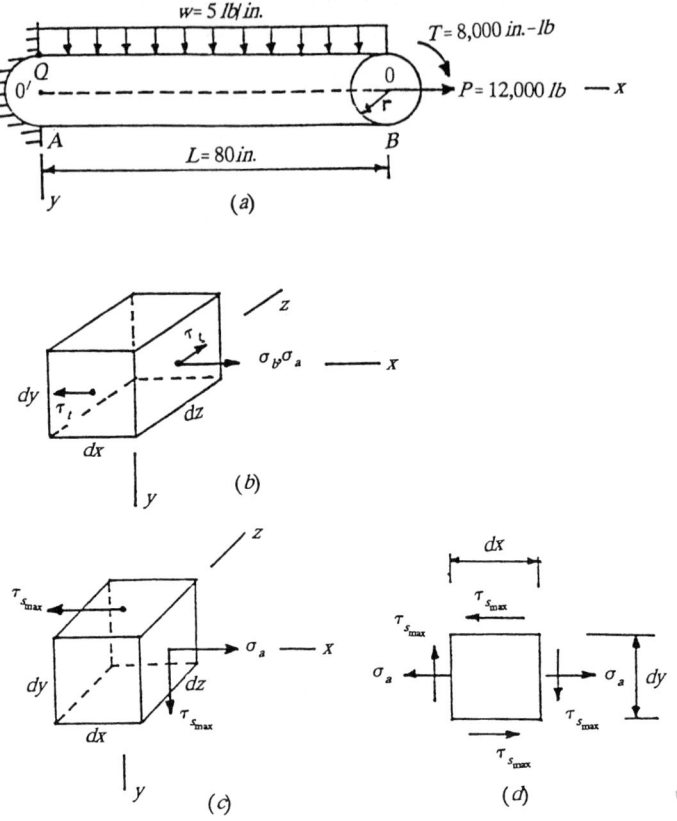

Figure 2.16. (a) Circular shaft subjected to bending, torsion, and axial force. (b) Stresses acting on an element located at point Q. (c) Stresses acting on an element located at point O'. (d) Usual two-dimensional representation of the element in Fig. 2.16c.

SOLUTION: The types of stresses acting on an infinitesimal element located at point Q in Fig. 2.16a are illustrated in Fig. 2.16b. The normal tensile stress σ_b is the maximum bending stress produced in the shaft, σ_a is the axial tensile normal stress produced by the axial force P, and τ_t is the maximum shear stress produced by the torsional moment T. The plane of the element that is normal to the x axis is defined as the x plane. On this basis, σ_a and σ_b are normal to this plane, and τ_t lies on the x plane. On the z plane, that is the plane of the element that is normal to the z axis, the shear stress τ_t is also present, because shear stresses on mutually perpendicular planes are equal (x and z planes are mutually perpendicular). On planes of the element that are parallel to x and z planes, the stresses are equal and opposite in sense. They are not shown on the element for simplicity purposes. On planes perpendicular to the y axis the stresses are zero.

The cross-sectional area A of the shaft is

$$A = \pi r^2 = \pi(1)^2 = \pi$$

The polar moment of inertia J_p is

$$J_p = \frac{\pi r^4}{2} = \frac{\pi(1)^4}{2} = \frac{\pi}{2}$$

The moment of inertia I_z about the z axis is

$$I_z = \frac{J_p}{2} = \frac{\pi}{4}$$

The bending moment M_A at the fixed end A is

$$M_A = (5)(80)(40) = 32{,}000 \text{ in.-lb}$$

Thus, the bending stress σ_b at point Q is

$$\sigma_b = \frac{M_A r}{I_z} = \frac{(32{,}000)(1)}{\pi/4} = 40{,}744 \text{ psi} \quad \text{(tension)}$$

At the same point, the stress σ_a produced by P is

$$\sigma_a = \frac{P}{A} = \frac{12{,}000}{\pi} = 3{,}820 \text{ psi} \quad \text{(tension)}$$

The shear stress τ_t produced at point Q by the torsional moment T may be obtained by using Eq. (2.24). Thus, we have

$$\tau_t = \frac{Tr}{J_p} = \frac{(8{,}000)(1)}{\pi/2} = 5{,}093 \text{ psi}$$

which is directed as shown in Fig. 2.16b. Note that σ_a and σ_b may be superimposed, yielding a total normal stress $\sigma = 44{,}564$ psi at point Q. Point Q is subjected to the highest normal tensile stress in the shaft that is produced by the combined loading.

At point O', which is the center of the circular cross section at the fixed end A, the stresses produced on an infinitesimal element of sides dx, dy, and dz are shown in Fig. 2.16c. The bending stress σ_b and the shear stress τ_t caused by the torsional moment T are both zero at O'. We have, however, at this point, the normal tensile stress σ_a that is produced by the axial force P, and the maximum shear stress $\tau_{s\max}$ that is produced by the shear force V acting at the cross section. The internal forces and moments acting at a cross section a distance x from support A are shown in Fig. 2.17a.

Both σ_a and $\tau_{s\max}$ are acting on the x plane of the element in Fig. 2.16c. On the y plane, we also have $\tau_{s\max}$ acting, because the x and y planes are mutually perpendicular. On planes parallel to the x and y planes, which are the opposite sides of the element, the stresses are equal and opposite in sense. On the remaining sides of the element, the stresses are zero. A customary representation of this two-dimensional state of stress of point O' is shown in Fig. 2.16d.

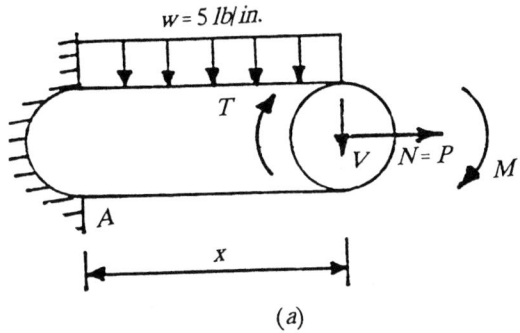

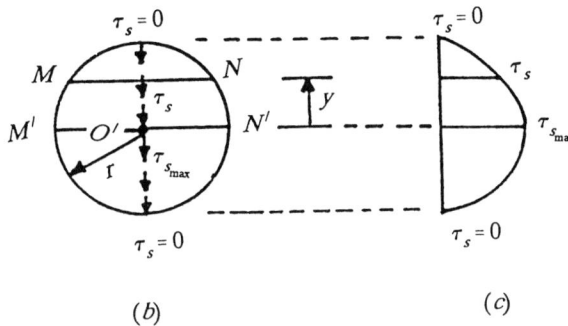

$$\tau_s = \frac{4V}{3A}\left(1 - \frac{y^2}{r^2}\right) \qquad \tau_{s\max} = \frac{4V}{3A}$$

Figure 2.17. (a) Internal forces and moments acting at a cross section of distance x from support A. (b) Shear stress τ_s at a circular section caused by the shear force V. (c) Variation of the shear stress τ_s at a circular cross section.

The numerical value of the normal stress σ_a due to P is

$$\sigma_a = \frac{P}{A} = \frac{12{,}000}{\pi} = 3{,}820 \text{ psi} \quad \text{(tension)}$$

For a circular cross section, the shear stress τ_s produced by V varies in accordance with the parabolic law shown in Fig. 2.17c. Thus, at a distance y from point O' in Fig. 2.17b, the shear stress τ_s may be determined from the following equation:

$$\tau_s = \frac{4V}{3A}\left(1 - \frac{y^2}{r^2}\right) \tag{2.113}$$

The maximum shear stress $\tau_{s\max}$ occurs when $y = 0$ in Eq. (2.113), yielding

$$\tau_{s\max} = \frac{4V}{3A} \tag{2.114}$$

Thus, the maximum shear stress produced by the shear force V is at the centroidal axis where the bending stress σ_b is zero.

It should be pointed out, however, that for circular cross sections the shear stress τ_s is not distributed uniformly along lines MN, or $M'N'$, in Fig. 2.17b, as it usually occurs for other types of cross sections (for example rectangular). A more exact analysis, based on the theory of elasticity, shows that at M' the shear stress is $1.23V/A$, and it is $1.38V/A$ at the center O' of the cross section. Equation (2.114) yields $1.33V/A$, a difference of -3.6 percent. For most practical situations, the value obtained from Eq. (2.114) is sufficiently accurate.

At the fixed support A, the shear force V is

$$V = (5)(80) = 400 \text{ lb}$$

Thus, from Eq. (2.114),

$$\tau_{s\max} = \frac{(4)(400)}{(3)(\pi)} = 170 \text{ psi}$$

which is directed as shown in Fig. 2.16c.

The subject of shear stresses τ_s, produced when a member is subjected to bending, is discussed in greater detail in the following section.

2.5 SHEAR STRESSES ASSOCIATED WITH THE BENDING OF MEMBERS

Consider the simply supported beam illustrated in Fig. 2.18a, which is loaded by a distributed load w as shown. The cross section of the member has the

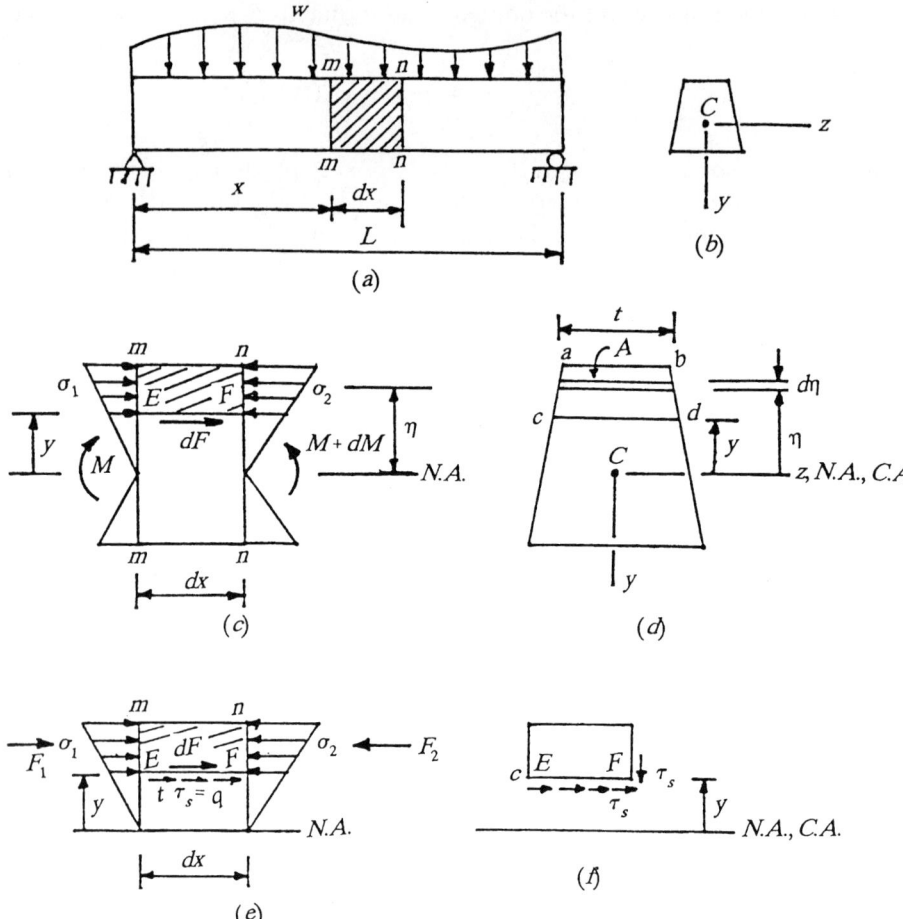

Figure 2.18. (*a*) Simply supported beam loaded as shown. (*b*) Cross section of the beam. (*c*) Element of the beam of length dx. (*d*) Cross section showing area *abcd*. (*e*) Free-body diagram of a portion of the element. (*f*) Element showing shear stress τ_s at a distance y from the neutral axis.

shape shown in Fig. 2.18*b*, with point C as its centroid. In Fig. 2.18*c* we show the bending stress distributions on the sides of an element of length dx, which is taken apart from the complete beam between section *mm* and *nn* in Fig. 2.18*a*. At a distance y from the neutral axis, the free-body diagram of the shaded portion of the element is shown in Fig. 2.18*e*.

On the left side of the element, the bending stress distribution produces a resultant force equal to F_1, and on the right side of the element the resultant force is F_2. If the bending moment at section *mm* is M, and if it is $(M + dM)$ at section *nn*, as shown in Fig. 2.18*c*, then forces F_1 and F_2 in Fig. 2.18*e* will

SHEAR STRESSES ASSOCIATED WITH THE BENDING OF MEMBERS

not be equal. The difference $F_2 - F_1$ must be balanced by the force dF along side EF in Fig. 2.18e, or Fig. 2.18c. Thus, in order to satisfy static equilibrium, we must have

$$dF = F_2 - F_1 \tag{2.115}$$

The bending stress σ at any distance η from the neutral axis, Fig. 2.18c, or Fig. 2.18e, is given by the equation

$$\sigma = \frac{M\eta}{I} \tag{2.116}$$

Thus, by consulting Figs. 2.18d and 2.18e, we have

$$F_1 = \int_A \sigma_1 t\, d\eta = \int_A \frac{M\eta t}{I}\, d\eta \tag{2.117}$$

$$F_2 = \int_A \sigma_2 t\, d\eta = \int_A \frac{(M + dM)\eta t}{I}\, d\eta \tag{2.118}$$

where σ_1 is the bending stress at a distance η on the left side of the element, σ_2 is the bending stress at a distance η on the right side of the element, t is the thickness of the cross section at any distance η, and A is the area $abcd$ of the cross section in Fig. 2.18d. The moments M and $(M + dM)$ are shown in Fig. 2.18c.

By substituting Eqs (2.117) and (2.118) into Eq. (2.115), we find

$$\begin{aligned} dF &= \int_A \frac{(M + dM)\eta t}{I}\, d\eta - \int_A \frac{M\eta t}{I}\, d\eta \\ &= \frac{dM}{I} \int_A \eta t\, d\eta \\ &= \frac{dM}{I} Q_A \end{aligned} \tag{2.119}$$

where

$$Q_A = \int_A \eta t\, d\eta \tag{2.120}$$

is the first moment of the area $abcd$, Fig. 2.18d, about the neutral axis.

From Fig. 2.18e, along side EF, the shear flow q is

$$q = \tau_s t \tag{2.121}$$

116 STRESS AND DEFORMATION

where τ_s is the shear stress along EF, and t is the thickness of the cross section at the distance y from the neutral axis, which is length cd in Fig. 2.18d. Thus,

$$dF = q\,dx = \tau_s t\,dx \tag{2.122}$$

By substituting Eq. (2.122) into Eq. (2.119), we find

$$\tau_s t\,dx = \frac{dM}{I} Q_A$$

or

$$\tau_s = \frac{Q_A}{It} \cdot \frac{dM}{dx} \tag{2.123}$$

Since the shear force V acting at the cross section is

$$V = \frac{dM}{dx} \tag{2.124}$$

we have

$$\tau_s = \frac{VQ_A}{It} \tag{2.125}$$

Equation (2.125) may be used to determine the shear stress τ_s at any distance y from the neutral axis, which is produced by the shear force V acting at the cross section when a member is subjected to bending. This is justified because, as shown in Fig. 2.18f, shear stresses on mutually perpendicular planes are equal. Equation (2.125) is also general, and it can be applied to various shapes of cross sections.

The following examples illustrate the application of the above theory to practical problems.

Example 2.12: The structural member in Fig. 2.19a is loaded by a uniformly distributed load $w = 5$ lb/in. and a force $P = 10{,}000$ lb, located as shown in the figure. Determine the maximum shear stress $\tau_{s_{\max}}$, and the largest numerical value of the normal stress at the fixed end A of the member.

SOLUTION: The bending moment M_A, at the section just to the right of support A, is

$$M_A = (10{,}000)(11.5) - (5)(100)(50)$$
$$= 90{,}000 \text{ in.-lb} \quad \text{(counterclockwise)}$$

SHEAR STRESSES ASSOCIATED WITH THE BENDING OF MEMBERS 117

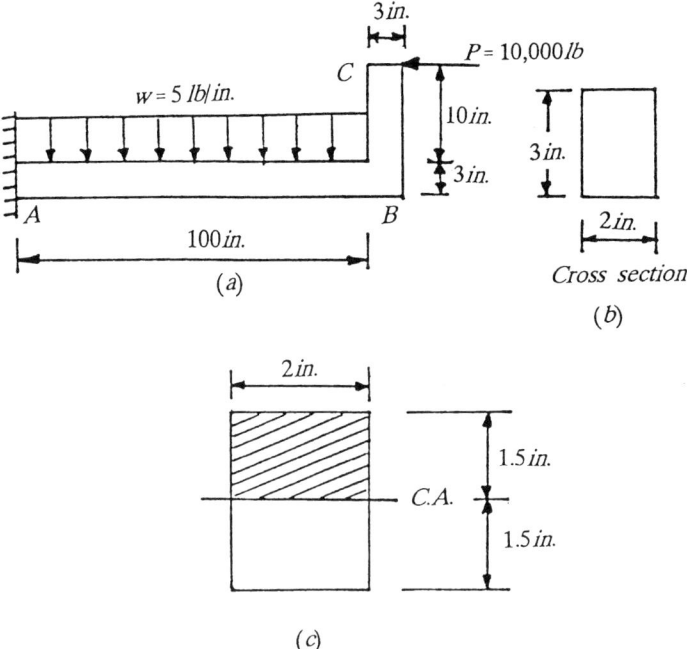

Figure 2.19. (a) Structural member loaded as shown. (b) Cross-sectional shape of the member. (c) Centroidal axis of the cross section.

On this basis, the top of the member at A will be subjected to a compressive bending stress, and its bottom will be subjected to a tensile bending stress.

The moment of inertia I about the centroidal axis is

$$I = \frac{(2)(3)^3}{12} = 4.5 \text{ in.}^4$$

Thus, the maximum numerical value of the bending stress $\sigma_{b\max}$ is

$$\sigma_{b\max} = \pm \frac{(90{,}000)(1.5)}{4.5} = \pm 30{,}000 \text{ psi}$$

The axial force acting at the cross section at A is compressive, and it is equal to P. Therefore, the axial compressive stress σ_a produced by P is

$$\sigma_a = -\frac{P}{A} = -\frac{10{,}000}{(2)(3)} = -1{,}667 \text{ psi}$$

118   STRESS AND DEFORMATION

The maximum numerical value of the normal stress σ at A is at the top of the member and has the value

$$\sigma = -30{,}000 - 1{,}667$$
$$= -31{,}667 \text{ psi} \quad \text{(compressive)}$$

The shear force V_A, at the section just to the right of A, is

$$V_A (5)(100) = 500 \text{ lb} \quad \text{(downward)}$$

The maximum shear stress $\tau_{s_{max}}$ produced by the shear force V_A will occur at the centroidal axis. Thus, Q_A in Eq. (2.125) would be the first moment of the shaded area in Fig. 2.19c about the centroidal axis. On this basis, we have

$$Q_A = (2)(1.5)(0.75) = 2.25 \text{ in.}^3$$

From Eq. (2.125), we obtain

$$\tau_{s_{max}} = \frac{V_A Q_A}{It} = \frac{(500)(2.25)}{(4.5)(2)}$$
$$= 125 \text{ psi} \quad \text{(downward)}$$

Example 2.13: Rework the problem in Example 2.12, by assuming that the structural member has the cross-sectional shape shown in Fig. 2.20b.

SOLUTION: From Fig. 2.20a, the moment M_A about the fixed end A of the member is

$$M_A = (10{,}000)(11.28) - (5)(100)(50)$$
$$= 87{,}800 \text{ in.-lb} \quad \text{(counterclockwise)}$$

By applying statics, the centroidal axis of the cross section of the beam is determined, and it is located as shown in Fig. 2.20b.

The moment of inertia I about the neutral axis is

$$I = \frac{(3)(0.2)^3}{12} + (3)(0.2)(1.18)^2 + \frac{(0.1)(5)^3}{12} + (5)(0.1)(1.42)^2$$
$$= 2.887 \text{ in.}^4$$

At the fixed end A, the compressive normal stress σ_A at the top of the cross

SHEAR STRESSES ASSOCIATED WITH THE BENDING OF MEMBERS 119

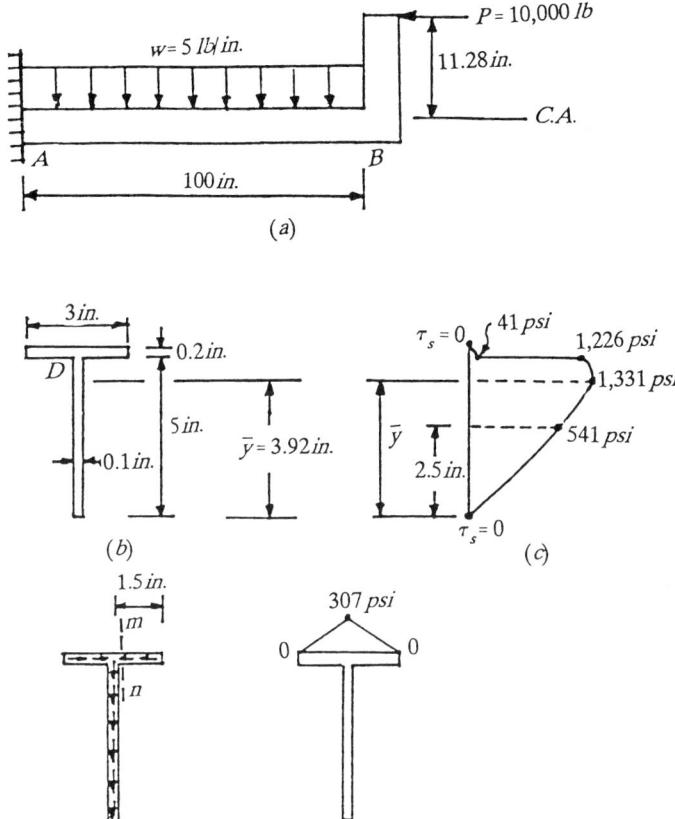

Figure 2.20. (*a*) Structural member loaded as shown. (*b*) Cross-sectional shape of the member. (*c*) Shear stress variation in the vertical direction. (*d*) Shear stress distribution at the cross section of the member. (*e*) Shear stress variation at the cross section of the web.

section is

$$(\sigma_A)_{\text{top}} = \frac{(87,800)(1.28)}{(2.887)}$$

$$= 38,928 \text{ psi} \quad (\text{compression})$$

At the bottom of the same cross section, the tensile normal stress σ_A is

$$(\sigma_A)_{\text{bottom}} = \frac{(87,800)(3.92)}{(2.887)}$$

$$= 119,216 \text{ psi} \quad (\text{tension})$$

120   STRESS AND DEFORMATION

We note here that the numerical value of the tensile stress is 3.06 times higher than that of the compressive stress at the same cross section.

The maximum shear stress $\tau_{s_{max}}$ is located at the neutral axis, and it is given by the equation

$$\tau_{s_{max}} = \frac{V_{max} Q_{A_{max}}}{It}$$

We have

$$Q_{A_{max}} = (3.92)(0.1)\left(\frac{3.92}{2}\right) = 0.7683 \text{ in.}^3$$

$$V_{max} = (5)(100) = 500 \text{ lb}$$

On this basis, we have

$$\tau_{s_{max}} = \frac{(500)(0.7683)}{(2.887)(0.1)}$$

$$= 1{,}331 \text{ psi} \quad \text{(downward)}$$

At point D in Fig. 2.20b, we have

$$Q = (3)(0.2)(1.18) = 0.708 \text{ in.}^3$$

Thus, at the same point,

$$\tau_s = \frac{(500)(0.708)}{(2.887)(3)} = 41 \text{ psi} \quad \text{(downward)}$$

$$\tau_s = \frac{(500)(0.708)}{(2.887)(0.1)} = 1{,}226 \text{ psi} \quad \text{(downward)}$$

We have two values of τ_s at D, because just above D the thickness $t = 3$ in., and just below D the thickness $t = 0.1$ in. The shear stress variation at the cross section is shown plotted in Fig. 2.20c.

Along the cross-sectional area of the flange, the shear stress τ_s is distributed as shown in Fig. 2.20d, and has the linear variation shown in Fig. 2.20e. At the center of the flange, cross section mn in Fig. 2.20d, the shear stress τ_s may be obtained again from the general equation:

$$\tau_s = \frac{VQ}{It}$$

In this case, the quantity Q may be determined by taking the first moment about the neutral axis of the cross-sectional area to the right of the section mn.

MAXIMUM AND MINIMUM NORMAL AND SHEAR STRESSES

That is

$$Q = (0.2)(1.5)(1.18) = 0.354 \text{ in.}^3$$

Thus,

$$\tau_s = \frac{(500)(0.354)}{(2.887)(0.2)} = 307 \text{ psi}$$

The above discussion shows that Eq. (2.125) may be used to determine shear stress produced by bending for many cross-sectional shapes. The theory is general.

2.6 MAXIMUM AND MINIMUM NORMAL AND SHEAR STRESSES

At any point in a stressed body, its state of stress may be represented by stress elements, as discussed in Section 2.4 and illustrated in Figs. 2.16a through 2.16d. Such stress elements are rectangular parallelepipeds, and their rectangular sides are rectangular planes that are normal to an x, y, z coordinate system of axes as shown in Fig. 2.16c. A three-dimensional state of stress is represented by the element shown in Fig. 2.21a. At the x plane the normal stress σ_x and the shear stresses τ_{xy} and τ_{xz} are acting, and at the z plane the normal stress σ_z and shear stresses τ_{zx} and τ_{zy} are acting. At the y plane we have the normal stress σ_y and the shear stresses τ_{yx} and τ_{yz}. At the parallel planes the normal and shear stresses are equal in magnitude and opposite in sense.

For the three-dimensional case of stress, we have all together three normal stresses and six shear stresses acting on the three sides of the element. However, from the theory of elasticity, we know that shear stresses on mutually perpendicular planes are equal. On this basis, we have $\tau_{xy} = \tau_{yx}$, $\tau_{yz} = \tau_{zy}$, and $\tau_{xz} = \tau_{zx}$. Thus, in a three-dimensional state of stress, a total of six unknown stresses need to be determined, namely, σ_x, σ_y, σ_z, τ_{xy}, τ_{yz}, and τ_{zx}.

A two-dimensional state of stress at a point, usually referred to as plane stress, is shown in Fig. 2.21b. In this case, we have the normal stresses σ_x and σ_y acting on the x and y planes, respectively, and the shear stresses τ_{xy} and τ_{yx} acting as shown in the figure. Since $\tau_{xy} = \tau_{yx}$, a total of three unknown stresses, namely, σ_x, σ_y, and τ_{xy}, are associated with a plane stress problem. Such a state of stress is usually represented by the flat element shown in Fig. 2.21c. Note that for a plane stress problem involving the x and y planes, the stresses σ_z, τ_{yz}, and τ_{xz} are equal to zero. Axially loaded members, shafts subjected to torsional moments, and beams in bending are examples of plane stress problems. Such problems are discussed in preceding sections of this chapter. When we describe a shear stress, the first subscript indicates the plane on which the stress is acting, and the second subscript indicates its direction. For

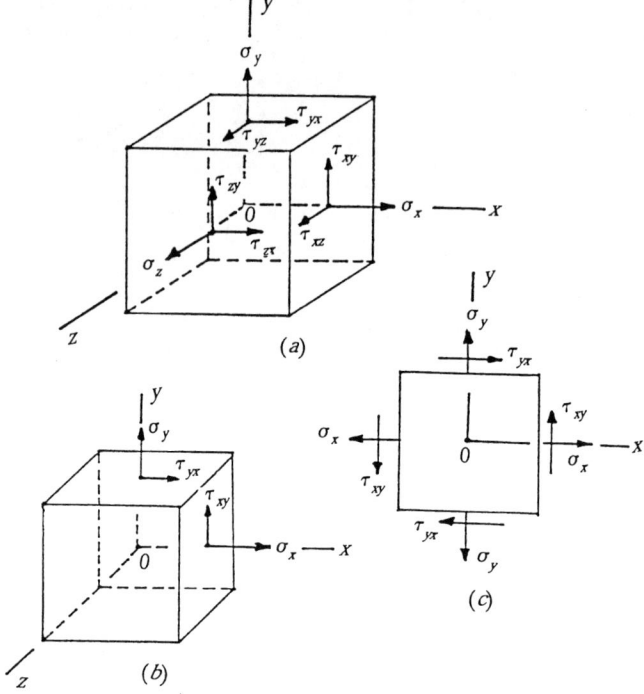

Figure 2.21. (a) Three-dimensional state of stress at a point. (b) Two-dimensional stress, or plane stress, at a point. (c) Usual representation of an element subjected to plane stress.

example, the shear stress τ_{xy} is located on the x plane and it is parallel to the y axis. Only one subscript, x, is used for a stress normal to the x plane, indicating that this stress is parallel to the x axis.

If we know the state of plane stress shown in Fig. 2.21c, the state of plane stress at another orientation x_1, y_1 of the element, as shown in Fig. 2.22a, may be determined by using transformation equations. In other words, if we know the stresses σ_x, σ_y, and τ_{xy} in Fig. 2.21c, we can determine the stresses $\sigma_{x_1}, \sigma_{y_1}$, and $\tau_{x_1 y_1}$ at any angle θ from the x axis, Fig. 2.22a, by using the following transformation equations:

$$\sigma_{x_1} = \frac{\sigma_x + \sigma_y}{2} + \frac{\sigma_x - \sigma_y}{2} \cos 2\theta + \tau_{xy} \sin 2\theta \qquad (2.126)$$

$$\sigma_{y_1} = \frac{\sigma_x + \sigma_y}{2} - \frac{\sigma_x - \sigma_y}{2} \cos 2\theta - \tau_{xy} \sin 2\theta \qquad (2.127)$$

$$\tau_{x_1 y_1} = -\frac{\sigma_x - \sigma_y}{2} \sin 2\theta + \tau_{xy} \cos 2\theta \qquad (2.128)$$

MAXIMUM AND MINIMUM NORMAL AND SHEAR STRESSES   123

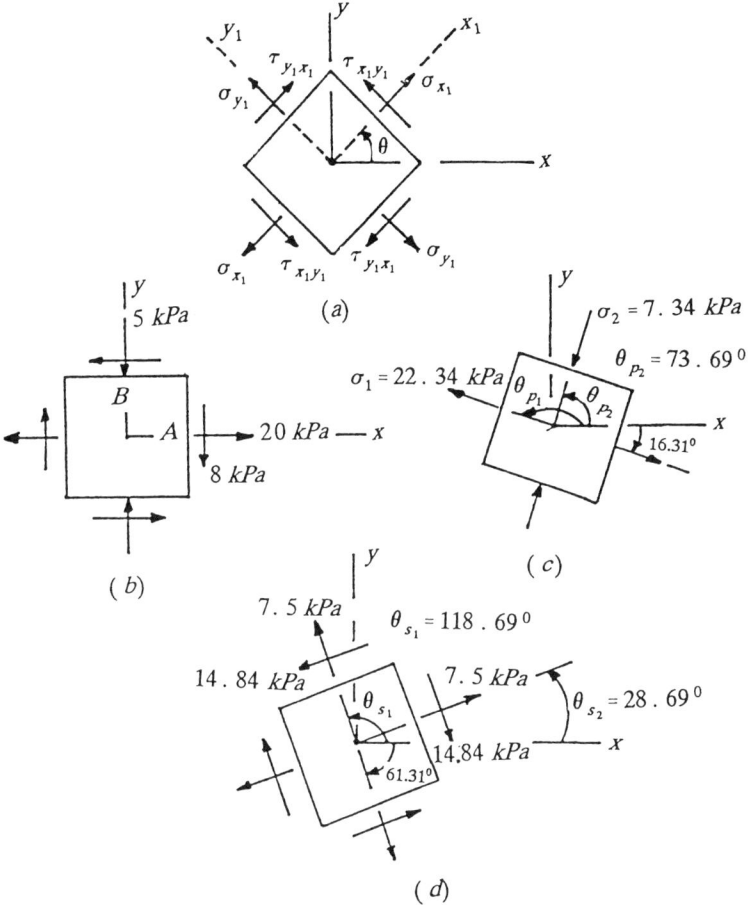

Figure 2.22. (a) Stresses on a plane stress element rotated at an angle θ for the x axis. (b) Element subjected to plane stress. (c) Location of principal stresses. (d) Location of maximum and minimum shear stresses.

The above three equations may be found in texts on elementary mechanics of solids [38]. We also know that the sum of the normal stresses σ_x and σ_y at any orientation θ is an invariant quantity (constant). Thus, we have the identity

$$\sigma_{x_1} + \sigma_{y_1} = \sigma_x + \sigma_y \qquad (2.129)$$

Equations (2.126) through (2.128) indicate that stresses σ_{x_1}, σ_{y_1}, and $\tau_{x_1 y_1}$ are functions of the angle θ and, consequently, they vary as θ changes. Therefore, there should be an orientation θ that makes the normal stresses maximum and minimum, known as principal stresses, and an orientation θ that provides the maximum and minimum shear stresses. Starting with σ_{x_1}, Eq.

(2.126), and setting equal to zero the derivative of σ_{x_1} with respect to θ, we find

$$\frac{d\sigma_{x_1}}{d\theta} = -(\sigma_x - \sigma_y)\sin 2\theta + 2\tau_{xy}\cos 2\theta = 0 \qquad (2.130)$$

or

$$\tan 2\theta_p = \frac{2\tau_{xy}}{\sigma_x - \sigma_y} \qquad (2.131)$$

The angle θ_p in Eq. (2.131) defines the orientation of the principal planes where the principal stresses are acting. Since two values of θ_p may be obtained from Eq. (2.131), two values that differ by 90°, one value of θ_p would be between 0 and 90°, and the other one between 90° and 180°. One of these two angles locates the principal plane with the maximum principal stress, and the other one locates the principal plane with the minimum principal stress. If the maximum and minimum principal stresses are denoted as σ_1 and σ_2, respectively, they can be determined from the following equation:

$$\sigma_{1,2} = \frac{\sigma_x + \sigma_y}{2} \pm \sqrt{\left(\frac{\sigma_x - \sigma_y}{2}\right)^2 + \tau_{xy}^2} \qquad (2.132)$$

The plus sign in Eq. (2.132) gives the algebraically larger principal stress σ_1, and the minus sign provides the minimum principal stress σ_2. This is extremely important information for the design engineer, because for design purposes only the largest value is usually needed.

It should be pointed out, however, that we do not know which value of θ_p from Eq. (2.131) corresponds to σ_1 and which one corresponds to σ_2. This question can be resolved by substituting one value of θ_p in Eq. (2.126) and noting whether σ_{x_1} will recognize σ_1 or σ_2, thus correlating the principal stresses with the principal angles. This is acceptable, because Eq. (2.126) applies for all values of θ.

We also note that Eq. (2.130) is identical to Eq. (2.128), indicating that the shear stresses on principal planes are zero; that is, on the planes where the normal stresses are maximum or minimum, the shear stresses on those planes are zero. This again is very important information for the design engineer, who needs to know the maximum value of the normal stress for the design of his or her infrastructural element. We also know that

$$\sigma_{x_1} + \sigma_{y_1} = \sigma_x + \sigma_y = \sigma_1 + \sigma_2 \qquad (2.133)$$

In a similar manner, by considering Eq. (2.128), we have

$$\frac{d\tau_{x_1y_1}}{d\theta} = (\sigma_x - \sigma_y)\cos 2\theta - 2\tau_{xy}\sin 2\theta = 0$$

or

$$\tan 2\theta_s = \frac{\sigma_x - \sigma_y}{2\tau_{xy}} \qquad (2.134)$$

The angle θ_s in Eq. (2.134) defines the orientation of the planes of maximum and minimum shear stresses. The two values of θ_s that can be obtained from Eq. (2.134) define the planes where the maximum and minimum values of the shear stress $\tau_{x_1 y_1}$ occur. These two values of θ_p differ by 90° as it was for the principal normal stresses.

By using trigonometry, we can prove that

$$\theta_s = \theta_p \pm 45° \qquad (2.135)$$

which indicates that the planes of maximum shear stress occur at 45° to the principal planes. If θ_{s1} and θ_{s2} are used to denote the angles where the maximum and minimum shear stresses, respectively, are located, and if θ_{p1} and θ_{p2} are used to denote the angles where the maximum and minimum principal stresses, respectively, are located, we find that

$$\theta_{s1} = \theta_{p1} - 45° \qquad (2.136)$$

and

$$\tau_{max} = \sqrt{\left(\frac{\sigma_x - \sigma_y}{2}\right)^2 + \tau_{xy}^2} \qquad (2.137)$$

Also, the minimum shear stress τ_{min} is

$$\tau_{min} = -\sqrt{\left(\frac{\sigma_x - \sigma_y}{2}\right)^2 + \tau_{xy}^2} \qquad (2.138)$$

If we use Eq. (2.132) and subtract the expression of σ_2 from the expression of σ_1, then compare the result with Eq. (2.137), we find

$$\tau_{max} = \frac{\sigma_1 - \sigma_2}{2} \qquad (2.139)$$

On each of the planes of maximum and minimum shear stress, however, the normal stress is equal to the average of the normal stresses acting on the x and y planes. If this stress is denoted as σ_{aver}, we have

$$\sigma_{aver} = \frac{\sigma_x + \sigma_y}{2} \qquad (2.140)$$

126 STRESS AND DEFORMATION

and it acts on both the plane of maximum shear stress and the plane of minimum shear stress.

The following example illustrates the application of the above theory.

Example 2.14: An element in plane stress is subjected to the stresses shown in Fig. 2.22b. (a) Determine the principal stresses and show them on a sketch of a properly oriented element. (b) Determine the maximum and minimum shear stresses and sketch them on a properly oriented element.

SOLUTION

Calculation of Principal Stresses: By using Eq. (2.131), we find

$$\tan 2\theta_p = \frac{2\tau_{xy}}{\sigma_x - \sigma_y} = \frac{(2)(-8)}{20 - (-5)} = -0.64$$

Thus,

$$2\theta_p = 147.38° \qquad \theta_p = 73.69°$$
$$2\theta_p = 327.38° \qquad \theta_p = 163.69°$$

By substituting $\theta_p = 73.69°$ into Eq. (2.126), we find

$$\sigma_{x_1} = \frac{20 - 5}{2} + \frac{20 + 5}{2}\cos(147.38°) - 8\sin(147.38°)$$

$$= 7.5 - 10.528 - 4.313$$

$$= -7.34 \text{ kPa}$$

In a similar manner, by using $\theta_p = 163.69°$, Eq. (2.126) yields $\sigma_{x_1} = 22.338$ kN. On this basis, the principal stresses and their corresponding principal angles are as follows:

$$\sigma_1 = 22.338 \text{ kPa} \qquad \theta_{p1} = 163.69°$$
$$\sigma_2 = -7.341 \text{ kPa} \qquad \theta_{p2} = 73.69°$$

Note that θ_{p1} and θ_{p2} differ by 90°. Also, we have

$$\sigma_1 + \sigma_2 = \sigma_x + \sigma_y = 15 \text{ kPa}$$

An alternative approach to determine the principal stresses would be to use Eq. (2.132). By substituting the values of σ_x, σ_y, and τ_{xy} into this equation, we find

$$\sigma_{1,2} = \frac{20-5}{2} \pm \sqrt{\left(\frac{20+5}{2}\right)^2 + (-8)^2}$$

$$= 7.50 \pm 14.84$$

Thus, the principal stresses σ_1 and σ_2 are

$$\sigma_1 = 22.34 \, \text{kPa} \qquad \sigma_2 = -7.34 \, \text{kPa}$$

The angles θ_{p_1} and θ_{p_2}, which define the planes where σ_1 and σ_2 are acting, may be determined from Eq. (2.131), as discussed earlier. Since we do not know which θ_p corresponds to σ_1 or σ_2, we use Eq. (2.126) and one value of θ_p that will yield either σ_1 or σ_2. In this manner, correlation between the angles and the principal stresses is established.

The properly oriented element, where the σ_1 and σ_2 principal stresses are acting, is shown in Fig. 2.22c.

Calculation of Maximum and Minimum Shear Stresses: The maximum and minimum shear stresses may be determined from Eqs. (2.137) and (2.138). These two equations yield

$$\tau_{\text{max}} = 14.84 \, \text{kPa}$$

$$\tau_{\text{min}} = -14.84 \, \text{kPa}$$

The angle θ_{s_1} locating the plane where τ_{max} is acting can be determined from Eq. (2.136). Thus,

$$\theta_{s_1} = \theta_{p_1} - 45° = 163.69° - 45° = 118.69°$$

The minimum shear stress τ_{min} is acting on the plane where

$$\theta_{s_2} = 118.69° - 90° = 28.69°$$

From Eq. (2.140), the normal stress acting on the planes of τ_{max} and τ_{min} is

$$\sigma_{\text{aver}} = \frac{\sigma_x + \sigma_y}{2} = \frac{20 - 5}{2} = 7.5 \, \text{kPa}$$

The properly oriented element, where τ_{max} and τ_{min} are acting, is shown in Fig. 2.22d.

2.7 MOHR'S CIRCLE FOR PLANE STRESS

The transformation equations given by Eqs. (2.126) and (2.128) may be transformed into a graphical form known as Mohr's circle, after the German engineer Otto Mohr (1835–1918). It is a very useful and practical method for the determination of principal stresses and maximum and minimum shear stresses for both two-dimensional and three-dimensional states of stress. In this section, only the two-dimensional application of Mohr's circle is discussed.

We rewrite Eqs. (2.126) and (2.128) as follows:

$$\sigma_{x_1} - \frac{\sigma_x + \sigma_y}{2} = \frac{\sigma_x - \sigma_y}{2} \cos 2\theta + \tau_{xy} \sin 2\theta \qquad (2.141)$$

$$\tau_{x_1 y_1} = -\frac{\sigma_x - \sigma_y}{2} \sin 2\theta + \tau_{xy} \cos 2\theta \qquad (2.142)$$

By squaring both sides of each equation given by Eqs. (2.141) and (2.142) and adding them together, we obtain

$$\left(\sigma_{x_1} - \frac{\sigma_x + \sigma_y}{2}\right)^2 + \tau_{x_1 y_1}^2 = \left(\frac{\sigma_x - \sigma_y}{2}\right)^2 + \tau_{xy}^2 \qquad (2.143)$$

or

$$(\sigma_{x_1} - \sigma_{\text{aver}})^2 + \tau_{x_1 y_1}^2 = R^2 \qquad (2.144)$$

where

$$\sigma_{\text{aver}} = \frac{\sigma_x + \sigma_y}{2} \qquad (2.145)$$

$$R = \sqrt{\left(\frac{\sigma_x - \sigma_y}{2}\right)^2 + \tau_{xy}^2} \qquad (2.146)$$

Equation (2.144) is the equation of a circle where σ_{x_1} and $\tau_{x_1 y_1}$ are its coordinates, R is its radius, and its center is defined by the coordinates $\sigma_{x_1} = \sigma_{\text{aver}}$ and $\tau_{x_1 y_1} = 0$.

The Mohr's circle may be constructed with the normal stress σ_{x_1} as the abscissa, and the shear stress $\tau_{x_1 y_1}$ as the ordinate. We consider here the first form of Mohr's circle with σ_{x_1} positive to the right and $\tau_{x_1 y_1}$ positive downward. On this basis the angle 2θ on the Mohr's circle is positive when counterclockwise, which is the same with the positive direction used in defining the transformation equations given in the preceding section.

For the plane stress element shown in Fig. 2.23a, the Mohr's circle may be constructed as shown in Fig. 2.23b. The center C of the circle has the

MOHR'S CIRCLE FOR PLANE STRESS   129

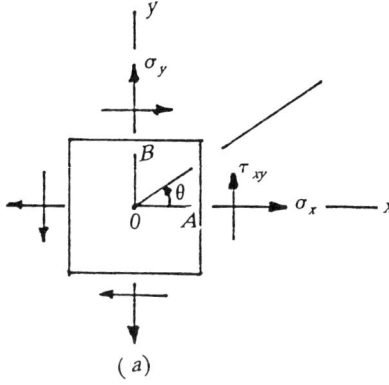

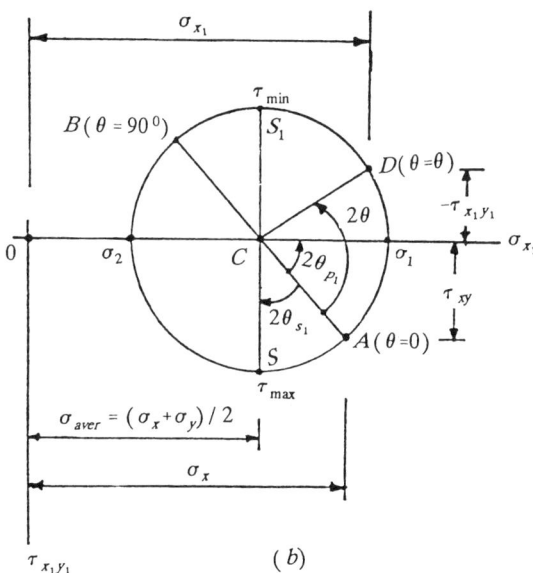

Figure 2.23. (a) Element in plane stress. (b) Mohr's circle for plane stress.

coordinates

$$\sigma_{x_1} = \sigma_{aver} = \frac{\sigma_x + \sigma_y}{2} \quad \text{and} \quad \tau_{x_1 y_1} = 0$$

Point A on the Mohr's circle represents the stress condition of the x plane in Fig. 2.23a where $\theta = 0$, and has the coordinates $\sigma_{x_1} = \sigma_x$ and $\tau_{x_1 y_1} = \tau_{xy}$. Point B on this circle represents the stress condition on the y plane of the element in

Fig. 2.23a where $\theta = 90°$, and has the coordinates $\sigma_{x_1} = \sigma_y$ and $\tau_{x_1y_1} = \tau_{xy}$. Thus, line AB in Fig. 2.23b passes through the center C of the Mohr's circle and becomes a diameter of the circle where points A and B represent the stresses on planes 90° to each other, but 180° apart on the Mohr's circle. With point C as the center and AB as the diameter, we draw the circle as shown in Fig. 2.23b. The radius R of the circle is the length of line CA, and the intersects σ_1 and σ_2 of the circle with the σ_{x_1} axis define the maximum and minimum normal stresses, referred to as principal stresses. Point S on the circle provides the plane where τ_{max} is located, and the plane of τ_{min} is represented by point S_1 on the circle. Note that the normal stress on planes S and S_1 is equal to σ_{aver} and that τ_{max} is equal to the radius R of the circle. Also, note that the angle $2\theta_{p_1}$ in the Mohr's circle locates the plane of the principal stress σ_1, and $2\theta_{s_1}$ locates the plane where τ_{max} is acting. A plane at an angle θ from the x axis in Fig. 2.23a is represented by point D in the Mohr's circle, and it is defined by the coordinates σ_{x_1} and $-\tau_{x_1y_1}$.

The following example illustrates the application of Mohr's circle for plane stress.

Example 2.15: Rework Example 2.14 by using Mohr's circle of plane stress.

SOLUTION: We first locate the center C of the Mohr's circle by using Eq. (2.145); that is,

$$\sigma_{aver} = \frac{\sigma_x + \sigma_y}{2} = \frac{20 - 5}{2} = 7.5 \, \text{kPa}$$

Figure 2.24 shows the location of point C. Then we locate point A in the circle by using the coordinates $\sigma_{x_1} = 20\,\text{kPa}$ and $\tau_{x_1y_1} = -8\,\text{kPa}$. These coordinates are the stresses acting on the x plane in Fig. 2.22b. The angle θ at A is zero. With CA as the radius, we draw the Mohr's circle as shown in Fig. 2.24. Point B is also shown in the same figure with coordinates $\sigma_{x_1} = 5\,\text{kPa}$ and $\tau_{x_1y_1} = 8\,\text{kPa}$, representing the stress condition of the y plane in Fig. 2.22b.

The radius R of the circle is

$$R = \sqrt{(12.5)^2 + (8)^2} = 14.84 \, \text{kPa}$$

The angle $2\theta_{p_1}$, defining the position of the maximum principal stress σ_1 shown in the figure, is

$$\tan 2\theta_{p_1} = \frac{-8}{12.5} = -0.64$$

or

$$2\theta_{p_1} = -32.62° \qquad \theta_{p_1} = -16.31°$$

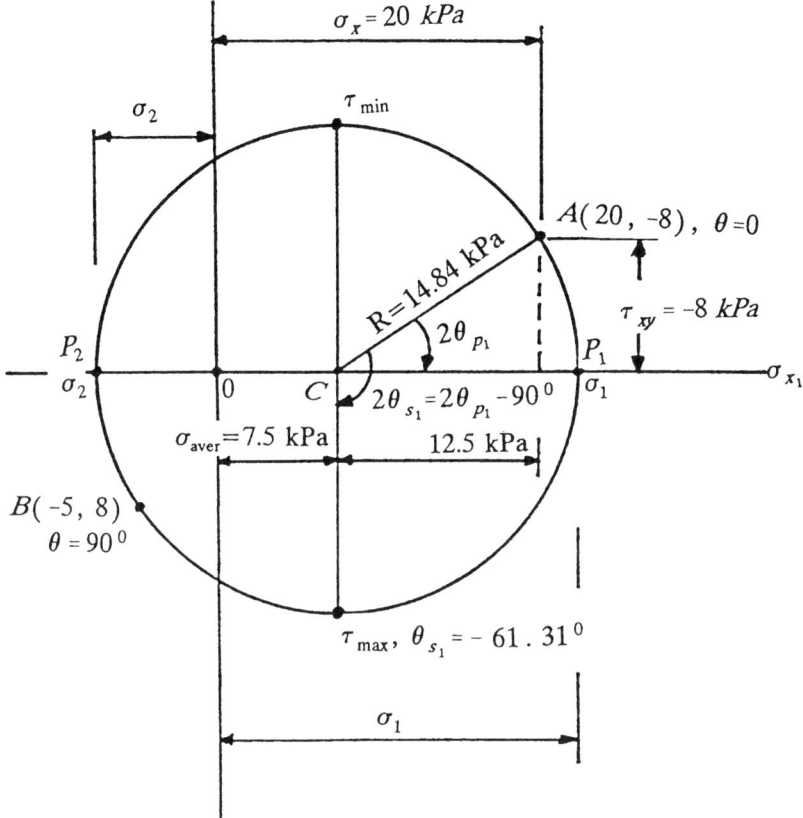

Figure 2.24. Mohr's circle of stress for the plane element in Fig. 2.22b.

Since $CA = CP_1$, the maximum principal stress σ_1 is

$$\sigma_1 = OC + CP_1 = 7.5 + 14.84 = 22.34 \text{ kPa}$$

The minimum principal stress σ_2 is

$$\sigma_2 = OP_2 = -14.84 + 7.5 = -7.34 \text{ kPa}$$

The maximum shear stress τ_{max} is equal to the radius R of the Mohr's circle and it is located as shown in the figure. Thus,

$$\tau_{max} = R = 14.84 \text{ kPa}$$

The minimum shear stress τ_{min} is also shown in the figure, and it is equal to

132   STRESS AND DEFORMATION

-14.84 kPa. The angle $2\theta_{s_1}$ locating τ_{max} is

$$2\theta_{s_1} = 2\theta_{p_1} - 90 = -32.62° - 90° = -122.62°$$
$$\theta_{s_1} = -61.31°$$

The properly oriented elements for the principal stresses and maximum and minimum shear stresses are shown in Figs. 2.22c and 2.22d, respectively.

PROBLEMS

2.1 For the stepped beam loaded as shown in Fig. P2.1: (a) determine the magnitude and location of its maximum normal stress; (b) determine the total elongation of the member; and (c) plot the variation of the axial deformation of the member starting from end A and ending at end C. The width of the beam is 2 in. and its modulus of elasticity $E = 10 \times 10^6$ psi.

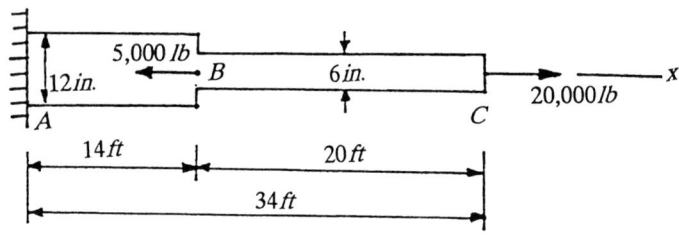

Figure P2.1.

2.2 For the indicated loading condition, design the stepped beam in Fig. P2.1 so that its maximum normal stress does not exceed 1,200 psi. Assume that the depth of the member is as shown in the figure.

2.3 A rigid bar ABC is hinged at A and supported by a steel wire at point B, as shown in Fig. P2.3. If a force $P = 500$ kN is applied at end C, determine the cross-sectional area of the wire so that its length change does not exceed 3 mm. The modulus of elasticity E of the wire is 200 GPa.

2.4 The structural steel trusses in Fig. P2.4 are loaded as shown. For the indicated cross-sectional areas, determine the length changes for the lettered members of each truss. The modulus of elasticity $E = 30 \times 10^6$ psi.

2.5 A long doubly tapered cantilever beam is loaded with an axial load P at its free end, as shown in Fig. P2.5. Derive a formula for the elongation

PROBLEMS 133

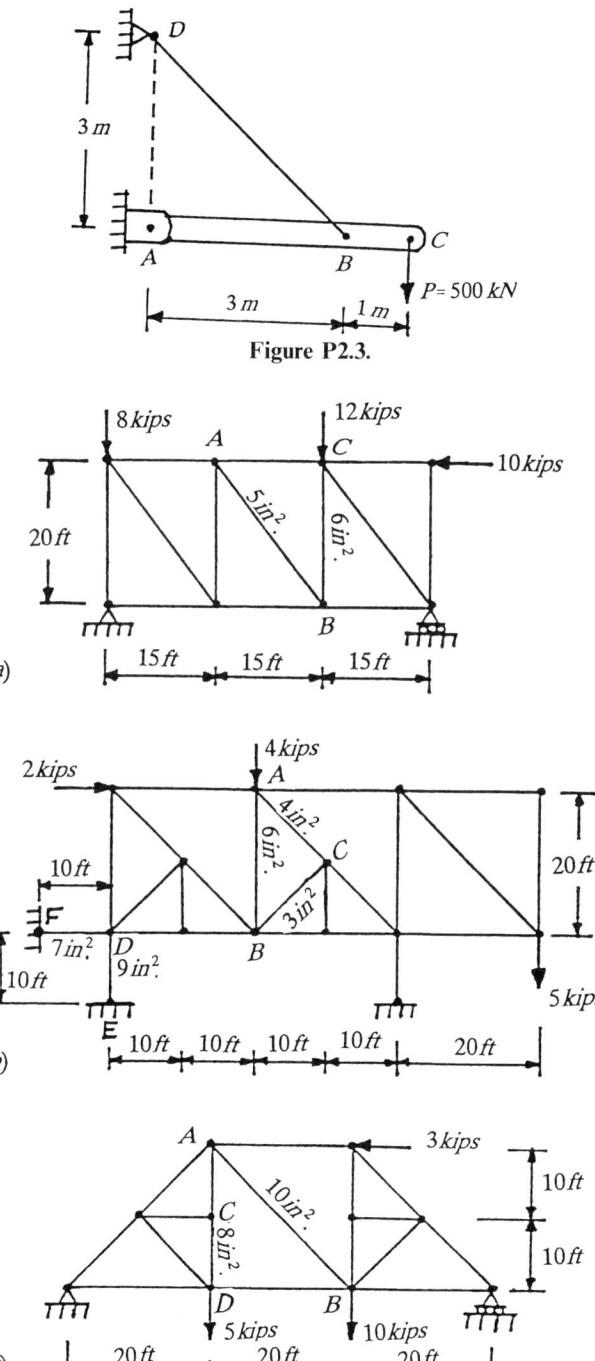

Figure P2.3.

(a)

(b)

(c)

Figure P2.4.

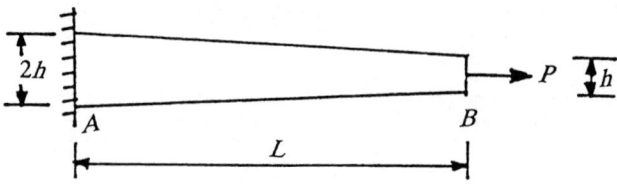

Figure P2.5.

δ of the beam. The width b is constant throughout the length of the member.

2.6 A steel uniform cantilever beam is loaded at its free end by a torsional moment, as shown in Fig. P2.6a. The cross-sectional shape of the member may have only one of the shapes shown in the figure. For each case, determine: (a) the maximum shear stress in the member caused by the applied torque; (b) the angular twist per unit of length; and (c) the total angular twist of the member. The shear modulus $G = 11 \times 10^6$ psi.

2.7 A stepped circular bar in torsion has the dimensions shown in Fig. P2.7, and is loaded at its free ends by a torque T. Determine the allowable torque T if the angle of twist between the ends of the bar is not to exceed 0.03 radians. The shear modulus of elasticity $G = 80$ GPa.

2.8 For the stepped circular bar in Fig. P2.7, determine the allowable torque T if the shear stress in the bar is not allowed to exceed 30 MPa.

2.9 Solve Problem 2.7 by assuming that the bar is hollow with an inside diameter of 30 mm throughout the entire length of the bar. Compare the results.

2.10 Solve Problem 2.8 by assuming that the bar is hollow with an inside diameter of 30 mm throughout the entire length of the bar. Compare the results.

2.11 Repeat Problem 2.7 by assuming that part AB of the bar is hollow with an inside diameter of 30 mm throughout the entire length AB of the bar. Compare the results.

2.12 For the cantilever beam in Example 2.4, determine the magnitude of the applied torque T so that the maximum shear stress does not exceed 8,000 psi.

2.13 The solid circular steel shaft in Fig. P2.13 is loaded by a torque $T = 10$ kip-ft located as shown in the figure. The rod AB is a rigid pointer fastened to the end of the shaft. Determine: (a) the maximum shear stress in the shaft; (b) the distance point A of the rod moves, measured from its load-free position.

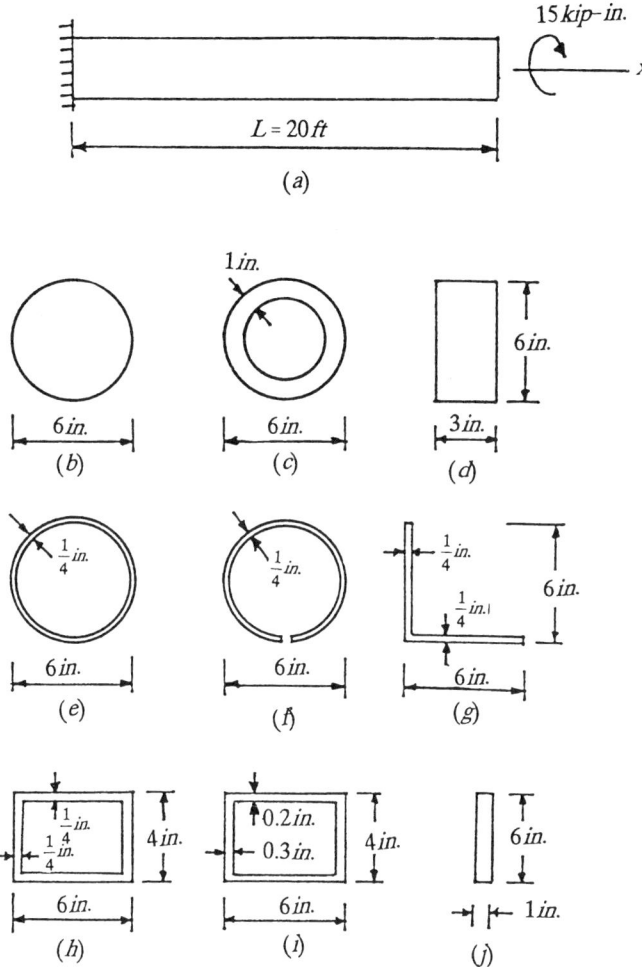

Figure P2.6.

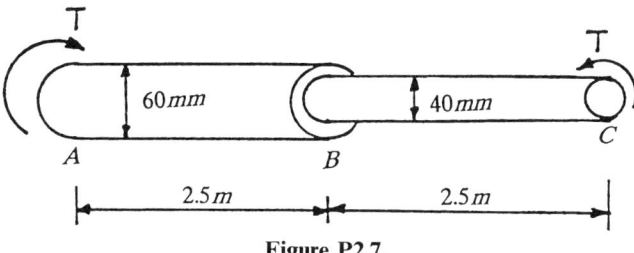

Figure P2.7.

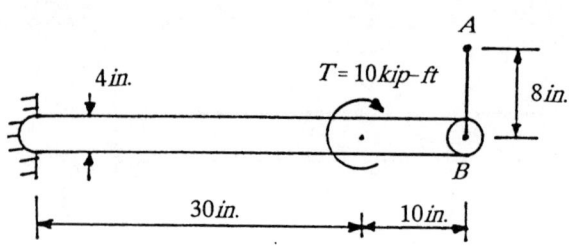

Figure P2.13.

2.14 The solid circular stepped steel shaft in Fig. P2.14 is subjected to a torque T as shown. If the allowable shear stress is 10,000 psi and the maximum allowable angle of twist of the shaft is 0.06 rad, determine the maximum allowable value of the torque T.

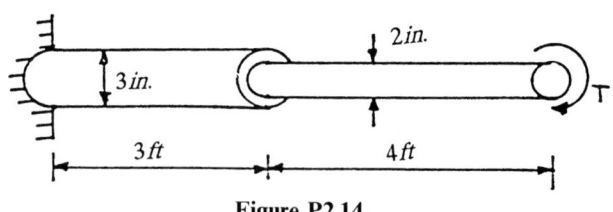

Figure P2.14.

2.15 For the uniform simply supported beam loaded as shown in Fig. 2.13a, assume that the equation of its elastic line y is given by the following:

$$y = C_1 x^5 + C_2 x^4 + C_3 x^3 + C_4 x^2 + C_5 x + C_6$$

By utilizing Eq. (2.52a) in Section 2.3 and setting up the required six end conditions, determine the constants C_1, C_2, C_3, C_4, C_5, and C_6. Also, write the expressions for the deflection y and slope y' of the member.

2.16 Solve Problem 2.15 by using positive signs for all the derivatives of the deflection y and first quadrant coordinates.

2.17 By using the double integration method, determine the equation of the elastic line for each member loaded as shown in Fig. P2.17. Also, for each member, determine the slope and deflection θ_B, y_B, and so on, as indicated in the figure. The stiffness EI is constant.

2.18 Repeat Problem 2.17 by using the last expression in Eq. (2.52a), and applying integration with appropriate boundary conditions regarding the evaluation of the constants of integration.

2.19 For the cantilever beam loaded as shown in Fig. P2.19a, determine the maximum numerical value of the normal stress acting on the beam. The cross section of the beam is shown in Fig. P2.19b.

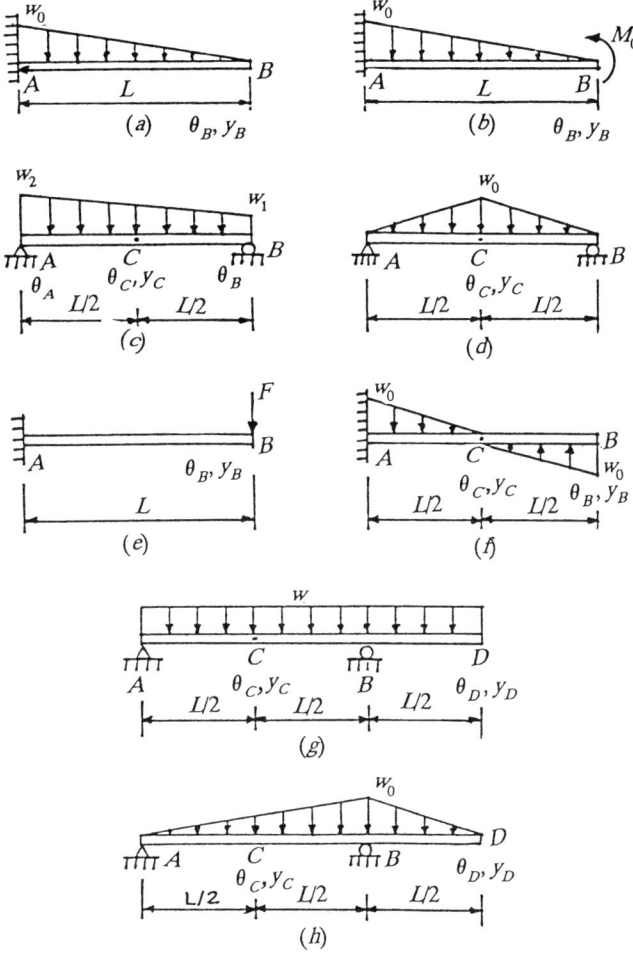

Figure P2.17.

2.20 A beam assembly is shown in Fig. P2.20a. Beam *BDE* is welded to the simply supported beam, as shown in the figure. If the yield in tension for the beams is 6×10^8 Pa, determine the depth h of the beam *ABC* for an allowable stress of 4×10^8 Pa. The cross section of the uniform beam *ABC* is rectangular, as shown in Fig. P2.20b.

2.21 For the uniform simply supported beam loaded as shown in Fig. P2.21a, the allowable normal stresses at the cross section (Fig. P2.21b) under the applied load P are 4,000 psi for tension and 6,000 psi for compression. Determine the maximum value of the load P that can be applied to the member. The cross section of the member is shown in Fig. P2.21b.

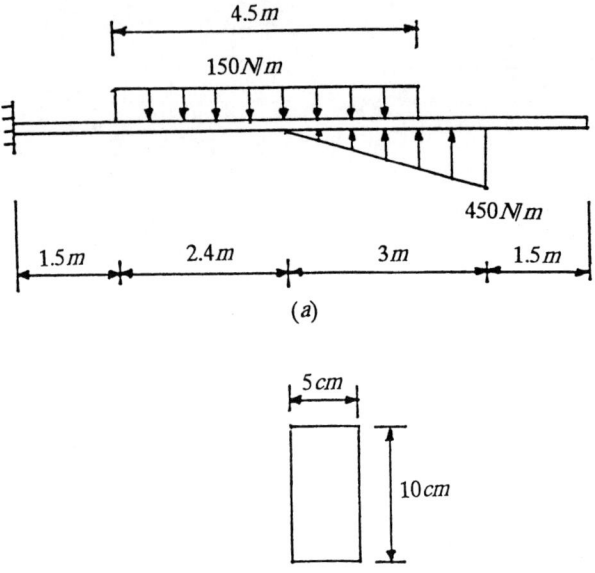

Figure P2.19.

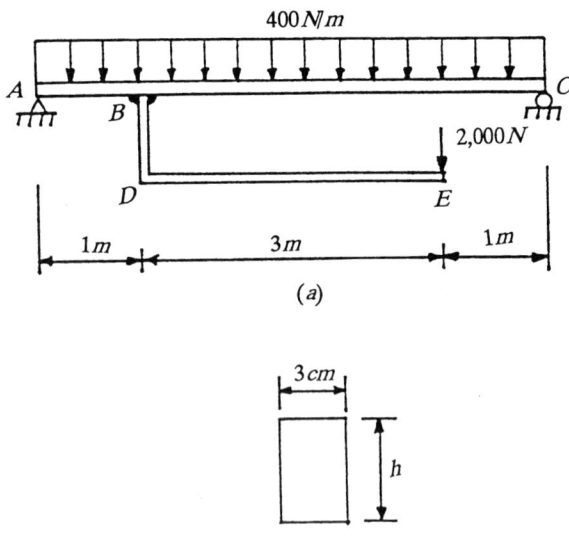

Figure P2.20.

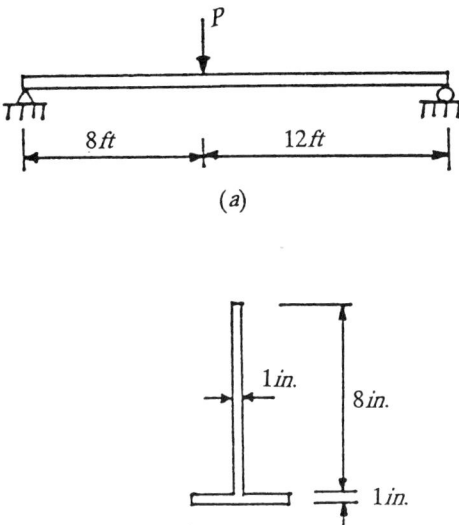

Figure P2.21.

2.22 The simply supported beam in Fig. P2.22 is loaded as shown. If we assume that the allowable normal stress is 18,000 psi, determine an appropriate cross section of the member that will satisfy this requirement. Consider: (a) a rectangular cross section; (b) a wide-flange cross section; and (c) a T section such as the one in Fig. P2.21b of appropriate dimensions. A repetitive procedure may be used.

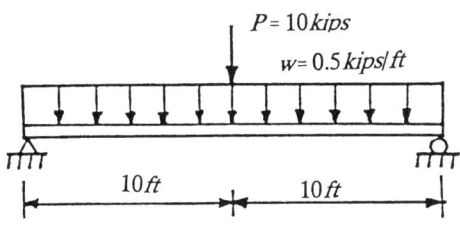

Figure P2.22.

2.23 A wooden cantilever beam is loaded as shown in Fig. P2.23a. Its cross-sectional shape is as shown in Fig. P2.23b, and its modulus of elasticity $E = 1,200$ ksi. Determine the normal stress distribution at the fixed end of the member. Consider the effect of the vertical deflection of the member.

140 STRESS AND DEFORMATION

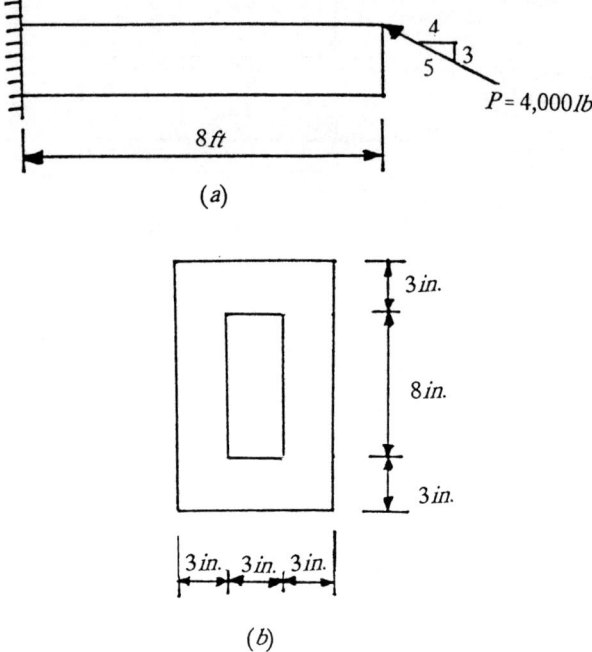

Figure P2.23.

2.24 For the short column loaded as shown in Fig. P2.24, determine the normal stresses at points *A*, *B*, *C*, and *D*.

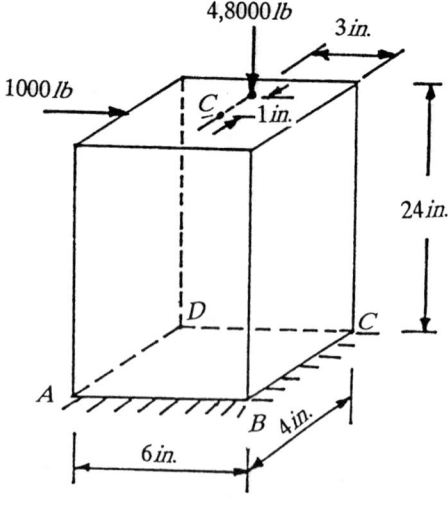

Figure P2.24.

PROBLEMS

2.25 A solid circular cantilever beam is loaded by a bending moment and an axial force, as shown in Fig. P2.25. If the allowable stress in tension is 120 MPa, determine the required diameter d of the beam. Neglect the weight of the beam.

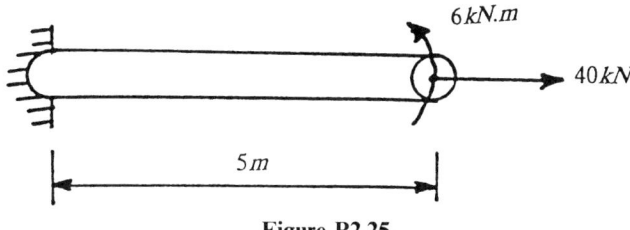

Figure P2.25.

2.26 For the circular shaft shown in Fig. 2.16a of Example 2.11, determine its radius r so that the normal stress at point Q does not exceed 20,000 psi.

2.27 For the circular shaft in Fig. 2.16a of Example 2.11, determine its radius r so that the shear stress at point Q does not exceed 4,000 psi.

2.28 For Problem 2.19, determine the maximum numerical value of the shear stress acting on the beam.

2.29 For the steel rod loaded as shown in Fig. P2.29, determine and show on a sketch the principal stresses and the maximum shear stress at the top surface near the support.

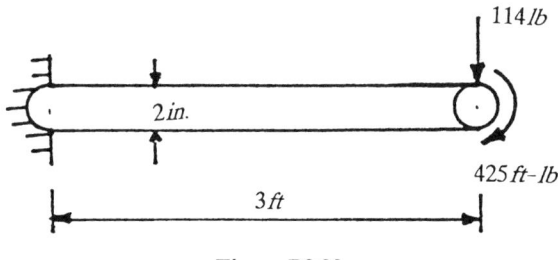

Figure P2.29.

2.30 The laminated simply supported beam loaded as shown in Fig. P2.30a is made up of three planks that are glued together to make the cross section shown in Fig. P2.30b. Determine the maximum load P that can be applied to the beam if the allowable shear stress at the glued joint is 0.5 MPa and the allowable bending stress is 12 MPa. Neglect the weight of the beam.

2.31 A uniform simply supported beam is loaded as shown in Fig. P2.31a, and has the cross-sectional shape shown in Fig. P2.31b. Determine the

142 STRESS AND DEFORMATION

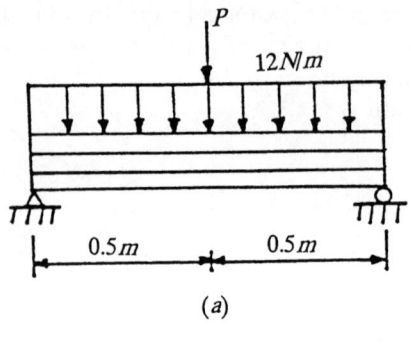

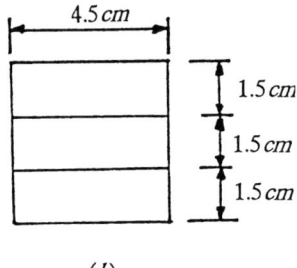

Figure P2.30.

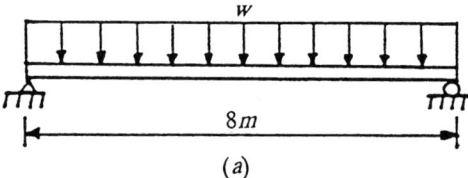

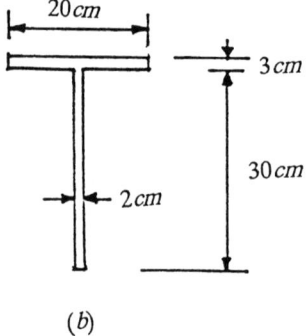

Figure P2.31.

maximum uniform load w that can be applied to the member if the allowable bending stress $\sigma_{al} = 110$ MPa and the allowable shear stress $\tau_{al} = 50$ MPa.

2.32 For the simply supported beam in Fig. P2.21 of Problem 2.21, determine the maximum load P that can be applied to the member if the allowable shear stress $\tau_{al} = 1,000$ psi.

2.33 Repeat Problem 2.22 if the allowable shear stress in the member is 8,000 psi.

2.34 An element in plane stress is subjected each time at the states of plane stress shown in Fig. P2.34. By applying the procedure discussed in Section 2.6, determine: (a) the normal and shear stresses at the angles $\theta = 30°, 110°, 230°,$ and $330°$; (b) the principal stresses and their location; (c) the maximum and minimum shear stress. Show all results on properly oriented elements. Counterclockwise θ is positive.

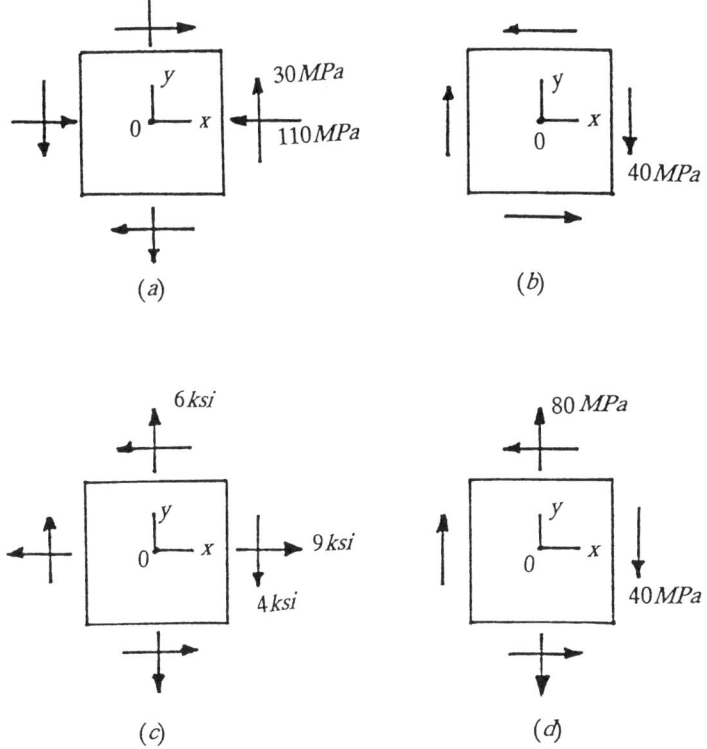

Figure P2.34.

2.35 The thin plate in Fig. P2.35 is subjected to plane stress consisting of stresses σ_x, σ_y, and τ_{xy}, as shown in the figure. If stress $\sigma_x = 25$ MPa and the normal tensile stresses on planes oriented at $\theta = 30°$ and $70°$ are 40 MPa each, determine the stresses σ_y and τ_{xy}.

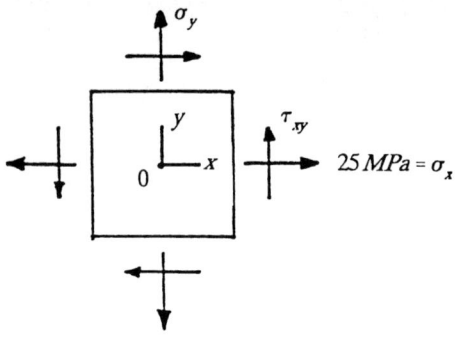

Figure P2.35.

2.36 Solve Problem 2.34 by using Mohr's circle for plane stress.

2.37 For the plane stress elements in Fig. P2.37, construct for each case the Mohr's circle for plane stress, and sketch on properly oriented elements the principal stresses and maximum and minimum shear stresses. What are the stresses on a plane oriented at an angle $\theta = 60°$?

2.38 A block of material is 20 in. long in the horizontal direction, 3 in. wide, and 6 in. high. It is subjected to a vertical pressure of 500 psi and a horizontal axial tension of 6,000 psi. By considering a plane through the block that has a slope of 3 vertical to 4 in the axial direction, determine the normal and shearing stresses on the plane by using Mohr's circle for plane stress.

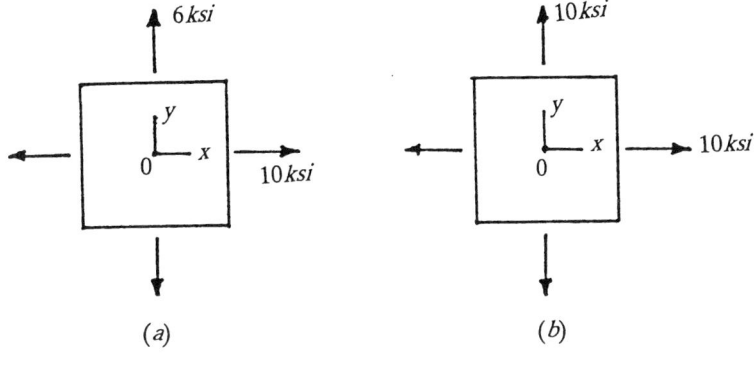
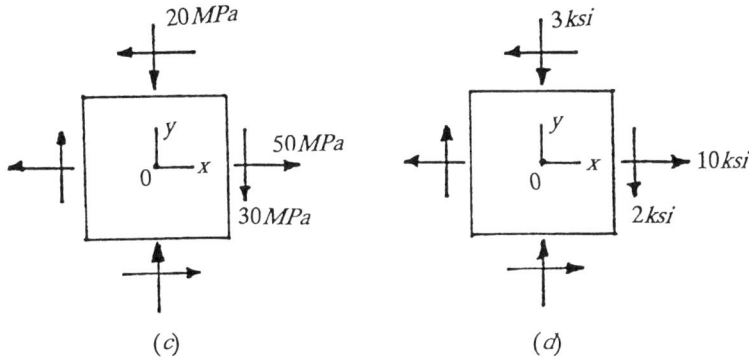

Figure P2.37.

3 Infrastructural Dynamics

3.1 INTRODUCTION

In the design of infrastructural elements such as highway bridges, buildings, tall stacks, machine foundations, and so on, the dynamic nature of the applied loading must be considered. In many cases, such a dynamic analysis can produce reasonable results by considering equivalent static loads and converting the dynamic problem into an equivalent statics problem where statics analysis can be used. In other cases, however, the actual dynamic problem must be solved, and reasonable idealized systems are often used to carry out such a dynamic analysis.

In this chapter, the dynamic analysis and design of infrastructural elements and components such as beams, frames, buildings, and highway bridges are carried out. Various types of dynamic loadings are considered, and appropriate idealized systems are developed to carry out an accurate dynamic analysis. Closed-form as well as numerical solutions are used, and the advantages, and/or possible disadvantages of such solutions are carefully examined. This chapter also includes an introduction to practical design and analysis for earthquakes by using available earthquake codes.

3.2 THE SIMPLEST INFRASTRUCTURAL DYNAMICS PROBLEM

The simplest infrastructural dynamics problem may be considered to be the one shown in Fig. 3.1a, consisting of a mass m attached to a linear spring of spring constant k. The constant k is defined as the force that is required to deform the spring by an amount equal to unity. We assume here that the mass m is large compared to the mass of the spring, thus making it possible to neglect the mass of the spring in the solution of the spring-mass problem.

The mass m is subjected to a force $F(t)$, which is a function of time. It can be a harmonic force, a periodic nonharmonic force, a force of some general arbitrary variation with respect to time, or it can be a random force. A harmonic variation of $F(t)$ is shown in Fig. 3.1b, a periodic nonharmonic force is shown in Fig. 3.1c, and a nonperiodic force of some general type is shown in Fig. 3.1d. An example of a random force $F(t)$ is the one shown in Fig. 1.27, which is produced by a rather strong earthquake. Since randomness implies unpredictability, a random force would have an irregular form in terms of both time and amplitude, and it is usually called nondeterministic. A force $F(t)$ can

148 INFRASTRUCTURAL DYNAMICS

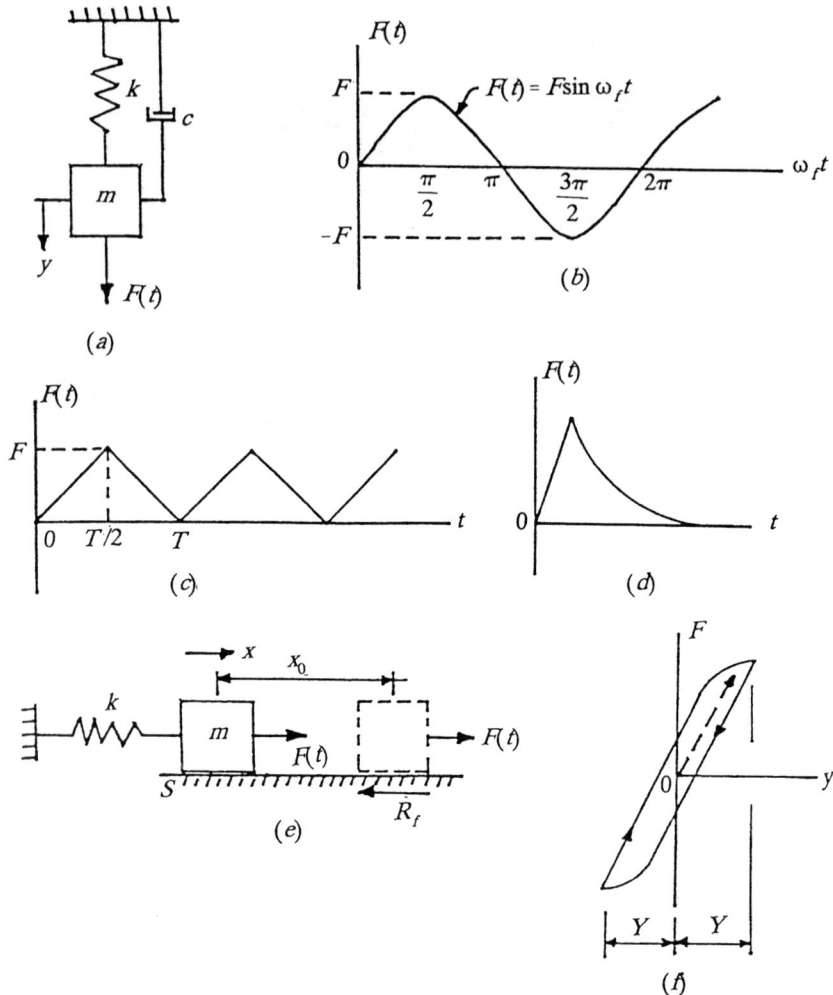

Figure 3.1. (*a*) Spring-mass system. (*b*) Harmonic force, (*c*) Periodic nonharmonic force. (*d*) Nonperiodic force of a general type. (*e*) Spring-mass system with Coulomb damping. (*f*) Hysteresis loop.

be also impulsive. Impulsive forces are forces of large magnitude with an instantaneous time duration.

The motion of the spring-mass system in Fig. 3.1*a* may take place under the influence of some kind of damping that opposes the motion. The damping is usually called viscous if the motion is executed through some kind of air or liquid medium, or if the parts of the system are moving through lubricated surfaces. In such cases, the resisting damping force $R_d(t)$ is usually assumed to be proportional to the velocity. The spring-mass system in Fig. 3.1*a* is assumed to be under the influence of viscous damping, where c is the viscous damping coefficient defined as the resisting force that is developed per unit of velocity. Thus, for viscously damped systems, we have

$$R_d(t) = -c\dot{y} \tag{3.1}$$

where $\dot{y}$ is the velocity of the mass m with respect to time t.

If the mass m is rubbing on a dry surface during motion, it said to be under the influence of Coulomb, or dry friction, damping, and the resisting force R_f is

$$R_f = \pm \mu N \tag{3.2}$$

where N is the force normal to the dry surface S, and μ is the coefficient of kinetic friction of the material of the surface S. For example, the motion of the spring-mass system in Fig. 3.1e is under the influence of Coulomb damping, because m is rubbing against the dry surface S during motion.

Another form of damping is the hysteresis damping, also known as solid or structural damping, which is the result of internal friction in the material of the body during its oscillatory motion. For example, consider a cycle of motion of the spring-mass system and plot the force-displacement diagram shown in Fig. 3.1f; the area within the loop of this diagram represents the amount of energy ΔU transformed into heat, per cycle of motion, due to internal friction in the material. Experiments have shown that ΔU can be obtained from the following approximate expression:

$$\Delta U = k\pi c_0 Y^2 \tag{3.3}$$

where c_0 is the dimensionless constant of the material for solid damping, and k is the spring constant. The rate of reduction of the amplitude in this case depends on the size of the area under the hysteresis loop in Fig. 3.1f. Rubber-type materials are considered to be good dampers and exhibit a much wider loop when compared to metallic materials.

3.2.1 Practical Applications of the Spring-Mass System

There are many useful practical applications of the single-degree spring-mass system in Fig. 3.1a. Many infrastructural elements such as beams, frames, buildings, foundations, concrete slabs, and so on, that have continuous mass and elasticity, may be designed and analyzed successfully if they are idealized as single-degree spring-mass system. Such important idealizations are briefly discussed below.

Beams: A simply supported beam, such as the one shown in Fig. 3.2a, may be idealized as shown in Fig. 3.2b or Fig. 3.2c. The beam is assumed to support a weight W, the weight is assumed to be heavy compared to the weight of the beam, and the weight is acted upon by a dynamic force $F(t)$, as shown. For

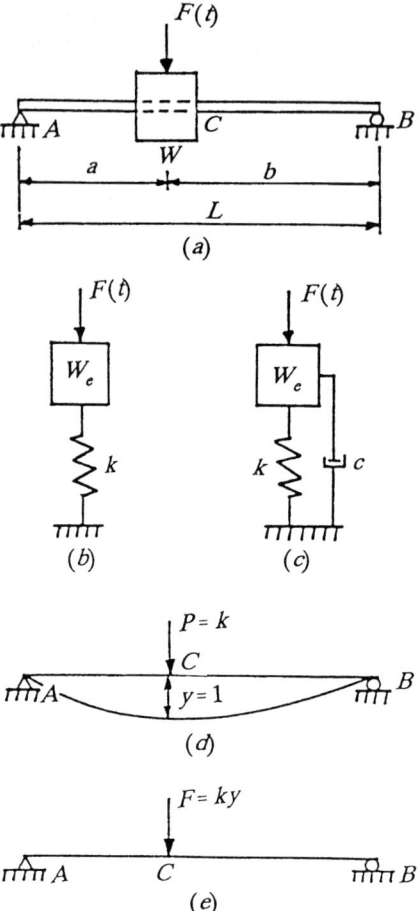

Figure 3.2. (a) Simply supported beam supporting a weight W and loaded as shown. (b) Undamped idealized spring-mass system. (c) Damped idealized spring-mass system. (d) Calculation of spring stiffness k. (e) Simply supported beam loaded with $F = ky$.

example, the heavy weight W may be a mechanical system supported by the beam, and the dynamic force may be a force that is produced by the moving parts of the machine. A centrifugal force from rotating machinery such as gears, machines used to shred logs, and shredders used to crush waste are examples of such dynamic forces.

The spring constant k in Fig. 3.2b represents the stiffness of the beam at point C, and is defined as the vertical force P at C, as shown in Fig. 3.2d, that is required to produce a vertical deflection y equal to unity. The weight W_e is the weight supported by the beam, and $F(t)$ is the dynamic force. The weight of the beam may be neglected if it is small, but, if desired, a certain portion of it [5, 37, 39] may be incorporated in the weight W_e of the idealized spring-mass system. A proper idealization would be to make W_e equal to the heavy weight W plus half the weight of the beam [5, 37, 39].

The idealized system in Fig. 3.2b is a reasonable one because the displacement of the weight produced by the dynamic force $F(t)$ represents very accurately the vertical displacement at C of the beam in Fig. 3.2a, and the spring force ky of the idealized spring-mass system is the force $F = ky$ acting on the beam at point C, as shown in Fig. 3.2e. Static analysis can be used to solve the beam in Fig. 3.2e, because the magnitude of the dynamic force $F = ky$ can be determined by solving the idealized spring-mass system. The solution is very accurate, because the maximum moment in the beam occurs at point C, and the exact beam stiffness at C is represented by the spring constant k. Thus, for the maximum force $F_{max} = ky_{max}$, we obtain the maximum dynamic moment at point C of the member, and, consequently, its maximum bending stress. On this basis, the beam in Fig. 3.2a may be designed for maximum allowable bending stress and maximum allowable vertical deflection by solving the problems in Figs. 3.2b and 3.2e.

Damping may also be taken into consideration by solving the idealized spring-mass system in Fig. 3.2c. On this basis, the analysis of the beam in Fig. 3.2e will incorporate damping, because in this case, ky is determined by using the idealized spring-mass system in Fig. 3.2c where c is the coefficient of viscous damping.

A similar procedure may be used to derive idealized spring-mass systems of beams of various boundary conditions and dynamic loadings $F(t)$.

Example 3.1: The beam in Fig. 3.3a is fixed at both ends and supports a mechanical system of weight $W = 230$ kN. The dynamic force $F(t)$ that is produced during its operation is a harmonic force $F \cos \omega_f t$, where $F = 150$ kN. The beam is made out of structural steel of modulus of elasticity $E = 210 \times 10^6$ kPa, and its uniform cross-sectional moment of inertia $I = 624 \times 10^{-6}$ m^4. Determine an idealized spring-mass system that can be used for the computation of the maximum bending stress and maximum deflection of the member. Neglect damping and the weight of the beam.

SOLUTION: From mechanics of solids, the vertical deflection at point C in Fig. 3.3b, produced by the vertical load P, is given by the expression

$$y = \frac{PL^3}{192EI} \quad (3.4)$$

When $y = 1$, we have

$$P = k = \frac{192EI}{L^3} = \frac{(192)(210)(10)^6(624)(10)^{-6}}{10^3}$$

$$= 25{,}159.68 \text{ kN/m}$$

which is the spring stiffness of the beams at point C.

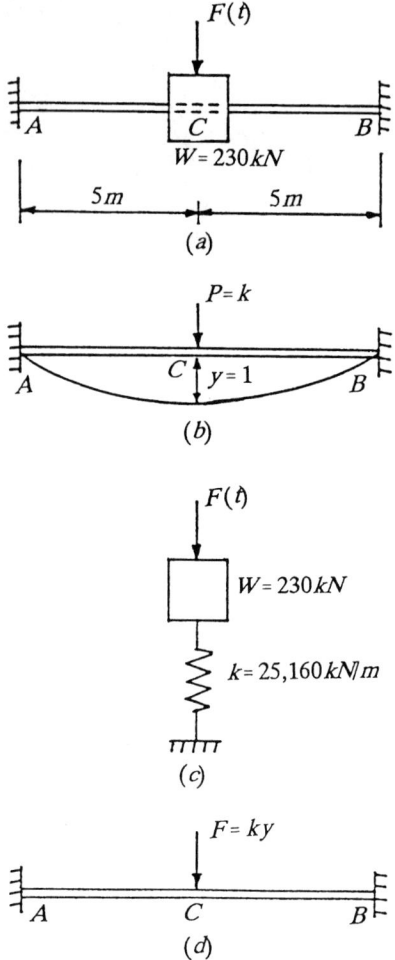

Figure 3.3. (a) Uniform beam fixed at both ends and loaded as shown. (b) Calculation of spring stiffness k of the member. (c) Idealized one-degree spring-mass system. (d) Static application of the load $F = ky$ at point C of the member.

The idealized single-degree spring-mass system is shown in Fig. 3.3c. The deflection y of the spring produced by the dynamic force $F(t)$ would be identical to the vertical deflection at point C of the original member in Fig. 3.3a. At any time t, the force ky of the spring in Fig. 3.3c can be applied as a static load at point C of the member, as shown in Fig. 3.3d. In the same figure, the maximum bending moment due to $F = ky$ occurs simultaneously at points A, B, and C, and it can be determined by using available tables or elementary methods of mechanics of solids [38, 6]. This topic is discussed at length later in the text. With known maximum moment, the maximum bending stress may be determined as discussed in Section (2.3). The maximum load ky_{max} is obtained when the deflection y of the idealized spring becomes y_{max}. On this basis, the member in Fig. 3.3a may be designed for maximum bending stress and maximum vertical deflection requirements.

Frames and Buildings: Many one-story framed structures subjected to dynamic excitations may be idealized and analyzed dynamically as single-degree of freedom spring-mass systems. Consider, for example, a one-story building that has the floor plotted as shown in Fig. 3.4a. The columns and the girders are

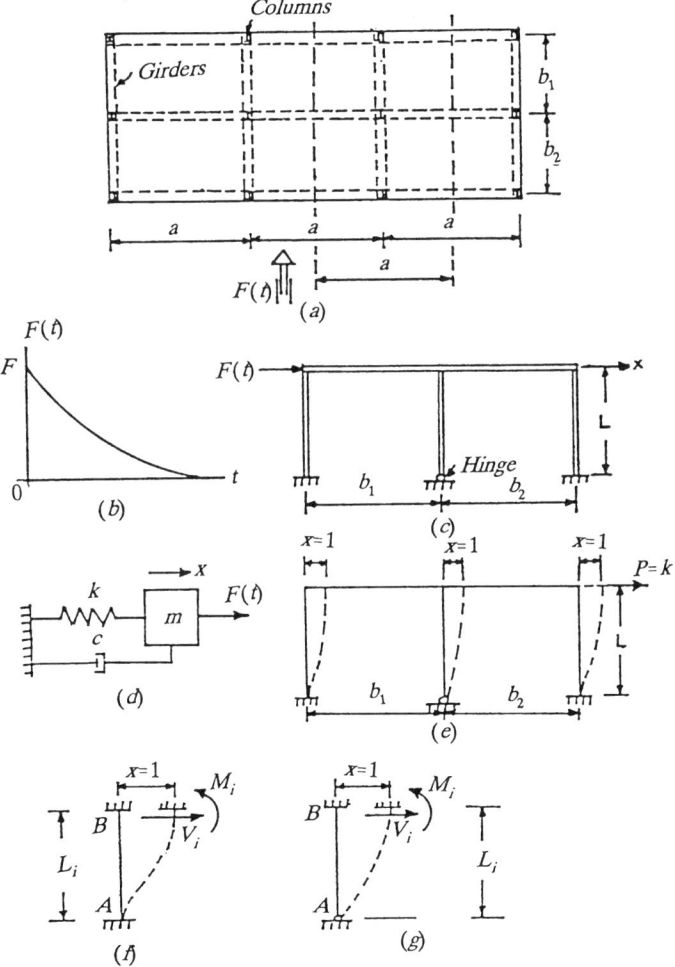

Figure 3.4. (a) Floor plan of a one-story building. (b) Time variation of $F(t)$. (c) Interior bay of the building. (d) Idealized one-degree spring-mass system. (e) Stiffness k of an interior bay. (f) Stiffness k_i of a fixed-fixed column. (g) Stiffness k_i of a hinged-fixed column.

arranged as shown in the figure, and standard structural steel wide-flange sections may be used for the construction of the building. If the side of the building is subjected to a dynamic force $F(t)$ that has a time variation such as the one in Fig. 3.4b, a single-degree spring-mass system may be determined that can provide a reasonable dynamic analysis concerning the design of the building. If the bays of the building are fairly uniform, such as the ones shown in Fig. 3.4a, its dynamic analysis can be carried out by considering only an interior bay of the building with the associated wall and floor areas, as shown in Fig. 3.4c. The dynamic force $F(t)$ has the time variation shown in Fig. 3.4b, and it is assumed to be a concentrated dynamic force located at the top of the front columns of the building.

If the dynamic force is distributed throughout the side of the building, reasonable idealizations can be made that provide equivalent concentrated dynamic forces $F(t)$ acting at the tops of the columns of the building.

The bay in Fig. 3.4c may be idealized as a one-degree spring-mass system having the elements shown in Fig. 3.4d. The mass m of the spring-mass system is composed of the total mass of roof and girders of the bay, plus half the mass of its walls and columns. Since the mass of the wall and columns is often small compared to the mass of the roof and girders, it is usually neglected. The dynamic force $F(t)$ acting on the mass m is the force $F(t)$ in Fig. 3.4c; it has the time variation shown in Fig. 3.4b. The stiffness k of the spring-mass system represents the stiffness at the top of the frame of the bay in Fig. 3.4c. It is defined as the horizontal force P at the top of the frame that produces the unit horizontal displacement shown in Fig. 3.4e.

If the girders are very stiff compared to the columns, a situation that is often encountered in practice, the tops of the columns may be assumed as fixed, as shown in Figs. 3.4f and 3.4g. On this basis, the stiffness k of the bay frame at its top may be obtained by determining the stiffness at the top of each column and adding them together. The stiffness k_i of such a column represents the amount of shear V_i that is developing at the fixed end B, when this end is displaced by a horizontal displacement $x = 1$, as shown in Figs. 3.4f and 3.4g. For the column in Fig. 3.4f, where the end A is also fixed, the stiffness k_i is given by the equation

$$k_i = \frac{12(EI)_i}{L_i^3} \tag{3.5}$$

For the column in Fig. 3.4g, where end A is hinged, we have

$$k_i = \frac{3(EI)_i}{L_i^3} \tag{3.6}$$

On this basis, the stiffness k of the spring in Fig. 3.4d is

$$k = \sum_{i=1}^{n} k_i \tag{3.7}$$

where n is the number of the columns. Note that the lengths L_i of the columns of the bay may be different.

When the dynamic force $F(t)$ is applied to the bay in Fig. 3.4c, the top of each column will be displaced horizontally by an amount x, which is the same for all columns. This displacement is the maximum displacement of the frame and is represented by the displacement x of the mass m in Fig. 3.4d. Thus, for each column, the shear force V_i at the top of each column is given by the equation

$$V_i = k_i x \tag{3.8}$$

and the total shear V at the top of the frame is

$$V = \sum_{i=1}^{n} k_i x$$

Since V is equal to the force kx developing in the spring of the idealized system in Fig. 3.4d, we have

$$V = kx = \sum_{i=1}^{n} k_i x \tag{3.9}$$

Thus, the dynamic analysis of the idealized system in Fig. 3.4d provides an accurate evaluation of the horizontal displacement x of the top of the bay frame in Fig. 3.4c. With known x, the shear forces V_i at the top of each column, as well as the total shear V of the bay frame, may be evaluated as discussed above. With known V_i, simple statics indicate that the moment M_i for the column in Fig. 3.4f is given by the equation

$$M_i = \frac{V_i L_i}{2} \tag{3.10}$$

For the column in Fig. 3.4g, where end A is hinged, we have

$$M_i = V_i L_i \tag{3.11}$$

With known moments, the maximum stresses at the top of each column may be evaluated as discussed in Section 2.3 of this text. In each column, this stress would reach its maximum value when the horizontal displacement x becomes maximum.

Example 3.2: The frame shown in Fig. 3.5a represents an interior bay of a single-story building with columns of different lengths. The width of the bay is 25 ft, and the weights of the walls and roof, by including the weight of the

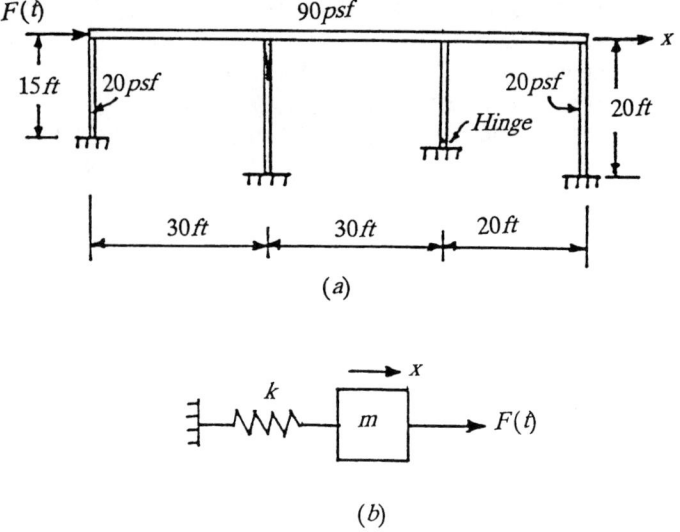

Figure 3.5. (a) Bay frame of a one-story building. (b) Idealized one-degree spring-mass system of the bay frame to be used for dynamic analysis.

girders and columns, is as shown in the figure. The first and third columns of the frame have W10 × 30 wide-flange sections with $I = 170$ in.^{4}, and the sections of the second and fourth columns are W10 × 60 with $I = 341$ in.^{4} Determine an idealized one-degree spring-mass system that is appropriate for the dynamic analysis of the bay frame. Neglect damping. The modulus of elasticity $E = 30 \times 10^6$ psi.

SOLUTION: The idealized one-degree spring-mass system is as shown in Fig. 3.5b. The displacement x of the mass m represents the horizontal displacement x at the top of the columns as shown in Fig. 3.5a. The mass m of the idealized system is obtained by considering the total weight of the roof and half the weight of the walls and columns, and dividing the total weight W by the acceleration of gravity g. On this basis, we have

$$W = (80)(25)(90 + \tfrac{1}{2}(15)(25)(20) + \tfrac{1}{2}(20)(25)(20) + \tfrac{1}{2}(60)(20) + \tfrac{1}{2}(30)(20)$$
$$= 180{,}000 + 3{,}750 + 5{,}000 + 600 + 300$$
$$= 189{,}650 \text{ lb}$$
$$= 189.65 \text{ kips}$$

Note that the contribution of the walls and columns is only 5.36 percent compared to the weight of the roof, and it could be neglected in the dynamic analysis of the bay frame.

The mass m of the spring-mass system, by assuming $g = 384.6 \text{ in./sec}^2$, is

$$m = \frac{W}{g} = \frac{189.65}{384.6} = 0.4931 \frac{\text{kip-sec}^2}{\text{in.}}$$

We assume now that the girders of the frame are infinitely stiff compared to the columns. On this basis the stiffnesses of the first, second, and fourth columns of the frame may be obtained by using Eq. (3.5), and Eq. (3.6) may be used for the stiffness of the third column. Thus, the spring stiffness k of the idealized system is

$$k = \frac{(12)(30)(10)^6(170)}{(15 \times 12)^3} + \frac{(2)(12)(30)(10)^6(341)}{(20 \times 12)^3} + \frac{(3)(30)(10)^6(170)}{(15 \times 12)^3}$$

$$= 10{,}494 + 17{,}760 + 2{,}623$$

$$= 30{,}877 \text{ lb/in.}$$

$$= 30.877 \text{ kips/in.}$$

The force $F(t)$ acting on the mass m of the idealized system is the dynamic force $F(t)$ acting at the top of the bay frame in Fig. 3.5a.

The idealized system obtained in Fig. 3.5b would be suitable for the dynamic analysis of the bay frame.

3.3 SOLUTION OF THE SIMPLEST INFRASTRUCTURAL DYNAMICS PROBLEM

In Section 3.2, it was shown that beams, frames, buildings, and many other infrastructural elements, may be very accurately analyzed dynamically by using idealized one-degree spring-mass systems. It was also shown how such idealized spring-mass systems may be derived. In this section, the solution of single-degree spring-mass systems will be obtained. We start the discussion by considering the viscously damped spring-mass system shown in Fig. 3.6a, which is subjected to a dynamic force $F(t)$. At some vertical displacement y, the free-body diagram of mass m is as shown in Fig. 3.6b, where $c\dot{y}$ is the resistance caused by the viscous damping, ky is the spring force, $m\ddot{y}$ is the inertia force, and $F(t)$ is the applied dynamic force. By using Newton's second law of motion, we find

$$m\ddot{y} = -c\dot{y} - ky + F(t)$$

or

$$m\ddot{y} + c\dot{y} + ky = F(t) \tag{3.12}$$

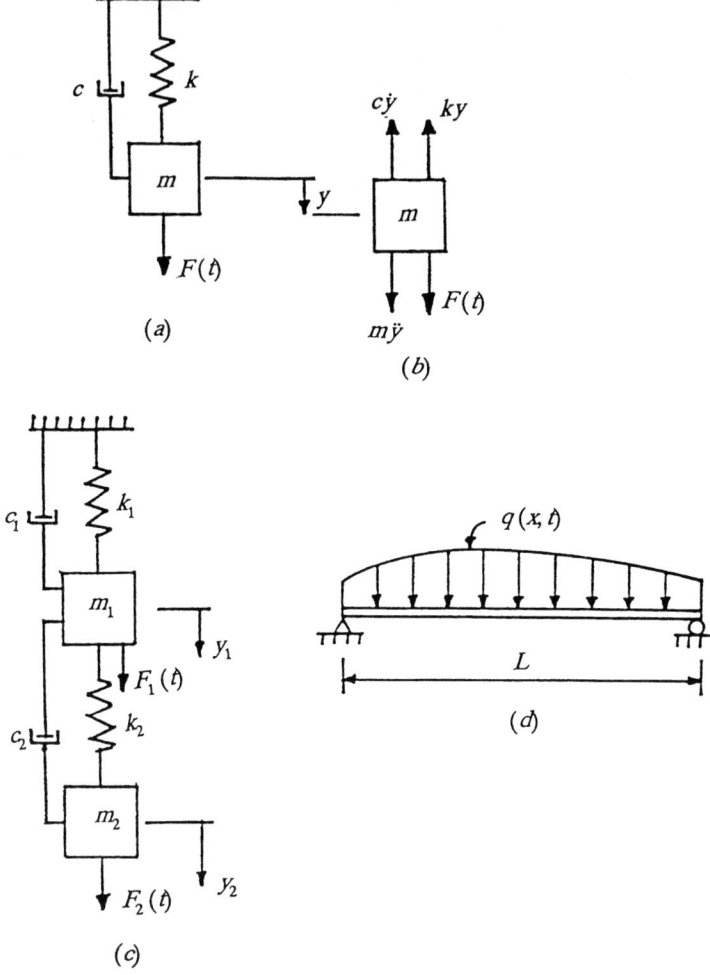

Figure 3.6. (*a*) Viscously damped single-degree spring-mass system. (*b*) Free-body diagram of mass *m*. (*c*) Two-degree spring-mass system. (*d*) System with infinite degrees of freedom.

Equation (3.12) is the differential equation of motion of the spring-mass system in Fig. 3.6a. Its solution yields the displacement y, velocity $\dot{y}$, and acceleration $\ddot{y}$, of the mass m at any time t. Note that the mass m is restricted to move only in the vertical direction y and, consequently, the position of mass m at any time t can be defined completely by the displacement y. Such systems are called single-degree of freedom spring-mass systems, because one coordinate is sufficient to locate the position of the mass m at any time t. If more coordinates are required, then the degrees of freedom of the system are

SOLUTION OF THE SIMPLEST INFRASTRUCTURAL DYNAMICS PROBLEM 159

increasing. In general, the number of degrees of freedom of a system is specified by the number of coordinates that are required to define the position of the moving parts, or particles, of the system. For example, if the two masses m_1 and m_2 in Fig. 3.6c are restricted to move only in the vertical direction, then we need a vertical displacement y_1 to define the position of m_1 during motion, and a vertical displacement y_2 to locate m_2 during motion. Thus, this system has two degrees of freedom. The simply supported beam in Fig. 3.6d has infinite degrees of freedom, because we need an infinite number of coordinates y to define the vertical displacement of the infinite number of particles along the length L of the member.

The solution of Eq. 3.12 largely depends upon the type and character of the dynamic force $F(t)$. It may be harmonic, periodic, or of some general type, as shown in Figs. 3.1b, 3.1c, and 3.1d, respectively.

If $F(t) = 0$, Eq. (3.12) yields

$$m\ddot{y} + c\dot{y} + ky = 0 \tag{3.13}$$

Equation (3.13) represents the free viscously damped vibration of the spring-mass system. If damping is also absent, then c in Eq. (3.13) is zero. Thus, we have

$$m\ddot{y} + ky = 0 \tag{3.14}$$

Equation (3.14) characterizes the free undamped vibration of the spring-mass system.

3.3.1 Free Undamped Vibration

In this case, we need to solve the differential equation given by Eq. (3.14). The vibratory motion represented by this equation is harmonic, and the amplitude y of the mass m may be expressed as follows:

$$y = y_0 \cos \omega t \tag{3.15}$$

where y_0 is the maximum amplitude of the vertical vibration of the mass m, and ω is its free undamped frequency of vibration.

We also have

$$\dot{y} = -y_0 \omega \sin \omega t \tag{3.16}$$

and

$$\begin{aligned}\ddot{y} &= -y_0 \omega^2 \cos \omega t \\ &= -\omega^2 y\end{aligned} \tag{3.17}$$

or

$$\ddot{y} + \omega^2 y = 0 \tag{3.18}$$

By substituting Eqs. (3.15) and (3.17) into Eq. (3.14), we find

$$-my_0\omega^2 \cos \omega t + ky_0 \cos \omega t = 0$$

or

$$\omega = \sqrt{\frac{k}{m}} \tag{3.19}$$

Equation (3.19) yields the free undamped frequency of vibration of the one-degree spring-mass system in radians per second (rps). In cycles per second (cps), or in hertz (Hz), the frequency f is

$$f = \frac{\omega}{2\pi} = \frac{1}{2\pi}\sqrt{\frac{k}{m}} \tag{3.20}$$

The period of vibration τ, which is defined as the time required to complete one cycle of motion, is

$$\tau = \frac{1}{f} = 2\pi\sqrt{\frac{m}{k}} \tag{3.21}$$

From Eq. (3.19), we find

$$k = m\omega^2 \tag{3.22}$$

On this basis, by substituting for k in Eq. (3.14), we obtain

$$\ddot{y} + \omega^2 y = 0 \tag{3.23}$$

Equation (3.23) is a second-order, homogeneous differential equation, and its well-known solution $y(t)$ is

$$y(t) = C_1 \sin \omega t + C_2 \cos \omega t \tag{3.24}$$

where the constants C_1 and C_2 may be determined by using the initial conditions of the motion. For example, if at $t = 0$ we have $y = y_0$, Eq. (3.24) yields

$$y_0 = C_1 \sin(0) + C_2 \cos(0)$$

SOLUTION OF THE SIMPLEST INFRASTRUCTURAL DYNAMICS PROBLEM

or

$$C_2 = y_0 \tag{3.25}$$

If $\dot{y} = \dot{y}_0$ at $t = 0$, the first derivative of Eq. (3.24) with respect to time t yields

$$\dot{y}_0 = C_1 \omega \cos(0) - C_2 \omega \sin(0)$$

or

$$C_1 = \frac{\dot{y}_0}{\omega} \tag{3.26}$$

With known C_1 and C_2, the solution given by Eq. (3.24) yields

$$y(t) = \frac{\dot{y}_0}{\omega} \sin \omega t + y_0 \cos \omega t \tag{3.27}$$

If the initial velocity $\dot{y}_0$ at $t = 0$ is also zero, Eq. (3.27) yields $y(t) = y_0 \cos \omega t$, which is identical to Eq. (3.15).

The most general solution of $y(t)$ may be obtained if at some time $t = t_0$ the initial displacement and initial velocity are y_0 and $\dot{y}_0$, respectively. By applying these two initial conditions as explained above, we find

$$C_1 = y_0 \sin \omega t_0 + \frac{\dot{y}_0}{\omega} \cos \omega t_0 \tag{3.28}$$

$$C_2 = y_0 \cos \omega t_0 - \frac{\dot{y}_0}{\omega} \sin \omega t_0 \tag{3.29}$$

On this basis, Eq. (3.24) yields

$$y(t) = y_0 \cos \omega(t - t_0) + \frac{\dot{y}_0}{\omega} \sin \omega(t - t_0) \tag{3.30}$$

Differentiation of Eq. (3.30) with respect to time t yields the velocity $\dot{y}(t)$ with respect to time t. That is

$$\dot{y}(t) = -y_0 \omega \sin \omega(t - t_0) + \dot{y}_0 \cos \omega(t - t_0) \tag{3.31}$$

Equations (3.24) and (3.30) may be also expressed in terms of a constant Y and the phase angle ϕ by using the relations

$$C_1 = y_0 = Y \sin \phi \tag{3.32}$$

$$C_2 = \frac{\dot{y}_0}{\omega} = Y \cos \phi \tag{3.33}$$

With this in mind, Eq. (3.30) yields

$$y(t) = Y \sin[\omega(t - t_0) + \phi] \tag{3.34}$$

and, by differentation,

$$\dot{y}(t) = Y \omega \cos[\omega(t - t_0) + \phi] \tag{3.35}$$

where

$$Y = \sqrt{C_1^2 + C_2^2} = \sqrt{y_0^2 + \left(\frac{\dot{y}_0}{\omega}\right)^2} \tag{3.36}$$

$$\tan \phi = \frac{C_1}{C_2} = \frac{Y \sin \phi}{Y \cos \phi} = \frac{\omega y_0}{\dot{y}_0} \tag{3.37}$$

Example 3.3: Determine the free undamped frequency of vibration and the free undamped period of vibration for the idealized spring-mass system in Fig. 3.3c. Assume that the dynamic force $F(t)$ is zero.

SOLUTION: The free undamped frequency of vibration ω is

$$\omega = \sqrt{\frac{k}{m}} = \sqrt{\frac{(25{,}160)(9.81)}{(230)}}$$

$$= 32.7586 \text{ rps}$$

In units of hertz, the undamped frequency f is

$$f = \frac{\omega}{2\pi} = \frac{32.7586}{2\pi} = 5.2137 \text{ Hz}$$

From Eq. (3.21), the period of vibration τ, is

$$\tau = \frac{1}{f} = \frac{1}{5.2137} = 0.1918 \text{ sec}$$

3.3.2 Free Vibration with Viscous Damping

This involves the solution of the differential equation given by Eq. (3.13). We write this equation again here:

$$m\ddot{y} + c\dot{y} + ky = 0 \tag{3.38}$$

We assume its solution $y(t)$ as follows:

$$y(t) = Ae^{pt} \tag{3.39}$$

where A and p are constants.

SOLUTION OF THE SIMPLEST INFRASTRUCTURAL DYNAMICS PROBLEM

By substituting Eq. (3.39) into Eq. (3.38), we find

$$mp^2 + cp + k = 0 \tag{3.40}$$

which yields the two roots of p as follows:

$$p_{1,2} = -\frac{c}{2m} \pm \sqrt{\left(\frac{c}{2m}\right)^2 - \frac{k}{m}} \tag{3.41}$$

Examination of the two terms under the radical of Eq. (3.41) leads us to draw the following conclusions:

1. If the sum of the terms under the radical is zero, then

$$\left(\frac{c}{2m}\right)^2 = \frac{k}{m} = \omega^2 \tag{3.42}$$

or

$$\frac{c}{2m} = \omega \tag{3.43}$$

where ω is the undamped frequency of the system. Under this condition, Eq. (3.41) yields $p_{1,2} = -c/2m$, which represents a transition from oscillatory to nonoscillatory configuration known as the condition of critical damping. That is, if the motion of the mass m starts from the position y_0 with velocity $\dot{y}_0$, the mass will return to rest without oscillation. The value of the damping factor c that produces this condition is defined as the critical damping factor and is denoted by the symbol c_c. Thus, from Eq. (3.42), the critical damping factor c_c is

$$c_c = 2\sqrt{km} = 2m\omega \tag{3.44}$$

This situation does not usually occur in practice.

2. When the sum under the radical is positive, we have the inequality

$$\left(\frac{c}{2m}\right)^2 > \omega^2 \tag{3.45}$$

or

$$\frac{c}{2m} > \omega \tag{3.46}$$

which is known as the condition of overdamping, indicating that the motion is aperiodic with real and negative values of p_1 and p_2.

3. We also have the condition of underdamping, or light damping, when the sum of the terms under the radical is negative. On this basis, we have

$$\left(\frac{c}{2m}\right)^2 < \omega^2 \tag{3.47}$$

or

$$\frac{c}{2m} < \omega \tag{3.48}$$

In this case, the spring mass system will vibrate with decreasing amplitude and eventually the vibratory motion will cease to exist. Most infrastructural systems are under the influence of light damping.

Similar conclusions may be made if we introduce the damping ratio ζ, which is defined by the expression

$$\zeta = \frac{c}{c_c} \tag{3.49}$$

On this basis, we have

$$\frac{c}{2m} = \zeta\omega \tag{3.50}$$

$$\left(\frac{c}{2m}\right)^2 - \frac{k}{m} = \omega^2(\zeta^2 - 1) \tag{3.51}$$

and Eq. (3.41) yields

$$p_{1,2} = (-\zeta \pm \sqrt{\zeta^2 - 1})\omega \tag{3.52}$$

Equation (3.52) shows that the condition of critical damping occurs when $\zeta = 1$, overdamping occurs when $\zeta > 1$, and underdamping, or light damping, will take place when $\zeta < 1$. This is very useful information for the design engineer who deals with the dynamic performance of infrastructural systems.

By using Eqs. (3.39) and (3.41), the solution $y(t)$ of Eq. (3.38) is found to be

$$y(t) = \exp\left[-\left(\frac{c}{2m}\right)\right]\left(A_1 \exp\left[\sqrt{\left(\frac{c}{2m}\right)^2 - \frac{k}{m}}\right]t \right.$$
$$\left. + A_2 \exp\left\{-\left[\sqrt{\left(\frac{c}{2m}\right)^2 - \frac{k}{m}}\right]t\right\}\right) \tag{3.53}$$

The constants A_1 and A_2 may be determined by using the initial conditions of the motion. For example, if at $t = 0$ the initial displacement is y_0 and the initial velocity is $\dot{y}_0$, we find

$$A_1 = y_0 - \frac{\dot{y}_0 - y_0 p_1}{p_2 - p_1} \tag{3.54}$$

$$A_2 = \frac{\dot{y}_0 - y_0 p_1}{p_2 - p_1} \tag{3.55}$$

If we apply the Euler's expression

$$\exp\left\{\pm\left[i\sqrt{\frac{k}{m} - \left(\frac{c}{2m}\right)^2}\right]t\right\} = \cos\left[\sqrt{\frac{k}{m} - \left(\frac{c}{2m}\right)^2}\right]t \pm i\sin\left[\sqrt{\frac{k}{m} - \left(\frac{c}{2m}\right)^2}\right]t \tag{3.56}$$

the trigonometric form of Eq. (3.53) is

$$y(t) = \exp\left[-\left(\frac{c}{2m}\right)t\right]\left\{C_1 \cos\left[\sqrt{\frac{k}{m} - \left(\frac{c}{2m}\right)^2}\right]t + C_2 \sin\left[\sqrt{\frac{k}{m} - \left(\frac{c}{2m}\right)^2}\right]t\right\} \tag{3.57}$$

where the constants C_1 and C_2 may be determined by applying the initial conditions of the motion.

By using Eqs. (3.50) and (3.51), the solution given by Eq. (3.57) may be expressed in terms of the damping ratio ζ in the following manner:

$$y(t) = \exp(-\zeta\omega t)[C_1 \cos(\omega\sqrt{1-\zeta^2})t + C_2 \sin(\omega\sqrt{1-\zeta^2})]t \tag{3.58}$$

From this equation we conclude that the damped frequency ω_d of the spring-mass system is

$$\omega_d = \omega\sqrt{1-\zeta^2} \tag{3.59}$$

Thus, the damped period of vibration τ_d of the spring-mass system is

$$\tau_d = \frac{2\pi}{\omega_d} = \frac{2\pi}{\omega\sqrt{1-\zeta^2}} \tag{3.60}$$

By applying the initial conditions that at $t = 0$ the initial displacement is y_0 and the initial velocity is $\dot{y}_0$, we find

$$C_1 = y_0 \tag{3.61}$$

$$C_2 = \frac{\dot{y}_0 + \zeta\omega y_0}{\omega_d} \tag{3.62}$$

By substituting into Eq. (3.58), we find

$$y(t) = \exp(-\zeta \omega t)\left(y_0 \cos \omega_d t + \frac{\dot{y}_0 + \zeta \omega y_0}{\omega_d} \sin \omega_d t\right) \quad (3.63)$$

which shows that the first term is proportional to $\cos \omega_d t$ and depends only on y_0, while the second term is proportional to $\sin \omega_d t$ and depends on both y_0 and $\dot{y}_0$.

In terms of a constant Y and the phase angle ϕ_d, Eq. (3.63) may be written as follows:

$$y(t) = Y \exp(-\zeta \omega t) \sin(\omega_d t + \phi_d) \quad (3.64)$$

where

$$Y = \sqrt{C_1^2 + C_2^2} = \sqrt{y_0^2 + \frac{(\dot{y}_0 + \zeta \omega y_0)^2}{\omega_d^2}} \quad (3.65)$$

$$\phi_d = \tan^{-1}\left(\frac{C_2}{C_1}\right) = \tan^{-1}\left(\frac{\dot{y}_0 + \zeta \omega y_0}{\omega_d y_0}\right) \quad (3.66)$$

Equation (3.64) represents a pseudoharmonic motion with an exponentially decaying amplitude $Y \exp(-\zeta \omega t)$ and phase angle ϕ_d. The rate of decay of the amplitude depends on the amount of damping ζ of the spring-mass system.

The rate of decay of the motion may be also evaluated by considering the logarithmic decrement δ, which is defined as the natural logarithm of the ratio of any two successive amplitudes of the motion. It may be obtained by using the expression

$$\delta = \zeta \omega \tau_d \quad (3.67)$$

or the expression

$$\delta = \frac{2\pi \zeta}{\sqrt{1 - \zeta^2}} \quad (3.68)$$

By comparing Eq. (3.67) with the term $Y \exp(-\zeta \omega t)$ of Eq. (3.64), which represents the amplitude decay of the motion, we note that

$$Y \exp(-\zeta \omega t) = \frac{Y}{\exp(\delta t / \tau_d)} \quad (3.69)$$

Equation (3.69) illustrates the influence of δ on the oscillatory motion of consecutive amplitudes.

SOLUTION OF THE SIMPLEST INFRASTRUCTURAL DYNAMICS PROBLEM 167

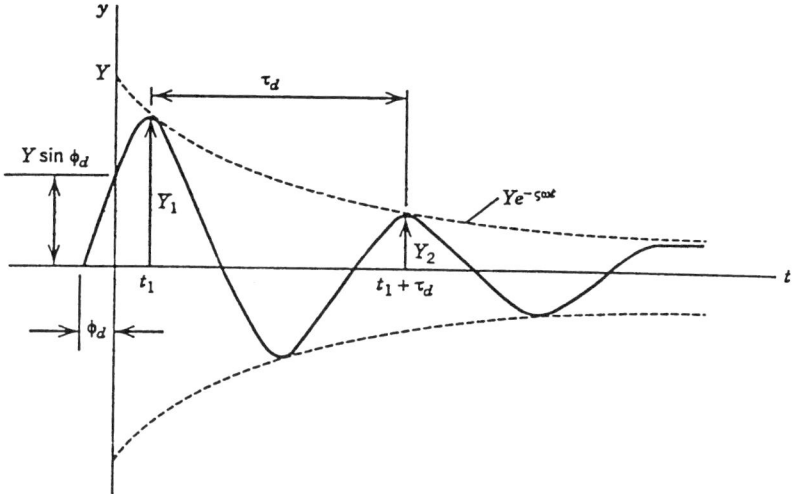

Figure 3.7. Graphical representation of Eq. (3.64).

For very light damping, say $\zeta < 0.30$, the denominator of Eq. (3.68) is approximately equal to unity. On this basis, the logarithmic decrement δ may be obtained by using the simpler expression

$$\delta = 2\pi\zeta \qquad (3.70)$$

A graphical representation of Eq. (3.64) is shown in Fig. 3.7.

Example 3.4: The frame shown in Fig. 3.5a is vibrating freely ($F(t) = 0$) under the influence of viscous damping with damping ratio $\zeta = 0.15$. By using its idealized one-degree spring-mass system, determine the free damped and undamped frequencies, the damped and undamped periods of vibration, the critical damping factor, the viscous damping constant c, and the logarithmic decrement δ.

SOLUTION: The idealized one-degree spring-mass system of the frame is determined in Example 3.2, and it is as shown in Fig. 3.8. The dynamic force $F(t)$ is assumed to be zero.
  The undamped free frequency ω is

$$\omega = \sqrt{\frac{k}{m}} = \sqrt{\frac{30.877}{0.4931}} = 7.9132 \text{ rps}$$

In hertz, this frequency is

$$f = \frac{\omega}{2\pi} = 1.2594 \text{ Hz}$$

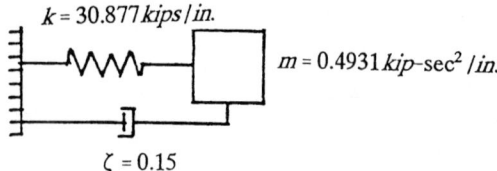

Figure 3.8. Idealized one-degree spring-mass system for the frame in Fig. 3.5a.

The low magnitude of this frequency indicates that the frame is very flexible in the horizontal direction, as would be expected. Therefore, the free frequency of vibration of an infrastructural component structure provides a measure of its flexibility. The lower the free frequency, the more flexible the structure is. This frequency concept can be used as a guidance by the practicing design engineer in design work.

The critical damping factor c_c is

$$c_c = 2m\omega = (2)(0.4931)(7.9132)$$
$$= 7.804 \text{ kip} \cdot \text{sec/in.}$$

Thus, the viscous damping constant c is

$$c = c_c \zeta = (7.804)(0.15) = 1.171 \text{ kip} \cdot \text{sec/in.}$$

The damped free frequency of vibration ω_d is

$$\omega_d = \omega\sqrt{1 - \zeta^2} = 7.9132\sqrt{1 - (0.15)^2}$$
$$= 7.8237 \text{ rps}$$

The damped frequency ω_d is just 1.131 percent lower compared to the undamped frequency ω.

The undamped period of vibration τ is

$$\tau = \frac{2\pi}{\omega} = \frac{2\pi}{7.9132} = 0.794 \text{ sec}$$

and the damped period of vibration τ_d is

$$\tau_d = \frac{2\pi}{\omega_d} = \frac{2\pi}{7.8237} = 0.8031 \text{ sec}$$

Thus, τ_d is only 0.882 percent higher compared to τ.

The amplitude of vibration, however, is much more sensitive to damping. For example, if $\zeta = 0$ and $t = 0$, we have $\exp(-\zeta\omega t) = 1$, and for $\zeta = 0.15$ and $t = \tau = 0.794$ sec, we have $\exp(-\zeta\omega t) = 1/\exp[(0.15)(7.9132)(0.794) = 0.3897$. This shows that 15 percent damping in the system produces 61.03 percent reduction in amplitude from pick-to-pick during one cycle of motion. Again, this is a very important information for the practicing engineer when designing a structural system for dynamic response.

The logarithmic decrement δ may be obtained by using Eq. (3.68). We have

$$\delta = \frac{2\pi\zeta}{\sqrt{1-\zeta^2}} = \frac{2\pi(0.15)}{\sqrt{1-(0.15)^2}} = 0.9532$$

If we use Eq. (3.70), we find

$$\delta = 2\pi\zeta = 2\pi(0.15) = 0.9425$$

which is 1.12 percent lower compared to the value obtained by using Eq. (3.68).

3.4 RESPONSE DUE TO HARMONIC EXCITATIONS

Consider the one-story frame in Fig. 3.9a, and let it be assumed that it is acted on by a harmonic force $F(t) = F\cos\omega_f t$, where F is its maximum amplitude and ω_f is its forced frequency. By following the procedure discussed in Section 3.2, the idealized one-degree spring-mass system may be determined, and it has the elements shown in Fig. 3.9b. This system represents an excellent idealization for the frame, provided that the purpose of the analysis and design of the frame is to determine maximum bending stresses and maximum horizontal displacement. This is an important accomplishment, because the dynamic solution of a rather complicated frame problem reduces to a solution of a simple one-degree spring-mass system, which provides excellent results for the practicing engineer in the design of the frame for dynamic response. In other words, the indicated dynamic analysis of a system with infinite degrees of freedom may be carried out satisfactorily by using the idealized spring-mass system in Fig. 3.9b.

By using the free-body diagram of the mass m shown in Fig. 3.9c and applying Newton's second law of motion, the differential equation of the idealized spring-mass system is as follows:

$$m\ddot{x} + c\dot{x} + kx = F\cos\omega_f t \tag{3.71}$$

If there is no damping, the viscous damping constant $c = 0$, and we have the following differential equation:

$$m\ddot{x} + kx = F\cos\omega_f t \tag{3.72}$$

170   INFRASTRUCTURAL DYNAMICS

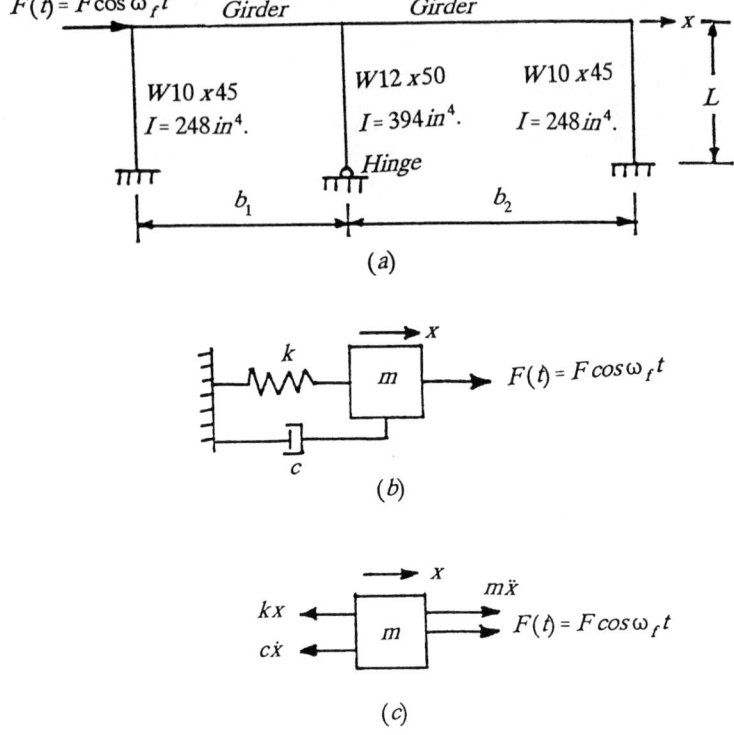

Figure 3.9. (a) One-story frame subjected to a harmonic force as shown. (b) Idealized one-degree spring-mass system for the frame. (c) Free-body diagram of mass m.

3.4.1 Closed-Form Solution of the Differential Equation

We start here with the solution of Eq. (3.71). Such a solution consists of two parts. The first part $x_c(t)$ is the solution of the homogeneous equation $m\ddot{x} + c\dot{x} + kx = 0$, and the second one is the particular solution $x_p(t)$ that satisfies Eq. (3.71). Thus, the complete solution $x(t)$ is

$$x(t) = x_c(t) + x_p(t) \tag{3.73}$$

The solution $x_c(t)$ is already known since it is derived in Section 3.3 of the text, Eq. (3.58), and it is as follows:

$$x_c(t) = \exp(-\zeta\omega t)[C_1 \cos(\omega\sqrt{1-\zeta^2})t + C_2 \sin(\omega\sqrt{1-\zeta^2})t] \tag{3.74}$$

where C_1 and C_2 are constants that can be determined from the initial

RESPONSE DUE TO HARMONIC EXCITATIONS   171

conditions of the motion. If preferred, we can use the solution

$$x_c(t) = C \exp(-\zeta \omega t) \sin(\omega_d t + \phi_d) \tag{3.75}$$

where ϕ_d is the phase angle, C is a constant, and

$$\omega_d = \omega \sqrt{1 - \zeta^2} \tag{3.76}$$

is the viscously damped free frequency of vibration.

The particular solution $x_p(t)$ is based on the harmonic forcing function, and it can have the form

$$x_p(t) = A \cos \omega_f t + B \sin \omega_f t \tag{3.77}$$

or the form

$$x_p(t) = X \cos(\omega_f t - \phi) \tag{3.78}$$

The constant X and phase angle ϕ are related to the constants A and B as follows:

$$X = \sqrt{A^2 + B^2} \tag{3.79}$$

$$\tan \phi = \frac{B}{A} \tag{3.80}$$

If we use the solution given by Eq. (3.77), carry out the required differentiations with respect to time, and substitute into Eq. (3.71), we find

$$(-mA\omega_f^2 + c\omega_f B + kA) \cos \omega_f t + (-mB\omega_f^2 - c\omega_f A + kB) \sin \omega_f t = F \cos \omega_f t \tag{3.81}$$

Equation (3.81) will be satisfied for all values of time t if we match the expressions of the coefficients of both $\cos \omega_f t$ and $\sin \omega_f t$ terms on the right-hand and left-hand sides of the equation. On this basis, we have

$$-mA\omega_f^2 + c\omega_f B + kA = F \tag{3.82}$$

$$-mB\omega_f^2 - c\omega_f A + kB = 0 \tag{3.83}$$

where Eq. (3.82) matches the coefficients of $\cos \omega_f t$ on each side of Eq. (3.81), while Eq. (3.83) matches the coefficients of $\sin \omega_f t$. Note that $\sin \omega_f t$ on the right-hand side of Eq. (3.81) is zero, and, consequently, its coefficient is zero.

Simultaneous solution of Eqs. (3.82) and (3.83) yields

$$A = \frac{F(k - m\omega_f^2)}{(k - m\omega_f^2)^2 + (c\omega_f)^2} \qquad (3.84)$$

$$B = \frac{Fc\omega_f}{(k - m\omega_f^2)^2 + (c\omega_f)^2} \qquad (3.85)$$

Also, by substituting into Eqs. (3.79) and (3.80), we find

$$X = \frac{F}{\sqrt{(k - m\omega_f^2)^2 + (c\omega_f)^2}} \qquad (3.86)$$

$$\tan \phi = \frac{c\omega_f}{k - m\omega_f^2} \qquad (3.87)$$

In terms of the damping ratio ζ and the ratio ω_f/ω, where ω is the free undamped vibration of the spring-mass system, Eqs. (3.86) and (3.87) are written as follows:

$$X = \frac{F}{k} \cdot \frac{1}{\sqrt{[1 - (\omega_f/\omega)^2]^2 + [2\zeta(\omega_f/\omega)]^2}} \qquad (3.88)$$

$$\tan \phi = \frac{2\zeta\omega_f/\omega}{1 - (\omega_f/\omega)^2} \qquad (3.89)$$

If we use Eqs. (3.78) and (3.88), the particular solution is written as

$$x_p(t) = \frac{F}{k} \cdot \frac{\cos(\omega_f t - \phi)}{\sqrt{[1 - (\omega_f/\omega)^2]^2 + [2\zeta(\omega_f/\omega)]^2}} \qquad (3.90)$$

If we use Eqs. (3.75) and (3.90), the complete solution $x(t)$ is given by the equation

$$x(t) = x_c(t) + x_p(t)$$

$$= C \exp(-\zeta\omega t) \sin(\omega_d t + \phi_d) + \frac{F}{k} \cdot \frac{\cos(\omega_f t - \phi)}{\sqrt{[1 - (\omega_f/\omega)^2]^2 + [2\zeta(\omega_f/\omega)]^2}} \qquad (3.91)$$

In Eq. (3.91), the phase angle ϕ is given by Eq. (3.89), and the constant C and phase angle ϕ_d may be determined from the initial conditions of the motion.

In Eq. (3.90), the term F/k is usually called the static deflection and it is denoted by the symbol X_{st}. It represents the static deflection of the mass m in

Fig. 3.9b, which is produced by the gradual application of the force F. The ratio X/X_{st} is called the magnification factor and it is denoted by the Greek letter Γ. Thus, we have

$$\Gamma = \frac{X}{X_{st}} = \frac{1}{\sqrt{[1 - (\omega_f/\omega)^2]^2 + [2\zeta(\omega_f/\omega)]^2}} \qquad (3.92)$$

The factor Γ provides the value by which X_{st} should be multiplied in order to obtain X. It depends on the damping ratio ζ and the ratio ω_f/ω. When $\omega_f = \omega$, the phenomenon of resonance occurs, which is the case where the frequency ω_f of the dynamic force coincides with the free undamped frequency ω of the spring-mass system. In such a case, the value of Γ depends on the amount of damping ζ in the system. If $\zeta = 0$, Γ will reach infinity with time. The influence of ζ on the factor Γ is shown in Fig. 3.10, which provides a family of curves of Γ versus the ratio ω_f/ω for various values of ζ.

If there is no damping in the system ($\zeta = 0$), Eq. (3.85) shows that $B = 0$, and Eq. (3.90) yields

$$x_p(t) = \frac{F}{k} \cdot \frac{\cos \omega_f t}{\sqrt{[1 - (\omega_f/\omega)^2]^2}} \qquad (3.93)$$

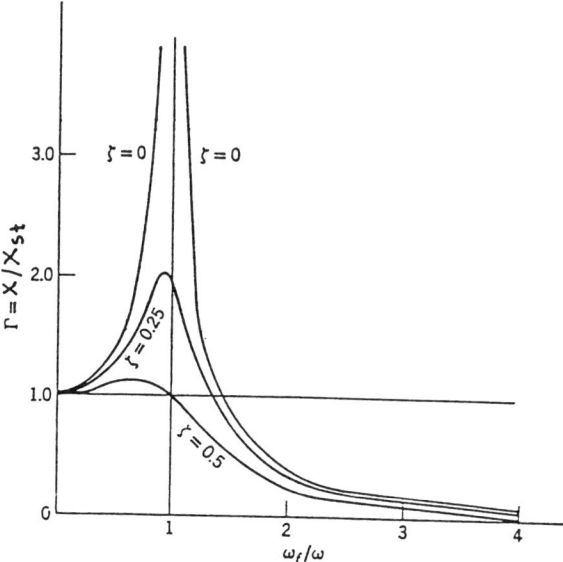

Figure 3.10. Family of curves representing the influence of ζ on the magnification factor Γ.

The homogeneous solution $x_c(t)$ of $m\ddot{x} + kx = 0$ is given by Eq. (3.24), which is as follows:

$$x_c(t) = C_1 \sin \omega t + C_2 \cos \omega t \qquad (3.94)$$

Thus, the complete solution $x(t)$ for the undamped spring-mass system is

$$x(t) = C_1 \sin \omega t + C_2 \cos \omega t + \frac{F}{k} \cdot \frac{\cos \omega_f t}{[1 - (\omega_f/\omega)^2]} \qquad (3.95)$$

By applying the initial condition that at time $t = 0$ the initial displacement is x_0 and the initial velocity is $\dot{x}_0$, we find that the constants C_1 and C_2 of Eq. (3.95) are as follows:

$$C_1 = \frac{\dot{x}_0}{\omega} \qquad (3.96)$$

$$C_2 = x_0 - \frac{F}{k} \cdot \frac{1}{[1 - (\omega_f/\omega)^2]} \qquad (3.97)$$

By substituting into Eq. (3.95), we find

$$x(t) = \frac{\dot{x}_0}{\omega} \sin \omega t + \left\{ x_0 - \frac{F}{k} \cdot \frac{1}{[1 - (\omega_f/\omega)^2]} \right\} \cos \omega t$$
$$+ \frac{F}{k} \cdot \frac{\cos \omega_f t}{[1 - (\omega_f/\omega)^2]} \qquad (3.98)$$

By examining Eq. (3.98), we note that the first two terms prescribe the free vibration of the system, and the third term provides the effect of the harmonic force $F \cos \omega_f t$. Note that $x(t)$ becomes infinite as t approaches infinity ($t \to \infty$). In practice, however, the system will fail long before the amplitude $x(t)$ becomes infinite.

With known dynamic displacement $x(t)$, the maximum bending moments and the maximum bending stresses in the frame shown in Fig. 3.9a may be determined as discussed in Section 3.2. The following example illustrates the application of the above theory.

Example 3.5: The steel frame in Fig. 3.9a is subjected to a concentrated harmonic force $F(t) = F \cos \omega_f t$ at the girder level, where $F = 15$ kips and $\omega_f = 11$ Hz. The weight of the girder and roof material is 1,200 lb/ft, $b_1 = 30$ ft, $b_2 = 40$ ft, and $L = 15$ ft. By assuming that the girders are infinitely stiff compared to the columns, determine the maximum horizontal displacement of the frame and the maximum bending stress in each column. The frame is under the influence of 8 percent viscous damping and the modulus of elasticity E of the steel columns is 30×10^6 psi. Perform the analysis by using an appropriate idealized one-degree spring-mass system and neglecting the weight of the columns and the associated wall areas.

SOLUTION: The idealized one-degree spring-mass system of the frame would have the form shown in Fig. 3.9b. The mass m represents the weight W of the roof and girders divided by the acceleration of gravity $g = 386.4$ in./sec². Thus,

$$W = (1{,}200)(70) = 84{,}000 \text{ lb}$$
$$= 84.00 \text{ kips}$$
$$m = \frac{W}{g} = \frac{84}{386.4} = 0.2174 \text{ kip} \cdot \text{sec}^2/\text{in.}$$

The stiffness k of the spring may be determined by using Eq. (3.5) for the first and third columns of the frame, and Eq. (3.6) for the second column. On this basis, we have

$$k = \frac{(2)(12)(30)(10)^3(248)}{(15 \times 12)^3} + \frac{(3)(30)(10)^3(294)}{(15 \times 12)^3}$$
$$= 30.62 + 6.08 = 36.70 \text{ kips/in.}$$

The free undamped frequency ω of the spring-mass system is

$$\omega = \sqrt{\frac{k}{m}} = \sqrt{\frac{36.70}{0.2174}} = 12.993 \text{ rps}$$

The critical damping c_c is

$$c_c = 2m\omega = (2)(0.2174)(12.993)$$
$$= 5.649 \text{ kip} \cdot \text{sec/in.}$$

Thus,

$$c = c_c \zeta = (5.649)(0.08) = 0.452 \text{ kip} \cdot \text{sec/in.}$$

If we consider the particular solution only, which represents the direct response of the frame caused by the application of the harmonic force, the maximum horizontal displacement X may be determined by using Eq. (3.88). This concept is often used in practical applications, because the effect of the free vibration dies out eventually since damping is present. On this basis, we have

$$x_{max} = X = \frac{15}{36.70} \cdot \frac{1}{\sqrt{[1-(11.00/12.993)^2]^2 + [(2)(0.08)(11.00/12.993)]^2}}$$
$$= (0.4087)(3.1845) = 1.3015 \text{ in.}$$

Note that $\Gamma = 3.1845$.

The maximum force, or resistance, R_{max}, developed in the spring of stiffness $k = 36.70$ kips/in., is

$$R_{max} = kx_{max} = (36.70)(1.3015)$$
$$= 47.7651 \text{ kips}$$

This is equal to the total shear force V_{max} at the top of the columns, and it is distributed to each column in proportion to their stiffness.

From Eq. (3.5), we find that the stiffnesses k_1 and k_3 of the first and third columns of the frame, respectively, are

$$k_1 = k_3 = 15.31 \text{ kips/in.}$$

and from Eq. (3.6), the stiffness k_2 of the second column of the frame is

$$k_2 = 6.08 \text{ kips/in.}$$

Thus, for the first and third columns of the frame, the shear forces V_1 and V_3, respectively, are

$$V_1 = V_3 = \frac{k_1}{k} V_{max} = \frac{15.31}{36.70}(47.7651)$$
$$= 19.926 \text{ kips}$$

The shear force V_2 in the second column is

$$V_2 = \frac{k_1}{k} V_{max} = \frac{6.08}{36.70}(47.7651)$$
$$= 7.9131 \text{ kips}$$

By using Eq. (3.10), the bending moments M_1 and M_3 at the top and bottom of the first and third columns of the frame, respectively, are

$$M_1 = M_3 = \frac{V_1 L}{2} = \frac{(19.926)(15)(12)}{2}$$
$$= 1,793.3 \text{ kip-in.}$$

By using Eq. (3.11), the moment M_2 at the top of the second column is

$$M_2 = V_2 L = (7.9131)(15)(12)$$

$$= 1{,}424.358 \text{ kip-in.}$$

Note that the moment at the base of the second column is zero, because the column is hinged at this end.

The maximum bending stresses in the first and third columns of the frame will occur at the top and bottom of each column, because M_1 and M_3 are maximum moments. With known M_1 and M_3, the maximum bending stress in each column may be determined by using Eq. (2.45) in Section 2.3. Since the depth h of the wide-flange cross section of these two columns is the same and it is equal to 10.10 in., the maximum bending stresses $\sigma_{1\max}$ and $\sigma_{3\max}$ in the first and third columns, respectively, are

$$\sigma_{1\max} = \sigma_{3\max} = \frac{M_1 y_{\max}}{I} = \frac{(1{,}793.3)(5.05)}{248}$$

$$= 36.5168 \text{ ksi}$$

For the second column of the frame, the depth h of the wide-flange cross section is 12.19 in., and the maximum bending stress $\sigma_{2\max}$ is

$$\sigma_{2\max} = \frac{M_2 y_{\max}}{I} = \frac{(1{,}424.358)(6.095)}{394}$$

$$= 22.0342 \text{ ksi}$$

The maximum horizontal displacement of the frame is at its top, and it is equal to 1.3015 in., as shown earlier in this problem.

The stresses $\sigma_{1\max}$, $\sigma_{2\max}$, and $\sigma_{3\max}$, and the maximum horizontal displacement $x_{\max}$ at the top of the frame are strictly the results produced by the application of the dynamic harmonic force $F(t) = F \cos \omega_f t$. We also have stresses that are produced by the dead weight of the frame and other possible static live loads supported by the frame. The stresses produced by such additional loads may be superimposed directly on the stresses produced by the harmonic force.

In the design of frames and buildings, safe limits are introduced for these stresses, and the total maximum stress in a frame or building should not exceed such limits. By total maximum stress we mean the maximum stress that is produced on the structure by considering all possible effects that can produce such stress. Thus, dynamic stresses, are only a part of the total picture.

3.5 RESPONSE DUE TO FORCE OF A GENERAL TYPE

In this section, we assume that the idealized one-degree spring-mass system is subjected to a dynamic force of some general time variation that does not repeat itself. Such forces could be the output from a conventional or nuclear explosion, the forces produced by the rotating parts of a mechanical system, or forces produced by strong gusty winds and other acts of nature such as hurricanes, tornados, and earthquakes, just to name a few. Two solution approaches are discussed in this section. The first one provides a closed-form solution of the differential equation representing the system, and the second one utilizes numerical solutions of the differential equation. Both methods have advantages and disadvantages, which will be pointed out during the discussion.

3.5.1 Closed-Form Solution of the Differential Equation

We consider here the spring-mass system shown in Fig. 3.9b, which is the idealized one-degree spring-mass system representing the one-story frame in Fig. 3.9a. In this discussion, however, we assume that $F(t)$ is a dynamic force of some general time variation, such as the one shown in Fig. 3.11a. By using the free-body diagram in Fig. 3.9c, the differential equation of motion of the spring-mass system may be written as follows:

$$m\ddot{x} + c\dot{x} + kx = F(t) \qquad (3.99)$$

In order to obtain a solution of Eq. (3.99), we subdivide the plot of the force $F(t)$ into an infinite number of impulses of time duration dT, as shown in Fig. 3.11a. The total response of the spring-mass system due to $F(t)$ is then considered to be the algebraic sum of the responses of all impulses. We initiate the procedure by considering the response produced by one impulse, such as the one shown in Fig. 3.11b. This impulse is suddenly applied to the mass m of the spring-mass system at time $t = T$, its time duration is dT, and its amplitude is $F(t)$. Since the time duration of the impulse is very small, we only need to determine the response of the spring-mass system at times $t \geqslant T$.

Under these conditions, the value F_{imp} of the impulse is

$$F_{imp} = F(T)\, dT \qquad (3.100)$$

and it is assumed to be applied suddenly to the spring-mass system. The velocity $d\dot{x}_0$ produced by the impulse may be determined by equating the linear momentum change $m\dot{x}_0$ to the linear impulse F_{imp}. On this basis, we have

$$m d\dot{x}_0 = F_{imp}$$

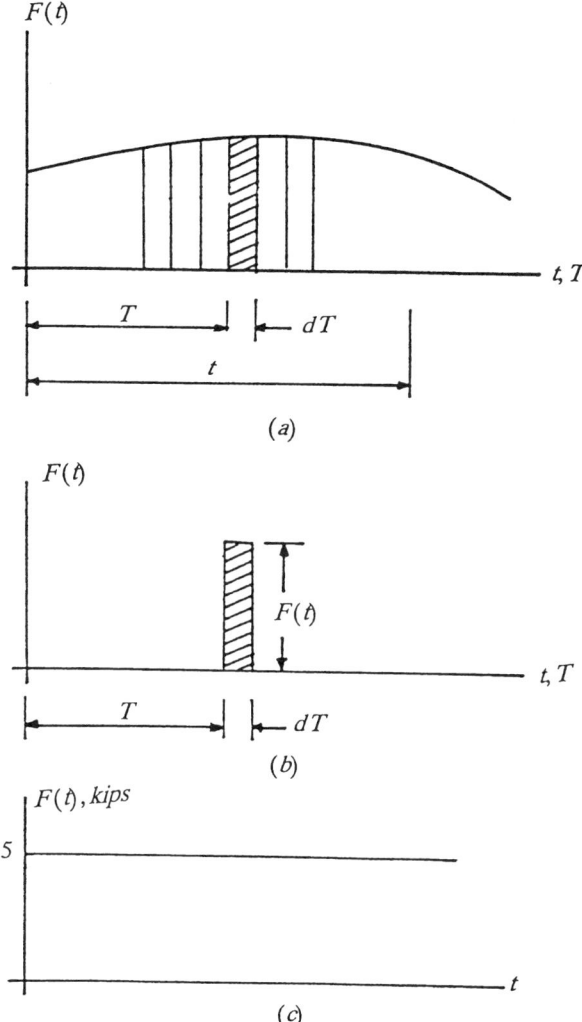

Figure 3.11. (a) Dynamic force of general time variation. (b) Impulse. (c) Dynamic force $F(t)$ applied suddenly and lasting indefinitely.

or

$$d\dot{x}_0 = \frac{F_{\text{imp}}}{m} \qquad (3.101)$$

where m is the mass of the spring-mass system. By using Eq. (3.100), we can write Eq. (3.101) as

$$d\dot{x}_0 = \frac{F(t)\, dT}{m} \qquad (3.102)$$

At times $t \geq T$, the dynamic force in Fig. 3.11b is zero, and the response $dx(t)$ of the system due to the impulse may be determined from the solution of the following differential equation:

$$m\ddot{x} + c\dot{x} + kx = F(t) \qquad (3.103)$$

The solution of Eq. (3.103) is obtained in Section 3.3.2, and it is given by Eq. (3.58). Since at time $t = T$ the initial displacement x_0 is zero and the initial velocity is given by Eq. (3.102), the displacement $dx(t)$ produced by the application of the impulse may be determined and is as follows:

$$dx(t) = \frac{F(T)\,dT}{m\omega_d} \exp[-\mu(t-T)] \sin \omega_d(t-T) \qquad (3.104)$$

where the damped frequency ω_d is given by the equation

$$\omega_d = \omega\sqrt{1-\zeta^2} \qquad (3.105)$$

and $\mu = \zeta\omega$.

The response $x_p(t)$ due to $F(t)$ may be determined by superimposing the responses of all impulses. Thus we have

$$x_p(t) = \frac{1}{m\omega_d} \int_0^t F(T) \exp[-\mu(t-T)] \sin \omega_d(t-T)\,dT \qquad (3.105a)$$

In order to complete the solution, the transient response $x_c(t)$, which is represented by the solution of the homogeneous equation given by Eq. (3.103), with $F(t) = 0$, should be added. If we start from time $t = 0$ with an initial displacement $x = x_0$ and initial velocity $\dot{x} = \dot{x}_0$, the solution may be obtained from Eq. (3.63) by replacing the y's with x's. Thus, the complete solution $x(t)$ is

$$x(t) = \exp(-\mu t)\left(x_0 \cos \omega_d t + \frac{\dot{x}_0 + \mu x_0}{\omega_d} \sin \omega_d t\right)$$
$$+ \frac{1}{m\omega_d} \int_0^t F(T) \exp[-\mu(t-T)] \sin \omega_d(t-T)\,dT \qquad (3.106)$$

It should be noted here that Eq. (3.106) is also known as Duhamel's integral or the convolution integral.

If damping is zero, Eq. (3.106) yields

$$x(t) = x_0 \cos \omega t + \frac{\dot{x}_0}{\omega} \sin \omega t + \frac{1}{m\omega} \int_0^t F(T) \sin \omega(t-T)\,dt \qquad (3.107)$$

RESPONSE DUE TO FORCE OF A GENERAL TYPE   181

For a given continuous force function $F(T)$, a closed-form solution for the response of the spring-mass system may be obtained from Eq. (3.107) when damping is zero, and from Eq. (3.106) when the spring-mass system is under the influence of viscous damping. For the more complicated functions of $F(t)$ such solutions are often very laborious, and numerical methods of analysis are employed. Such approaches may only include numerical techniques to carry out the required integrations, or well-known numerical methods of analysis may be used, as discussed later in this section.

Example 3.6: Solve the problem in Example 3.5 by assuming that $F(t)$ is a constant force of 15 kips that is applied suddenly to the frame and lasts indefinitely, as shown in Fig. 3.11c.

SOLUTION: The idealized one-degree spring-mass system representing the frame is derived in Example 3.5 and it is as shown in Fig. 3.12a. The horizontal

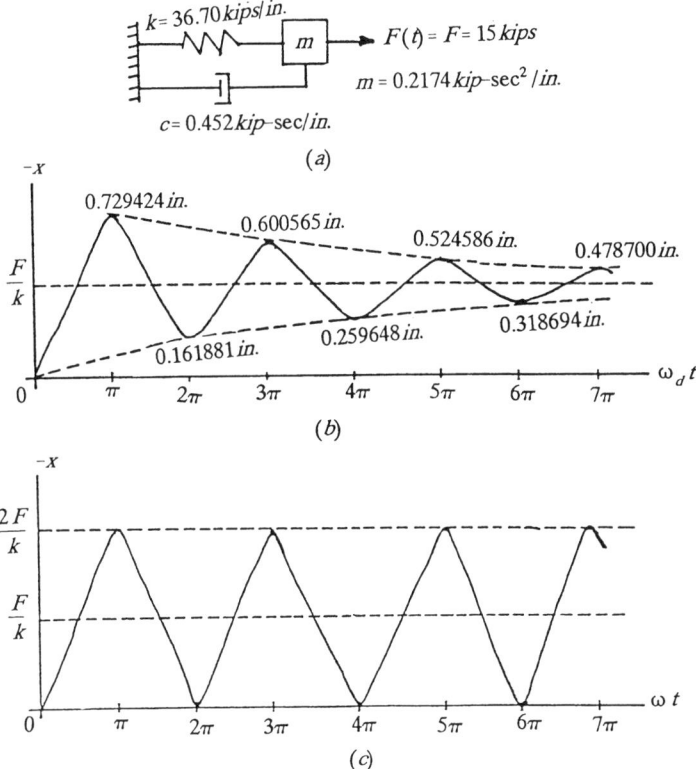

Figure 3.12. (*a*) Idealized one-degree spring-mass system for the frame in Example 3.5. (*b*) Plot of the exponentially decaying amplitude $x(t)$. (*c*) Plot of $x(t)$ when damping is zero.

displacement $x(t)$ of the top of the frame may be obtained by using Eq. 3.106. Since the initial displacement x_0 and initial velocity $\dot{x}_0$ are both zero, Eq. (3.106) yields

$$x(t) = \frac{1}{m\omega_d} \int_0^t F(T) \exp[-\mu(t-T)] \sin \omega_d(t-T) \, dT \qquad (3.108)$$

Since $F(t)$ is constant and equal to $F = 15$ kips, we have

$$x(t) = \frac{1}{m\omega_d} \int_0^t F \exp[-\mu(t-T)] \sin \omega_d(t-T) \, dT \qquad (3.109)$$

Integration of Eq. (3.109) yields

$$x(t) = \frac{F}{m[\mu^2 + \omega_d^2]} - \frac{F \exp(-\mu t)}{m\omega_d[\mu^2 + \omega_d^2]} (\mu \sin \omega_d t + \omega_d \cos \omega_d t) \qquad (3.110)$$

where $\mu = \zeta\omega$.

From Example 3.5 we know that $\omega = 12.993$ rps, and, consequently, the damped frequency ω_d is

$$\omega_d = \omega\sqrt{1-\zeta^2} = 12.993\sqrt{1-(0.08)^2}$$
$$= 12.951 \text{ rps}$$

Also, we have

$$\mu = \zeta\omega = (0.08)(12.993) = 1.03944$$

By substituting all known values into Eq. (3.110), we find

$$x(t) = \frac{15}{(0.2174)[(1.03944)^2 + (12.951)^2]}$$
$$- \frac{15 \exp(-1.03944t)}{(0.2174)(12.951)[(1.03944)^2 + (12.951)^2]}$$
$$\times [1.03944 \sin 12.951t + 12.951 \cos 12.951t]$$

or, after simplifying,

$$x(t) = 0.408730 - 0.031560 \exp(-1.03944t)$$
$$\times [1.03944 \sin 12.951t + 12.951 \cos 12.951t] \qquad (3.111)$$

We note that at $t = 0$ Eq. (3.111) yields $x(0) = 0$, indicating that it satisfies the initial condition of zero displacement at $t = 0$.

RESPONSE DUE TO FORCE OF A GENERAL TYPE   183

The graph of Eq. (3.111) for various values of $\omega_d t$ is shown plotted in Fig. 3.12b. It shows that the suddenly applied force F, illustrated in Fig. 3.11c, produces an exponentially decaying amplitude that forces the mass m of the idealized spring-mass system in Fig. 3.12a to vibrate about the static equilibrium position $x(t) = x_{st}$, where

$$x_{st} = \frac{F}{k} = \frac{15}{36.70} = 0.408719 \text{ in.} \qquad (3.112)$$

Because of the presence of damping, the vibration will eventually die out and the mass will come to rest at its static equilibrium position x_{st}, given by Eq. (3.112).

Equation (3.111) also represents the horizontal amplitude $x(t)$ of the top of the frame in Fig. 3.9a, when the suddenly applied dynamic force $F(t)$ is as shown in Fig. 3.11c. From Fig. 3.12b, we note that the maximum value of $x(t)$ is 0.729424 in., and it occurs at $\omega_d t = \pi$, or $t = 0.242575$ sec.

From Example 3.5, we note that the stiffnesses k_1 and k_3 of the first and third columns, respectively, of the frame are

$$k_1 = k_3 = 15.31 \text{ kips/in.}$$

and the stiffness k_2 of the second column of the frame is

$$k_2 = 6.08 \text{ kips/in.}$$

By using the largest value of $x(t)$ shown in Fig. 3.12b, the maximum resistance R_{max}, developed in the spring of stiffness k in Fig. 3.12a, is

$$R_{max} = kx_{max} = (36.70)(0.729424)$$
$$= 26.7699 \text{ kips}$$

This resistance also represents the total shear force V_{max} at the tops of the columns of the frame, and it is distributed to each column in proportion to their stiffness.

On this basis, for the first and third columns of the frame the shear forces V_1 and V_3 are, respectively,

$$V_1 = V_3 = \frac{k_1}{k} V_{max} = \frac{15.31}{36.70}(26.7699)$$
$$= 11.1675 \text{ kips}$$

and for the second column the shear force V_2 is

$$V_2 = \frac{k_2}{k} V_{max} = \frac{6.08}{36.70}(26.7699) = 4.4349 \text{ kips}$$

By using Eq. (3.10), the maximum bending moments M_1 and M_3 at the top and bottom of the first and third columns of the frame, respectively, are

$$M_1 = M_3 = \frac{V_1 L}{2} = \frac{(11.1675)(15)(12)}{2}$$

$$= 1{,}005.08 \text{ kip-in.}$$

From Eq. (3.11), the moment M_2 at the top of the second column is

$$M_2 = V_2 L = (4.4349)(15)(12) = 798.282 \text{ kip-in.}$$

With known moments, the maximum bending stresses $\sigma_{1\max}$ and $\sigma_{3\max}$ in the first and third columns of the frame, respectively, are

$$\sigma_{1\max} = \sigma_{3\max} = \frac{M_1 y_{\max}}{I} = \frac{(1{,}005.08)(5.05)}{248}$$

$$= 20.4663 \text{ ksi}$$

For the second column of the frame, we have

$$\sigma_{2\max} = \frac{M_2 y_{\max}}{I} = \frac{(798.282)(6.095)}{394.5}$$

$$= 12.3334 \text{ ksi}$$

As pointed out earlier, the maximum horizontal displacement of the frame is at its top and it is equal to 0.729424 in. We note here that the maximum stresses and maximum displacement produced by the forcing function $F(t)$ shown in Fig. 3.11c, are much lower than the analogous values produced by the application of the harmonic force $F \cos \omega_f t$ given in Example 3.5.

If damping is zero, Eq. (3.110) yields

$$x(t) = \frac{F}{m\omega^2} - \frac{F}{m\omega^2} \cos \omega t \tag{3.113}$$

or

$$x(t) = \frac{F}{k}(1 - \cos \omega t) \tag{3.114}$$

where ω is the free undamped frequency of vibration of the idealized spring-mass system.

By substituting the appropriate values of F, k, and ω in Eq. (3.114), we obtain

$$x(t) = 0.408719(\cos 12.993 t - 1) \tag{3.115}$$

The graph of Eq. (3.115) is shown in Fig. 3.12c, indicating again that the suddenly applied force F caused the system to vibrate about the equilibrium position $x(t) = x_{st}$. Since there is no damping, the mass of the idealized spring-mass system will vibrate about x_{st} indefinitely. The maximum amplitude x_{max} occurs at $\omega t = n\pi$, $n = 1, 2, 3, \ldots$, and is as follows:

$$x_{max} = \frac{2F}{k} = \frac{(2)(15)}{36.70} = 0.817438 \text{ in.}$$

The resulting maximum stresses in the columns of the frame may be determined as stated earlier.

3.5.2 Numerical Solution of the Differential Equation

In the first part of this section, a rigorous exact solution of the differential equation was used to determine the dynamic response of the idealized one-degree spring-mass system of a frame. Such solution, however, can only be obtained for force functions $F(t)$ that make the exact integration of Eq. (3.107) or Eq. (3.106) possible. In many practical cases, such as forces produced by conventional and nuclear explosions, for example, the force function $F(t)$ is rather arbitrary, and numerical methods of analysis are used to obtain dynamic responses.

An interesting numerical method, known as the acceleration impulse extrapolation method (AIEM), was developed at the Massachusetts Institute of Technology (MIT) during the late 1950s, in order to carry out the dynamic analysis that was required for a successful moon landing (5, 14, 37). Simple dynamic principles are used to formulate this method, but a very accurate and simple solution may be obtained for very complicated dynamics problems. In this section, the AIEM method is used to solve idealized one-degree spring-mass systems. The method, however, is simple and general, and it can be used to solve much more complicated problems, as discussed later in this text and other given references [37, 39].

The general discussion is initiated by considering an idealized one-degree spring-mass system, such as the one shown in Fig. 3.4d, which is under the influence of viscous damping of damping coefficient c. The differential equation of its motion may be derived in the usual way, and it is as follows:

$$m\ddot{x} + c\dot{x} + kx = F(t) \tag{3.116}$$

Solving for the acceleration $\ddot{x}$, we obtain

$$\ddot{x} = \frac{F(t) - c\dot{x} - kx}{m} \tag{3.117}$$

In the AIEM method, the acceleration curve in Fig. 3.13a is replaced by a series of equally spaced impulses that occur at times $t_0, t_1, \ldots, t_{i-1}, t_i, t_{i+1}, \ldots, t_n$. The time interval of each impulse is Δt, and at a time t_i, the magnitude of the acceleration impulse is $\ddot{x}_i(\Delta t)$. Between time stations the velocity $\dot{x}$ is constant and the displacement x is linear.

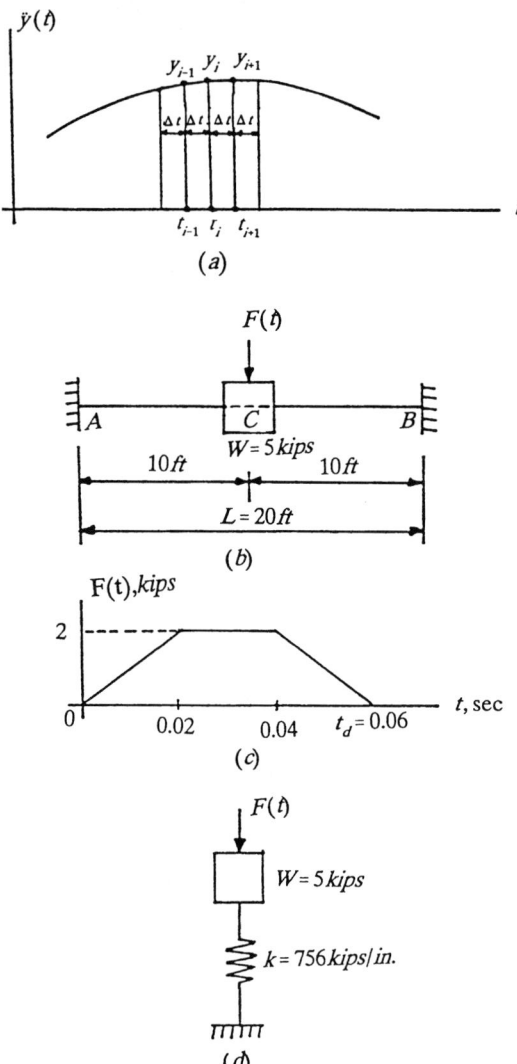

Figure 3.13. (a) Acceleration curve divided into a series of equally spaced impulses. (b) Fixed-fixed beam supporting operating machinery. (c) Time variation of $F(t)$. (d) Idealized one-degree spring-mass system.

The solution of Eq. (3.116) involves a step-by-step procedure that starts at time $t = 0$, where the displacement and velocity are known, and from there the displacement is extrapolated from one time station to the next.

If the displacements $x^{(i-1)}$ and $x^{(i)}$ at time stations $i - 1$ and i, respectively, are known, we can determine the acceleration $\ddot{x}^{(i)}$ at station i by using Eq. (3.117). We note, however, that the acceleration $\ddot{x}^{(i)}$ at station i depends on the velocity $\dot{x}^{(i)}$, which is not known. The original version of the method suggests that a reasonable approximate expression for the velocity $\dot{x}^{(i)}$ at station i is the following:

$$\dot{x}^{(i)} = \frac{x^{(i)} - x^{(i-1)}}{\Delta t} + \ddot{x}^{(i)}\left(\frac{\Delta t}{2}\right) \tag{3.118}$$

The second term on the right-hand side of Eq. (3.118) gives an estimate of the amount by which the average velocity of the preceding interval should be increased, in order to obtain the velocity at the next station.

By substituting Eq. (3.118) into Eq. (3.117) and simplifying, we obtain

$$\ddot{x}^{(i)} = \frac{2F(t)}{2m + c(\Delta t)} - \frac{2c[x^{(i)} - x^{(i-1)}]}{(\Delta t)[2m + c(\Delta t)]} - \frac{2kx^{(i)}}{2m + c(\Delta t)} \tag{3.119}$$

Equation (3.119) combines Eqs. (3.117) and (3.118), and it may be used here to determine the acceleration $\ddot{x}^{(i)}$ of the idealized one-degree spring-mass system at any station i. If damping is zero, Eq. (3.119) yields

$$\ddot{x}^{(i)} = \frac{F(t) - kx^{(i)}}{m} \tag{3.120}$$

as should be expected.

Between time stations i and $i + 1$, the average velocity $\dot{x}_{av}$ may be written as follows:

$$\dot{x}_{av} = \frac{x^{(i)} - x^{(i-1)}}{\Delta t} + \ddot{x}^{(i)}(\Delta t) \tag{3.121}$$

Equation (3.121) is also a reasonable expression to use in place of Eq. (3.118) if consistency is desired. If we substitute Eq. (3.121) into Eq. (3.117) and rearrange, we obtain

$$\ddot{x}^{(i)} = \frac{F(t)}{m + c(\Delta t)} - \frac{c[x^{(i)} - x^{(i-1)}]}{(\Delta t)[m + c(\Delta t)]} - \frac{kx^{(i)}}{m + c(\Delta t)} \tag{3.122}$$

Equation (3.122) may be also used to determine the acceleration $\ddot{x}^{(i)}$ at any station i.

188   INFRASTRUCTURAL DYNAMICS

At any time station $i + 1$, the displacement $x^{(i+1)}$ may be determined from the equation

$$x^{(i+1)} = x^{(i)} + \dot{x}_{av}(\Delta t) \qquad (3.123)$$

By substituting Eq. (3.121) into Eq. (3.123), we obtain

$$x^{(i+1)} = 2x^{(i)} - x^{(i-1)} + \ddot{x}^{(i)}(\Delta t)^2 \qquad (3.124)$$

Equation (3.124) may be used to determine the displacement $x^{(i+1)}$ at time station $i + 1$, when the preceding two displacements $x^{(i)}$ and $x^{(i-1)}$ at time stations i and $i - 1$, respectively, are known. Equation (3.119), or Eq. (3.122), may be used to determine the acceleration $\ddot{x}^{(i)}$ shown in Eq. (3.124).

Equation (3.124), however, cannot be used for the evaluation of the displacement $x^{(1)}$ at station 1, because at time $t = 0$, the displacement $x^{(0)}$ exists, but the displacement $x^{(0-1)}$ at time station $0 - 1$ does not exist. We can eliminate this difficulty if for the first time station the displacement $x^{(1)}$ is computed by using one of the following equations:

$$x^{(1)} = \tfrac{1}{6}[2\ddot{x}^{(0)} + \ddot{x}^{(1)}](\Delta t)^2 \qquad (3.125)$$

$$x^{(1)} = \frac{\ddot{x}^{(0)}}{2}(\Delta t)^2 \qquad (3.126)$$

In Equation (3.125) the acceleration during the first time interval is assumed to vary linearly, and in Eq. (3.126) the acceleration during the first time interval is assumed to be constant and equal to the initial value. Either one of these two equations is reasonable to use, but if the force $F(t)$ is zero at $t = 0$, then Eq. (3.125) must be used because $\ddot{x}^{(0)} = 0$ and Eq. (3.126) does not apply.

The following examples illustrate the application of the numerical method.

Example 3.7: Solve the frame problem in Example 3.6 by using the idealized one-degree spring-mass system shown in Fig. 3.12a, and compare the results.

SOLUTION: Since the idealized spring-mass system is under the influence of viscous damping, we may use Eq. (3.119) to determine the acceleration $\ddot{x}^{(i)}$ at any station i. Before we start the procedure, a decision should be made regarding the size of the time interval Δt. In practice, the size of Δt is often taken approximately equal to one-tenth of the smallest period of vibration of the spring-mass system. In our case here, the system has only one degree of freedom and, consequently, we have only one period of vibration. Thus, the damped period of vibration τ_d is

$$\tau_d = \frac{2\pi}{\omega_d} = \frac{2\pi}{12.951} = 0.485 \text{ sec}$$

On this basis, a reasonable value of Δt is 0.04 sec.

By substituting into Eq. (3.119), we obtain

$$\ddot{x}^{(i)} = \frac{(2)(15)}{(2)(0.2174) + (0.452)(0.04)} - \frac{(2)(0.452)[x^{(i)} - x^{(i-1)}]}{(0.04)[(2)(0.2174) + (0.452)(0.04)]}$$
$$- \frac{(2)(36.70)x^{(i)}}{(2)(0.2174) + (0.452)(0.04)}$$

or, after simplifying and rearranging,

$$\ddot{x}^{(i)} = 66.242713 - 211.976683 x^{(i)} + 49.902844 x^{(i-1)} \qquad (3.127)$$

At time $t = 0$, the displacement $x^{(0)}$ is zero. At the first time station, the displacement $x^{(1)}$ may be determined by using Eq. (3.126). The acceleration $\ddot{x}^{(0)}$ at $t = 0$ may be obtained by using Eq. (3.117) or Eq. (3.127). Since $x = \dot{x} = 0$ at $t = 0$, Eq. (3.117) yields

$$\ddot{x}^{(0)} = \frac{F(t)}{m} = \frac{15}{0.2174} = 68.997240 \text{ in./sec}^2$$

Thus, by substituting into Eq. (3.126), we find

$$x^{(1)} = \tfrac{1}{2}\ddot{x}^{(0)}(\Delta t)^2 = \frac{68.997240}{2}(0.04)^2$$
$$= 0.055198 \text{ in.}$$

The step-by-step procedure may be carried out as shown in Table 3.1. The first, second, and third terms of Eq. (3.127) are shown in Columns 3, 4, and 5, respectively, of the table, and the acceleration $\ddot{x}^{(i)}$ at any station i is shown in the sixth column of the table. In the seventh column, the product $\ddot{x}^{(i)}(\Delta t)^2$ of Eq. (3.124) appears, and the displacements $x^{(i)}$ at any station i are shown in the eighth column. Note that Eq. (3.126) is used to determine the displacement at the first time station. The remaining values of $x^{(i)}$ are determined by using Eq. (3.124), because the values of the displacements at the preceding two time stations are known, as well as the acceleration $\ddot{x}^{(i)}$ at any station i. For example, with known $x^{(1)}$, the second row of Table 3.1 can be completely determined. Thus, by using Eq. (3.124), the displacement $x^{(2)}$ at station 2 is

$$x^{(2)} = (2)(0.055198) - 0 + 0.087267$$
$$= 0.197663 \text{ in.}$$

With known $x^{(2)}$, the third row of Table 3.1 is determined, which makes it possible to determine the displacement $x^{(3)}$ at time station 3, and so on.

TABLE 3.1. Variation of the Displacement x of Mass m with Time t

(1) i	(2) t (sec)	(3) 66.242713	(4) $-211.976683 x^{(i)}$	(5) $49.902844 x^{(i-1)}$	(6) $\ddot{x}^{(i)}$	(7) $\ddot{x}^{(i)}(\Delta t)^2$	(8) $x^{(i)}$ (in.)
0	0	66.242713	0	0	66.242713	0.105988	0
1	0.04	66.242713	−11.700689	0	54.542024	0.087267	0.055198[a]
2	0.08	66.242713	−41.899947	2.754537	27.097303	0.043356	0.197663
3	0.12	66.242713	−81.289666	9.863946	−5.183007	−0.008293	0.383484
4	0.16	66.242713	−118.921463	19.136942	−33.541808	−0.053667	0.561012
5	0.20	66.242713	−145.177107	27.996094	−50.938300	−0.081501	0.684873
6	0.24	66.242713	−154.156439	34.177110	−53.736616	−0.085979	0.727233
7	0.28	66.242713	−144.910228	36.290995	−42.376520	−0.067802	0.683614

[a] Equation (3.126) is used for the displacement of the first time station.

RESPONSE DUE TO FORCE OF A GENERAL TYPE   191

Table 3.1 shows that the maximum displacement $x_{\max}$ occurs at time $t = 0.24$ sec and is equal to 0.727233 in. The exact value of this displacement is 0.729424 in., as shown in Fig. 3.12b. The difference is 0.30 percent. Better accuracy may be obtained by reducing the size of the time increment Δt.

The solution is also obtained by using Eq. (3.122) for the acceleration $\ddot{x}^{(i)}$ at any time station i. This equation yields

$$\ddot{x}^{(i)} = 63.699677 - 203.839986 x^{(i)} + 47.988109 x^{(i-1)} \quad (3.128)$$

The results are tabulated in Table 3.2. The maximum displacement occurs at $t = 0.24$ sec and is equal to 0.728159 in. This value is almost identical to the value obtained in Table 3.1 where Eq. (3.127) was used.

Example 3.8: The fixed-fixed wide-flange steel beam in Fig. 3.13b supports a heavy operating machinery of weight $W = 5$ kips located as shown in the figure. During operation, the machine transmits to the member a dynamic force $F(t)$ that varies as shown in Fig. 3.13c. By using a wide-flange beam cross section, design the member so that its vertical deflection due to $F(t)$ under the machine does not exceed 4 mils. Assume that damping is zero and perform the analysis and design by using an appropriate idealized one-degree spring-mass system. The modulus of elasticity $E = 30 \times 10^3$ ksi. Neglect the weight of the beam.

SOLUTION: Design dynamic criteria are often used in practice, and one such condition is deflection performance. Particular attention is given to designs that support operating machinery, sensitive scientific equipment, heavy equipment, and other similar structural and mechanical systems. Upper limits on deflection are usually established in order to avoid damage and also to protect the health and safety of people in the immediate environment.

The idealized one-degree spring-mass system for the beam in Fig. 3.13b may be determined by using the procedure discussed in Section 3.2. The deflection y of the member at its midpoint C is given by the following equation:

$$y = \frac{PL^3}{192EI} \quad (3.129)$$

When the deflection $y = 1$, we have

$$P = k = \frac{192EI}{L^3} \quad (3.130)$$

Equation (3.130) characterizes the stiffness of the beam at midpoint C.

We start the procedure by using a trial wide-flange section W24 × 68 with moment of inertia $I = 1814.5$ in.^{4} Thus, by substituting into Eq. (3.130), we find

$$k = \frac{(192)(30)(10)^3(1{,}814.5)}{(20 \times 12)^3} = 756 \text{ kips/in.}$$

TABLE 3.2. Variation of the Displacement x of Mass m with Time t

(1) i	(2) t (sec)	(3) 63.699677	(4) $-203.839986 x^{(i)}$	(5) $47.988109 x^{(i-1)}$	(6) $\ddot{x}^{(i)}$	(7) $\ddot{x}^{(i)}(\Delta t)^2$	(8) $x^{(i)}$ (in.)
0	0	63.699677	0	0	63.699677	0.101919	0
1	0.04	63.699677	−10.387584	0	53.312093	0.085299	0.050960[a]
2	0.08	63.699677	−38.162718	2.445474	27.982433	0.044772	0.187219
3	0.12	63.699677	−75.064075	8.984286	−2.380112	−0.003808	0.368250
4	0.16	63.699677	−111.189209	17.671621	−29.817911	−0.0477709	0.545473
5	0.20	63.699677	−137.589341	26.176218	−47.713446	−0.076342	0.674987
6	0.24	63.699677	−148.427920	32.391350	−52.336893	−0.083739	0.728159
7	0.28	63.699677	−142.197144	34.942973	−43.554494	−0.069687	0.697542

[a]Equation (3.126) is used to determine the displacement $x^{(i)}$ of the first time station.

The idealized one-degree spring-mass system is as shown in Fig. 3.13d. Its mass m is

$$m = \frac{W}{g} = \frac{5}{386} = 0.012953 \text{ kip-sec}^2/\text{in.}$$

which includes the mass of the operating machinery.

The differential equation of motion of the idealized system is

$$m\ddot{y} + ky = F(t) \tag{3.131}$$

Thus,

$$\ddot{y} = \frac{F(t) - ky}{m} \tag{3.132}$$

By substituting into Eq. (3.132), we find

$$\ddot{y}^{(i)} = 77.20 F(t) - 58{,}365 y^{(i)} \tag{3.133}$$

The free frequency of vibration ω of the idealized system is

$$\omega = \sqrt{\frac{k}{m}} = \sqrt{\frac{756}{0.012953}} = 241.59 \text{ rps}$$

The period of vibration τ is

$$\tau = \frac{2\pi}{\omega} = \frac{2\pi}{241.59} = 0.026 \text{ sec}$$

We start the procedure by using a time interval Δt equal to 0.004 sec. This is reasonable because it provides a sufficient number (15) of time stations within the time duration of $F(t)$ for an accurate solution, if the maximum response occurs within the time duration t_d of the dynamic force $F(t)$. Larger time intervals may be used if the maximum deflection y would occur outside t_d.

Since $F(t)$ is zero at $t = 0$, Eq. (3.125) is used to determine the displacement $y^{(1)}$ at the first time station. Note that the displacement at $t = 0$ is zero. Thus,

$$y^{(1)} = \tfrac{1}{6}[2\ddot{y}^{(0)} + \ddot{y}^{(1)}](\Delta t)^2$$
$$= \tfrac{1}{6}[0 + 77.20(0.40) - 58{,}365 y^{(1)}](0.004)^2$$
$$= 82.346672(10)^{-6} - 155{,}640(10)^{-6} y^{(1)}$$

TABLE 3.3. Variation of the Vertical Displacement y of the Member under the Machine

(1) i	(2) t (sec)	(3) 77.20F(t)	(4) $-58{,}365 y^{(i)}$	(5) $\ddot{y}^{(i)}$	(6) $\ddot{y}^{(i)}(\Delta t)^2 \times 10^{-3}$	(7) $y^{(i)} \times 10^{-3}$ (in.)
1	0	0	0	0	0	0
2	0.004	30.880	−4.159	26.721	0.428000	0.071256[a]
3	0.008	61.760	−33.298	28.462	0.455000	0.570512
4	0.012	92.640	−88.993	3.647	0.058000	1.524768
5	0.016	123.520	−148.073	−24.553	−0.393000	2.537024
6	0.020	154.400	−184.216	−29.816	−0.477000	3.156280
7	0.024	154.400	−192.519	−38.119	−0.610000	3.298536
8	0.028	154.400	−165.219	−10.819	−0.173000	2.830792

[a]Equation (3.125 is used for the first time station.

Solving for $y^{(1)}$ we obtain

$$y^{(1)} = \frac{82.346672(10)^{-6}}{1.15564} = 0.071256(10)^{-3} \text{ in.}$$

The step-by-step procedure is illustrated in Table 3.3. Note that column 3 contains the first term on the right-hand side of Fig. (3.133), column 4 contains the second term, and column 6 contains the last term on the right-hand side of Eq. (3.124) if $\ddot{x}^{(i)}$ is replaced by $\ddot{y}^{(i)}$. Table 3.3 shows that the maximum deflection y occurs at $t = 0.024$ sec, and is equal to 3.298536 mils. This is well below the limit deflection of 4 mils, indicating a satisfactory design. If we wish to get closer to 4 mils of deflection, the procedure may be repeated by using a smaller wide-flange beam cross section. The deflection limit of 4 mils, however, should not be exceeded.

3.6 VIBRATION OF HIGHWAY BRIDGES

Considerable discussion regarding the dynamics of highway bridges is given in Section 1.8, and some failures associated with highway bridges are briefly discussed in Section 1.2. Bridges are essential parts of an infrastructure system and deserve particular attention and a great deal of research work, in order to be able to maintain their structural integrity. In this section, vibrational aspects of highway bridges are examined.

A typical cross section of a highway bridge is shown in Fig. 3.14a, which involves an assembly consisting of the bridge deck and the girders. The girders are used to support the bridge deck and, consequently, the vehicular traffic

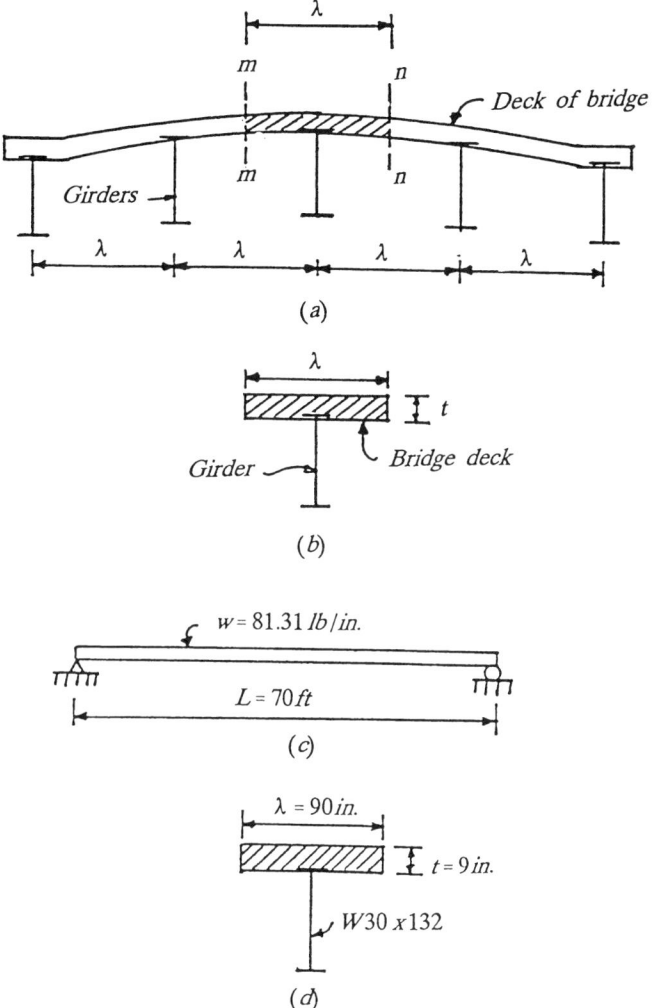

Figure 3.14. (*a*) Typical cross section of a highway bridge. (*b*) Composite girder-deck cross section. (*c*) Simply supported beam. (*d*) Cross section of the simply supported beam.

passing across the bridge deck. The vehicular traffic entering and leaving a bridge sets the bridge in vibratory motions that may influence its structural integrity and produce undesirable conditions. When a vehicle is entering the bridge it causes the bridge deck and girder system to respond dynamically. As discussed in Section 1.8, the bridge-vehicle system is associated with various types of vibratory motions as the vehicle moves along the length of the bridge, motions that can create potentially hazardous resonant vibrations. When the

196 INFRASTRUCTURAL DYNAMICS

vehicle leaves the bridge, the deck and girder system will continue to vibrate freely for an indefinite period of time, or until the motion is damped out, since some damping is always present in such vibratory motions.

The free frequencies of vibration of a highway bridge consisting of a concrete deck and steel girders, such as the one shown in Fig. 3.14a, may be determined with sufficient accuracy by considering one of its interior girders together with the associated portion of the bridge deck that is supported by the girder. This is shown in Fig. 3.14b. If we have shear connectors at the top of the girder, the bridge deck becomes rigidly connected to the girder and it will follow the girder during its free vibratory motion. Under this condition, we have what is known as composite bridge, which implies that there exists a composite action between the girders and the deck of the bridge. If shear connectors are not used, we have what is termed a noncomposite bridge, and the bridge deck is not always in contact with the girder during the vibratory motion. Only partial contact is usually present.

It is important to distinguish these two types of motion, because for a composite bridge the total weight of both deck and girder participates in the vibratory motion, and hence all of it must be considered in the vibration analysis of the bridge. In addition, for a composite bridge, the moment of inertia of the composite cross section, shown in Fig. 3.14b, must be used. If the deck is made out of concrete, then we have a cross section composed of two materials: concrete and steel. As shown later, methods are available to calculate moments of inertia of cross sections involving two materials.

For a noncomposite bridge, only part of the weight of the deck remains in contact with the girder during its free vibratory motion. Experience and experimental investigations have shown that an accurate vibration analysis may be carried out by using only 50 percent of the weight of the deck and 100 percent of the weight of the girder. On the other hand, since there is no composite action between deck and girder, only the moment of inertia of the girder should be considered in the analysis. In addition, experimental investigations and experience have shown [40] that the dynamic modulus of the steel is the same as that used for static analysis, while the dynamic modulus of concrete is larger, and is equal to 5×10^6 psi.

The following examples illustrate the application of the theory by including important observations and concluding remarks regarding the free vibration of highway bridges.

Example 3.9: A single-span composite highway bridge has the cross section shown in Fig. 3.14a. The thickness t of the concrete deck is 9 in., the spacing λ of the steel girders is 90 in., and the size of the wide-flange section is W30 × 132 with moment of inertia equal to 5,770 in.^{4} The steel modulus of elasticity $E_s = 30 \times 10^6$ psi, and the concrete dynamic modulus $E_c = 5 \times 10^6$ psi. Determine the free undamped fundamental frequency of vibration of the highway bridge when its length $L = 70$ ft. Assume that the ends of each girder are simply supported.

SOLUTION: From the theory of vibrations [5, 37, 39], the undamped fundamental frequency of vibration ω for a uniform simply supported single-span beam is given by the following equation:

$$\omega = \frac{\pi^2}{L^2}\sqrt{\frac{EI}{m}} \qquad (3.134)$$

where m is the mass per unit of length of the member.

By considering the cross section in Fig. 3.14b, the weight w per unit of length of the member is

$$w = \lambda t \gamma(1) + \frac{132}{12} = (90)(9)(86.806)(10)^{-3}(1) + 11$$

$$= 70.31 + 11 = 81.31 \text{ lb/in.}$$

The specific weight γ per cubic inch of concrete is considered to be equal to 86.806×10^{-3} lb/in.^{3} Thus, the mass m per unit of length of the beam is

$$m = \frac{w}{g} = \frac{81.31}{386} = 0.2106 \text{ lb} \cdot \text{sec}^2/\text{in.}$$

The undamped fundamental frequency of vibration of the highway bridge may be determined by using the uniform single-span beam in Fig. 3.14c, which has the cross section shown in Fig. 3.14d, and applying Eq. (3.134). In order to calculate the moment of inertia I_s of the composite cross section about its centroidal, or neutral, axis, we transform the area of the concrete part of the cross section, which is the deck, into an equivalent steel area by using the ratio η between E_c and E_s. For practical applications, it is reasonable to neglect the steel reinforcement of the bridge deck in this transformation. On this basis, we have

$$\eta = \frac{E_c}{E_s} = \frac{5 \times 10^6}{30 \times 10^6} = \frac{1}{6}$$

Thus, the equivalent steel area $(A_c)_s$ is

$$(A_c)_s = \tfrac{1}{6} A_c = \tfrac{1}{6}(90)(9) = (15)(9)$$

$$= 135 \text{ in.}^2$$

Only the λ dimension of the concrete should be changed, and the transformed cross section, which is now all steel, is shown in Fig. 3.15.

Since the neutral and the centroidal axes coincide, their location may be determined by taking first moments about the bottom of the steel girder. On

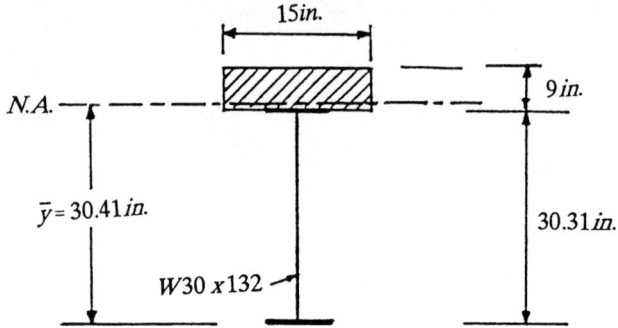

Figure 3.15. Transformed cross section.

this basis, we have

$$\bar{y} = \frac{(15)(9)(30.31 + 9/2) + (38.9)(30.31/2)}{(15)(9) + 38.9} = 30.41 \text{ in.}$$

Note that the area of the steel girder is 38.9 in.²

The moment of inertia I_s of the transformed section about the centroidal axis is

$$I_s = \frac{(15)(9)^3}{12} + (15)(9)(4.4)^2 + 5{,}770 + (38.9)(15.155)^2$$

$$= 18{,}347.46 \text{ in.}^4$$

Thus, by using Fig. 3.134), we find

$$\omega = \frac{\pi^2}{L^2}\sqrt{\frac{EI}{m}} = \frac{\pi^2}{(70 \times 12)^2}\sqrt{\frac{(30)(10)^6(18{,}347.46)}{0.2106}}$$

$$= 22.61 \text{ rps}$$

In units of cycles per second (cps), or hertz, named after the German mathematician, Heinrich Hertz, the free undamped fundamental frequency f of the bridge is

$$f = \frac{\omega}{2\pi} = \frac{22.61}{2\pi} = 3.60 \text{ Hz}$$

Previous experimental investigations by the Michigan Department of Transportation [41] have shown that the free undamped fundamental frequency of a well-designed highway bridge should be 5.5 Hz or larger. The above frequency of 3.60 Hz does not conform with such design criteria. Vibration design criteria are important for bridges, because their rigidity, or flexibility, is extremely dependent on the associated natural frequency of vibration of a bridge. Examination of Eq. (3.134) reveals that the frequency ω is inversely proportional to the square of the length of the beam, directly proportional to the square root of its stiffness EI, and inversely proportional to the square root of its mass m. Since the flexibility of a bridge increases as the frequency ω decreases, its length L should be closely controlled, and it should be in line with appropriate frequency design criteria.

If, for the given problem, we wish to retain the frequency of 3.60 Hz, but double its length L, a critical situation is developing. To compensate for such an increase in L, we must increase the stiffness EI of the member 16 times. Since E is constant, we have to make $I = (16)(18{,}347.46) = 293{,}559$ in.^{4} On the other hand, the increase in EI is accompanied with an increase in mass m, indicating that the ratio EI/m must be increased 16 times. Therefore, there exists a maximum value on the span length of a bridge, a maximum that controls its design when frequencies are considered. The limitations that could be imposed on the span length of a bridge would largely depend on the assigned limits of its free fundamental frequency of vibration. Such standards must be developed in the future in order to secure the structural integrity of our highway bridges. Static analysis alone is not sufficient.

Example 3.10: Repeat the bridge problem in Example 3.9 by assuming that there is no composite action between the bridge deck and the girders and compare the results.

SOLUTION: In this case, only half of the weight of the bridge deck is considered to be participating in the vibrational motion of the girders. Thus,

$$w = \frac{\lambda t\gamma(1)}{2} + \frac{132}{12} = \frac{(90)(9)(86.806)(10)^{-3}(1)}{2} + 11$$

$$= 35.16 + 11 = 46.16 \text{ lb/in.}$$

$$m = \frac{w}{g} = \frac{46.16}{386} = 0.1196 \text{ lb} \cdot \text{sec}^2/\text{in.}$$

Also, in this case, only the moment of inertia of the girder is considered.

By applying Eq. (3.134), we have

$$\omega = \frac{\pi^2}{L^2}\sqrt{\frac{EI}{m}} = \frac{\pi^2}{(70 \times 12)^2}\sqrt{\frac{(30)(10)^6(5770)}{0.1196}}$$

$$= 16.8277 \text{ rps}$$

or

$$f = \frac{\omega}{2\pi} = \frac{16.8277}{2\pi} = 2.68 \text{ Hz}$$

Thus, if we do not have composite action between the deck and the girders, the frequency f would be 25.56 percent lower compared to the composite bridge. This indicates that composite action between deck and girders made the bridge a great deal stiffer.

3.7 VIBRATION ANALYSIS OF CONTINUOUS HIGHWAY BRIDGES

In this section we are concerned with the vibration analysis and characteristics of multispan highway bridges. Since exact methods of analysis are not usually available for such types of bridges, numerical and approximate methods of analysis are usually employed. Exact methods of analysis usually involve the solution of the appropriate differential equation of motion, or they may involve the utilization of energy concepts and principles that can provide exact solutions. Numerical and approximate methods of analysis that are often used for the solution of structural and mechanical systems are the finite element and finite difference methods, the method of equivalent systems (which is particularly convenient for the solution of complex linear and nonlinear problems), transfer matrices, Rayleigh's, Stodola's, and Myklestad's methods, and so on. Complete information regarding these methods is given by Fertis [3, 5, 37, 39].

The method of Stodola with an iteration procedure is used in this section for the determination of free undamped frequencies of vibration of highway bridges.

3.7.1 Stodola's Method and Iteration Procedure

Utilization of the method of Stodola requires the employment of concentrated, or lumped, masses and influence (or deflection) coefficients. On this basis, the required equations of motion would be in terms of concentrated masses and influence or deflection coefficients. A step-by-step procedure is used here to derive the required equations of motion by utilizing the three-span continuous beam in Fig. 3.16a. The beam is of continuous mass and elasticity, but it may also include the mass of additional weights that are attached to the member and participate in its free vibratory motion. Such additional masses must also be included in its vibration analysis. The step-by-step procedure may start as follows:

Step 1

We subdivide the members in Fig. 3.16a into n segments and then lump the weight of each segment at the center of the segment, as shown. Also include any weights that are securely attached to the beam. The segments are not required to be of the same length.

Step 2

Divide each lumped weight W_i ($i = 1, 2, 3, \ldots, n$) by the acceleration of gravity g to obtain the lumped masses m_i ($i = 1, 2, 3, \ldots, n$), as shown in Fig. 3.16b.

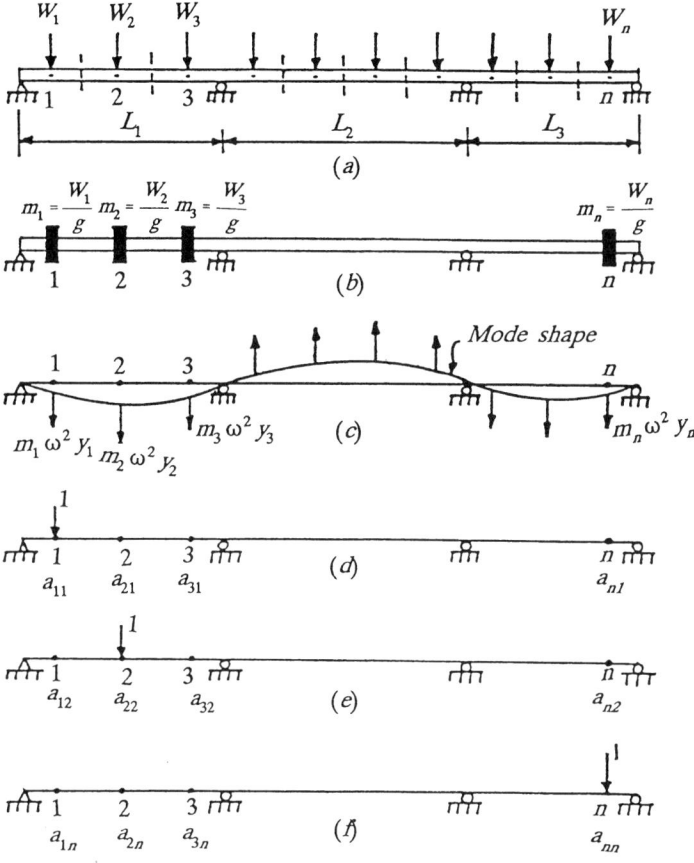

Figure 3.16. (*a*) Three-span continuous beam with lumped weights. (*b*) Lumped masses $m_1, m_2, m_3, \ldots, m_n$. (*c*) Mode shape with corresponding inertia forces. (*d*) Unit load at point 1. (*e*) Unit load at point 2. (*f*) Unit load at point *n*.

Step 3

Assume that the member vibrates in one of its modes of vibration, such as the one shown in Fig. 3.16c. The inertia forces

$$m_1\omega^2 y_1, m_2\omega^2 y_2, m_3\omega^2 y_3, \ldots, m_n\omega^2 y_n$$

are acting in the direction of the mode shape as shown in the same figure. They are acting downward in the first span since we have a downward deflection in this span, they are acting in the upward direction in the second span to match the upward direction of the deflection, and so on. The deflections under the mass concentration points $1, 2, 3, \ldots, n$ are designated as $y_1, y_2, y_3, \ldots, y_n$, respectively. The static application of the inertia loads in the direction shown in Fig. 3.16c, should produce the indicated mode shape, provided that such inertia loads are known.

Step 4

We apply unit loads at points $1, 2, \ldots, n$, as shown in Figs. 3.16d, 3.16e, and 3.16f, respectively, and we determine the indicated a_{ij} deflection coefficients. For example, if we apply a unit load at point 1, as shown in Fig. 3.16d, we determine the deflections $a_{11}, a_{21}, a_{31}, \ldots, a_{n1}$, at points $1, 2, 3, \ldots, n$, respectively, produced by the unit load. Any known methods of analysis can be used for this purpose. See also Fertis [6]. Note that the first subscript of a defines the point and the second one shows the location of the unit load. That is, a_{21} means the deflection at point 2 due to a unit load at point 1.

Step 5

The deflections $y_1, y_2, y_3, \ldots, y_n$ under the mass concentration points $1, 2, 3, \ldots, n$, respectively, caused by the inertia forces are written as follows:

$$y_1 = a_{11}(m_1\omega^2 y_1) + a_{12}(m_2\omega^2 y_2) + a_{13}(m_3\omega^2 y_3) + \cdots + a_{1n}(m_n\omega^2 y_n) \quad (3.135)$$

$$y_2 = a_{21}(m_1\omega^2 y_1) + a_{22}(m_2\omega^2 y_2) + a_{23}(m_3\omega^2 y_3) + \cdots + a_{2n}(m_n\omega^2 y_n) \quad (3.136)$$

$$y_3 = a_{31}(m_1\omega^2 y_1) + a_{32}(m_2\omega^2 y_2) + a_{33}(m_3\omega^2 y_3) + \cdots + a_{3n}(m_n\omega^2 y_n) \quad (3.137)$$

$$\vdots$$

$$y_n = a_{n1}(m_1\omega^2 y_1) + a_{n2}(m_2\omega^2 y_2) + a_{n3}(m_3\omega^2 y_3) + \cdots + a_{nn}(m_n\omega^2 y_n) \quad (3.138)$$

Step 6

We write Eqs. (3.135) through (3.138) in matrix notation as follows:

$$\begin{bmatrix} y_1 \\ y_2 \\ y_3 \\ \vdots \\ y_n \end{bmatrix} = \omega^2 \begin{bmatrix} a_{11}m_1 & a_{12}m_2 & a_{13}m_3 & \cdots & a_{1n}m_n \\ a_{21}m_1 & a_{22}m_2 & a_{23}m_3 & \cdots & a_{2n}m_n \\ a_{31}m_1 & a_{32}m_2 & a_{33}m_3 & \cdots & a_{3n}m_n \\ \vdots & \vdots & \vdots & & \vdots \\ a_{n1}m_1 & a_{n2}m_2 & a_{n3}m_3 & \cdots & a_{nn}m_n \end{bmatrix} \begin{bmatrix} y_1 \\ y_2 \\ y_3 \\ \vdots \\ y_n \end{bmatrix} \quad (3.139)$$

or, in short form notation,

$$\{y\} = \omega^2 [M]\{y\} \quad (3.140)$$

Step 7

Solve the matrix equation given by Eq. (3.139), or Eq. (3.140), by applying an iteration procedure. The iteration procedure may be initiated by assuming arbitrarily that $y_1 = y_2 = y_3 = \cdots = y_n = 1$ in the right-hand column of Eq. (3.139) and performing the matrix multiplication. If preferred, other arbitrary values for $y_1, y_2, y_3, \ldots, y_n$ may be used. The result of the matrix multiplication is a column matrix that is normalized by using as a common factor any one of its elements, preferably the smallest one. The elements of the normalized column matrix represent a better approximation of the amplitudes $y_1, y_2, y_3, \ldots, y_n$ of the fundamental mode. The procedure is repeated until the amplitudes $y_1, y_2, y_3, \ldots, y_n$ of the last trial are approximately equal to the corresponding ones obtained in the preceding trial. These amplitudes characterize the shape of the fundamental mode. The fundamental frequency ω in radians per second is obtained by substituting the computed values of $y_1, y_2, y_3, \ldots, y_n$ into any one of the expressions given by Eqs. (3.135) through (3.138) and solving for ω.

The iteration procedure converges very rapidly, and it usually requires three to five repetitions for an excellent approximation of the fundamental frequency and its corresponding mode shape. Higher frequencies of vibration may be also determined by using Stodola's method. See Fertis [5, 37, 39].

The application of Stodola's method for the computation of the fundamental frequency of vibration of continuous highway bridges is given below.

3.7.2 Vibration of Highway Bridges Using Stodola's Method

The method of Stodola and the iteration procedure is applied here for the computation of the fundamental frequency of vibration of continuous highway bridges. The procedure is best demonstrated by using an example.

Example 3.11: The girder of a continuous three-span highway bridge located in the state of Michigan is shown in Fig. 3.17a. It is a composite bridge and the uniform weight w of the girder and deck per unit of length is equal to 75 lb/in. The calculated moment of inertia of the composite section is 58×10^3 in.⁴ and the steel modulus of elasticity $E_s = 30 \times 10^6$ psi. By using the method of Stodola and the iteration procedure, determine the fundamental free frequency of vibration of the bridge. Neglect damping.

SOLUTION: Each span of the beam in Fig. 3.17a is divided into three equal segments, and the weight of each segment is lumped at the center of the segment, as shown in Fig. 3.17b. On this basis we have nine concentrated weights located at points 1 through 9. Between weight concentration points the beam is assumed to be massless, but it retains its elastic properties. If we divide each weight by the acceleration of gravity g, we obtain lumped masses. The degrees of freedom of the member are now reduced to nine, because we only need nine vertical displacements $y_1, y_2, \ldots, y_9$ to locate the position of the nine masses during vibration. The initial system in Fig. 3.17a has infinite degrees of

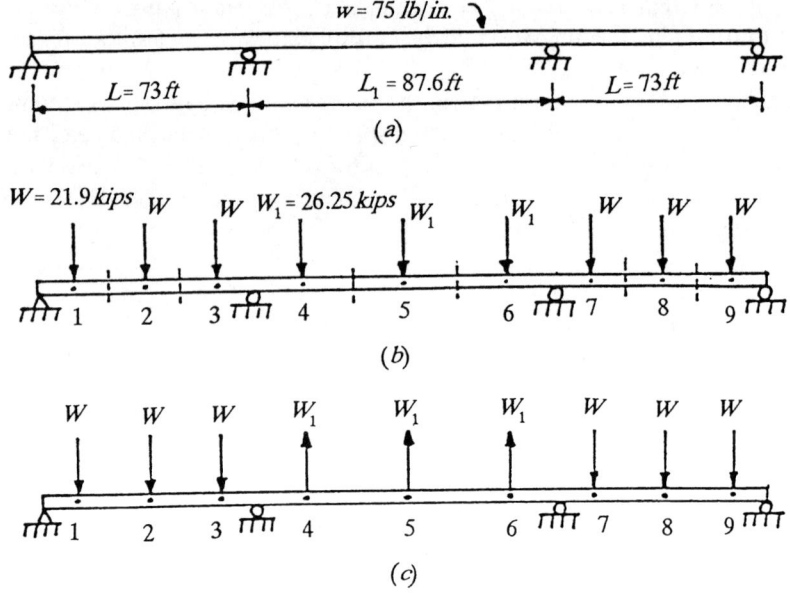

Figure 3.17. (a) Three-span continuous bridge girder. (b) Lumped weights. (c) Lumped weights assumed to act in the direction of the fundamental mode.

freedom, because we need an infinite number of coordinates y to locate the position of all particles of the member during vibration.

Since the shape of the fundamental mode of vibration is similar to the one shown in Fig. 3.16c, we rearrange the weights W and W_1 in Fig. 3.17b so that they are acting in the directions shown in Fig. 3.17c. In other words, we assume that the lumped weights are acting in the direction of the fundamental mode, which is the same as the direction of the inertia forces $m_i\omega^2 y_i$. Doing so is similar to assuming that the lumped weights are the inertia forces producing an approximate shape of the fundamental mode and, consequently, that they must be acting in the direction of the mode shape in the same way as the inertia forces do. The iteration procedure, however, leads to the correct mode shape.

The a_{ij} deflection coefficients may be determined by applying a unit load at each of the weight concentration points and calculating the resulting deflections at all nine points. Any known methods from the theory of structures or mechanics may be used for this purpose. See Fertis [6, 39]. Available computer programs may be also used for this purpose. The deflection coefficients for this problem are determined by using the conjugate beam method and they are tabulated as shown in Table 3.4. For example, the coefficient a_{11} in this table is $L^3/186.7EI$, the coefficient $a_{31} = a_{13} = L^3/382EI$, and so on. Note that $a_{ij} = a_{ji}$ in accordance with the Maxwell's theorem of reciprocity.

By substituting into Eq. (3.139) and simplifying, we have

$$\begin{bmatrix} y_1 \\ y_2 \\ y_3 \\ y_4 \\ y_5 \\ y_6 \\ y_7 \\ y_8 \\ y_9 \end{bmatrix} = \frac{\omega^2 L^3}{52.70 EIg} \begin{bmatrix} 6.19 & 8.65 & 3.02 & 2.75 & 3.57 & 1.21 & 0.53 & 0.78 & 0.34 \\ 8.65 & 17.10 & 2.84 & 6.38 & 8.35 & 2.81 & 1.23 & 1.80 & 0.78 \\ 3.02 & 2.84 & 4.51 & 4.34 & 5.65 & 1.90 & 0.83 & 1.23 & 0.53 \\ 2.29 & 5.60 & 3.61 & 7.06 & 9.93 & 3.66 & 1.61 & 2.36 & 1.02 \\ 2.97 & 6.91 & 4.71 & 10.40 & 26.25 & 10.40 & 4.71 & 6.91 & 2.97 \\ 1.02 & 2.36 & 1.61 & 3.66 & 9.93 & 7.06 & 3.61 & 5.60 & 2.29 \\ 0.53 & 1.23 & 0.83 & 1.90 & 5.65 & 4.34 & 4.51 & 2.84 & 3.02 \\ 0.78 & 1.80 & 1.23 & 2.81 & 8.35 & 6.71 & 2.84 & 17.10 & 8.65 \\ 0.34 & 0.78 & 0.53 & 1.21 & 3.57 & 2.75 & 3.02 & 8.65 & 6.22 \end{bmatrix} \begin{bmatrix} y_1 \\ y_2 \\ y_3 \\ y_4 \\ y_5 \\ y_6 \\ y_7 \\ y_8 \\ y_9 \end{bmatrix}$$

(3.141)

The iteration procedure is initiated by assigning values for $y_1, y_2, y_3, \ldots, y_9$ on the right-hand column of Eq. (3.141). We may start with the static deflections produced by the lumped weights directed as shown in Fig. 3.17c, or we may simply assume that $y_1 = y_2 = y_3 = \cdots = y_9 = 1.00$. For simplicity purposes, we assume that $y_1 = y_2 = y_3 = \cdots = y_9 = 1.00$. This is reasonable and permissible in starting the first trial procedure. By substituting into Eq. (3.141) and performing the matrix operation, we obtain

$$\begin{bmatrix} y_1 \\ y_2 \\ y_3 \\ y_4 \\ y_5 \\ y_6 \\ y_7 \\ y_8 \\ y_9 \end{bmatrix} = \frac{\omega^2 L^3}{52.70 EIg} \begin{bmatrix} 26.74 \\ 49.94 \\ 24.85 \\ 37.14 \\ 76.23 \\ 37.14 \\ 24.85 \\ 49.94 \\ 26.74 \end{bmatrix} = \frac{24.85 \omega^2 L^3}{52.70 EIg} \begin{bmatrix} 1.076 \\ 2.010 \\ 1.000 \\ 1.495 \\ 3.068 \\ 1.495 \\ 1.000 \\ 2.010 \\ 1.076 \end{bmatrix}$$

(3.142)

Equation (3.142) provides a better approximation of the amplitudes $y_1, y_2, y_3, \ldots, y_9$ of the fundamental mode. The procedure may be repeated by

TABLE 3.4. Deflection Coefficients for the Continuous Beam in Fig. 3.17c

Point of Unit Load	Deflection at Point									
	1	2	3	4	5	6	7	8	9
1	$\dfrac{L^3}{186.7EI}$	$\dfrac{L^3}{133.4EI}$	$\dfrac{L^3}{382EI}$	$-\dfrac{L^3}{503EI}$	$-\dfrac{L^3}{388EI}$	$-\dfrac{L^3}{1,137EI}$	$\dfrac{L^3}{2,185EI}$	$\dfrac{L^3}{1,484EI}$	$\dfrac{L^3}{3,430EI}$
2	$\dfrac{L^3}{133.4EI}$	$\dfrac{L^3}{67.2EI}$	$\dfrac{L^3}{406EI}$	$-\dfrac{L^3}{206EI}$	$-\dfrac{L^3}{167EI}$	$-\dfrac{L^3}{489EI}$	$\dfrac{L^3}{939EI}$	$\dfrac{L^3}{642EI}$	$\dfrac{L^3}{1,477EI}$
3	$\dfrac{L^3}{382EI}$	$\dfrac{L^3}{406EI}$	$\dfrac{L^3}{256EI}$	$-\dfrac{L^3}{319EI}$	$-\dfrac{L^3}{239EI}$	$-\dfrac{L^3}{717EI}$	$\dfrac{L^3}{1,394EI}$	$\dfrac{L^3}{940EI}$	$\dfrac{L^3}{2,185EI}$
4	$-\dfrac{L^3}{503EI}$	$-\dfrac{L^3}{206EI}$	$-\dfrac{L^3}{319EI}$	$\dfrac{L^3}{199EI}$	$\dfrac{L^3}{138EI}$	$\dfrac{L^3}{375EI}$	$-\dfrac{L^3}{717EI}$	$-\dfrac{L^3}{489EI}$	$-\dfrac{L^3}{1,137EI}$
5	$-\dfrac{L^3}{388EI}$	$-\dfrac{L^3}{167EI}$	$-\dfrac{L^3}{239EI}$	$\dfrac{L^3}{138EI}$	$\dfrac{L^3}{52.7EI}$	$\dfrac{L^3}{138EI}$	$-\dfrac{L^3}{239EI}$	$-\dfrac{L^3}{167EI}$	$-\dfrac{L^3}{388EI}$
6	$-\dfrac{L^3}{1,137EI}$	$-\dfrac{L^3}{489EI}$	$-\dfrac{L^3}{717EI}$	$\dfrac{L^3}{375EI}$	$\dfrac{L^3}{138EI}$	$\dfrac{L^3}{199EI}$	$-\dfrac{L^3}{319EI}$	$-\dfrac{L^3}{206EI}$	$-\dfrac{L^3}{503EI}$
7	$\dfrac{L^3}{2,185EI}$	$\dfrac{L^3}{939EI}$	$\dfrac{L^3}{1,394EI}$	$-\dfrac{L^3}{717EI}$	$-\dfrac{L^3}{239EI}$	$-\dfrac{L^3}{319EI}$	$\dfrac{L^3}{256EI}$	$\dfrac{L^3}{406EI}$	$\dfrac{L^3}{382EI}$
8	$\dfrac{L^3}{1,484EI}$	$\dfrac{L^3}{642EI}$	$\dfrac{L^3}{940EI}$	$-\dfrac{L^3}{489EI}$	$-\dfrac{L^3}{167EI}$	$-\dfrac{L^3}{206EI}$	$\dfrac{L^3}{406EI}$	$\dfrac{L^3}{67.2EI}$	$\dfrac{L^3}{133.4EI}$
9	$\dfrac{L^3}{3,430EI}$	$\dfrac{L^3}{1,477EI}$	$\dfrac{L^3}{2,185EI}$	$-\dfrac{L^3}{1,137EI}$	$-\dfrac{L^3}{388EI}$	$-\dfrac{L^3}{503EI}$	$\dfrac{L^3}{382EI}$	$\dfrac{L^3}{133.4EI}$	$\dfrac{L^3}{186.7EI}$

VIBRATION ANALYSIS OF CONTINUOUS HIGHWAY BRIDGES 207

using again Eq. (3.141) and the amplitudes given by Eq. (3.142), yielding

$$\begin{bmatrix} y_1 \\ y_2 \\ y_3 \\ y_4 \\ y_5 \\ y_6 \\ y_7 \\ y_8 \\ y_9 \end{bmatrix} = \frac{44.00\omega^2 L^3}{52.70 EIg} \begin{bmatrix} 1.055 \\ 2.081 \\ 1.000 \\ 1.620 \\ 3.528 \\ 1.620 \\ 1.000 \\ 2.081 \\ 1.055 \end{bmatrix} \qquad (3.143)$$

With two additional repetitions, Eq. (3.141) converges to the following values for the amplitudes $y_1, y_2, y_3, \ldots, y_9$:

$$\begin{bmatrix} y_1 \\ y_2 \\ y_3 \\ y_4 \\ y_5 \\ y_6 \\ y_7 \\ y_8 \\ y_9 \end{bmatrix} = \frac{47.50\omega^2 L^3}{52.70 EIg} \begin{bmatrix} 1.037 \\ 2.056 \\ 1.000 \\ 1.595 \\ 3.441 \\ 1.595 \\ 1.000 \\ 2.056 \\ 1.037 \end{bmatrix} \qquad (3.144)$$

The fundamental frequency of vibration of the highway bridge may be determined by using any one of the nine equations represented by Eq. (3.144). By using the third equation, we have

$$y_3 = \frac{47.50\omega^2 L^3}{52.70 EIg} y_3$$

or, solving for ω, we find

$$1 = \frac{47.50\omega^2 L^3}{52.70 EIg} \quad (1)$$

$$\omega = \sqrt{\frac{52.70 EIg}{47.50 L^3}} \qquad (3.145)$$

By substituting the values of E, I, L, and g in Eq. (3.145), we find

$$\omega = \sqrt{\frac{52.70(30 \times 10^6)(58 \times 10^3)386}{(47.50 \times 10^3)(73 \times 12)^3}}$$

$$= 33.294 \text{ rps}$$

or

$$f = \frac{\omega}{2\pi} = \frac{33.294}{2\pi} = 5.30 \text{ Hz}$$

This is in very close agreement with the experimental value obtained for the highway bridge by the Michigan Department of Transportation.

3.8 PRACTICAL DESIGN FOR EARTHQUAKES USING AVAILABLE CODES

Practical design considerations regarding earthquakes are given in Section 1.9, and various methodologies regarding the design of structures to resist their catastrophic effects are briefly introduced. One of these methodologies involves the utilization of carefully prepared engineering design codes. The development of such codes is based on experience gained from previous observations of earthquake damage, such as the Los Angeles Earthquake discussed in Section 1.10, as well as on current developments in the field of earthquake design. As stated earlier, these codes are intended to provide a simple but adequate design for many types of structures. For more specialized cases, however, a more elaborate analysis is required. In this section, we consider the Structural Engineers Association of California (SEAOC) code for the design of building structures.

The procedure for lateral force requirements in the utilization of the SEAOC code is based only on the first mode of the structure. This is justifiable, because experience has shown that, for many engineering structures, the contribution of the higher modes is of secondary importance. If we can assume a characteristic shape for the first mode, we make it possible to transform the maximum condition of response into a set of equivalent static forces, and the actual design may then be carried out by using static analysis.

Conceptually, the SEAOC recommendation is evolved by using the following two expressions:

$$V = KCW \qquad (3.146)$$

$$C = \frac{0.05}{\tau^{1/3}} \qquad (3.147)$$

where V is the total dynamic base shear, W is the total weight of the building, τ is the natural period of vibration of the first mode, and K is a coefficient that

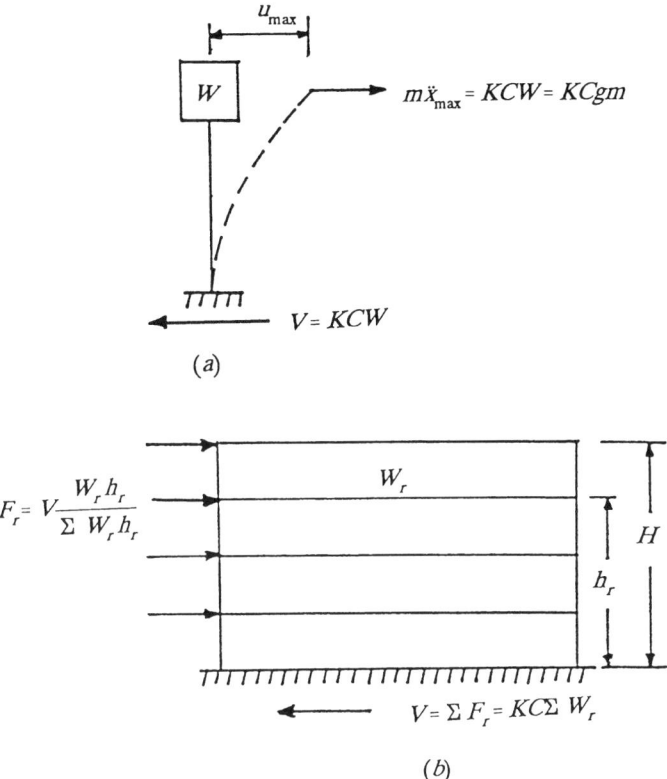

Figure 3.18. (a) Seismic coefficients. (b) SEAOC recommendations for earthquake forces on buildings.

varies between 0.67 and 1.50. The coefficient K reflects the ability of the structure to deform into the plastic range. For moment resisting frames that are fairly ductile, a value of 0.67 may be used for K, and the value of 1.33 may be used for less ductile arrangements, such as those consisting of concrete walls.

The seismic coefficient C represents an equivalent maximum acceleration expressed as a function of the acceleration of gravity g. This is shown in Fig. 3.18a, where the inertia force $m\ddot{x}_{max}$ is written as

$$m\ddot{x}_{max} = KCW = KCgm \tag{3.148}$$

Equation (3.148) shows that when C is multiplied by the weight W of the building, it yields the maximum horizontal inertia force.

By considering the mass m_r of a lumped-mass system, the maximum inertia force F_r of the mass m_r in the considered mode is taken to be as follows:

$$F_r = m_r \ddot{A}_m \phi_r \tag{3.149}$$

where $\ddot{A}_m$ is the maximum modal acceleration and ϕ_r is the coordinate of the characteristic shape of the mass m_r. The base shear force V may be expressed as

$$V = \ddot{A}_m \sum^r m_r \phi_r \qquad (3.150)$$

indicating that it should be equal to the sum of all inertia forces acting on a building in order to satisfy dynamic equilibrium.

By eliminating $\ddot{A}_m$ from Eqs. (3.149) and (3.150), we obtain

$$F_r = \frac{m_r \phi_r}{\Sigma^r m_r \phi_r} V \qquad (3.151)$$

The SEAOC recommendation indicates that ϕ_r should be taken as

$$\phi_r = \frac{h_r}{H} \qquad (3.152)$$

where h_r is the height above ground of mass m_r and H is the total height of the building.

By substituting Eq. (3.152) into Eq. (3.151), we find

$$F_r = \frac{m_r(h_r/H)}{\Sigma^r m_r(h_r/H)}$$

or

$$F_r = \frac{W_r h_r}{\Sigma^r W_r h_r} \qquad (3.153)$$

Equation (3.153) represents the SEAOC recommendation for the distribution of the lateral force. In this equation, W_r is the weight of the mass m_r; that is, $W_r = m_r g$, where g is the acceleration of gravity. A set of F forces acting on a building is shown in Fig. 3.18b. On this basis, the building may be analyzed by assuming that these forces are statically applied.

Note that Eq. (3.153) is obtained by assuming that the fundamental mode shape of the building is linear with zero amplitude at the base of the building. This is a reasonable assumption for practical applications, because the combined effects of the shear distortion in the stories of the frame and the change in length of the columns caused by the overall bending of the building produce a shape that is a close approximation to the shape of a straight line.

The following example illustrates the application of the preceding theory.

Example 3.12: For the three-story building shown in Fig. 3.19, determine the floor forces F_1, F_2, and F_3 and the design values of the story shears. The weights of the three floors are $W_1 = W_2 = 4{,}000$ kips and $W_3 = 2{,}000$ kips, and

PRACTICAL DESIGN FOR EARTHQUAKES USING AVAILABLE CODES 211

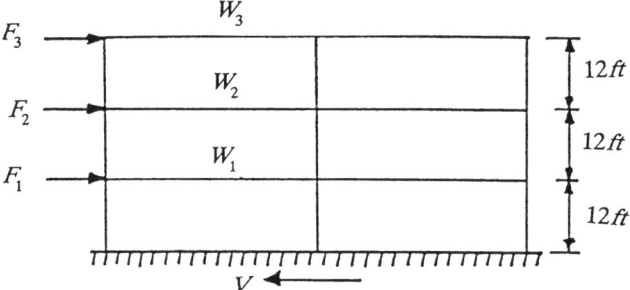

Figure 3.19. Three-story building.

the total weight W of the building is 10,000 kips. Assume that the building is constructed with moment resisting frames and that the period of vibration τ of the fundamental mode is 0.8 sec.

SOLUTION: For moment resisting frames, we can assume that $K = 0.67$. From Eq. (3.147), the seismic coefficient C is

$$C = \frac{0.05}{\tau^{1/3}} = \frac{0.05}{(0.8)^{1/3}} = 0.0539$$

Thus, the total base shear V is

$$V = KCW = (0.67)(0.0539)(10,000)$$
$$= 361 \text{ kips}$$

By following the distribution given by Eq. (3.153), the floor forces are as follows:

$$F_1 = \frac{(4,000)(12)}{(4,000)(12) + (4,000)(24) + (2,000)(36)} (361)$$
$$= \frac{17.328(10)^6}{216(10)^3}$$
$$= 80.22 \text{ kips}$$

$$F_2 = \frac{(4,000)(24)}{(216)(10)^3} (361)$$
$$= 160.44 \text{ kips}$$

$$F_3 = \frac{(2,000)(36)}{(216)(10)^3} (361)$$
$$= 120.33 \text{ kips}$$

212 INFRASTRUCTURAL DYNAMICS

Therefore the design values of the story shears are 361 kips for the first story, $361 - 80.22 = 280.78$ kips for the second story, and $280.78 - 160.44 = 120.34$ kips for the third story.

If an elastic analysis of the building was performed based on a response spectrum of a rather strong earthquake, the shear so obtained would be much larger than the one obtained using the SEAOC code recommendations obtained above. However, this should be expected, because the elastic analysis is known to be very conservative. In using the code recommendations, we should keep in mind that for a strong earthquake it is expected that the structure would undergo considerable plastic distortion.

PROBLEMS

3.1 The uniform steel beams shown in Figs. P3.1a through P3.1e support operating equipment of weight W located as shown. The cross section of each beam is a wide-flange section of moment of inertia I. For each beam, determine an idealized one-degree spring-mass system by following the procedure discussed in Section 3.2. Neglect the weight of the beam and assume that damping is zero.

3.2 Repeat Problem 3.1 by assuming that the member moves under the influence of viscous damping of viscous damping coefficient c.

3.3 The frames shown in Figs. P3.3a through P3.3c represent interior bays of single-story buildings. The horizontal dynamic force acting at the top of each frame is $F(t)$, and it can vary with respect to time. The width of each bay is 25 ft, and the weights of the walls and roof, including the weights of the columns and girders, are shown in the figures. For each bay frame, determine an idealized one-degree spring-mass system that is appropriate for dynamic analysis. Assume that the girders are infinitely stiff compared to the columns and neglect damping. The modulus of elasticity $E = 30 \times 10^6$ psi, and the column sizes are shown in the figures.

3.4 Repeat Problem 3.3 by assuming that the building frames are under the influence of viscous damping of viscous damping coefficient c.

3.5 By using the idealized one-degree spring-mass systems obtained in Problem 3.1 and assuming that the dynamic force $F(t) = 0$, determine in each case the free undamped frequency of vibration and the period of vibration. Assume that $W = 20$ kips, $L = 24$ ft, $k = 120$ kips/in., $k_1 = 60$ kips/in., $k_2 = 200$ kips/in., $E = 30 \times 10^6$ psi, and $I = 2{,}000$ in.^{4}

3.6 Repeat Problem 3.5 by assuming that the beams are moving under the influence of 10 percent viscous damping. Compare the results. Also, determine the logarithmic decrement δ.

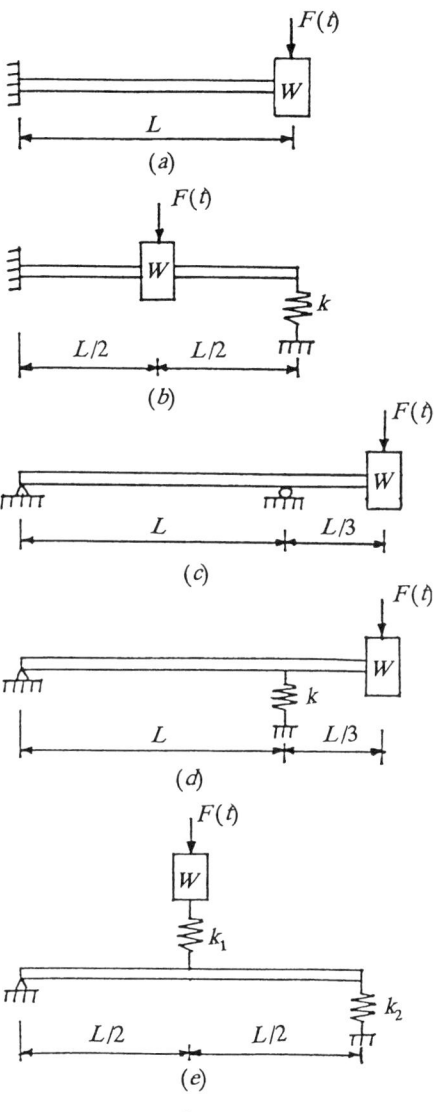

Figure P3.1.

3.7 Repeat Problem 3.5 by assuming that the beams are moving under the influence of 15, 20, 25, 30, and 35 percent viscous damping and compare the results.

3.8 By using the idealized one-degree spring-mass systems obtained for the building frames in Problem 3.3 and assuming that the dynamic force $F(t) = 0$, determine in each case the free undamped frequency of vibration and the period of vibration.

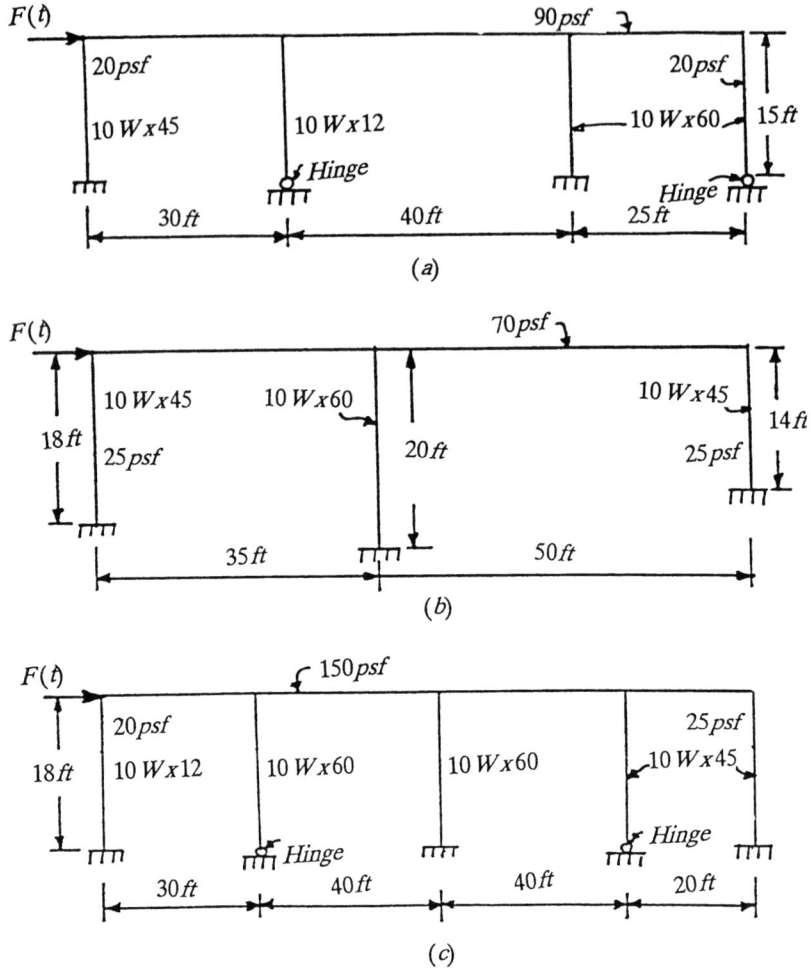

Figure P3.3.

3.9 Repeat Problem 3.8 by assuming that the building frames are vibrating under the influence of 10 percent viscous damping and compare the results. Also, determine the logarithmic decrement δ.

3.10 Repeat Problem 3.9 by assuming that the frames are moving under the influence of 5, 10, 15, and 20 percent viscous damping and compare the results.

3.11 The structural steel members shown in Fig. P3.1 support a weight $W = 70$ kN, located as shown. The dynamic force $F(t)$ acting on the weight W is a harmonic force $F(t) = F \cos \omega_f t$, where $F = 15$ kN and the frequency ω_f of the force is equal to 4π rps. By using an appropriate idealized one-degree spring-mass system, determine in each case the maximum displacement of the member and its maximum bending stress. The cross section of each member is rectangular with width $b = 20$ cm and depth $h = 40$ cm. The length $L = 8$ m, $k = 5{,}000$ kN/m, $k_1 = 10{,}000$ kN/m, $k_2 = 20{,}000$ kN/m, and $E = 200$ GPa. Neglect the weight of the members and assume that damping is zero.

3.12 Repeat Problem 3.11 by assuming that the structural members are under the influence of 10 percent viscous damping and compare the results.

3.13 Repeat Problem 3.11 when the harmonic dynamic force $F(t) = F \sin \omega_f t$, where $F = 15$ kN and the frequency $\omega_f = 4\pi$ and compare the results.

3.14 Repeat Problem 3.11 when $\omega_f = 2\pi, 3\pi, 5\pi, 6\pi$, and 7π rps, and compare the results.

3.15 Repeat Problem 3.12 by assuming that the frequency ω_f of the harmonic dynamic force coincides with the free frequency ω of the idealized spring-mass system and compared the results.

3.16 By utilizing in each case the idealized one-degree spring-mass system obtained for the building frames in Problem 3.3, determine the maximum horizontal displacement of each frame and the maximum bending stresses in its columns. The dynamic force $F(t)$ is a harmonic force of the form $F \cos \omega_f t$, where $F = 15$ kips and $\omega_f = 15$ rps.

3.17 Repeat Problem 3.16 when $\omega_f = 8, 10, 15, 20,$ and 25 rps and compare the results.

3.18 Repeat Problem 3.16 when the motion of the building frames is under the influence of 10 percent damping and compare the results.

3.19 Repeat Problem 3.18 for the case where the frequency ω_f of the harmonic force coincides with the free frequency of vibration ω of the idealized spring-mass system and compare the results.

3.20 Repeat Problem 3.16 when $F(t) = F \sin \omega_f t$, where $F = 15$ kips and $\omega_f = 15$ rps, and compare the results.

3.21 Repeat Problem 3.11 by assuming that the dynamic force $F(t)$ is a suddenly applied constant force of 15 kN of infinite duration. Use the rigorous solution given in Section 3.5.1.

3.22 Repeat Problem 3.21 by assuming that the idealized spring-mass system is under the influence of 10 percent viscous damping and compare the results.

3.23 Repeat Problem 3.21 by assuming that the dynamic force $F(t)$ has the time variations shown in Figs. P3.23a through P3.23d.

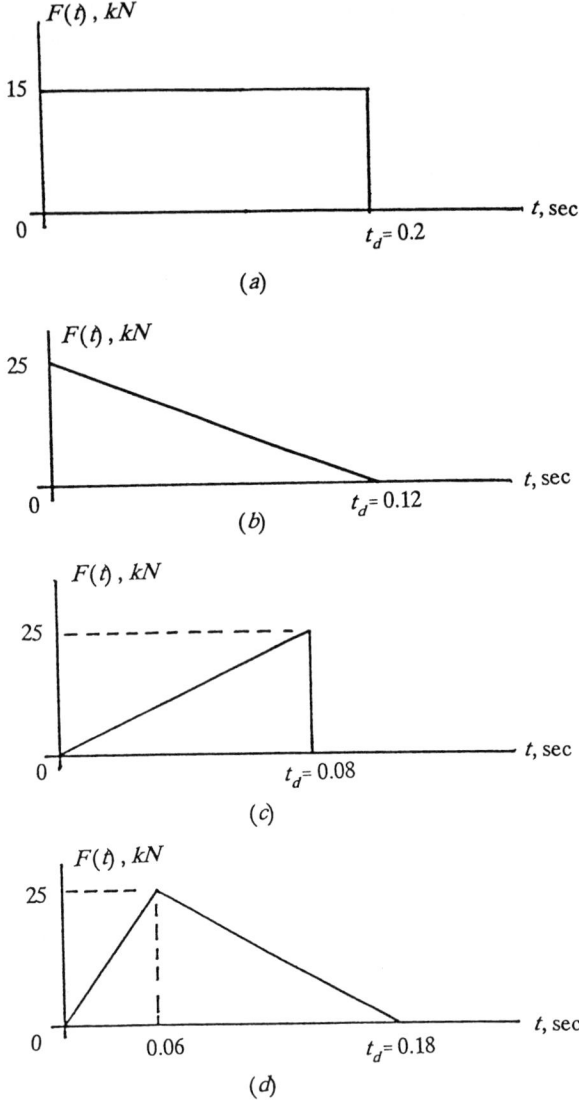

Figure P3.23.

3.24 Repeat Problem 3.22 by assuming that the dynamic force $F(t)$ has the time variations shown in Figs. P3.23a through P3.23d.

3.25 Repeat Problem 3.16 by assuming that the dynamic force $F(t)$ is a suddenly applied constant force of 15 kips that lasts indefinitely and compare results. Use the rigorous solution discussed in Section 3.5.1.

3.26 Repeat Problem 3.25 by assuming that the building frame is under the influence of 10 percent viscous damping and compare the results.

3.27 Repeat Problem 3.25 by assuming that the force $F(t)$ has the magnitude and time variations shown in Fig. P3.23.

3.28 Repeat Problem 3.27 by assuming that the building frame is under the influence of 8 percent viscous damping and compare the results.

3.29 Repeat Problem 3.11 by using the AIEM and compare the results.

3.30 Repeat Problem 3.12 by using the AIEM and compare the results.

3.31 Repeat Problem 3.14 by using the AIEM and compare the results.

3.32 Repeat Problem 3.16 by using the AIEM and compare the results.

3.33 Repeat Problem 3.18 by using the AIEM and compare the results.

3.34 Repeat Problem 3.21 by using the AIEM and compare the results.

3.35 Repeat Problem 3.22 by using the AIEM and compare the results.

3.36 Repeat Problem 3.23 by using the AIEM and compare the results.

3.37 Repeat Problem 3.24 by using the AIEM and compare the results.

3.38 Repeat Problem 3.25 by using the AIEM and compare the results.

3.39 Repeat Problem 3.26 by using the AIEM and compare the results.

3.40 Repeat Problem 3.27 by using the AIEM and compare the results.

3.41 Repeat Problem 3.28 by using the AIEM and compare the results.

3.42 Rework the highway bridge problem in Example 3.9 by assuming that $\lambda = 72$ in., $t = 8$ in., $L = 90$ ft, and the size of the girder section is W24 × 162. Neglect damping.

3.43 Design the highway bridge in Problem 3.42 so that its fundamental free frequency of vibration is 6 Hz. Select an appropriate wide-flange section for the girder and repeat the procedure if it is required.

3.44 Rework Problem 3.42 by assuming noncomposite action between girder and bridge deck and compare the results.

3.45 Rework Problem 3.43 by assuming noncomposite action between girder and bridge deck and compare the results.

3.46 Repeat Problem 3.42 by using Stodola's method and the iteration procedure and compare the results.

3.47 Rework Problem 3.44 by using Stodola's method and the iteration procedure and compare the results.

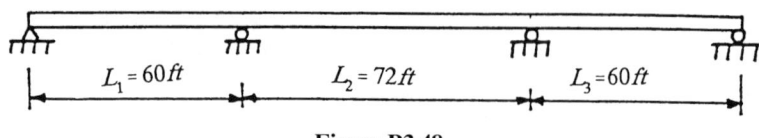

Figure P3.48.

3.48 An internal girder of a continuous three-span composite highway bridge and its associated bridge deck has the cross-sectional dimensions shown in Fig. 3.14d. The cross-sectional dimensions of the deck-girder system are constant throughout the length of the bridge and the lengths of each span are shown in Fig. P3.48. By using Stodola's method and the iteration procedure, determine the free fundamental frequency of vibration of the highway bridge and its corresponding mode shape. The modulus of elasticity E_s of steel is 30×10^6 psi and the dynamic modulus E_c of concrete is 5×10^6 psi. Neglect damping.

3.49 Repeat Problem 3.48 by assuming that there is no composite action between the deck and the girders and compare the results.

3.50 Rework Example 3.12 by assuming that $K = 1.33$ and compare the results.

3.51 For the four-story building in Fig. P3.51, determine the floor forces F_1, F_2, F_3, and F_4 and the design values of the story shears. The weights of the floors are $W_1 = W_2 = W_3 = 4{,}500$ kips and $W_4 = 2{,}250$ kips, and the total weight W of the building is 15,750 kips. Assume that the building is constructed with moment resisting frames and that the period of vibration τ of the fundamental mode is 1.2 sec.

3.52 Repeat Problem 3.51 by assuming that $K = 1.33$ and $\tau = 0.8$ sec and compare the results.

3.53 Repeat Problem 3.51 for $\tau = 0.5$, 1, and 1.5 sec and compare the results.

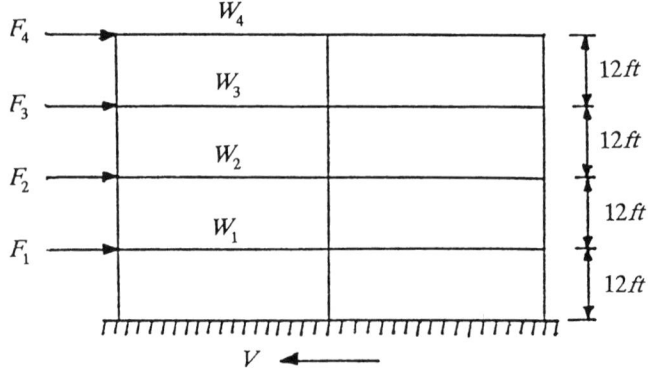

Figure P3.51.

4 Advanced Methods and Problems of Infrastructural Analysis

4.1 INTRODUCTION

Our structural and mechanical infrastructural elements are becoming increasingly complex today, and require more advanced technologies to cope with the associated structural and mechanical problems. Many such systems incorporate structural and mechanical problems. Many such systems incorporate structural components of arbitrary stiffness variations. These components can be subjected to hazardous environments such as blast, earthquake, and gusty winds, the elastic limit of their material may be exceeded, and deformations in general may be large. In this chapter, as well as in following chapters of this text, many of these problems are examined, and powerful methods of analysis for the solution of such problems are discussed. In this chapter, such methods of analysis would include the utilization of simplified equivalent systems for an accurate or exact solution of complex problems, the moment-area and conjugate-beam methods, the moment distribution method, and the three-moment equation. Both statically determinate and statically indeterminate practical problem are examined and solved. Energy methods, as well as the finite element and finite difference methods, are discussed in later chapters of the text.

4.2 MEMBERS OF ARBITRARY STIFFNESS VARIATIONS

Infrastructural systems are often composed of members where the moment of inertia I and the modulus of elasticity E may vary along their length. The design purpose in such variations is to increase the ability of the member to resist stress and deformation. For example, as pointed out in Section 3.6, longer span highway bridges are more flexible and thus more susceptible to undesirable deformations and general vibration and dynamic performance. Increased rigidity is often obtained by composite action between girders and deck, and also by using girders with linear or parabolic variation in their depth.

The modulus of elasticity of a member may also be variable along its length. For example, when the applied loading on a member is large, the stresses in the member may exceed the elastic limit of its material and the member will respond inelastically. On this basis, the modulus of elasticity E at cross sections along the length of the member will become variable and, consequently, the stiffness EI of the member will vary along its length. Such situations can easily occur when the infrastructural system responds to the action of an earthquake, nuclear and conventional explosions, and the acts of nature.

A very convenient method that can be used for both static and dynamic analysis of such problems is the author's method of the equivalent systems. The fundamental aspects of this method and theory are discussed in this section. See also Fertis [3, 5, 6, 37, 39].

4.2.1 Theory and Method of the Equivalent Systems

The theory and method of the equivalent systems [3, 5, 6] permit the replacement of a member of variable stiffness $E_x I_x$ with a member of uniform stiffness $E_1 I_1$ by the application of a theorem of equivalent systems, which reads as follows:

> A member of variable stiffness $E_x I_x$ can be replaced by an equivalent system of arbitrarily selected uniform stiffness $E_1 I_1$, where the deflection curve of the equivalent system is everywhere identical to the deflection curve of the original variable stiffness member.

The validity of this theorem can be proved by using known methods of mechanics of solids, and it applies to both statically determinate and statically indeterminate members. The theorem is also valid when the deformations of the variable stiffness member are large and when the member behaves inelastically [3, 5, 6]. In this section, the application of the theorem is limited to statically determinate variable stiffness members subjected to small deformations.

In the following analysis, a mathematical proof regarding the existence of such an equivalent system is first developed, based on the general differential equation of an elastic line, and then the procedure is simplified and used for the solution of infrastructural problems. In other words, the exact solution of the problem is first developed and then, based on the exact solution, a very accurate approximate method is derived that simplifies the solution of problems with any stiffness variation and applied loading conditions.

The proof of the theorem is initiated by considering the following general Euler–Bernoulli law of beam deformation:

$$\frac{y''}{[1 + (y')^2]^{3/2}} = -\frac{M_x}{E_x I_x} \qquad (4.1)$$

In Eq. (4.1), y'' is the curvature of the member, y' is the slope of its elastic line, M_x is the bending moment at any coordinate x, and $E_x I_x$ is the variable stiffness of the member. We assume here that both E_x and I_x are permitted to vary along the length of the member.

If the deformations are small, then y' in the denominator of Eq. (4.1) is small compared to one and it can be neglected. On this basis, Eq. (4.1) is written as follows:

$$y'' = -\frac{M_x}{E_x I_x} \qquad (4.2)$$

Equation (4.2) is linear since we assume that the deformations of the member are small.

If Eq. (4.2) is integrated twice, we find

$$y = \int \left[-\int \frac{M_x d_x}{E_x I_x} \right] dx + C_1 \int dx + C_2 \qquad (4.3)$$

where C_1 and C_2 are the constants of integration, which depend upon the boundary conditions of the member.

We express the variable stiffness $E_x I_x$ of the member as

$$E_x I_x = E_1 I_1 g(x) f(x) \qquad (4.4)$$

where $E_1 I_1$ is an arbitrarily selected reference value of the stiffness $E_x I_x$, $g(x)$ is a function representing the variation of E_x with reference value E_1, and $f(x)$ is the function representing the variation of I_x with reference value I_1. By substituting Eq. (4.4) into Eq. (4.3), we obtain

$$y = \frac{1}{E_1 I_1} \int \left[-\int \frac{M_x d_x}{g(x) f(x)} \right] dx + C_1 \int dx + C_2 \qquad (4.5)$$

If the beam of constant stiffness $E_1 I_1$ chosen has the same length and the same reference system of axes as the one represented by Eq. (4.5), its Euler–Bernoulli differential equation is

$$y_e'' = -\frac{M_e}{E_1 I_1} \qquad (4.6)$$

where y_e is its elastic line and M_e is its bending moment at any coordinate x. By differentiating Eq. (4.6) twice, we obtain

$$y_e = \frac{1}{E_1 I_1} \int \left[-\int M_e dx \right] dx + C_1' \int dx + C_2' \qquad (4.7)$$

where C_1' and C_2' are the constants of integration.

The elastic lines y and y_e represented by Eqs. (4.5) and (4.7), respectively, would be identical if

$$\text{(a)} \ C_1 = C_1' \qquad \text{(b)} \ C_2 = C_2' \qquad (4.8)$$

and

$$\int \left[-\int \frac{M_x d_x}{g(x) f(x)} \right] dx = \int \left[-\int M_e \, dx \right] dx \qquad (4.9)$$

The conditions in Eq. (4.8) may be satisfied if the members represented by Eqs. (4.5) and (4.7) have the same length and the same boundary conditions. Equation (4.9) will be satisfied if

$$M_e = \frac{M_x}{g(x) f(x)} \qquad (4.10)$$

This proves that, for a beam of variable stiffness $E_x I_x$, there exists an equivalent system of constant stiffness $E_1 I_1$ with boundary conditions and length the same as those of the variable stiffness member, but with its bending moment M_e at any cross section x given by Eq. (4.10). Thus, the bending moment diagram of the equivalent system of constant stiffness $E_1 I_1$ can be determined from Eq. (4.10), provided that $f(x)$ and $g(x)$ are known.

With known M_e, the shear force V_e and the applied load w_e of the equivalent system may be determined from the well-known expressions

$$V_e = \frac{d}{dx}(M_e) = \frac{d}{dx}\left[\frac{M_x}{g(x) f(x)} \right] \qquad (4.11)$$

$$w_e = -\frac{d}{dx}(V_e) = -\frac{d^2}{dx^2}\left[\frac{M_x}{g(x) f(x)} \right] \qquad (4.12)$$

If $g(x) = f(x) = 1$, then $E_x I_x = E_1 I_1$, and the equivalent system in this case is identical to the original system.

Only elastic deflections produced by bending are considered in the preceding derivation of an equivalent system. However, if deflections produced by shear are appreciable and need to be considered in the analysis, an equivalent system for shear deflection may also be obtained. See Fertis [6] and Fertis and Cunningham [42]. A wide range of applications of the theory and method of the equivalent systems to both linear and nonlinear complex statics and dynamics problems may be found in the author's work in References [3, 5, 6, 37, 39].

The following example illustrates the derivation of exact equivalent systems.

Example 4.1: A variable stiffness cantilever beam of length L is loaded as shown in Fig. 4.1a. The variation of its depth h_x is given by the expression

$$h_x = h_1 \left(\frac{L + 3x}{L} \right)^{1/3} \qquad (4.13)$$

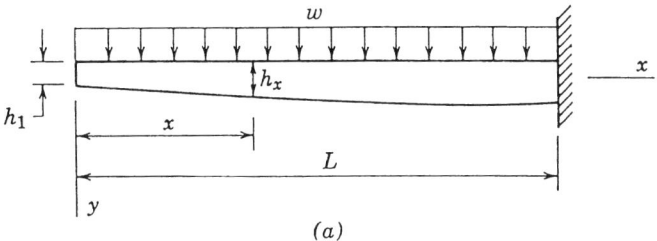

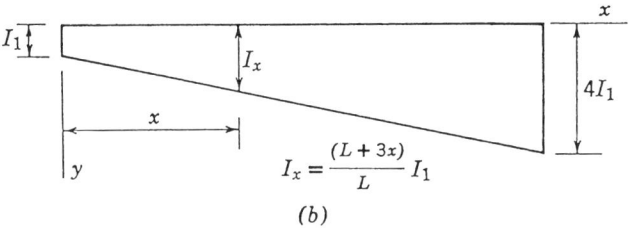

Figure 4.1. (a) Variable stiffness cantilever beam loaded as shown. (b) Moment of inertia variation of the member.

and its modulus of elasticity E is assumed to be constant throughout its length L, which implies that the function $g(x) = 1$. Determine an exact equivalent system of constant stiffness EI_1, where I_1 is the moment of inertia of the member at its free end. The width of the member is constant and equal to b.

SOLUTION: The moment of inertia I_x at a distance x is

$$I_x = \frac{bh_x^3}{12} = \left(\frac{L + 3x}{L}\right) I_1 \tag{4.14}$$

where

$$I_1 = \frac{bh_1^3}{12} \tag{4.15}$$

It is shown plotted in Fig. 4.1b. Thus, with $g(x) = 1$ and I_1 as the reference value of I_x, the moment of inertia function $f(x)$ is

$$f(x) = \frac{L + 3x}{L} \tag{4.16}$$

Note that $f(x)$ is linear.

By applying simple statics, the moment M_x of the variable stiffness member at any location x is

$$M_x = -\frac{wx^2}{2} \qquad (4.17)$$

By using Eq. (4.10), the moment M_e of the equivalent system of constant stiffness EI_1 is

$$M_e = \frac{M_x}{g(x)f(x)} = -\frac{wLx^2}{2L + 6x} \qquad (4.18)$$

Thus, by applying Eqs. (4.11) and (4.12) and carrying out the required differentiations, the shear force V_e and the applied load w_e of the equivalent system are

$$V_e = \frac{d}{dx}(M_e) = -\frac{4wL^2x + 6wLx^2}{(2L + 6x)^2} \qquad (4.19)$$

$$w_e = -\frac{d}{dx}(V_e) = \frac{8wL^3}{(2L + 6x)^3} \qquad (4.20)$$

If we make the assumption that $w = 1$ lb/in. and $L = 60$ in., the graphs of Eqs. (4.18), (4.19), and (4.20), are shown in Figs. 4.2c, 4.2d, and 4.2e, respectively. Thus, the equivalent system of uniform stiffness EI_1 is as shown in Fig. 4.2e, and its elastic line, its length, and its boundary conditions are identical to the analogous quantities of the variable stiffness member in Fig. 4.2a. The solution of the equivalent system in Fig. 4.2e yields deflections and rotations identical to the corresponding ones of the original variable stiffness member in Fig. 4.2a. Note that the applied loading w_e on the equivalent system is different in order to compensate for the change in stiffness.

4.2.2 Approximate Method of the Equivalent Systems

The exact method of the equivalent systems may become laborious for more complicated stiffness variations, and simplifications must be introduced for an easier solution of the problem. The practicing engineer does not usually require an exact value for the design, because he/she knows that exact values for complex problems are extremely difficult, and in many cases impossible to obtain. He/she usually requires a reliable value with reasonable and predictable accuracy. This difficulty may be removed in this case by introducing an accurate simplification in the derivation of the equivalent system of constant stiffness E_1I_1.

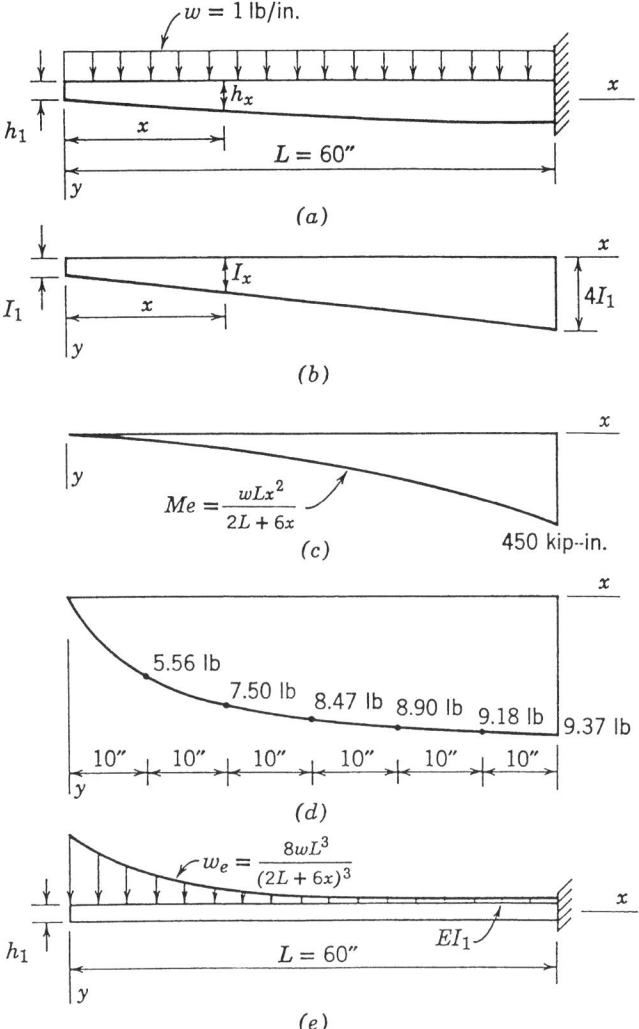

Figure 4.2. (a) Original member. (b) Moment of inertia variation. (c) M_e diagram of the equivalent system. (d) Equivalent shear force diagram V_e. (e) Equivalent system of constant stiffness EI_1.

Consider, for example, the variable stiffness cantilever beam in Fig. 4.3a and assume that the elastic modulus function $g(x) = 1$, and that the moment of inertia function $f(x) = I_x/I_1$ is as shown in Fig. 4.3b. This graph can be easily plotted by selecting a reference value I_1 and computing the values of $f(x) = I_x/I_1$ at a sufficient number of cross sections along the length of the member. At the same cross sections, the corresponding values of the bending

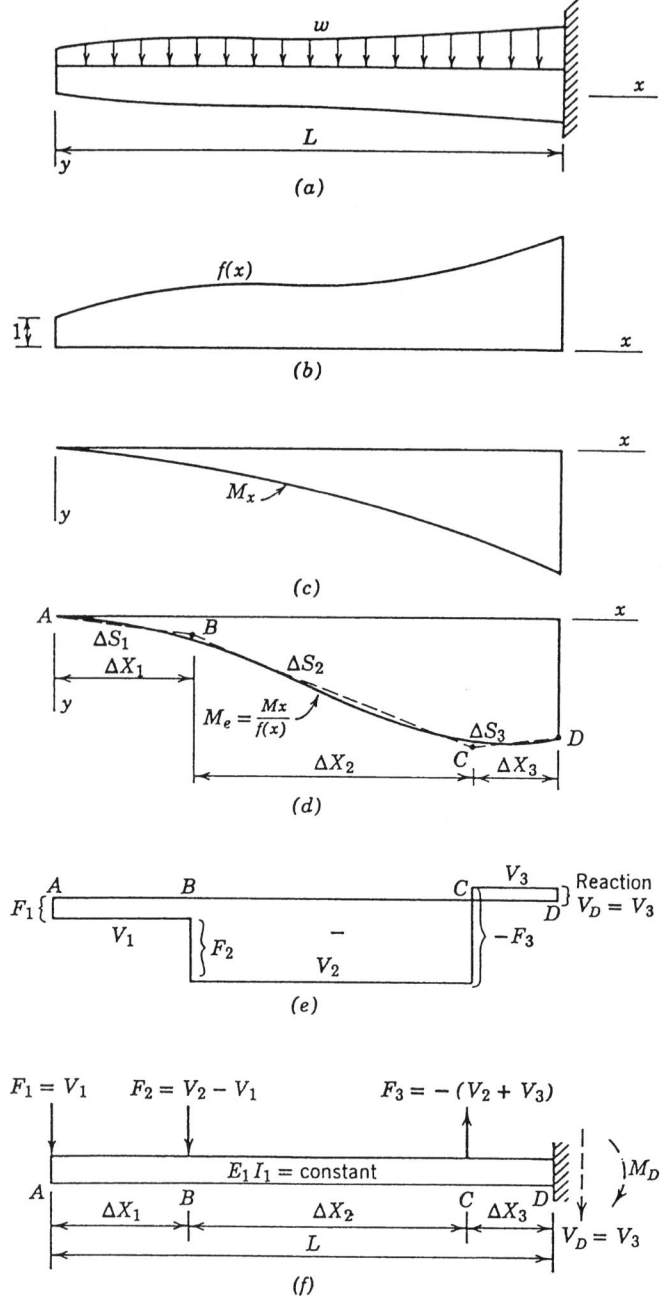

Figure 4.3. (*a*) Original variable stiffness member. (*b*) Variation of function $f(x)$. (*c*) Bending moment diagram M_x. (*d*) M_e diagram with its shape approximated with straight line segments. (*e*) Equivalent shear force diagram. (*f*) Equivalent system of constant stiffness.

moment M_x of the member in Fig. 4.3a are computed by using statics, and let it be assumed that the resulting diagram has the shape shown in Fig. 4.3c. The values of M_e of the equivalent system of constant stiffness $E_1 I_1$ may be determined by using Eq. 4.10, and they are plotted as shown by the solid line in Fig. 4.3d. That is, the ordinates of the M_e diagram are the computed values of M_x divided by the corresponding values of $f(x)$. Note that $g(x)$ is assumed to be equal to one.

The shape of the M_e diagram is now approximated by straight line segments of length ΔS_n ($n = 1, 2, 3, \ldots$), as shown by the dashed lines in Fig. 4.3d. It is not required for the ΔS_n segments to be of equal length, and their juncture points are usually located above or below the solid-line curve, so that the areas added to or subtracted from the M_e diagram are approximately balanced, as shown in Fig. 4.3d. The main purpose of this approximation is to approximate the shape of M_e with straight-line segments. The number of such straight-line segments for an excellent approximation depends on the shape of M_e. Usually three to five segments judiciously selected are sufficient.

The equivalent shear force diagram is drawn as shown in Fig. 4.3e. For example, the shear force V_2 is equal to the difference in bending moment between sections at C and B, divided by the length ΔX_2. The equivalent system of uniform stiffness $E_1 I_1$ is determined as shown in Fig. 4.3f, and it is acted upon by three concentrated loads. The slope and deflection of any point on the elastic line of the variable stiffness beam in Fig. 4.3a are very accurately determined by using the equivalent system in Fig. 4.3f and applying known methods of mechanics of solids, or existing formulas such as the ones given in Appendix A. For example, an approximation of the shape of the M_e diagram that yields two to four concentrated loads on the equivalent system produces less than one percent error in the computed deflections and rotations. If desired, better accuracy may be obtained if we approximate the shape of M_e with more straight-line segments.

If E and I are both uniform, then both $g(x)$ and $f(x)$ are equal to one, and the M_e diagram is identical to the M_x diagram of the original system. If desired, the approximation of its shape with straight-line segments can be carried out in the same way as for the variable stiffness beams, and an equivalent system loaded with simple concentrated loads may be obtained. This procedure is useful for uniform stiffness members with complicated loading conditions.

The above simplification in the derivation of an equivalent system has the advantage of being applicable to any member of variable stiffness, whether or not the variation of its stiffness and loading can be expressed as continuous functions of x. The following example illustrates the application of this method.

Example 4.2: For the variable stiffness cantilever beam of Example 4.1, determine an equivalent system of constant stiffness EI_1 by applying the approximate method of the equivalent systems. By using the equivalent system, determine the deflection and rotation at its free end.

SOLUTION: The original variable stiffness member is shown in Fig. 4.2a. The bending moment at a selected number of cross sections along the length of the member may be obtained by using statics, or by using Eq. (4.17) since such an equation is available for this problem. By selecting the moment of inertia I_1 at the free end of the member as the reference value, we calculate I_x in terms of I_1 at the selected cross sections where M_x is evaluated. Equation (4.14) may also be used for this purpose since it is available. The moment of inertia function $f(x)$ is the ratio I_x/I_1. The function $g(x)$ for this problem is assumed to be equal to one. With the help of Eq. (4.10), the values of $M_e = M_x/f(x)$ at the selected cross sections are computed, and they are plotted as shown by the solid line in Fig. 4.4a. This is the moment diagram M_e of the equivalent system of constant stiffness EI_1 and it is exact, because the exact values of M_x, $g(x)$, and $f(x)$ at the selected cross sections are used.

In Fig. 4.4a, the shape of the exact M_e diagram is approximated with three straight-line segments ΔS_1, ΔS_2, and ΔS_3, as shown by the dashed lines. The equivalent shear force diagram is shown in Fig. 4.4b, where

$$V_1 = \frac{\Delta M_B}{\Delta X_1} = -\frac{30}{10} = -3 \text{ lb}$$

$$V_2 = \frac{\Delta M_C - \Delta M_B}{\Delta X_2} = -\frac{127 - 30}{14} = -6.92 \text{ lb}$$

$$V_3 = \frac{\Delta M_D - \Delta M_C}{\Delta X_3} = -\frac{450 - 127}{36} = -8.98 \text{ lb}$$

The equivalent system of constant stiffness EI_1 is shown in Fig. 4.4c, where $F_1 = V_1 = 3$ lb, $F_2 = V_2 - V_1 = 3.92$ lb, and $F_3 = V_3 - V_2 = 2.06$ lb. The solution of the equivalent system for the computation of deflections and rotations at any point along the length of the member may be carried out by using well-known methods of mechanics, such as the moment-area method, or by using expressions such the ones provided in Appendix A and existing handbooks.

The deflection y_A at the free end A of the equivalent system in Fig. 4.4c is computed here by using the moment-area method, and is found to be

$$y_A = \frac{(491.9)(10)^3}{EI_1} \quad \text{(downward)}$$

The exact value of this deflection was also determined by solving the exact equivalent system in Fig. 4.2e, yielding

$$y_{A\text{exact}} = \frac{(489)(10)^3}{EI_1} \quad \text{(downward)}$$

MEMBERS OF ARBITRARY STIFFNESS VARIATIONS   229

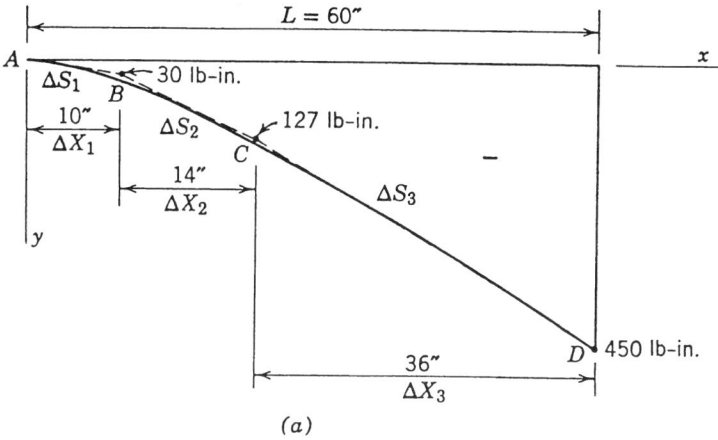

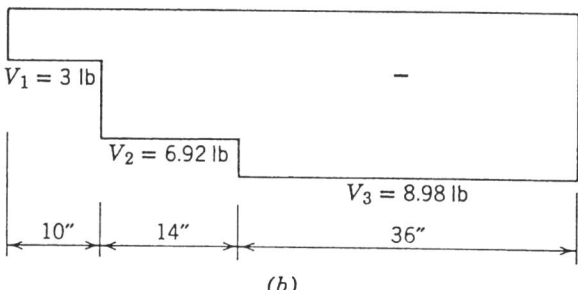

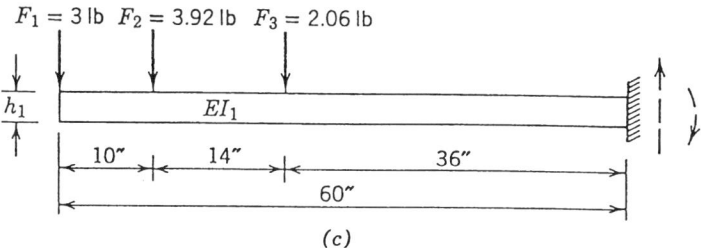

Figure 4.4. (a) M_e diagram with its shape approximated with three straight-line segments. (b) Equivalent shear force diagram. (c) Equivalent system of constant stiffness EI_1.

The error is

$$\text{Error} = \frac{2{,}900}{489{,}000} = \frac{1}{169} = 0.59\%$$

The rotation θ_A at the free end A of the equivalent system was also determined by using the moment-area method, yielding

$$\theta_A = \frac{(11.63)(10)^3}{EI_1} \quad \text{(counterclockwise)}$$

The exact solution yields, for all practical purposes, identical results.

4.3 EQUIVALENT SYSTEMS FOR HIGHWAY BRIDGES WITH GIRDERS OF VARIABLE DEPTH

The management organizations that govern roadway infrastructure systems in the United States fall under the general guidance of the Federal Highway Administration (FHWA), part of the Department of Transportation (DOT). FHWA is the successor organization to the Bureau of Public Roads (BPR), which was established in 1918 as part of the Department of Agriculture. The name of FHWA dates to 1970 after BPR was absorbed into the Department of Transportation (DOT), which was formed in 1967. The construction and maintenance of Federal aid and state roads is primarily the responsibility of the state departments of transportation, with headquarters in the capital of each state of the union.

Standards, guidelines, and helpful publications regarding the design of highway bridges became the responsibility of the American Association of State Highway and Transportation Officials (AASHTO). This group began, in 1921, as the Committee on Bridges and Structures of the American Association of State Highway Officials. Though in nonprinted form, a complete specification was available in 1926 and it was revised in 1928, at which time it was used by the bridge engineering profession. The first edition of the Standard Specifications for Highway Bridges was published in 1931, followed by many editions thereafter.

In-depth bridge inspections by the state highway departments began in 1968 with the development and printing of appropriate inspection forms. The immediate result of the 1968–1969 inspections was a complete structure inventory, along with identification and correlation of the most serious problems, and the adoption of a detailed periodic inspection procedure on a nationwide basis. See also References [43, 44]. Roads, streets, and bridges constitute the essential fabric of the nation's transportation infrastructure system, and their 20-year needs for the future amount to about $2 trillion. This constitutes the largest part of the needs in our infrastructure crisis.

EQUIVALENT SYSTEMS FOR HIGHWAY BRIDGES 231

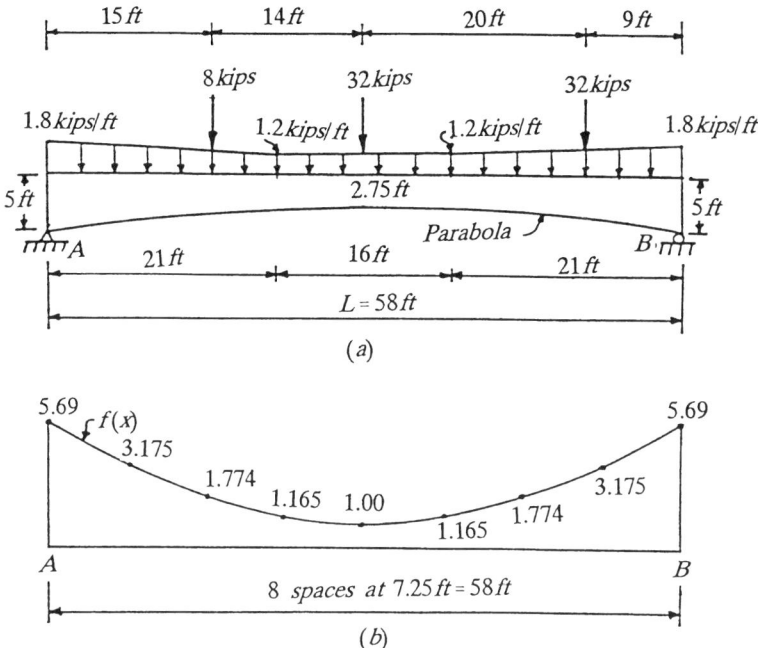

Figure 4.5. (*a*) Bridge girder of parabolic variation in depth. (*b*) Variation of moment of inertia function $f(x)$.

This section outlines the application of the method of the equivalent systems to the infrastructure problem of analyzing single-span highway bridges with girders of variable stiffness. The utilization of this method greatly simplifies the analysis of such complex problems. Multispan continuous highway bridges with variable stiffness girders are discussed in later sections of this chapter.

The following example illustrates the application of the method.

Example 4.3: The single-span beam in Fig. 4.5a is a girder of an existing highway bridge in the state of Michigan. It is a reinforced concrete girder of parabolic variation in depth, and has the dimensions shown in the figure. The distributed weight on the girder includes the dead weight of the girder and the associated dead weight of the bridge deck. The concentrated weights represent a standard H-S truck, in accordance with AASHTO standard specifications, and they are located as shown. By considering the moment of inertia I at the center of the girder as reference, the variation of $f(x)$ is as shown in Fig. 4.5b. By using the approximate method of the equivalent systems discussed in Section 4.2.2, determine an accurate equivalent system of constant stiffness EI_1 that is loaded with equivalent concentrated loads. By using the equivalent system, determine the deflection of the girder at a distance $x = 31$ ft from support A. The modulus E is constant and $EI_1 = 460 \times 10^6$ kip-in.^{2}

TABLE 4.1. Values of M_x, $f(x)$, and M_e at Various Values of x from Support A

(1) x (ft)	(2) M_x (kip-ft)	(3) $f(x)$	(4) $M_e = M_x/f(x)$ (kip-ft)
0	0	5.690	0
7.25	447.51	3.175	140.95
14.50	811.38	1.774	457.37
21.75	1,048.24	1.165	899.78
29.00	1,216.80	1.000	1,216.80
36.75	1,090.29	1.165	935.87
43.50	899.39	1.774	506.98
50.75	579.66	3.175	182.57
58.00	0	5.690	0

SOLUTION: Starting from end A in Fig. 4.5a, the bending moments M_x of the original variable stiffness girder are calculated at various locations x by using statics. The values of x with origin at A are shown in the first column of Table 4.1. The second column of the table shows the calculated values of M_x corresponding to the indicated locations x, and the third column gives the values of the function $f(x)$ at the same points. Dividing the values of M_x by the corresponding values of $f(x)$, we obtain the values of the moment M_e of the equivalent system of constant stiffness EI_1. They are shown in the fourth column of the table.

The exact moment diagram M_e of the equivalent system of constant stiffness EI_1 is shown plotted by the solid line in Fig. 4.6a. We approximate the shape of M_e with five straight-line segments, as shown by the dashed lines in the same figure. Note that the juncture points of the segments are located above or below the curve of the exact M_e, so that the areas added to the exact M_e or subtracted from it are approximately balanced. It is sufficient for practical purposes to do this balancing by inspection, because even rather crude approximations yield sufficiently accurate results for practical applications. The approximation of the shape of the exact M_e shown in Fig. 4.6a is excellent, and the error in deflection should be less than one percent.

The shear force diagram corresponding to the approximated M_e is shown in Fig. 4.6b. The 22.73-kip value of the shear force, for example, is obtained by dividing the moment of 250 kip-ft by 11 ft, and so on. The equivalent system of constant stiffness EI_1 is shown in Fig. 4.6c. The first concentrated load of 36.65 kips represents the difference of $(59.38 - 22.73)$ kips in shear shown in Fig. 4.6b, and so on. The deflections and rotation of the elastic line of the equivalent system should be closely identical to the corresponding ones of the initial system in Fig. 4.5a. The solution, however, of the equivalent system is much simpler. In fact, the formula given in Fig. A.1d of Appendix A may be

EQUIVALENT SYSTEMS FOR HIGHWAY BRIDGES   233

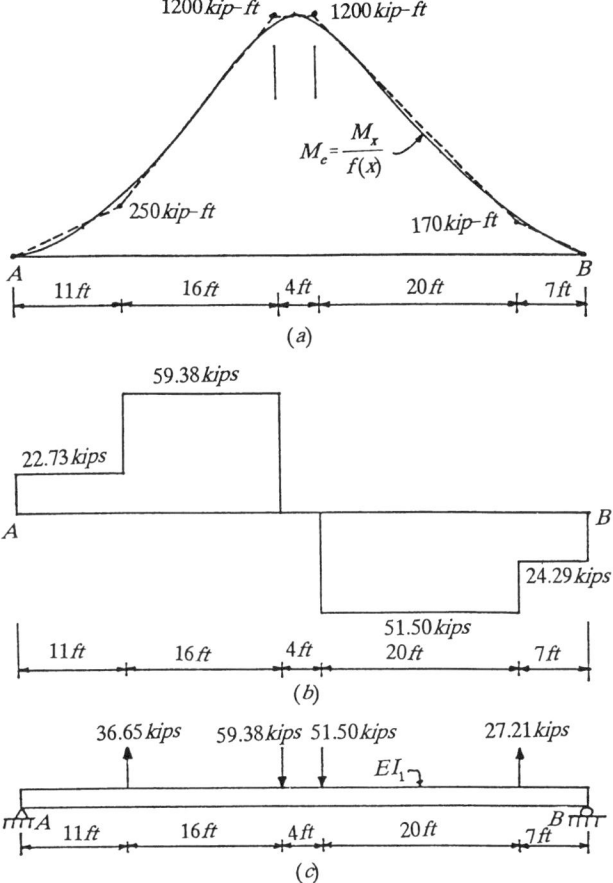

Figure 4.6. (a) Bending moment diagram M_e of the equivalent system approximated with five straight-line segments. (b) Equivalent shear force diagram. (c) Equivalent system of constant stiffness EI_1.

used to determine the vertical deflection of the equivalent system at any point along its length. At the distance $x = 31$ ft from support A, the application of this expression for the four concentrated loads yields

$$y_{x=31\,\text{ft}} = \frac{12^3}{(6)(460)(10)^6(58)} \{-(36.25)(11)(27)[58^2 - 11^2 - 27^2]$$

$$+ (59.38)(27)(27)[58^2 - 27^2 - 27^2]$$

$$+ (2)(51.50)(31)^2(27)^2 - (27.21)(7)(31)[58^2 - 7^2 - 31^2]\}$$

$$= -0.292171 + 0.890630 + 0.778923 - 0.150038$$

$$= 1.227344 \text{ in.}$$

This value should be in very close agreement with the value that would be obtained if the exact solution was used. An exact solution, however, would be much more difficult to obtain.

The vertical deflections at other points of the equivalent system may be obtained in a similar manner. Similar procedures may be also used for the solution of other challenging highway bridge problems.

4.4 STATICALLY INDETERMINATE VARIABLE STIFFNESS MEMBERS

When a member, or a bridge girder, of variable stiffness, is statically indeterminate, the method of the equivalent systems may be also used to carry out the required analysis. Consider for example, the statically indeterminate beam in Fig. 4.7a, and let it be assumed that it is required to determine an equivalent system of constant stiffness by applying the approximate method of the equivalent systems discussed in Sections 4.2.2 and 4.3. The solution is initiated by selecting the reaction R_A and the bending moment M_A at the end A as the redundants, as shown in Fig. 4.7b. We assume that the modulus function $g(x) = 1$ and that the moment of inertia function $f(x)$ has the arbitrary variation shown in Fig. 4.7c.

The actual moment diagrams of the cantilever beam in Fig. 4.7b, by parts, is shown in Fig. 4.7d. The exact equivalent moment diagram, by parts, is derived in the same manner as for statically determinate beams by using Eq. (4.10). It is shown by the solid lines in Fig. 4.8a. There are three equivalent moment diagrams shown in this figure; the one produced by the reaction R_A, one by the moment M_A, and the third one is the result of the applied loading on the member. The approximation of their shape with straight-line segments is shown by the dashed lines in the same figure.

For each approximated equivalent moment diagram shown in Fig. 4.8a, an equivalent system of uniform stiffness loaded with concentrated loads may be derived, as discussed in Section 4.2.2, and the results are assumed to be as shown in Figs. 4.8b, 4.8c, and 4.8d. The equivalent system in Fig. 4.8b corresponds to the approximated equivalent moment diagram $R_A x/f(x)$ in Fig. 4.8a, the one in Fig. 4.8c corresponds to $M_A/f(x)$, and the one in Fig. 4.8d corresponds to $M_{\text{Load}}/f(x)$.

The redundants R_A and M_A may be computed by utilizing the equivalent systems in Figs. 4.8b, 4.8c, and 4.8d and applying known methods of mechanics of solids, or existing formulas if such formulas are available. For example, if the deflections at end A in Figs. 4.8b, 4.8c, and 4.8d are designated, respectively, as δ'_A, δ''_A, and δ'''_A, and the rotations at the same end as θ'_A, θ''_A, and θ'''_A, then the boundary conditions at end A of the initial system in Fig. 4.7a will be satisfied if

$$\delta'_A + \delta''_A + \delta'''_A = 0 \tag{4.21}$$

STATICALLY INDETERMINATE VARIABLE STIFFNESS MEMBERS   235

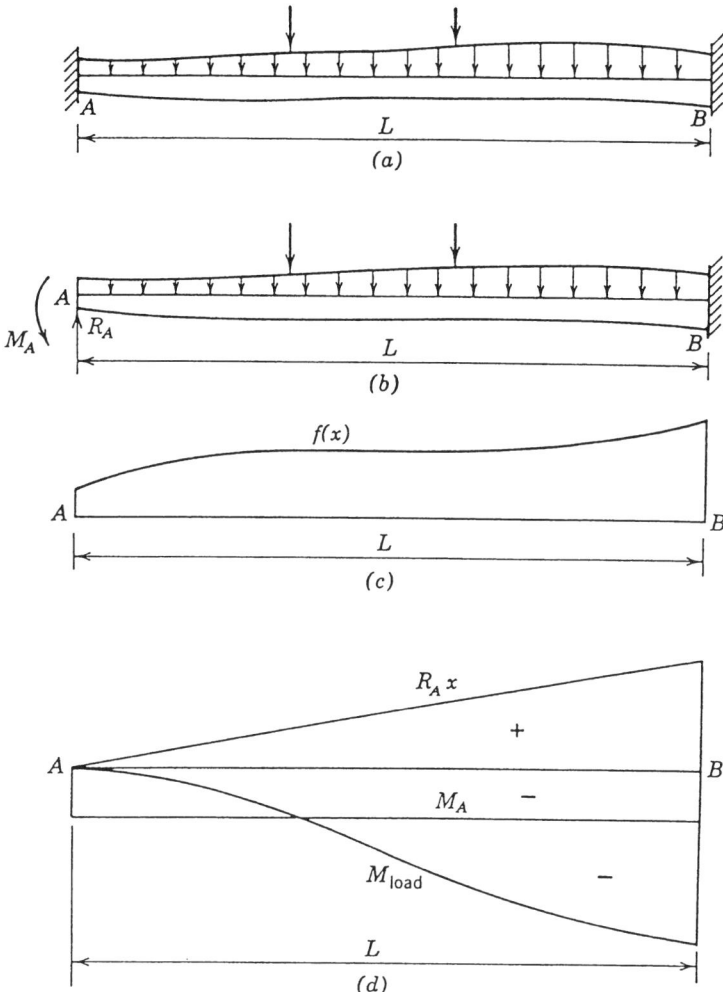

Figure 4.7. (*a*) Original statically indeterminate variable stiffness member. (*b*) Cantilever beam loaded with M_A, R_A, and applied loading. (*c*) Variation of the moment of inertia function $f(x)$. (*d*) Actual moment diagram by parts.

and

$$\theta'_A + \theta''_A + \theta'''_A = 0 \qquad (4.22)$$

Equations (4.21) and (4.22) are functions of R_A and M_A. Their simultaneous solution will yield the values of the reaction R_A and moment M_A at the fixed end A of the original system in Fig. 4.7*a*. With known redundants R_A and M_A, the values of the actual moments M_x at cross sections along the length of the

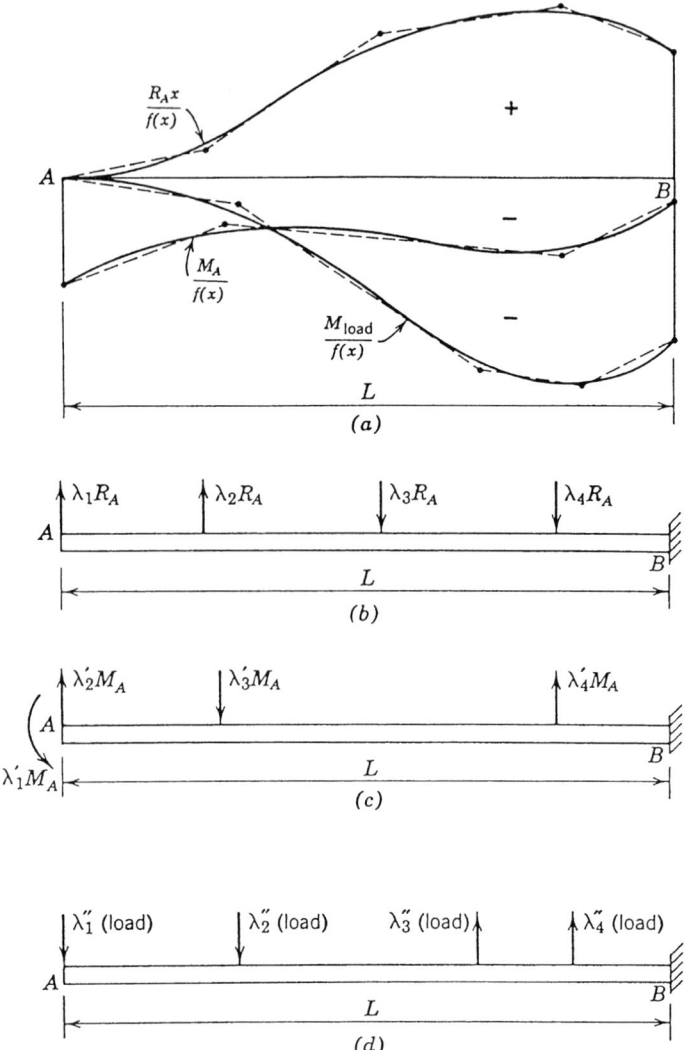

Figure 4.8. (a) Equivalent moment diagram approximated with straight-line segments. (b) Equivalent system for R_A. (c) Equivalent system for M_A. (d) Equivalent system for applied loading.

member in Fig. 4.7a can be computed by applying simple statics. Thus, by using the approximate method of the equivalent systems, an equivalent system of constant stiffness can be derived, as shown in Fig. 4.9. The equivalent system in Fig. 4.9d has the same length and boundary conditions as the original system in Fig. 4.7a, and approximately identical elastic lines. Thus, the deflections and rotations of any point on the elastic line of the variable stiffness member can

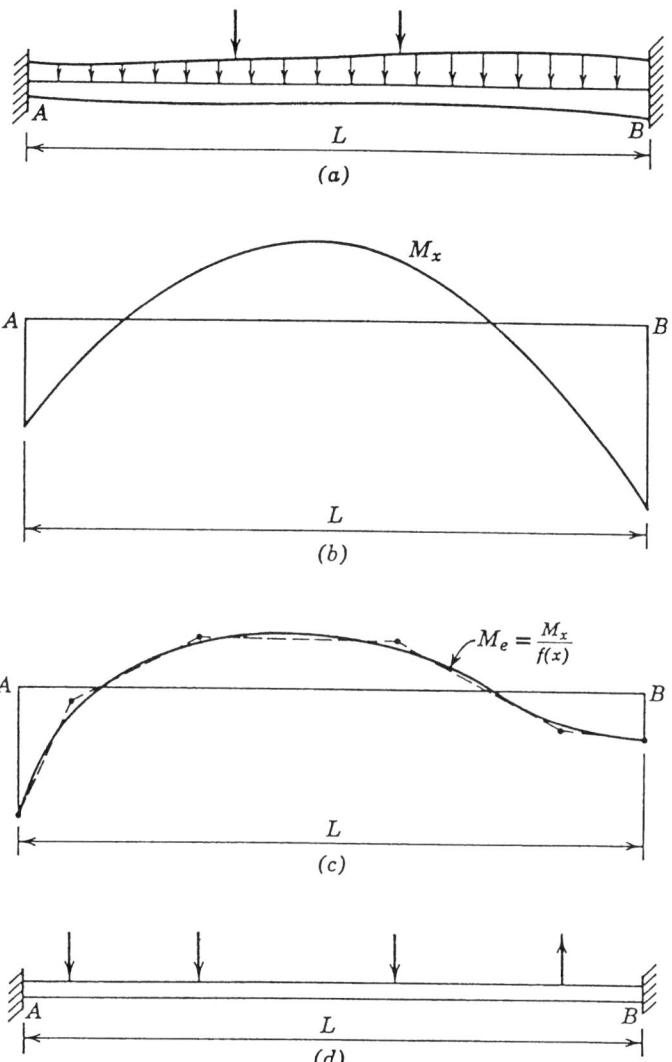

Figure 4.9. (*a*) Original variable stiffness system. (*b*) Moment diagram M_x of the original system. (*c*) Moment diagram M_e of the equivalent system with its shape approximated with straight-line segments. (*d*) Equivalent system of constant stiffness.

be computed by using the equivalent system in Fig. 4.9*d* and applying known methods of mechanics of solids. If preferred, the deflections and rotations of the original system may be determined by using the equivalent systems in Figs. 4.8*b*, 4.8*c*, and 4.8*d*, and superimposing the results.

The following examples illustrate the application of the preceding methodology.

Example 4.4: The depth of the fixed-fixed beam in Fig. 4.10a varies parabolically, and the applied loading on the beam is as shown. Its modulus of elasticity E is constant, and the moment of inertia is smallest at the center of the beam, and is equal to I_1. Determine an approximate equivalent system of constant stiffness EI_1, when $EI_1 = 460 \times 10^6$ kip-in.^{2} The variation $f(x)$ of I_x, with reference value I_1, is shown in Fig. 4.10b.

SOLUTION: By considering the reaction R_A and the moment M_A at end A as the redundants, Fig. 4.10c, the actual moment diagrams due to R_A, M_A, and

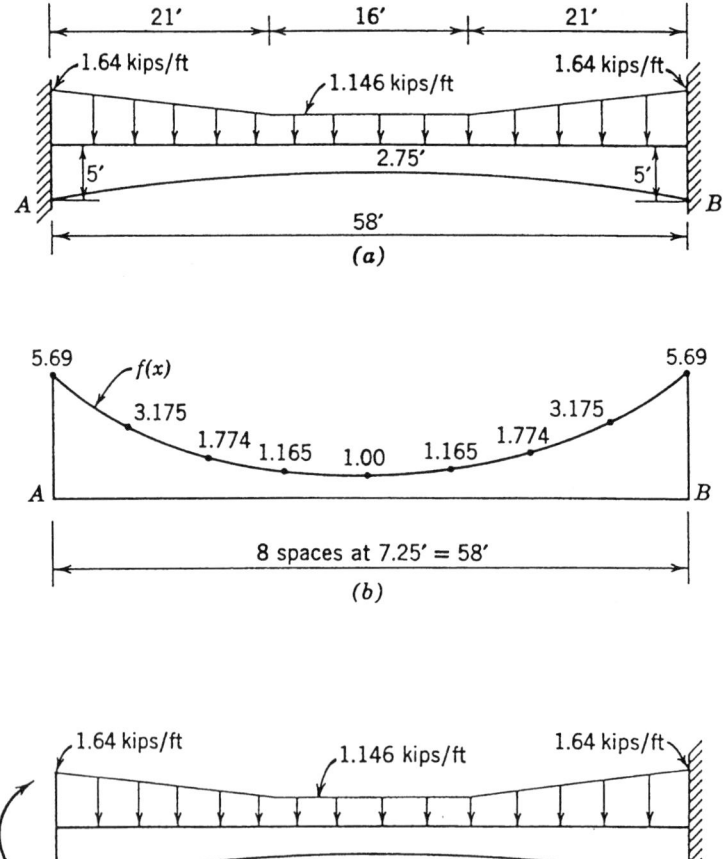

Figure 4.10. (a) Original statically indeterminate variable stiffness member. (b) Variation of moment of inertia function $f(x)$. (c) Cantilever beam loaded with the redundants R_A and M_A, and the applied loading.

STATICALLY INDETERMINATE VARIABLE STIFFNESS MEMBERS 239

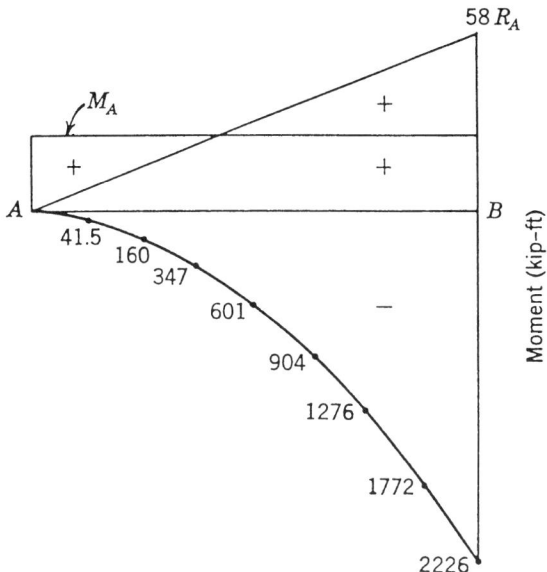

Figure 4.11. Moment diagram of the cantilever beam in Fig. 4.10c by parts.

distributed loading are shown in Fig. 4.11. It is the actual moment diagram of the cantilever variable stiffness beam, by parts, in Fig. 4.10c. When the ordinates of this diagram are divided by the corresponding values of $f(x)$ in Fig. 4.10b and their shapes are approximated with straight-line segments, the equivalent systems of uniform stiffness EI_1 for R_A, M_A, and distributed loading can be determined as discussed in Section 4.2.2. They are as shown in Figs. 4.12a, 4.12b, and 4.12c, respectively.

The deflections δ_1, δ_2, and δ_3 at end A in Figs. 4.12a, 4.12b, and 4.12c, respectively, and the corresponding rotations θ_1, θ_2, and θ_3, at the same end, are determined by using the equivalent systems and applying the moment-area method. Known formulas from Appendix A and available handbooks may be also used for this purpose; they are as follows:

$$EI_1\delta_1 = 34{,}244 R_A \qquad EI_1\theta_1 = 988.4 R_A$$
$$EI_1\delta_2 = 986.5 M_A \qquad EI_1\theta_2 = 33.94 M_A$$
$$EI_1\delta_3 = -898{,}000 \qquad EI_1\theta_3 = -23{,}470$$

The boundary conditions at end A of the original system in Fig. 4.10a are

$$\delta = \delta_1 + \delta_2 + \delta_3 = 0$$
$$\theta = \theta_1 + \theta_2 + \theta_3 = 0$$

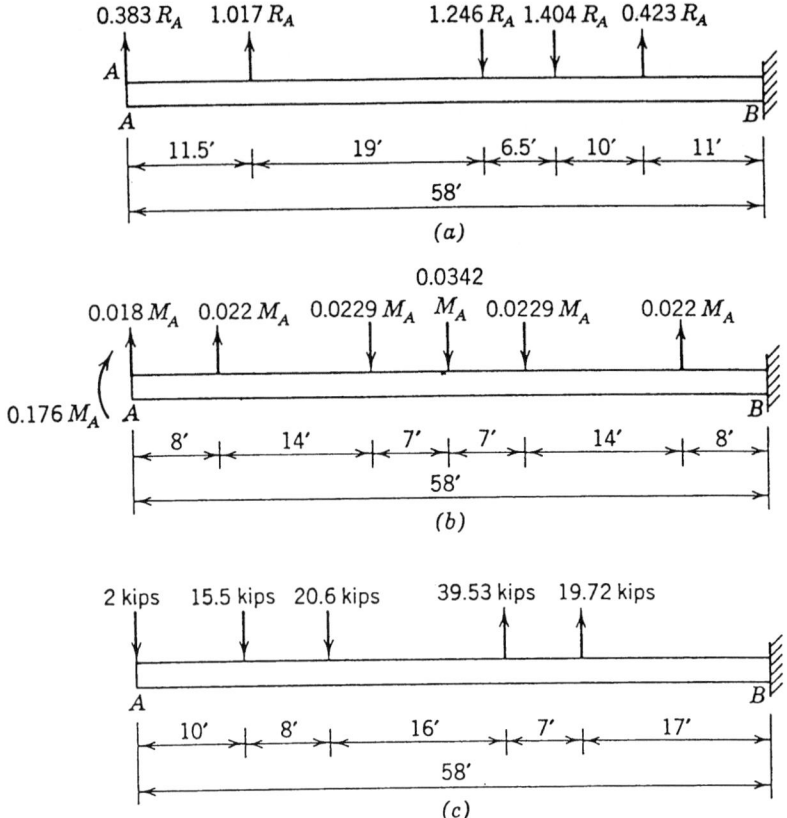

Figure 4.12. (a) Equivalent system for R_A. (b) Equivalent system for M_A. (c) Equivalent system for applied loading.

or, by substitution,

$$34{,}244 R_A + 986.5 M_A - 988{,}000 = 0$$
$$988.4 R_A + 33.94 M_A - 23{,}470 = 0$$

Simultaneous solution of the preceding two equations yields

$$R_A = 38.4 \text{ kips} \uparrow \qquad (4.23)$$

$$M_A = 427.0 \text{ kip-ft} \quad \text{(counterclockwise)} \qquad (4.24)$$

which are the actual reaction and moment at end A of the variable stiffness beam in Fig. 4.10a.

With known R_A and M_A, the values of the moments at cross sections along the length of the member in Fig. 4.10a can be obtained by applying simple

STATICALLY INDETERMINATE VARIABLE STIFFNESS MEMBERS   241

statics. When these moments are divided by the corresponding values of $f(x)$, the equivalent moments $M_e = M_x/f(x)$ of the equivalent system of constant stiffness EI_1 are determined and they are shown plotted by the solid line in Fig. 4.13a. In the same figure, the approximation of the shape of the M_e diagram with straight-line segments is also shown by the dashed line. The equivalent system of constant stiffness EI_1 is derived in the usual way, and it is shown in Fig. 4.13c. The deflections and rotations along the length of the equivalent system accurately approximate the corresponding values of the original system in Fig. 4.10a. If preferred, the deflections and rotations of the original system may be determined by solving the equivalent systems in Figs. 4.12a, 4.12b, and 4.12c, and superimposing the results.

Example 4.5: A statically indeterminate beam of variable stiffness is fixed at end B and elastically supported by a linear spring of constant $k_A = 400$ kips/in., as shown in Fig. 4.14a. Its uniformly distributed load $w = 100$ lb/in., its width is constant and equal to b, and its variable depth d_x at any distance x from support A is given by the expression

$$d_x = \left(\frac{L + 3x}{L}\right)^{1/3} d_1$$

By applying the approximate method of the equivalent systems, determine an approximate equivalent system of constant stiffness EI_1, where E is constant, and I_1 is the moment of inertia of the member at the elastic support A. The value of $EI_1 = 90 \times 10^6$ kip-in.^{2}

SOLUTION: If the fixed end moment M_B at end B is selected as the redundant, the problem to be solved is the variable stiffness elastically supported beam shown in Fig. 4.14b, which is loaded with the redundant M_B at end B and the applied loading $w = 100$ lb/in. In order to obtain a solution, we first apply the distributed load $w = 100$ lb/in., as shown in Fig. 4.15a, and proceed with the approximate method of the equivalent systems to determine the equivalent system of constant stiffness EI_1 shown in Fig. 4.15d.

The values of the actual moment M_x at various distances x from support A are shown in the third column of Table 4.2, and $I_x = f(x)I_1$ and M_e are shown in columns two and four, respectively, of the same table. The exact M_e is plotted as shown by the solid line in Fig. 4.15b. Note that $g(x) = 1$, because E is constant. The approximation of M_e with five straight-line segments, as shown by the dashed lines, leads to the equivalent system of constant stiffness EI_1 shown in Fig. 4.15d.

The value of the spring constant k_{A_e} of the equivalent system may be determined by satisfying the condition of equal deflection at the elastic support A for both the original and the equivalent systems. The deflection δ_A at support

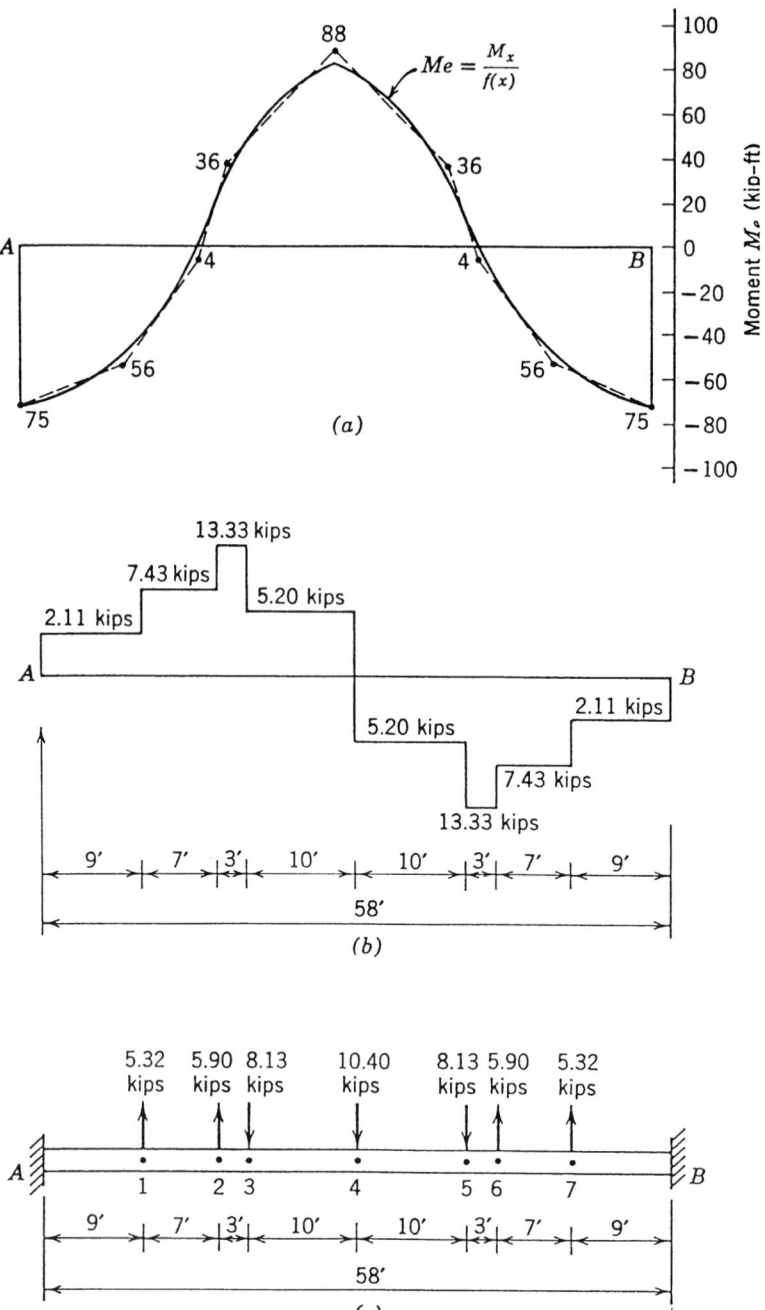

Figure 4.13. (a) Moment diagram M_e of the equivalent system of constant stiffness EI_1 approximated with straight-line segments. (b) Shear force diagram of the equivalent system. (c) Equivalent system of constant stiffness EI_1.

STATICALLY INDETERMINATE VARIABLE STIFFNESS MEMBERS   243

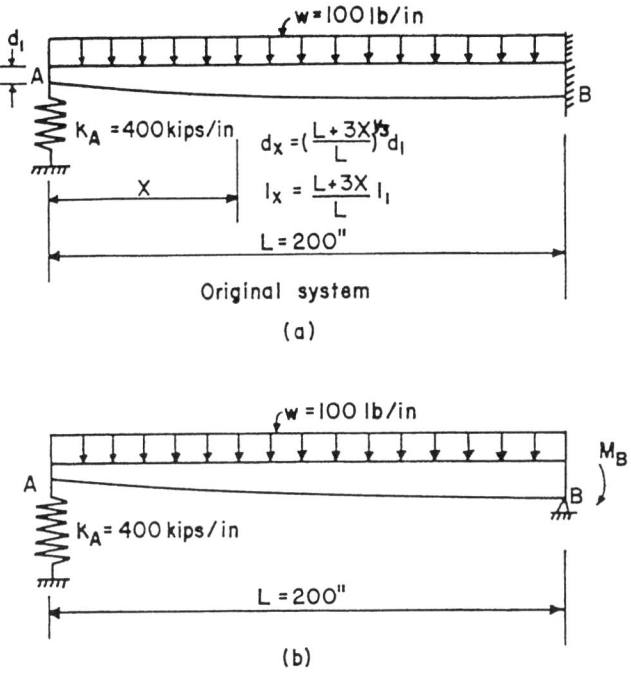

Figure 4.14. (*a*) Statically indeterminate elastically supported beam of variable stiffness. (*b*) Simply supported variable stiffness member loaded with the applied load *w* and the redundant moment M_B.

A of the original system in Fig. 4.15*a* is

$$\delta_A = \frac{X_A}{k_A} \tag{4.25}$$

where X_A is the elastic reaction at the end *A*.

The deflection δ_{A_e} of the equivalent system at end *A*, Fig. 4.15*d*, is

$$\delta_{A_e} = \frac{X_{A_e}}{k_{A_e}} \tag{4.26}$$

where X_{A_e} is the elastic reaction at end *A*. Thus, with

$$\delta_A = \delta_{A_e} \tag{4.27}$$

we find

$$\frac{X_A}{k_A} = \frac{X_{A_e}}{k_{A_e}} \tag{4.28}$$

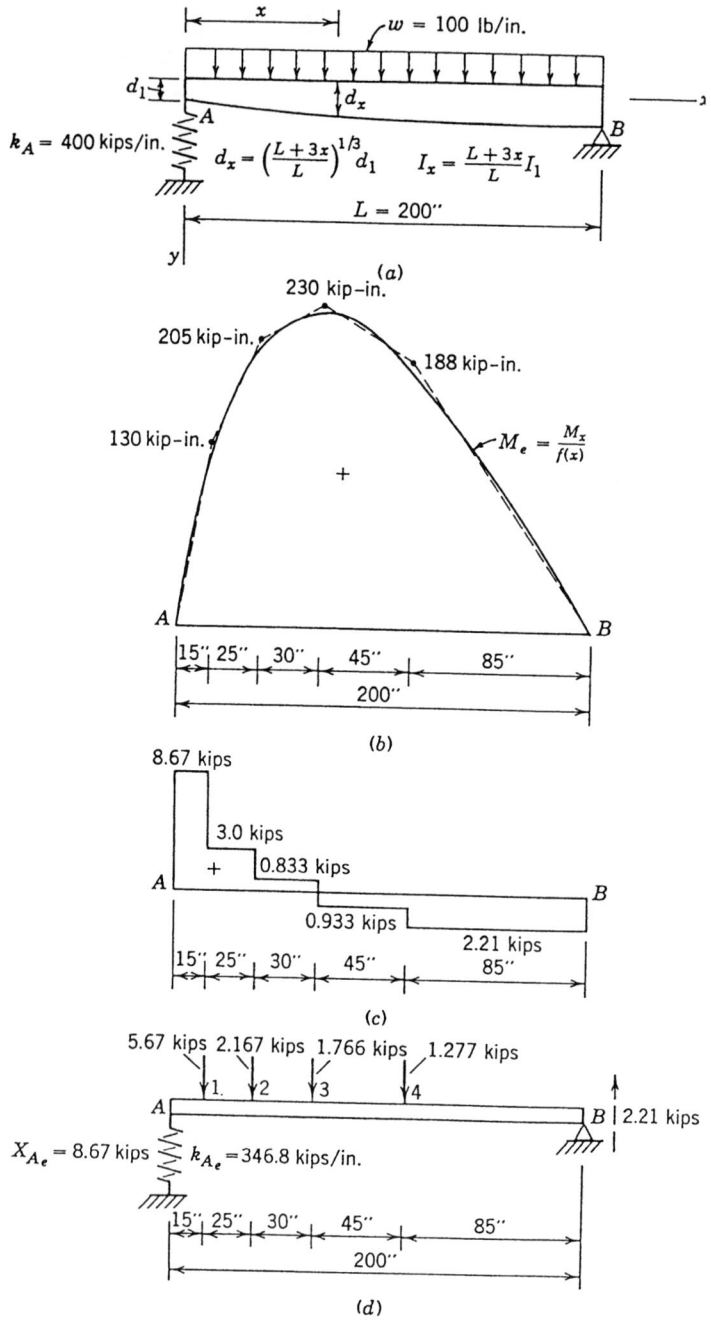

Figure 4.15. (*a*) Elastically supported variable stiffness beam loaded with the applied loading *w*. (*b*) M_e diagram with its shape approximated with five straight-line segments. (*c*) Equivalent shear force diagram. (*d*) Equivalent system of constant stiffness EI_1.

TABLE 4.2. Values of I_x, M_x, and M_e at Various Values of x Measured from Support A

(1) x (in.)	(2) I_x	(3) M_x (kip-in.)	(4) $M_e = M_x/f(x)$ (kip-in.)
0	I_1	0	0
25	$1.375I_1$	218.8	159.0
50	$1.750I_1$	375.0	214.0
75	$2.125I_1$	468.0	221.0
100	$2.500I_1$	500.0	320.0
125	$2.875I_1$	468.0	163.0
150	$3.250I_1$	375.0	115.0
175	$3.625I_1$	218.0	60.4
200	$4.000I_1$	0	0

or

$$k_{A_e} = k_A \frac{X_{A_e}}{X_A} \tag{4.29}$$

By applying statics, we find $X_A = 10$ kips and $X_{A_e} = 8.67$ kips. Thus,

$$k_{A_e} = k_A \frac{X_{A_e}}{X_A} = 400 \frac{(8.67)}{10}$$

$$= 346.8 \text{ kips/in.}$$

On this basis, the elastic line of the equivalent system in Fig. 4.15d would be closely identical to the elastic line of the original system in Fig. 4.15a.

We apply now the redundant moment M_B at end B, as shown in Fig. 4.16a. If we apply the approximate method of the equivalent systems as in the preceding case, the M_e diagram with solid line and its shape approximation with four straight-line segments with dashed lines are shown in Fig. 4.16b. This leads to the equivalent system of constant stiffness EI_1 shown in Fig. 4.16d. In this case, $X_A = 0.0050M_B$ and $X_{A_e} = 0.00367M_B$. Thus,

$$k_{A_e} = k_A \frac{X_{A_e}}{X_A} = (400) \frac{(0.00367M_B)}{(0.0050M_B)}$$

$$= 293.5 \text{ kips/in.}$$

If the symbols θ'_B and θ''_B are used to denote the rotations at end B in Figs. 4.15d and 4.16d, respectively, the total rotation θ_B at end B of the original

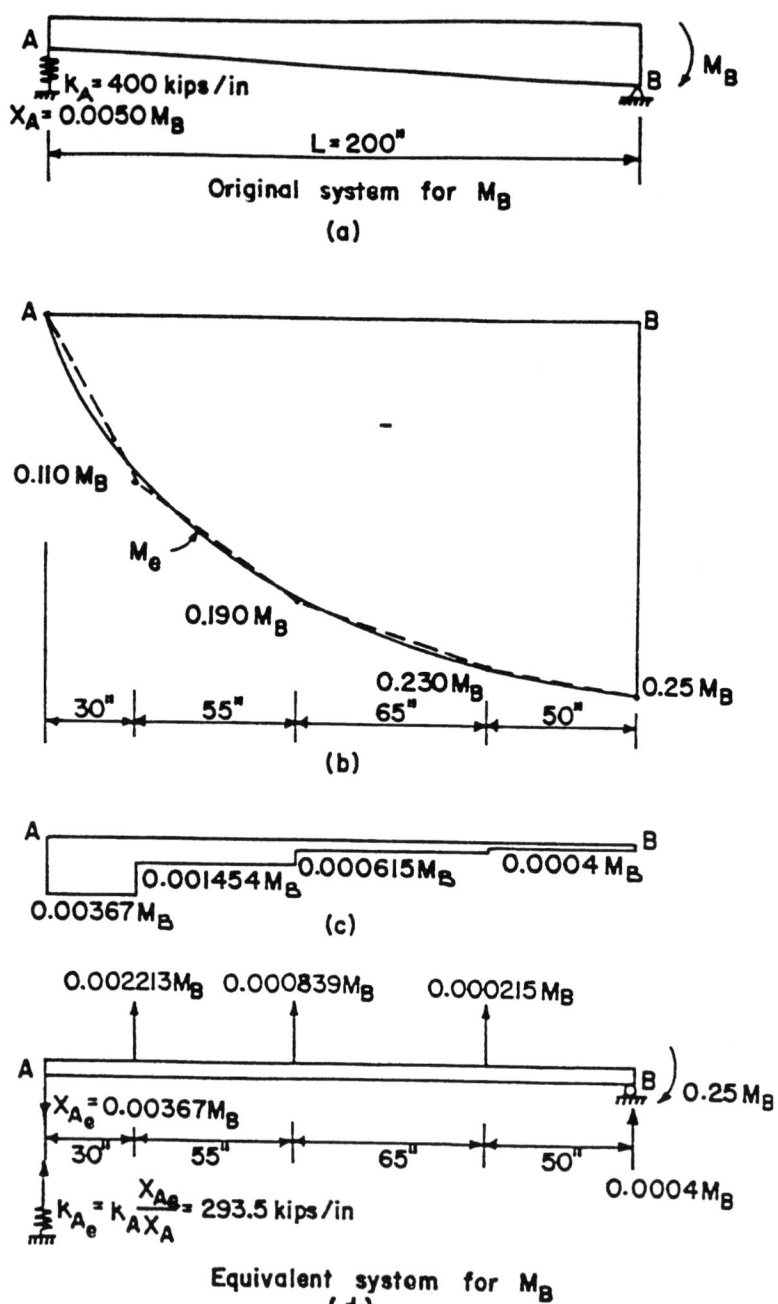

Figure 4.16. (*a*) Elastically supported variable stiffness beam loaded with the redundant moment M_B at B. (*b*) M_e diagram with its shape approximated with four straight-line segments. (*c*) Equivalent shear force diagram. (*d*) Equivalent system of constant stiffness EI_1.

system in Fig. 4.14a is

$$\theta_B = \theta'_B + \theta''_B = 0 \qquad (4.30)$$

The rotations θ'_B and θ''_B may be determined by using known methods of solid mechanics, or available formulas. They are as follows:

$$\theta'_B = \frac{0.25}{L} + \frac{10^6}{EI_1 L} \quad (2.1935)$$

$$\theta''_B = -\frac{0.0000125 M_B}{L} - \frac{M_B}{EI_1 L} \quad (4249.2)$$

By substituting into Eq. (4.30), we obtain the following equation in terms of M_B:

$$0.049372 - 0.000059713 M_B = 0 \qquad (4.31)$$

The solution of Eq. (4.31) yields

$$M_B = 827.0 \text{ kip-in.}$$

With known M_B, the elastic reaction X_A at support A of the original system in Fig. 4.14a may be determined by using simple statics, yielding

$$X_A = 5.865 \text{ kips}$$

The equivalent system of constant stiffness EI_1, which will have approximately the same elastic line as the original system in Fig. 4.14a, may now be determined by applying the approximate method of the equivalent systems in the usual way. The value of I_x, M_x, and M_e at various locations x from end support A are shown in Table 4.3. The equivalent moment diagram M_e is shown plotted by the solid line in Fig. 4.17a. The approximation of its shape, as shown by the dashed lines in the same figure, leads to the equivalent system of constant stiffness EI_1 shown in Fig. 4.17c. The equivalent spring constant k_{A_e} is

$$k_{A_e} = k_A \frac{X_{A_e}}{X_A} = (400) \frac{(9.00)}{(5.865)}$$

$$= 613.81 \text{ kips/in.}$$

The equivalent system in Fig. 4.17c has approximately the same elastic line as the original variable stiffness member in Fig. 4.14a. Thus, the equivalent system may be used to determine deflections and rotations of the original system. It is also correct, if preferred, to solve the equivalent systems in Figs. 4.15d and 4.16d and superimpose the results.

TABLE 4.3. Values of I_x, M_x, and M_e at Various Values of x Measured from Support A

(1) x (in.)	(2) I_x	(3) M_x (kip-in.)	(4) $M_e = M_x/f(x)$ (kip-in.)
0	I_1	0	0
25	$1.375I_1$	115.4	85.5
50	$1.750I_1$	168.5	96.3
75	$2.124I_1$	158.0	74.3
100	$2.500I_1$	86.5	34.6
125	$2.875I_1$	−49.0	−17.1
150	$3.250I_1$	−245.0	−75.3
175	$3.625I_1$	−504.2	−139.0
200	$4.000I_1$	−827.0	−207.0

4.5 MOMENT-AREA AND CONJUGATE-BEAM METHODS

In this section, a brief discussion on the very popular methods of moment area and conjugate beam is provided. Extensive discussion on the moment-area method may be also found in basic text dealing with mechanics of solids [38]. Both methods have been extensively used by practicing engineers for many centuries for the solution of beam-like infrastructural elements.

4.5.1 Moment-Area Method

The moment-area method is used extensively for the determination of deflections and rotations of straight members, and is based on the assumptions that the material of the member is continuous, homogeneous and isotropic, and that its elastic limit is not exceeded. Its application is limited to small deflections, about a principal axis, produced by bending only. The effect of shear on the deflection of the member may be neglected, provided that its length-to-depth ratio is large. If the member is curved, straight beam theory will provide sufficiently accurate results if the ratio of the radius of curvature of the member to its depth at any cross section is large. The method is also applicable to both statically determinate and statically indeterminate straight members. The development of the method is based on the following two theorems, or propositions:

Theorem 1: The first theorem states that the angle between two lines drawn tangent to an elastic curve at different points is equal, in radians, to the area of the M/EI diagram between the two points.

If we consider two points at distances x and $(x + dx)$ from the origin O of the elastic curve in Fig. 4.18a, the incremental angle $d\theta$ between the two

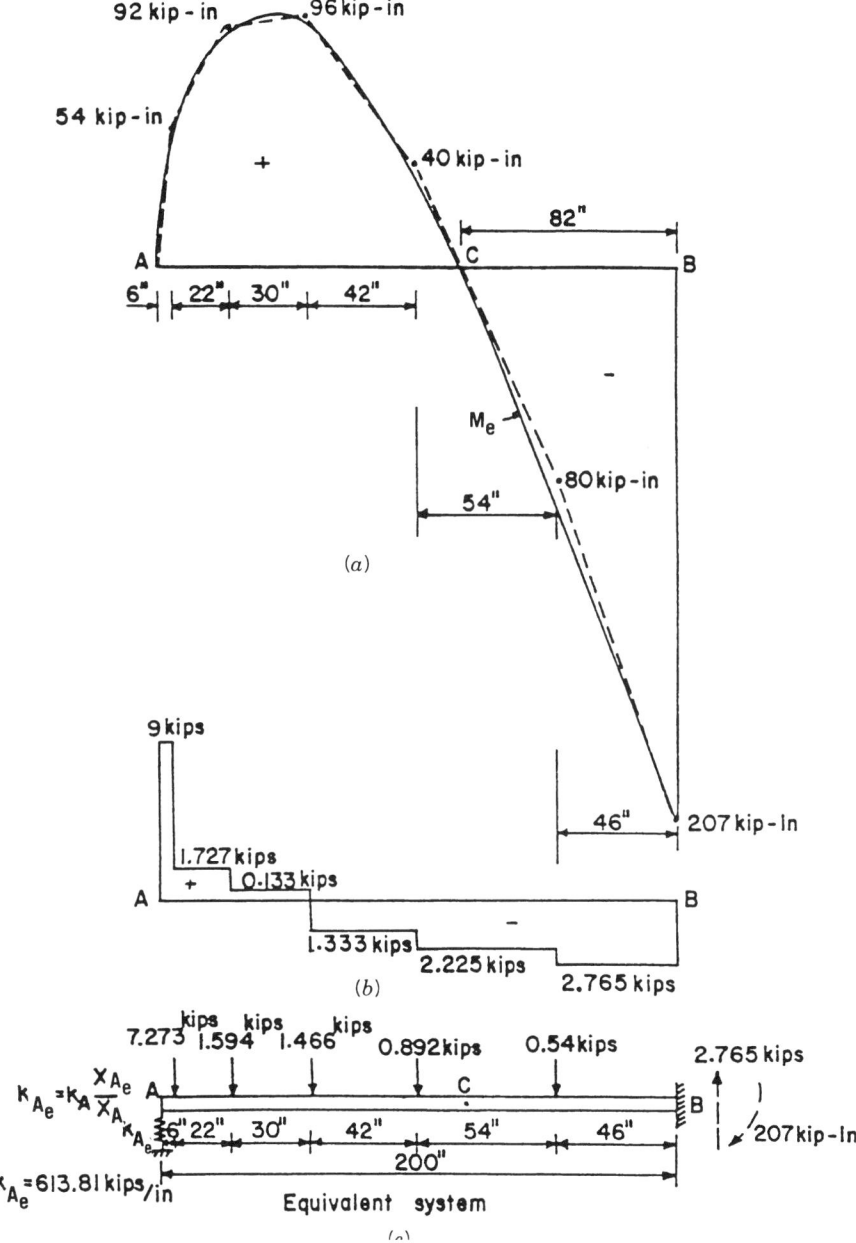

Figure 4.17. (*a*) Moment diagram M_e with its shape approximated with straight-line segments. (*b*) Equivalent shear force diagram. (*c*) Equivalent system of constant stiffness EI_1.

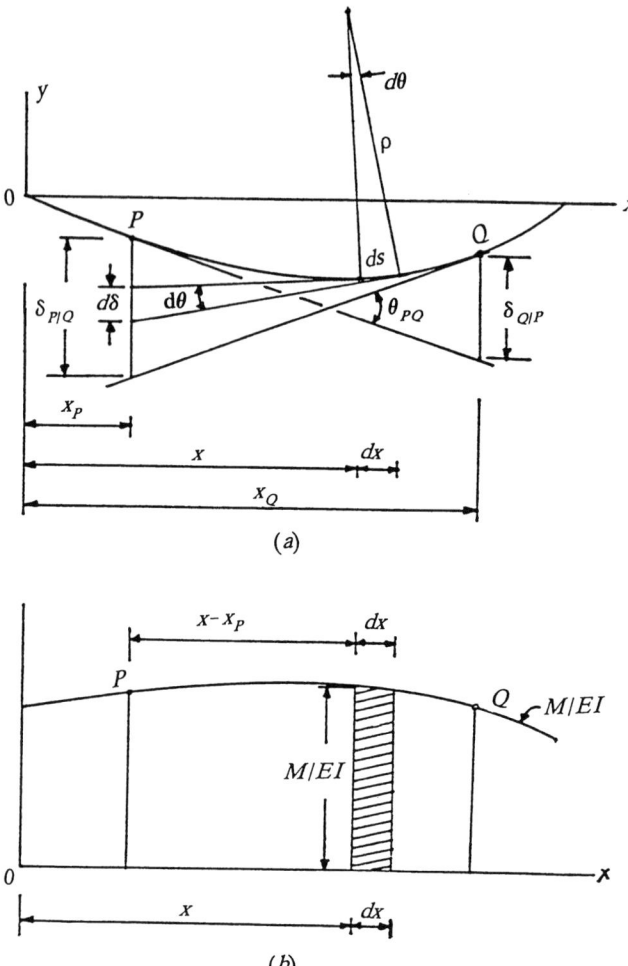

Figure 4.18. (a) Elastic curve of a straight member. (b) M/EI diagram of the member.

tangents drawn as shown at these two points of the elastic curve is equal to the shaded area of the M/EI diagram shown in Fig. 4.18b. Mathematically, it is expressed as follows:

$$d\theta = \frac{M d_x}{EI} \tag{4.32}$$

The angle θ_{PQ}, in radians, between the two tangents drawn at points P and Q in Fig. 4.18a is

$$\theta_{PQ} = \int_{x_P}^{x_Q} \frac{M d_x}{EI} \tag{4.33}$$

MOMENT-AREA AND CONJUGATE-BEAM METHODS   251

The area represented by the integral on the right-hand side of Eq. (4.33) is the M/EI area between points P and Q of the M/EI diagram in Fig. 4.18b. Thus, Eq. (4.33) represents the first theorem of the moment-area method.

Theorem 2: The second moment-area theorem states that the deflection of any point P on the elastic curve, with respect to a tangent drawn at point Q, is equal to the first moment of the M/EI area between points P and Q of the M/EI diagram, taken about P.

In Fig. 4.18a, the incremental contribution $d\delta$ to the deflection of point P resulting from the angle $d\theta$ between tangents to the deflection curve drawn at points of distances x and $(x + dx)$ from the origin O, is

$$d\delta = (x - x_p)d\theta \qquad (4.34)$$

By substituting Eq. (4.32) into Eq. (4.34), we obtain

$$d\delta = \frac{(x - x_p)M\,dx}{EI} \qquad (4.35)$$

In Eqs. (4.34) and (4.35), the distance $(x - x_p)$ is the moment arm of the incremental M/EI area shown by the shaded area in Fig. 4.18b. Thus, the deflection $\delta_{P/Q}$ of point P on the elastic curve, with respect to the tangent drawn at point Q, may be expressed as follows:

$$\delta_{P/Q} = \int_{x_P}^{x_Q} \frac{(x - x_p)M\,dx}{EI} \qquad (4.36)$$

Equation (4.36) represents the statement expressed by the second moment-area theorem. In a similar manner, the deflection $\delta_{Q/P}$ of point Q on the elastic curve, with respect to the tangent drawn at point P, is

$$\delta_{Q/P} = \int_{x_Q}^{x_P} \frac{(x_Q - x)M\,dx}{EI} \qquad (4.37)$$

The utilization of the preceding two theorems regarding the determination of actual deflections and rotations of statically determinate and statically indeterminate members is illustrated by the following practical examples.

Example 4.6: The beam shown in Fig. 4.19a is a reinforced concrete girder of a highway bridge, and it is loaded by a standard H-S truck with the axle loads located as shown in the same figure. By using the moment-area method, determine the rotations θ_A, θ_C, and θ_D at points A, C, and D, respectively, of the member, and the vertical deflections y_C and y_D at points C and D, respectively. The stiffness EI is constant throughout the length of the girder, and it is equal to 460×10^6 kip-in.^{2}

252 ADVANCED METHODS AND PROBLEMS OF INFRASTRUCTURAL ANALYSIS

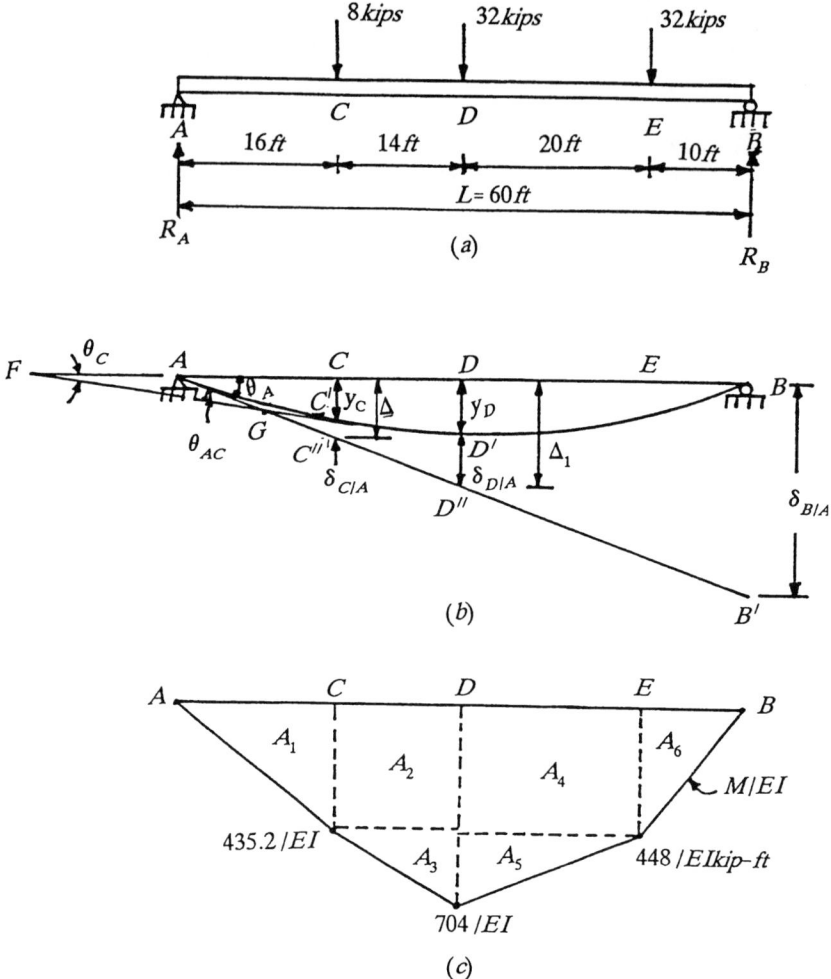

Figure 4.19. (a) Concrete girder of a highway bridge loaded as shown. (b) Deflection curve of the girder. (c) M/EI diagram.

SOLUTION: The first steps in applying the moment-area method are to draw the correct shape of the deflection curve and to plot the M/EI diagram of the girder. The correct shape of the deflection curve is shown in Fig. 4.19b. The bending moments at cross sections along the length of the girder may be determined by using simple statics. When these moments are divided by the constant stiffness EI, we obtain the M/EI diagram shown plotted in Fig. 4.19c.

By drawing a tangent at end A, as shown in Fig. 4.19b, the vertical distance $\delta_{B/A}$ of end B from this tangent may be calculated by using the second moment-area theorem. In accordance with this theorem, $\delta_{B/A}$ is equal to the first moment of the M/EI area between points A and B in Fig. 4.19c, taken about B. By subdividing this area into rectangles and triangles as shown in the figure and taking their first moments about point B, we have

$$\delta_{B/A} = \frac{1}{2}(16)\left(\frac{435.2}{EI}\right)\left[60 - \frac{(2)(16)}{3}\right] + (14)\left(\frac{435.2}{EI}\right)(37)$$

$$+ \frac{1}{2}(14)\left(\frac{704.0 - 435.2}{EI}\right)\left(30 + \frac{14}{3}\right) + (20)\left(\frac{448.0}{EI}\right)(20)$$

$$+ \frac{1}{2}(20)\left(\frac{704.0 - 448.0}{EI}\right)\left[10 + \frac{(2)(20)}{3}\right] + \frac{1}{2}(10)\left(\frac{448.0}{EI}\right)\left[\frac{(2)(20)}{3}\right]$$

$$= \frac{719{,}228.64}{EI}$$

The units of EI in this equation and in subsequent derivations should be kip-ft^2.

The rotation θ_A at the end A of the girder is

$$\theta_A = \frac{\delta_{B/A}}{L} = \frac{1}{EI}\left(\frac{719{,}228.64}{60}\right)$$

$$= \frac{11{,}987.144}{EI}$$

If we draw a tangent at point C of the elastic curve as shown in Fig. 4.19b, it intersects the horizontal line AB at point F, thus forming the triangle FAG shown in the same figure. Note that point G represents the point of intersection of the tangents to the deflection curve at points A and C, and θ_{AC} is the angle formed by these two tangents. By observing triangle FAG, we note that its outside angle θ_A is equal to the sum of the angles θ_C and θ_{AC}, where θ_C is the slope of the deflection curve at point C. Thus,

$$\theta_A = \theta_C + \theta_{AC}$$

or

$$\theta_C = \theta_A - \theta_{AC} \quad (4.38)$$

The angle θ_{AC} between the tangents at points A and C, in accordance with the first moment-area theorem, is equal to the M/EI area between points A and C in Fig. 4.19c. Thus,

$$\theta_{AC} = \frac{1}{2}(16)\left(\frac{435.2}{EI}\right) = \frac{3{,}481.6}{EI}$$

By substituting the values of θ_A and θ_{AC} into Eq. (4.38), we find

$$\theta_c = \frac{1}{EI}(11{,}987.144 - 3{,}481.6)$$

$$= \frac{8{,}505.544}{EI}$$

By referring again to Fig. 4.19b, we observe that the deflection y_C is

$$y_C = \Delta - \delta_{C/A} \tag{4.39}$$

where Δ is the vertical distance CC'' and $\delta_{C/A}$ is the distance $C'C''$, which is the vertical distance from point C' on the deflection curve to the tangent drawn at point A. In accordance with the second moment-area theorem, $\delta_{C/A}$ is the first moment of the M/EI area between points A and C in Fig. 4.19c, taken about C. Thus,

$$\delta_{C/A} = \frac{1}{2}(16)\left(\frac{435.2}{EI}\right)\left(\frac{16}{3}\right) = \frac{18{,}568.53}{EI}$$

The distance Δ can be determined by using the similar triangles ACC'' and ABB', yielding

$$\Delta = \frac{16}{60}\delta_{B/A} = \frac{16}{60}\left(\frac{719{,}228.64}{EI}\right)$$

$$= \frac{191{,}794.30}{EI}$$

Thus, from Eq. (4.39),

$$y_C = \frac{1}{EI}(191{,}794.30 - 18{,}568.53)$$

$$= \frac{173{,}225.77}{EI}$$

Equation (4.38) applies for any point C along the length of the girder. That is, if we consider point D, the rotation θ_D is

$$\theta_D = \theta_A - \theta_{AD} \tag{4.40}$$

where θ_{AD} is the angle between the tangent at points A and D and it is equal to the M/EI area between points A and D in Fig. 4.19c. That is,

$$\theta_{AD} = \frac{1}{2}(16)\left(\frac{435.2}{EI}\right) + (14)\left(\frac{435.2}{EI}\right) + \frac{1}{2}(14)\left(\frac{704.0 - 435.2}{EI}\right)$$

$$= \frac{13{,}337.6}{EI}$$

By substituting into Eq. (4.40), we find

$$\theta_D = \frac{1}{EI}[11{,}987.144 - 13{,}337.60]$$

$$= -\frac{1{,}350.46}{EI}$$

where the minus sign indicates a change in the slope direction.

By examining Fig. 4.19b, we find that the vertical deflection y_D at point D is

$$y_D = \Delta_1 - \delta_{D/A} \tag{4.40a}$$

where Δ_1 is equal to the distance DD'' and $\delta_{D/A}$ is distance $D'D''$ from point D' on the deflection curve to the tangent drawn at point A. Thus, $\delta_{D/A}$ is the first moment of the M/EI area between points A and D in Fig. 4.19c, taken about D. On this basis,

$$\delta_{D/A} = \frac{1}{2}(16)\left(\frac{435.2}{EI}\right)\left(14 + \frac{(16)}{3}\right) + (14)\left(\frac{435.2}{EI}\right)(7)$$

$$+ \frac{1}{2}(14)\left(\frac{704.0 - 435.2}{EI}\right)\left(\frac{14}{3}\right)$$

$$= \frac{118{,}741.22}{EI}$$

From the similarity of triangles ADD'' and ABB' in Fig. 4.19c, we find

$$\Delta_1 = \frac{30}{60}\delta_{B/A} = \frac{1}{2}\left(\frac{719{,}228.64}{EI}\right)$$

$$= \frac{359{,}614.32}{EI}$$

Substituting into Eq. (4.40a), we obtain

$$y_D = \frac{1}{EI}[359{,}614.32 - 118{,}741.2]$$

$$= \frac{240{,}873.10}{EI}$$

which is the vertical deflection of the bridge girder at point D.

256   ADVANCED METHODS AND PROBLEMS OF INFRASTRUCTURAL ANALYSIS

By taking into consideration that

$$EI = 460 \times 10^6 \text{ kip-in.}^2$$
$$= 3.1944 \times 10^6 \text{ kip-ft}^2$$

the final values of θ_A, θ_C, θ_D, y_C, and y_D are as follows:

$$\theta_A = 3.7525 \times 10^{-3} \text{ rad}$$
$$\theta_C = 2.6626 \times 10^{-3} \text{ rad}$$
$$\theta_D = -0.4228 \times 10^{-3} \text{ rad}$$
$$y_c = 0.054227 \text{ ft} = 0.6507 \text{ in.}$$
$$y_D = 0.075404 \text{ ft} = 0.9048 \text{ in.}$$

Example 4.7: The uniform beam in Fig. 4.20a supports an experimental testing machine at its free end C. During the operation of the machine, an equivalent static force F of 12 kN is produced. The nature of the experimental work is such that the deflection of the member at point C produced by the force F should not exceed 3 mm. Design the member so that the deflection limitation at point C is not violated. Neglect the dead weight of the testing machine and the dead weight of the member.

SOLUTION: The deflection curve of the member is shown in Fig. 4.20b, and the M/EI diagram is shown in Fig. 4.20c. At support B, the common tangent on the deflection curve is drawn as shown in Fig. 4.20b. In accordance with the moment-area second theorem,

$$\delta_{A/B} = \frac{1}{2}(8)\left(\frac{24}{EI}\right)\left[\frac{(2)(8)}{3}\right]$$
$$= \frac{512}{EI}$$

The units of EI are $kN \cdot m^2$.

Examining the deflection curve in Fig. 4.20b, we observe that the deflection y_C at point C is

$$y_C = \Delta + \delta_{C/B} \qquad (4.41)$$

where $\delta_{C/B}$ is the vertical distance $C''C'$ from point C'' on the deflection curve to the tangent drawn at B. In accordance with the moment-area second

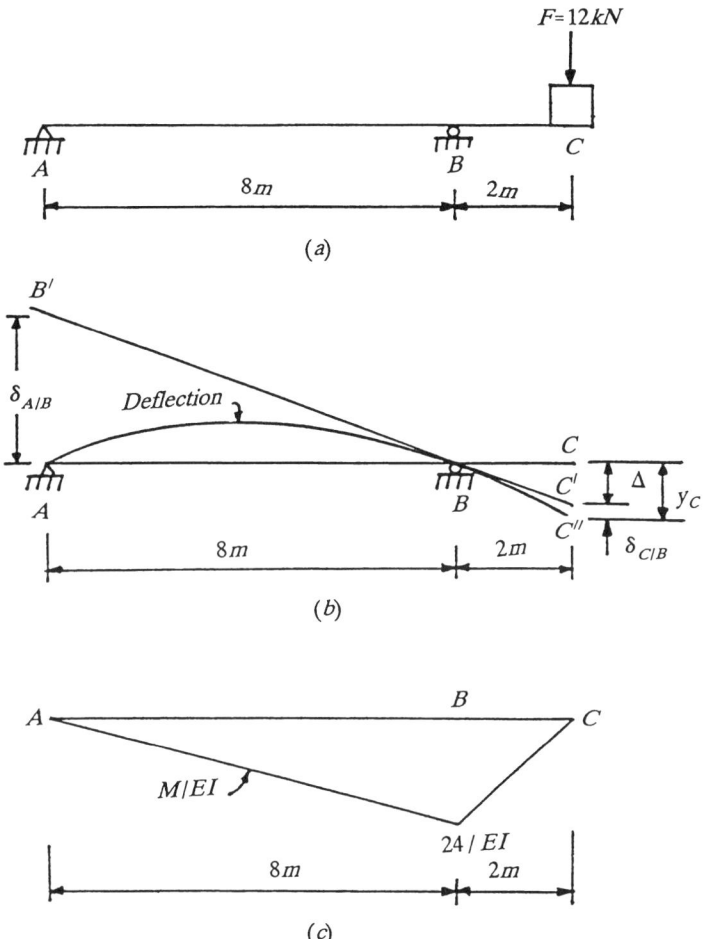

Figure 4.20. (a) Uniform beam loaded as shown. (b) Shape of the deflection curve. (c) M/EI diagram.

theorem,

$$\delta_{C/B} = \frac{1}{2}(2)\left(\frac{24}{EI}\right)\left(\frac{4}{3}\right)$$

$$= \frac{32}{EI}$$

From the similarity of triangles ABB' and CBC', we find

$$\Delta = \frac{\delta_{A/B}}{4} = \frac{512}{4EI} = \frac{128}{EI}$$

By substituting into Eq. (4.41), we find

$$y_c = \frac{1}{EI}[128 + 32] = \frac{160}{EI}$$

In order to satisfy the design requirement of $y_C \leqslant 3$ mm, we should have

$$0.003 m = \frac{160}{EI}$$

or

$$EI = \frac{160}{0.003} = 53{,}333.33 \text{ kN} \cdot \text{m}^2 \qquad (4.42)$$

which is the value of the stiffness EI that is required for the member to conform with the design limit on the deflection y_C.

If the material of the member is steel with elastic modulus $E = 207 \times 10^6$ kPa, Eq. (4.42) yields

$$(20710)^6 \left(\frac{\text{kN}}{\text{m}^2}\right) I = 53{,}333.33 \text{ kN} \cdot \text{m}^2$$

or, solving for I, we find

$$I = 0.257649 \times 10^{-3} \text{ m}^4$$
$$= 25{,}764.9 \text{ cm}^4$$

If we select a rectangular steel cross section of constant width $b = 15$ cm and of constant depth $d = 28$ cm, we find that $I = 27{,}440$ cm^4. On this basis, the member conforms with design requirements regarding the deflection y_C.

4.5.2 Conjugate-Beam Method

The conjugate-beam method may be considered as an extension of the moment-area method, since in both methods the M/EI diagram is used to determine rotations and displacements of members. In this method, the boundary conditions of the real beam are satisfied as support conditions on the conjugate beam and, consequently, the rotations and displacements determined by the conjugate-beam method are the true ones. In the moment-area method, the true rotations and deflections are obtained by geometric considerations regarding the deflection curve, and application of the two moment-area theorems.

In the application of the conjugate-beam method, we utilize a fictitious beam, known as conjugate beam, which satisfies both boundary and loading conditions of the real beam. Both real and conjugate beams have the same length and are in static equilibrium. The loading on the conjugate beam is the M/EI diagram of the real beam. With the M/EI as loading, the following two statements may be made regarding the conjugate beam:

Statement 1: The slope at any point along the length of the real beam is represented as a shear force at the corresponding point of the conjugate beam.

Statement 2: The deflection at any point along the length of the real beam is represented as moment at the corresponding point of the conjugate beam. In other words, the moment diagram of the conjugate beam is the deflection curve of the real beam.

Therefore, on the basis of the preceding two statements, the boundary conditions of the conjugate beam represent the boundary conditions of the real beam in terms of deflection and rotation. For example, a fixed support on a real beam permits no rotation or deflection of the support. Consequently, the corresponding support on the conjugate beam will satisfy the boundary conditions of the real beam if no shear or bending moment acts at the support. This means that a fixed support on a real beam is represented by a free end on the conjugate beam. By following the same analogy, a free end on a real beam is represented by a fixed support on the conjugate beam, because a free end permits both rotation and deflection and, consequently, the conjugate beam will have both shear and moment at this end. Fig. 4.21 illustrates corresponding boundary conditions for real and conjugate beams. It should be noted that the conjugate beam is always statically determinate.

The sign convention for the M/EI loading on the conjugate beam is such that a positive M/EI diagram is a downward loading on the beam, and vice versa, as shown in Fig. 4.22.

The following example illustrates the application of the method.

Example 4.8: For the uniform beam loaded as shown in Fig. 4.23a, determine the deflection and rotation of point A by using the conjugate-beam method. The stiffness EI is constant.

SOLUTION: Without regard to loading, the conjugate beam for the real beam in Fig. 4.23a is shown in Fig. 4.23b. In Fig. 4.23c, the conjugate beam is loaded with the M/EI loading that is produced by the application of the distributed load w acting on the real beam, and in Fig. 4.23d, the conjugate beam is loaded with the M/EI loading caused by the application of the concentrated load P at the free end A of the real beam. Deflections and rotations at any point of the real beam may be determined by calculating the corresponding moment and shear for the conjugate-beam problems in Figs.

260   ADVANCED METHODS AND PROBLEMS OF INFRASTRUCTURAL ANALYSIS

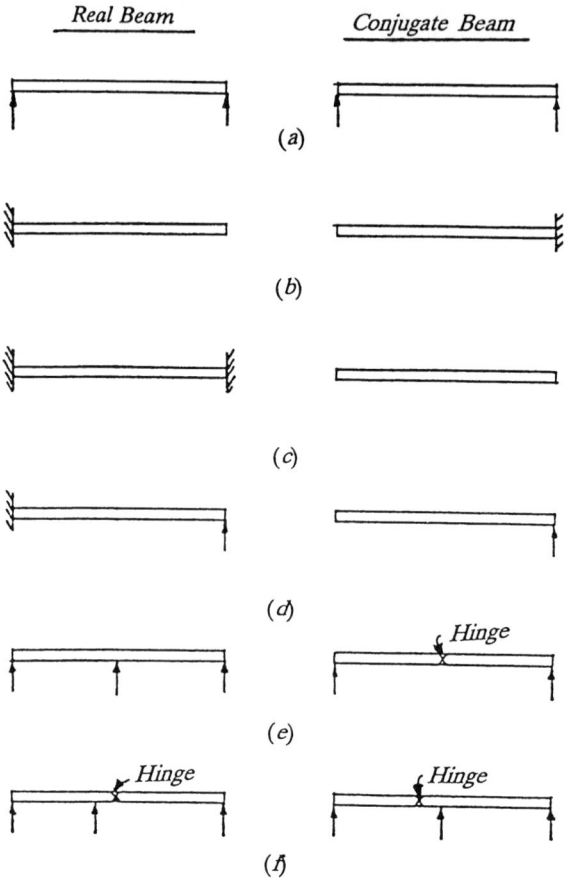

Figure 4.21. Corresponding real and conjugate-beam support conditions without regard to loading.

4.23c and 4.23d and superimposing the results. In this problem it is required to determine the deflection and rotation of point A of the real beam. Thus, for the deflection δ_A at A, we calculate the moments at A of the conjugate beams in Figs. 4.23c and 4.23d and we superimpose the results. For the rotation θ_A at A, we calculate the shear forces at A of the conjugate beams in Figs. 4.23c and 4.23d and we superimpose the results. The principles of statics are used to do this work.

By considering Fig. 4.23c and taking the sum of moments about B of all the forces to the right of B, we have

$$\Sigma M_B = 0 = 2LR'_2 - \frac{2}{3}(2L)\frac{wL^2}{2EI}(L)$$

MOMENT-AREA AND CONJUGATE-BEAM METHODS   261

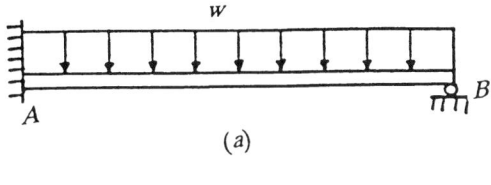

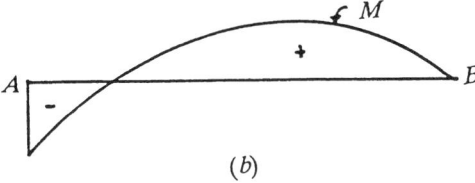

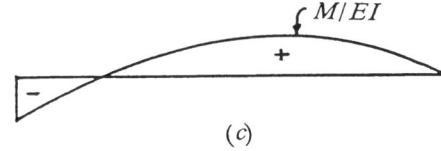

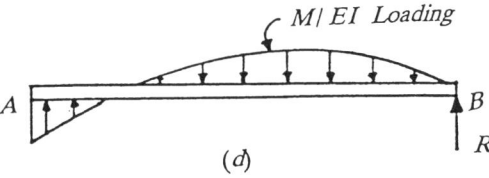

Figure 4.22. (*a*) Real beam. (*b*) Moment diagram of the real beam. (*c*) M/EI diagram. (*d*) Conjugate beam with M/EI loading and reaction.

or

$$R'_2 = \frac{wL^3}{3EI} \tag{4.43}$$

By considering the conjugate beam in Fig. 4.23*d* and performing a similar operation about *B*, we have

$$\Sigma M_B = 0 = 2LR''_2 - \frac{1}{2}\frac{PL}{EI}(2L)\frac{1}{3}(2L)$$

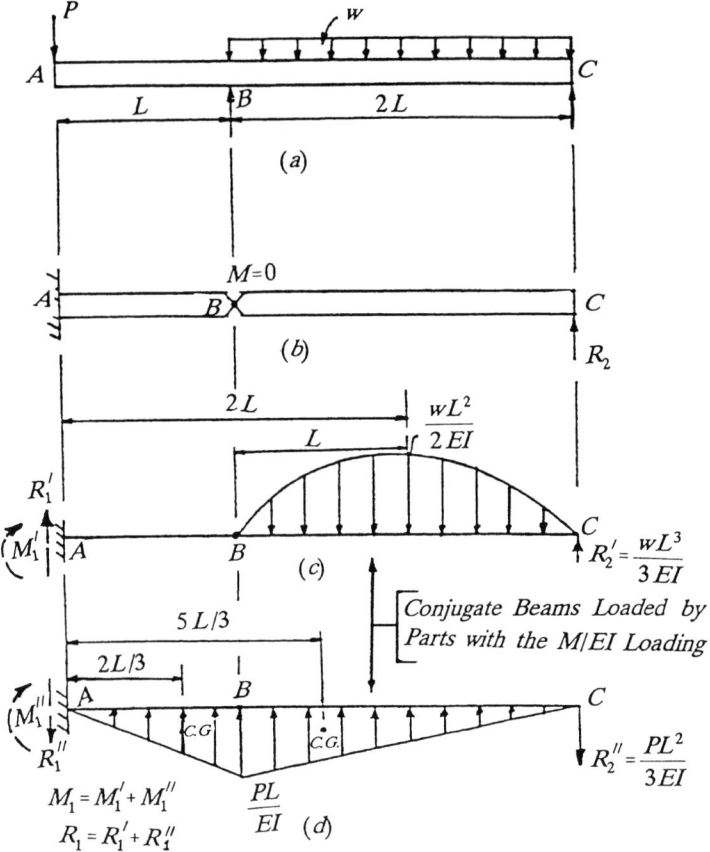

Figure 4.23. (a) Real beam loaded as shown. (b) Conjugate beam without regard to loading. (c) Conjugate beam loaded with the M/EI loading caused by the distributed load w. (d) Conjugate beam loaded with the M/EI loading produced by the application of the concentrated load P.

or

$$R_2'' = \frac{PL^2}{3EI} \tag{4.44}$$

Since the shear force of the conjugate beam at point A is the rotation θ_A of the real beam at A, we have

$$\theta_A = R_1 = R_1' + R_1'' \tag{4.45}$$

By considering the conjugate beams in Figs. 4.23c and 4.23d and applying static equilibrium in the vertical direction, we find

$$R'_1 = \frac{2}{3}\frac{wL^2}{2EI}(2L) - \frac{wL^3}{3EI} \qquad (4.46)$$

$$R''_1 = \frac{PL^2}{3EI} - \frac{1}{2}\frac{PL}{EI}(3L) \qquad (4.47)$$

By substituting Eqs. (4.46) and (4.47) into Eq. (4.45) and rearranging, we find

$$\theta = -\frac{7PL^2}{6EI} + \frac{wL^3}{3EI} \qquad (4.48)$$

It was stated earlier that the moment at A on the conjugate beam is the deflection δ_A of the real beam at point A. By using the conjugate beams in Figs. 4.23c and 4.23d and applying statics, we find

$$\delta_A = M_A = \frac{wL^3}{3EI}(3L) - \frac{2}{3}\frac{wL^2}{2EI}(2L)(2L) - \frac{PL^2}{3EI}(3L)$$

$$+ \frac{1}{2}\frac{PL}{EI}(L)\frac{2L}{3} + \frac{1}{2}\frac{PL}{EI}(2L)\frac{5L}{3}$$

or

$$\delta_A = \frac{PL^3}{EI} - \frac{wL^4}{3EI} \qquad (4.49)$$

4.5.3 Statically Indeterminate Members

The application of the moment-area and conjugate-beam methods to statically indeterminate problems is illustrated by the following examples:

Example 4.9: By applying the moment-area method, determine the moment M_C at the fixed end C of the statically indeterminate beam loaded as shown in Fig. 4.24a. The stiffness EI is constant.

SOLUTION: We select the moment M_C at the end C of the beam as the redundant. On this basis, the original member in Fig. 4.24a reduces to a simply supported beam with overhang that is loaded with a concentrated load P at point A and a bending moment M_C at support C. The moment M_C may be determined by calculating the rotation θ_C in Fig. 4.24b and setting it equal to zero. This satisfies the boundary condition of zero rotation at point C of the original member, and also provides an equation that can be used to determine M_C.

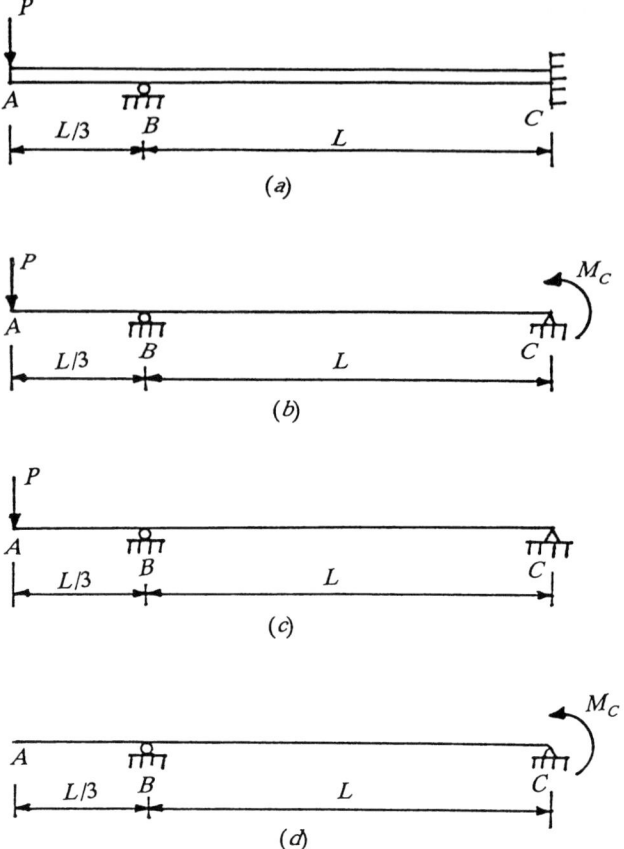

Figure 4.24. (*a*) Statically indeterminate beam loaded as shown. (*b*) Simply supported beam with overhang loaded by M_C and P. (*c*) Simply supported beam with overhand loaded with P. (*d*) Simply supported beam with overhang loaded with M_C.

For convenience, the simply supported beam with the overhang is loaded separately by P and M_C as shown in Figs. 4.24*c* and 4.24*d*, respectively. These two problems are now statically determinate, and the procedure discussed in Section 4.5.1 may be used to calculate the rotation at C. By solving the problem in Fig. 4.24*c*, we find

$$\theta'_C = \frac{PL^2}{18EI} \quad \text{(clockwise)} \tag{4.50}$$

Solution of the problem in Fig. 4.24*d* yields

$$\theta''_C = \frac{M_C L}{3EI} \quad \text{(counterclockwise)} \tag{4.51}$$

By applying superposition, we find

$$\theta_C = \theta'_C + \theta''_C = 0$$

or, by substitution, we obtain

$$-\frac{PL^2}{18EI} + \frac{M_C L}{3EI} = 0 \qquad (4.52)$$

Counterclockwise rotations were assumed positive in writing Eq. (4.52). By solving Eq. (4.52) for M_C, we find

$$M_C = \frac{PL}{6} \qquad (4.53)$$

Example 4.10: Solve the problem in Example 4.9 by using the conjugate-beam method.

SOLUTION: We consider again the moment at the fixed end C as the redundant. This reduces the original member into a simply supported beam with an overhang loaded as shown in Fig. 4.24b. The conjugate-beam method as discussed in Section 4.5.2 may now be used to solve the problems in Figs. 4.24c and 4.24d, and to calculate the rotations at the support C. For example, the conjugate beam of the problem in Fig. 4.24d, loaded with the M/EI loading, is shown in Fig. 4.25. By taking moments about the hinge B, we have

$$\Sigma M_B = 0 = -R''_C L + \frac{1}{2}\left(\frac{M_C}{EI}\right)(L)\frac{2L}{3}$$

or

$$\theta''_C = R''_C = \frac{M_C L}{3EI} \quad \text{(counterclockwise)}$$

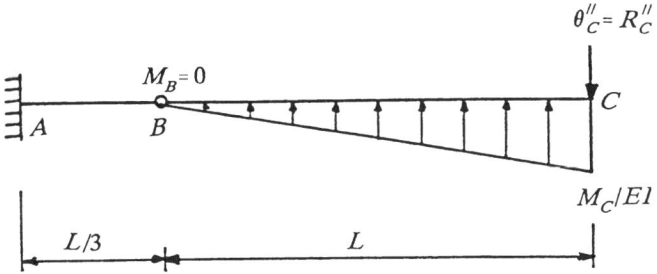

Figure 4.25. Conjugate beam for the problem in Fig. 4.24d loaded with the M/EI loading.

This is identical to the value obtained in Eq. (4.51) by using the moment-area method.

If we solve the problem in Fig. 4.24c by using the conjugate-beam method, we obtain

$$\theta'_C = R'_C = \frac{PL^2}{18EI} \quad \text{(clockwise)}$$

which is the same as Eq. (4.50). Thus,

$$\theta_C = \theta'_C + \theta''_C = 0$$

or

$$-\frac{PL^2}{18EI} + \frac{M_C L}{3EI} = 0$$

Solution for M_C yields

$$M_C = \frac{PL}{6}$$

which is identical to Eq. (4.53).

4.6 MOMENT DISTRIBUTION METHOD

The method of moment distribution was originated and developed by Professor Hardy Cross, and it was published in the Transactions of the American Society of Civil Engineers (ASCE) in 1932 [45]. Originally Professor Cross applied the method to continuous structures over rigid supports. The joints of the structure were limited to rotation without translation, but later on, modifications on the method were made to include joint translation. In this section, the application of the method is limited to structures on rigid supports with joints that are permitted to rotate but that do not translate.

Initially, all joints of the structure are assumed to be locked against rotation, which implies that each member of the structure is fixed at its ends. The fixed-end moments acting at the ends of each member are determined from that member's individual loading conditions. The sign convention used in this method designates clockwise moments on either end of the method as positive. At each locked joint, the algebraic sum of the fixed-end moments will not be zero. If all joints of the structure were instantaneously released from their hypothetically locked positions, the unbalanced moment at each joint would be distributed throughout the structure so that the final configuration of the

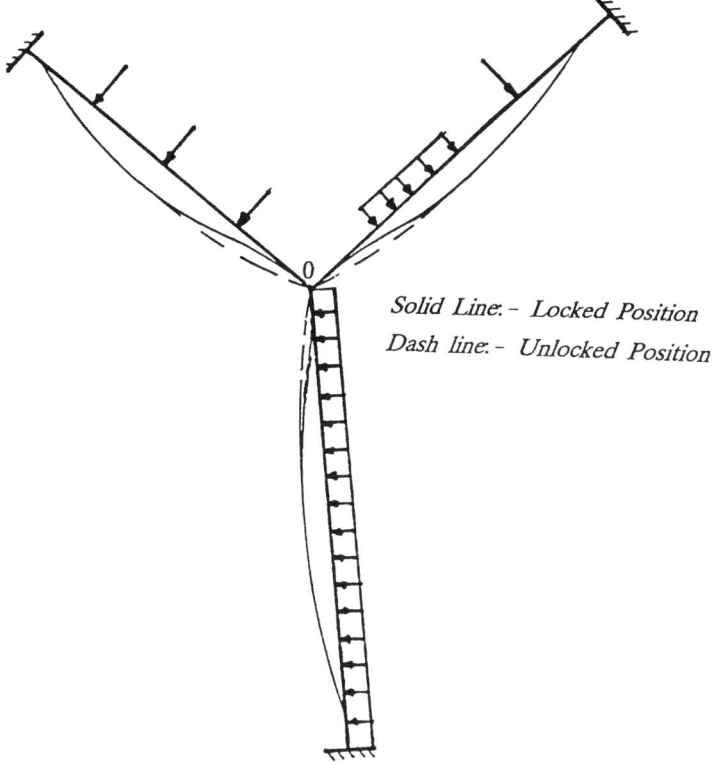

Figure 4.26. Locked and unlocked configuration of a joint of a structure.

structure would satisfy the conditions of continuity and equilibrium. This result is materialized in the moment distribution method by releasing one joint at a time, defining its effect on the adjacent joints, relocking the joint in the new position, and repeating the procedure for every joint in the structure until equilibrium is established.

The unbalanced moment at each joint is distributed to the members meeting at the joint in accordance with their stiffness when the joint is unlocked. Figure 4.26 exemplifies a joint of a structure, where the solid lines represent the deflected positions of the loaded members in their hypothetically locked positions, and the dashed lines illustrate the effect of unlocking the joint. In other words, the dashed deflection curves are the deflection configurations of the members when each member is considered simply supported at the joint and fixed at the opposite end. Thus, the stiffness factor K of a member may be defined as the magnitude of the moment that is required to produce unit rotation at its simply supported end, while its other end is fixed. This is illustrated in Fig. 4.27.

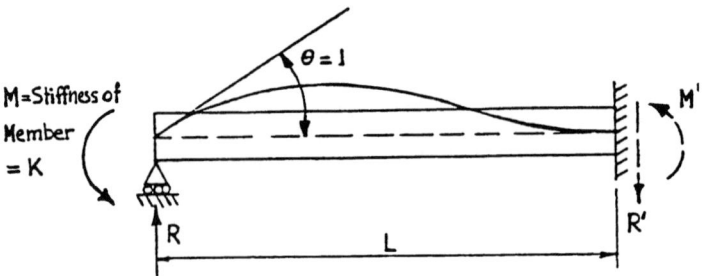

Figure 4.27. Member subjected to unit rotation at the simply supported end, while its opposite end is fixed.

In Fig. 4.27, the ratio M'/M is termed the carry-over factor C of the member. Carry-over factors are used to calculate the magnitudes of the moments transmitted to the fixed ends during the process of unlocking a joint. The proportion of the unbalanced moment carried by each member at the joint in its unlocked position is defined by the distribution factor D. The distribution factor of a member at the joint is defined as the ratio of the stiffness factor of the member to the sum of the stiffness factors of all the members meeting at the joint. Thus, for a joint composed of n members, the distribution factor D_i for the ith member is

$$D_i = \frac{K_i}{K_1 + K_2 + \cdots + K_i + \cdots + K_n} \qquad (4.54)$$

When a member meeting at a joint of a structure is of uniform cross section, such as the one in Fig. 4.27, the stiffness factor K is given by the following expression:

$$M_{\theta=1} = K = \frac{4EI}{L} \qquad (4.55)$$

The moment M' at the fixed end of the member in Fig. 4.27 is

$$M' = \frac{2EI}{L} \qquad (4.56)$$

Consequently, the carry-over factor C is

$$C = \frac{M'}{M} = \frac{1}{2} \qquad (4.57)$$

If the stiffness EI of the member varies along its length, its stiffness factor K

should be determined by using appropriate methods of mechanics and structures. The author's method of the equivalent systems, as discussed in earlier sections, will simplify the procedure.

The following step-by-step procedure may be used in the application of the moment distribution method:

Step 1: Determine the fixed-end moments for each member of the structure based on its applied loading.

Step 2: Calculate the stiffness factor K for each member using the EI/L value if the member is uniform. Use appropriate methods of analysis, such as the method of the equivalent system, if the stiffness of the member is variable.

Step 3: Use a carry-over factor $C = \frac{1}{2}$ if the member is uniform. For variable stiffness members, this factor would have to be calculated in accordance with the ratio M'/M of the end moment.

Step 4: Determine distribution factors at each joint.

Step 5: Beginning with the joint having the largest unbalanced moment, reverse its sign and distribute it to the members of the joint in accordance with their distribution factor.

Step 6: Carry over one-half of the distributed moment on each member to the other end if the member is of uniform cross section. If the member is of variable cross section, use the calculated value of the carry-over factor C.

Step 7: Repeat the distribution for each joint until the unbalanced moment residue is negligible.

Step 8: The distributed moments are algebraically added to the original fixed-end moment on each member to obtain the correct internal moment acting on the member at the joint.

The following examples illustrate the application of the moment distribution method to practical problems.

Example 4.11: A continuous beam with overhangs is loaded by distributed loads, as shown in Fig. 4.28a. Determine the internal moments at supports B, C, and D by using the moment distribution method. The cross section of the member is uniform throughout its length.

SOLUTION: The fixed-end moments for each span, in units of kip-ft, are calculated by using methods discussed in preceding sections, or handbook formulas if such formulas are available. They are shown in Fig. 4.28b. Since EI is constant throughout the length of the continuous beam, the stiffness factors are functions only of the lengths of the spans. Thus, the stiffness factor of each span is equal to the reciprocal of its length. Their values are given in Fig. 4.28a. The distribution factors are shown boxed in Fig. 4.28b, and the moment distribution procedure is carried out as shown in the same figure. The arrows

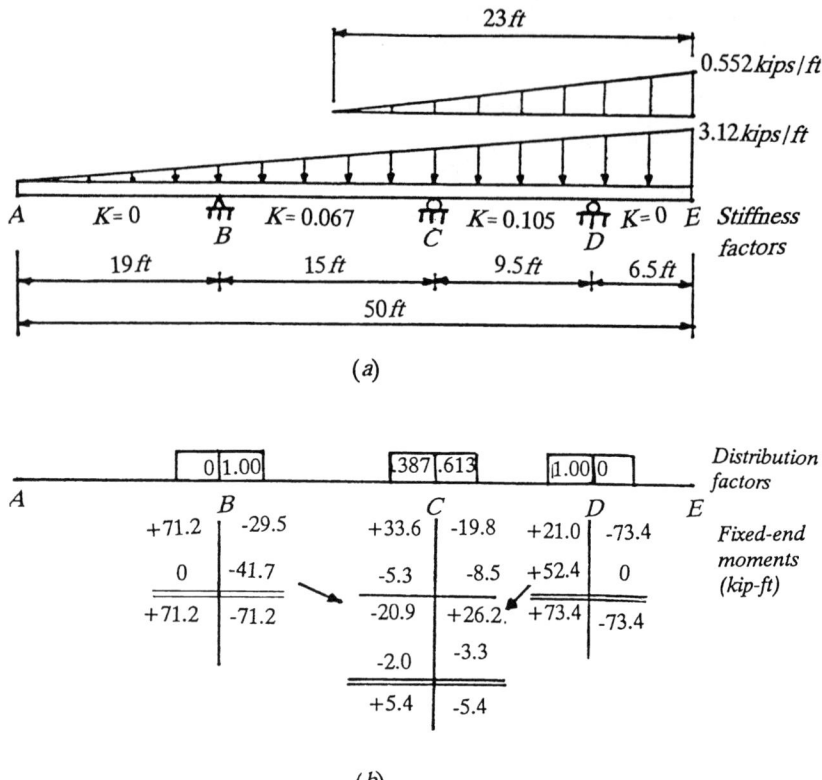

Figure 4.28. (a) Continuous beam with overhangs loaded as shown. (b) Moment distribution procedure.

in the figure point out the moments carried over to the other end of each member in accordance with the carry-over factor. The double straight lines are used to indicate that the moments at each joint are balanced. Thus, the internal moment at support B is ± 71.2 kip-ft, it is ± 5.4 kip-ft at support C, and ± 73.4 kip-ft at support D.

Example 4.12: A reinforced concrete culvert forms a closed frame structure and has the dimensions shown in Fig. 4.29a. If the two sides of the culvert are loaded with a distributed load, as shown in the figure, determine the internal moments at each corner of the structure by using the moment distribution method.

SOLUTION: By considering the vertical side BC of the culvert as a fixed-fixed beam, the fixed-end moments M_{F_B} and M_{F_C} at ends B and C, respectively, may

MOMENT DISTRIBUTION METHOD   271

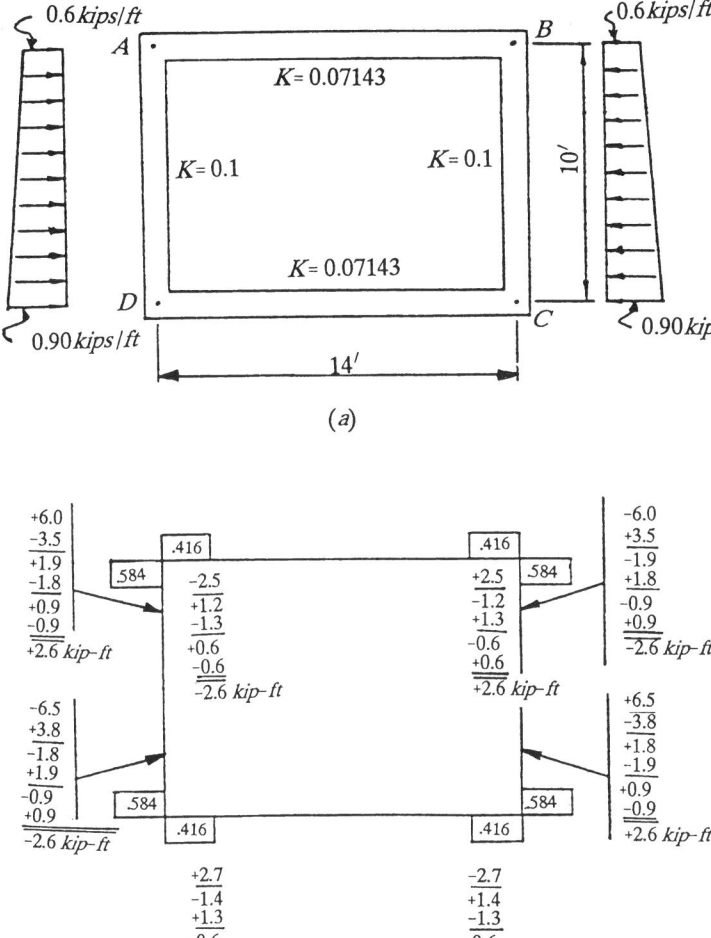

Figure 4.29. (*a*) Reinforced concrete culvert loaded as shown. (*b*) Moment distribution procedure.

be determined from the following equations:

$$M_{F_B} = -\frac{w_1 L^2}{12} - \frac{w_2 L^2}{30} \qquad (4.58)$$

$$M_{F_C} = \frac{w_1 L^2}{12} + \frac{w_2 L^2}{20} \qquad (4.59)$$

where $w_1 = 0.6$ kips/ft and $w_2 = (0.9 - 0.6) = 0.3$ kips/ft. The first term in Eqs. (4.58) and (4.59) provides the fixed-end moments produced by the uniformly distributed load w_1, and the second term provides the fixed-end moments produced by the triangular load w_2.

By substituting into Eqs. (4.58) and (4.59), we find

$$M_{F_B} = -\frac{(0.6)(10)^2}{12} - \frac{(0.3)(10)^2}{30} = -6.0 \text{ kip-ft}$$

$$M_{F_C} = \frac{(0.6)(10)^2}{12} + \frac{(0.3)(10)^2}{20} = 6.5 \text{ kip-ft}$$

On the side AD of the culvert, we have $M_{F_A} = -M_{F_B}$ and $M_{F_D} = -M_{F_C}$. Since the stiffness EI is constant on all sides of the culvert, the stiffness factor K of each side of the culvert is equal to the reciprocal of its length. These values are shown in Fig. 4.29a.

The distribution factors are determined by using Eq. (4.54), and they are shown in attached rectangular boxes in Fig. 4.29b. The moment distribution procedure is carried out as shown in the same figure. At the corners of the culvert, the internal moments are equal to ± 2.6 kip-ft.

Example 4.13: The symmetrical frame shown in Fig. 4.30 is loaded in a way that yields the fixed-end moments shown at the joints of the frame. The stiffness factor K for each member of the frame is also shown in Fig. 4.30. By using the moment distribution method, determine the end moments of the members of the frame.

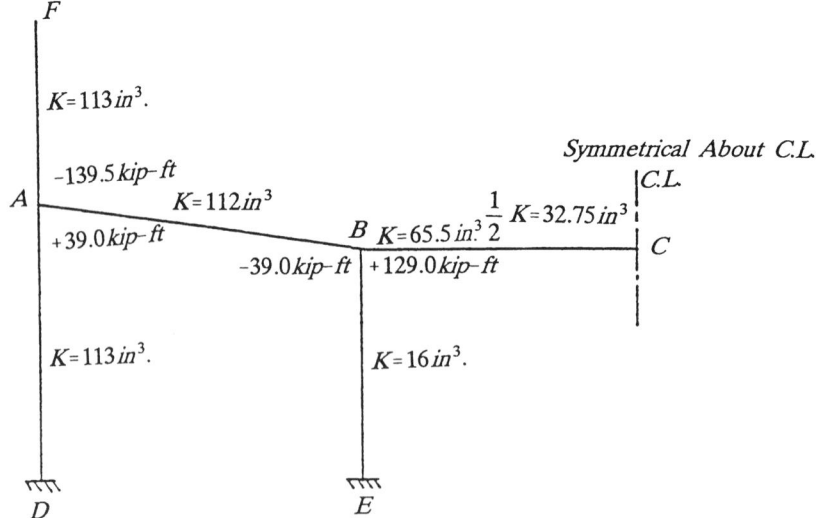

Figure 4.30. Symmetrical frame.

MOMENT DISTRIBUTION METHOD FOR CONTINUOUS BRIDGE GIRDERS

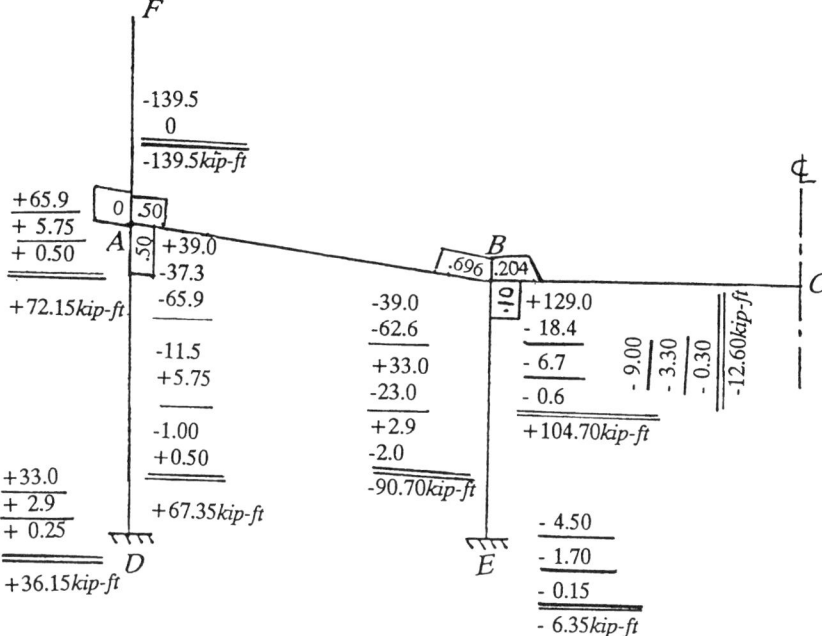

Figure 4.31. Moment distribution procedure for the symmetrical frame in Fig. 4.30.

SOLUTION: By using Eq. (4.54), the distribution factors of the members at each joint are determined, and they are shown enclosed in small boxes in Fig. 4.31. It should be noted that the stiffness of member BC, used in the determination of distribution factors at joint B, is taken as one-half the stiffness of the entire member, because of symmetry. Consideration of symmetry reduces the work that would be required if the entire structure were used.

The moment distribution procedure is carried out as shown in Fig. 4.31. The end moments of each member of the frame are as shown in the same figure.

4.7 MOMENT DISTRIBUTION METHOD FOR CONTINUOUS BRIDGE GIRDERS OF VARIABLE DEPTH

For structures with members of variable stiffness EI, the fixed end moments and moment distribution factors can be determined rather easily by using the approximate method of the equivalent systems. Consider, for example, the continuous beam of variable depth shown in Fig. 4.32a. With joints B, C, D, and E locked against rotation, the fixed-end moments at the ends of the individual spans can be determined by considering each span as shown in Figs. 4.32b and 4.32e, and applying the approximate method of the equivalent systems, as discussed in Sections 4.2 and 4.4.

274 ADVANCED METHODS AND PROBLEMS OF INFRASTRUCTURAL ANALYSIS

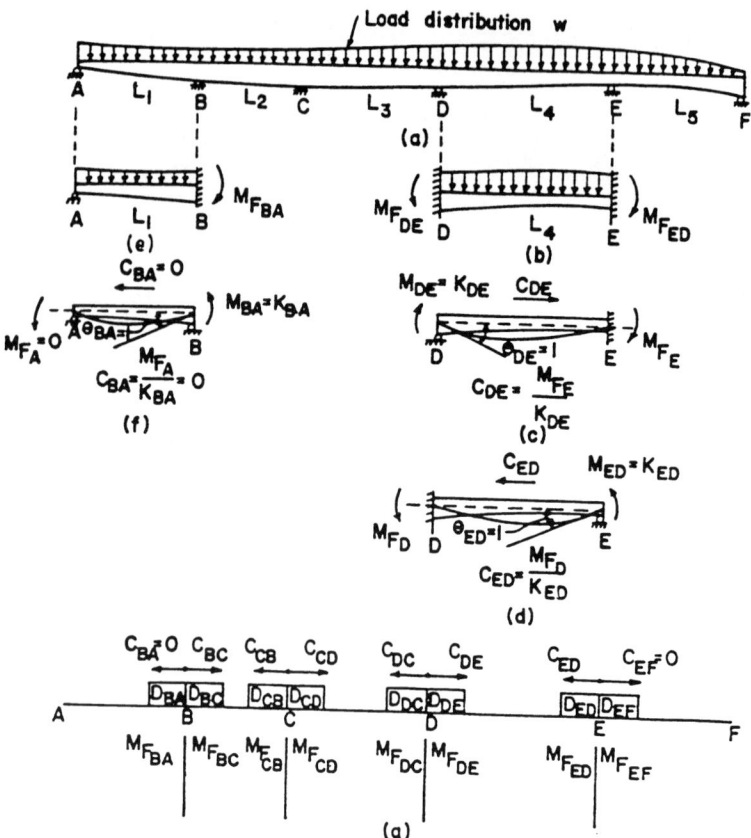

Figure 4.32. (a) Variable depth girder. (b) Span DE with fixed ends. (c) and (d) Stiffness and carry-over factors for span DE. (e) Span AB with end B fixed. (f) Stiffness and carry-over factors for span AB. (g) Schematic arrangement for moment distribution procedure.

The moment distribution factors are defined in the same way as for continuous beams of uniform EI. Thus, the stiffness factors would be computed by using the method of the equivalent systems and solving each span for the moment required to produce unit rotation at the simply supported end, as illustrated in Figs. 4.32c, 4.32d, and 4.32f. The carry-over factors are also shown in the same figures. Since EI is variable, the value of a carry-over factor is usually other than $\frac{1}{2}$. Upon computation of all stiffness factors, the distribution factors are determined in the same way as for continuous beams of uniform EI. By arranging the preliminary information as shown in Fig. 4.32g, the moment distribution procedure is carried out in the same way as for uniform beam, and it will yield the true negative internal moments at supports B, C, D, and E in Fig. 4.32a.

MOMENT DISTRIBUTION METHOD FOR CONTINUOUS BRIDGE GIRDERS

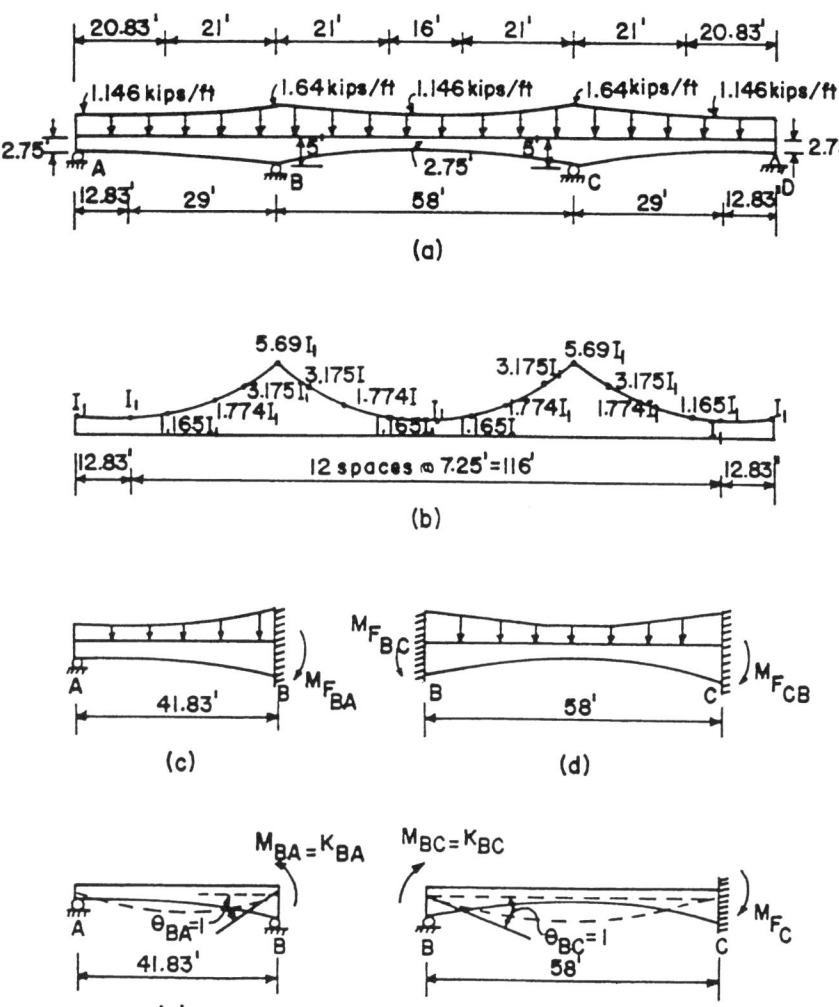

Figure 4.33. (*a*) Bridge girder of variable depth. (*b*) Stiffness variation of the girder. (*c*) Span *AB* fixed at end *B*. (*d*) Span *BC* with fixed ends. (*e*) Span *AB*. (*f*) Span *BC*.

The following example illustrates the application of the methodology.

Example 4.14: The three-span continuous concrete girder of an existing highway bridge in the state of Michigan is loaded as shown in Fig. 4.33*a*. The variation of the moment of inertia along its length is shown in Fig. 4.33*b*, where the reference value I_1 is the smallest moment of inertia of the girder. The modulus of elasticity E is constant and $EI_1 = 460 \times 10^6$ kip-in.^{2} By using the moment distribution method, determine the internal moments at supports B and C.

276   ADVANCED METHODS AND PROBLEMS OF INFRASTRUCTURAL ANALYSIS

SOLUTION: With joints B and C fixed against rotation, the fixed-end moments of spans AB, BC, and CD are determined by using the method of the equivalent systems, as discussed in Sections 4.2 and 4.4, to solve the problems in Figs. 4.33c and 4.33d. The solution of these two problems yields

$$M_{F_{BA}} = +392.0 \text{ kip-ft}$$
$$M_{F_{BC}} = -427.0 \text{ kip-ft}$$
$$M_{F_{CB}} = +427.0 \text{ kip-ft}$$
$$M_{F_{CD}} = -392.0 \text{ kip-ft}$$

The stiffness factors are determined by using the same method and solving the problems in Figs. 4.33e and 4.33f. The computations involved will not require appreciable amounts of additional work, because most of the required calculations for these problems are available from the computation of the fixed-end moments. The results are

$$K_{BA} = K_{CD} = \frac{EI_1}{6.28}$$

$$K_{BC} = K_{CB} = \frac{EI_1}{5.50}$$

Thus, the distribution factors are as follows:

$$D_{BA} = \frac{K_{BA}}{K_{BA} + K_{BC}} = 0.466$$

$$D_{BC} = 1.000 - 0.466 = 0.534$$

$$D_{CB} = D_{BC} = 0.534$$

$$D_{CD} = D_{BA} = 0.466$$

From Figs. 4.33e and 4.33f, the carry-over factors are

$$C_{BA} = C_{CD} = 0$$

$$C_{BC} = \frac{M_{Fc}}{K_{BC}} = 0.665$$

$$C_{CB} = C_{BC} = 0.665$$

The moment distribution procedure is shown in Fig. 4.34. The negative moments M_B and M_C at supports B and C, respectively, are

$$M_B = M_C = 417.0 \text{ kip-ft}$$

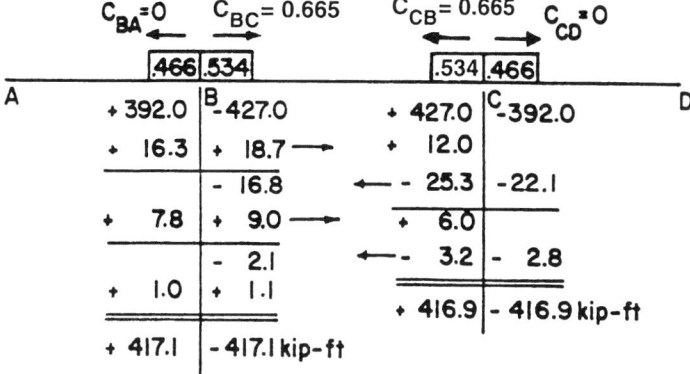

Figure 4.34. Moment distribution procedure for the variable thickness bridge girder.

With M_B and M_C known, the actual moments at cross sections of the beam in Fig. 4.33a may be determined by using statics.

4.8 THE THREE-MOMENT EQUATION

An interesting method that can be used for the analysis of continuous beams is known as the three-moment equation. For any two consecutive spans of a continuous beam, we can write an equation, known as the three-moment equation, that relates the three unknown internal moments at the three supports of the two spans. For a given continuous-beam problem, the number of such equations that can be written would be equal to the number of the unknown internal moments. Thus, simultaneous solution of this system of algebraic equations yields the values for the unknown internal moments. With known internal moments, each span of the continuous beam can be treated as a simply supported beam that is loaded with the applied loading and the calculated end internal moments.

We derive the three-moment equation by considering the continuous beam in Fig. 4.35a that is loaded arbitrarily as shown in the same figure. The span lengths can have different dimensions, and the moment of inertia can vary from span to span, but it is assumed to be constant within the span. We now consider the two consecutive spans AB and BC, and we draw the free-body diagrams shown in Figs. 4.35b and 4.35c. The purpose is to write an equation that relates the three moments M_A, M_B, and M_C. Since the deflection curve has a common slope at support B, this equation can be derived by satisfying the following equation:

$$\theta'_B = -\theta''_B \tag{4.60}$$

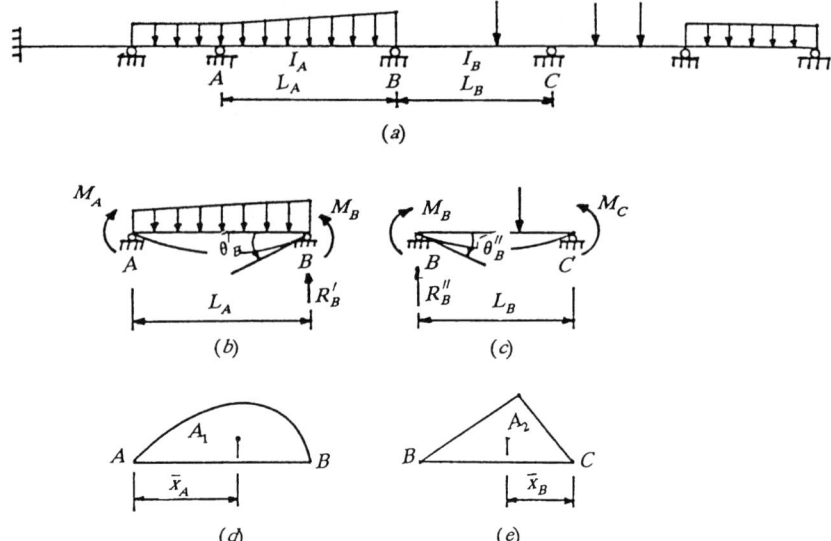

Figure 4.35. (a) Continuous beam loaded as shown. (b) Free-body diagram of span AB. (c) Free-body diagram of span BC. (d) Moment diagram of the applied loading on span AB. (e) Moment diagram of the applied loading on span BC.

where θ'_B is the slope of member AB at point B and θ''_B is the slope of member BC at point B.

The derivation of the three-moment equation may be completed by writing the expressions for θ'_B and θ''_B and substituting into Eq. (4.60). Then, with some rearrangement of terms, the appropriate form of the three-moment equation is obtained.

The expression for θ'_B is as follows:

$$\theta'_B = \frac{M_A L_A}{6EI_A} + \frac{M_B L_A}{3EI_A} + \frac{A_1 \bar{x}_A}{EI_A L_A} \qquad (4.61)$$

The first term on the right-hand side of Eq. (4.61) represents the contribution of M_A to the slope at B in Fig. 4.35b, the second term provides the contribution of M_B, and the third term is the slope at B produced by the applied loading on the span AB in Fig. 4.35b. This last part is obtained by considering the moment diagram of the applied loading shown in Fig. 4.35d, and applying the second moment-area theorem as discussed in Section 4.5.1. A_1 in this figure is the total area of the moment diagram, and $\bar{x}_A$ is the distance from the centroid of the area A_1 to the end A.

In a similar manner, an expression for θ''_B can be written that includes the contributions of the moments M_B and M_C, and the contribution of the applied loading on span *BC*. See Figs. 4.35*c* and 4.35*e*. This equation is

$$\theta_B'' = \frac{M_B L_B}{3EI_B} + \frac{M_C L_B}{6EI_B} + \frac{A_2 \bar{x}_B}{EI_B L_B} \tag{4.62}$$

Note that A_2 is the area of the moment diagram in Fig. 4.35*e* produced by the applied loading on span *BC*, and $\bar{x}_B$ is the distance from the centroid of A_2 to the end support *C*.

By substituting Eqs. (4.61) and (4.62) into Eq. (4.60) and rearranging, we obtain the following equation:

$$M_A \left(\frac{L_A}{I_A}\right) + 2M_B \left(\frac{L_A}{I_A} + \frac{L_B}{I_B}\right) + M_C \left(\frac{L_B}{I_B}\right)$$
$$= -\frac{6A_1 \bar{x}_A}{I_A L_A} - \frac{6A_2 \bar{x}_B}{I_B L_B} \tag{4.63}$$

Equation (4.63) is known as the three-moment equation, because it relates the three internal moments M_A, M_B, and M_C.

If the moment of inertia is the same throughout the length of the continuous beam, Eq. (4.63) simplifies as follows:

$$M_A L_A + 2M_B(L_A + L_B) + M_C L_B = -\frac{6A_1 \bar{x}_A}{L_A} - \frac{6A_2 \bar{x}_B}{L_B} \tag{4.64}$$

Equation (4.64) reduces to the following equation if all spans of the continuous beam have the same length *L*:

$$M_A + 4M_B + M_C = -\frac{6}{L^2}(A_1 \bar{x}_A + A_2 \bar{x}_B) \tag{4.65}$$

By examining the continuous beam in Fig. 4.35*a*, we note that five equations such as the one given by Eq. (4.63) may be written, involving the five internal moments at the five internal supports of the beam, and the fixed-end moment at its fixed-end support. Therefore, we have five equations with six unknowns, not enough equations to solve for all unknown moments. This situation can be corrected by replacing the fixed support by an additional span having an infinite moment of inertia, or, if preferred, by an additional span of zero length. Then an additional three-moment equation may be written by using the additional span and the one next to it. In this manner, we have six equations involving six unknown moments. If the other end of the continuous beam is also fixed, the procedure may be repeated by replacing the fixed end with an additional span having infinite moment of inertia.

The following example illustrates the application of the method.

Example 4.15: The two-span continuous beam shown in Fig. 4.36a is loaded as shown. By using the three-moment equation, determine the bending moments and the reactions at the supports. The modulus of elasticity E and moment of inertia I are constant throughout the length of the member. Also, determine the vertical deflection at the center of the first span when $E = 206 \times 10^6$ kPa and $I = 0.416 \times 10^{-3}$ m^4.

SOLUTION: Since the continuous beam in Fig. 4.36a is fixed at end A, we introduce an additional span AD of length $L = 0$, or moment of inertia $I = \infty$, as shown in Fig. 4.36b. Three-moment equations may be written by using the

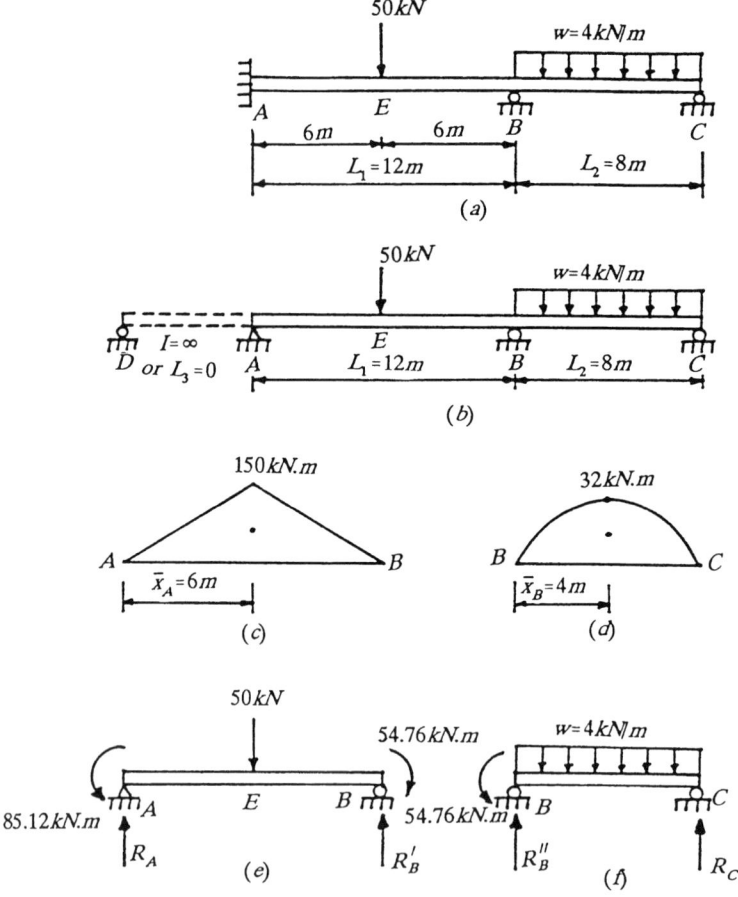

Figure 4.36. (a) Uniform continuous beam loaded as shown. (b) Continuous beam with additional span of infinite moment of inertia. (c) Moment diagram A_1 for span AB. (d) Moment diagram A_2 for span BC. (e) Free-body diagram for span AB. (f) Free-body diagram for span BC.

pair of spans AD and AB, and the pair AB and BC. By applying Eq. (4.64) and using the first pair of spans with span length AD equal to zero, we find

$$M_D(0) + 2M_A(12) + M_B(12) = -\frac{6A_1(6)}{12}$$

or

$$24M_A + 12M_B = -3A_1 \tag{4.66}$$

where A_1 is the area of the moment diagram in Fig. 4.36c.
From Fig. 4.36c, the area A_1 is

$$A_1 = \tfrac{1}{2}(12)(150) = 900$$

By substituting for A_1 in Eq. (4.66), we find

$$24M_A + 12M_B = -2{,}700 \tag{4.67}$$

By using Eq. (4.64), the pair of spans AB and BC yields

$$M_A(12) + 2M_B(12 + 8) + 0 = -\frac{6A_1(6)}{12} - \frac{6A_2(4)}{8}$$

or

$$12M_A + 40M_B = -3A_1 - 3A_2 \tag{4.68}$$

The areas A_1 and A_2 in Eq. (4.68) are the areas of the moment diagrams in Figs. 4.36c and 4.36d, respectively. They are:

$$A_1 = \tfrac{1}{2}(12)(150) = 900$$
$$A_2 = \tfrac{2}{3}(8)(32) = 170.67$$

By substituting A_1 and A_2 into Eq. (4.68), we find

$$12M_A + 40M_B = -3{,}212 \tag{4.69}$$

Simultaneous solution of Eqs. (4.67) and (4.69) yields

$$M_A = -85.12 \text{ kN} \cdot \text{m}$$
$$M_B = -54.76 \text{ kN} \cdot \text{m}$$

By comparing with the sign used in Figs. 4.35b and 4.35c to derive the three-moment equation, we conclude that

$$M_A = 85.12 \text{ kN} \cdot \text{m} \quad \text{(counterclockwise)}$$
$$M_B = 54.76 \text{ kN} \cdot \text{m} \quad \text{(clockwise for span } AB \text{ at } B)$$
$$M_B = 54.76 \text{ kN} \cdot \text{m} \quad \text{(counterclockwise for span } BC \text{ at } B)$$

The free-body diagrams of spans AB and BC are shown in Figs. 4.36e and 4.36f.

By using the free-body diagrams in Figs. 4.36e and 4.36f and applying statics, the reactions at the supports are calculated, and they are as follows:

$$R_A = 27.53 \text{ kN} \uparrow$$
$$R_B = R'_B + R''_B = 22.47 + 22.85$$
$$= 45.32 \text{ kN} \uparrow$$
$$R_C = 9.15 \text{ kN} \uparrow$$

Deflections and rotations at any point along the length of the continuous beam in Fig. 4.36a, may be determined by using the free-body diagrams in Fig. 4.36e and 4.36f and applying the methods discussed in Sections 1.7 and 4.5, or by using known handbook formulas. See also Appendix A. For example, by using formulas, the vertical deflection y_E at the center of span AB is

$$y_E = \frac{(50)(12)^3}{48EI} - \frac{(85.12)(12)^2}{16EI} - \frac{(54.76)(12)^2}{16EI}$$

$$= \frac{1}{EI}(1{,}800.00 - 766.08 - 492.84)$$

$$= \frac{541.08}{EI}$$

If $E = 206 \times 10^6$ kPa and $I = 0.416 \times 10^{-3}$ m^4, then y_E is

$$y_E = \frac{541.08}{(206)(10)^6 \, (0.416)(10)^{-3}}$$

$$= 6.3139 \times 10^{-3} \text{ m}$$

$$= 6.3139 \text{ mm}$$

Deflections and rotations at other points of the continuous beam may be determined in a similar manner.

4.9 DYNAMIC RESPONSE OF RIGID FRAMES COMPOSED OF MEMBERS OF VARIABLE STIFFNESS

In Chapter 3, the building frames were assumed to be composed of members of uniform stiffness EI. The stiffness, however, could vary between members, but it would stay uniform within the length of any member of a frame. This is a rather important restriction, because for many structures it is often required to vary the stiffness EI within the length of a member. From the architect's point of view, such stiffness variations can create unique architectural effects that please the eye of an observer. From the engineer's point of view, the purpose is rather practical. The stiffness of a member will usually vary in accordance with established strength criteria, with the ultimate purpose directed toward a lighter and safer structure.

If the members of a frame are of variable stiffness, the procedures regarding the analysis for dynamic response are still similar to those used for frames with uniform members, but the computations involved become more difficult and tedious. The approximate method of the equivalent systems, as discussed in preceding sections of this chapter, provides an excellent simplification regarding the analysis of such structural problems. In the following analysis, the general procedures and assumptions used in Chapter 3 are assumed to apply, but the stiffness EI within the length of a member of a frame is permitted to be variable.

Consider, for example, the building frame in Fig. 4.37a, and let it be assumed that the stiffness EI of its members is variable. If the stiffness of the girders in comparison to that of the columns is large, it would be reasonable to assume that the girders are infinitely rigid and neglect the small rotations at the tops of the columns. On this basis, reasonable results can be obtained by analyzing the frame as a two-degree of freedom spring-weight system, such as the one shown in Fig. 4.37b. In this figure, the weights W_1 and W_2 represent the weight of the frame lumped at the girder levels in exactly the same way as in Chapter 3. During motion, the displacements x_1 and x_2 of the weights in Fig. 4.37b will be equal to the horizontal displacements x_1 and x_2, respectively, of the frame in Fig. 4.37a, because the spring constants k_1 and k_2 are made equal to the total stiffness of the columns at the girder levels.

The spring constant k_2, for example, can be determined by considering each column of the second story of the frame as fixed at both ends, and applying at each column's top a displacement δ equal to unity, as shown in Fig. 4.37c. The equivalent of this problem is shown in Fig. 4.37d. In this figure, the purpose of the force $P = k$ and moment M is to produce at the top of the column a displacement $\delta = 1$, while the rotation θ is kept equal to zero. The spring constant k_2 in Fig. 4.37b is the sum of the k's of all the columns of the second story of the frame. In a similar manner, the k's at the top of the columns of the first story can be determined. Their sum will yield the spring constant k_1 in Fig. 4.37b.

284   ADVANCED METHODS AND PROBLEMS OF INFRASTRUCTURAL ANALYSIS

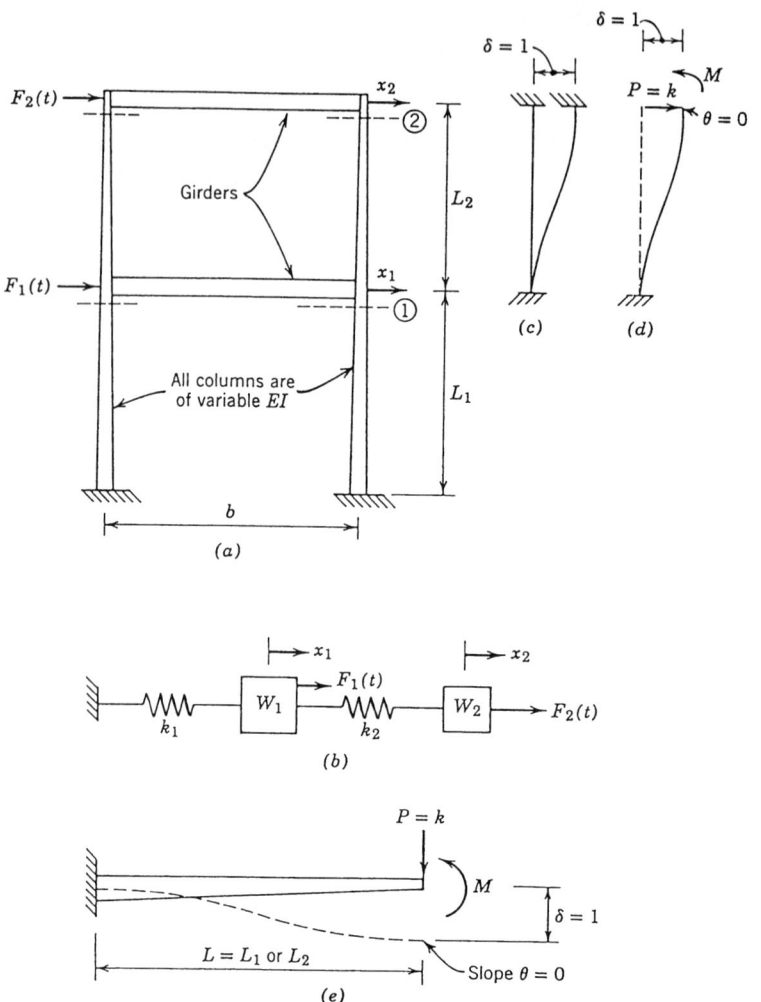

Figure 4.37. (*a*) Two-story frame with members of variable stiffness. (*b*) Idealized spring-mass system for the frame. (*c*) Second-story column. (*d*) Free-body diagram of a second-story column. (*e*) Free-body diagram for stiffness computations.

The computation of the k's of the variable stiffness columns becomes rather easy if the approximate method of the equivalent systems is used. The problem to be solved by this method is shown in Fig. 4.37e. In brief, the problem in Fig. 4.37e can be solved by applying first the load P and determining a constant stiffness equivalent system in the usual way. Then, by applying the moment M, a second equivalent system is obtained. Each equivalent system is solved for deflection and rotation at the free end. The algebraic sum of the two deflections and the two rotations should be equal to unity and zero, respectively. This

DYNAMIC RESPONSE OF RIGID FRAMES OF VARIABLE STIFFNESS 285

yields two equations with k and M as the unknowns. Simultaneous solution of these two expressions yields the value of k of the column.

The dynamic displacements x_1 and x_2, as well as the dynamic shear forces, moments, and stresses, can be computed by applying the AIEM discussed in Section 3.5, or by using other appropriate methods of analysis.

As an illustration, let it be assumed that it is required to determine the dynamic response of the two-story rigid frame building in Fig. 4.38, whose material properties, loading, and so on, are made uniform throughout the entire length of the building. In this case, the five rigid frames forming the building are of the form shown in Fig. 4.39a. The girders are assumed to be infinitely rigid compared to the columns. The moment of inertia I of the columns is variable, and it varies linearly from the elevation C to the elevation A. At C, the moment of inertia I_C of the columns is 70 in.^{4} and it is equal to 140 in.^{4} at A. The modulus of elasticity E is constant and equal to 30×10^6 psi. The entire building will be analyzed for dynamic response by considering only the interior frame in Fig. 4.39a together with the associated wall and floor areas.

The idealized two-degree of freedom spring-mass system is shown in Fig. 4.39d. The spring constant k_1 is the sum of the k's of the two columns of the first story of the frame in Fig. 4.39a. For each column, the k is determined by using the approximate method of the equivalent systems to solve the problem in Fig. 4.39c. In solving this problem, it should be noted that the boundary conditions that should be satisfied at end B are $\delta = 1$ and $\theta = 0$. The spring constant k_2 is the sum of the k's of the columns in the second story. Each k is determined by using the method of the equivalent systems and solving the problem in Fig. 4.39b. The results are

$$k_1 = 8.32 \text{ kips/in.}$$
$$k_2 = 20.10 \text{ kips/in.}$$

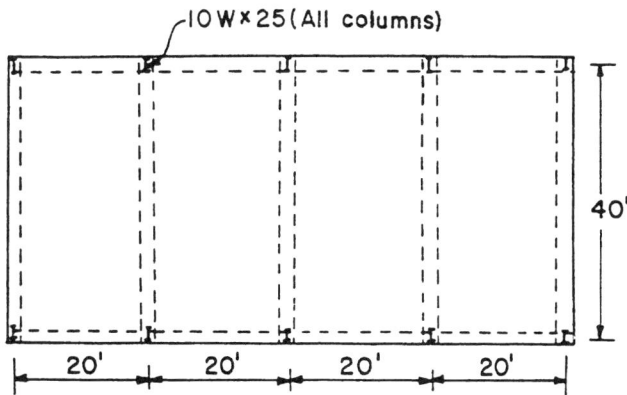

Figure 4.38. Top view of a two-story building.

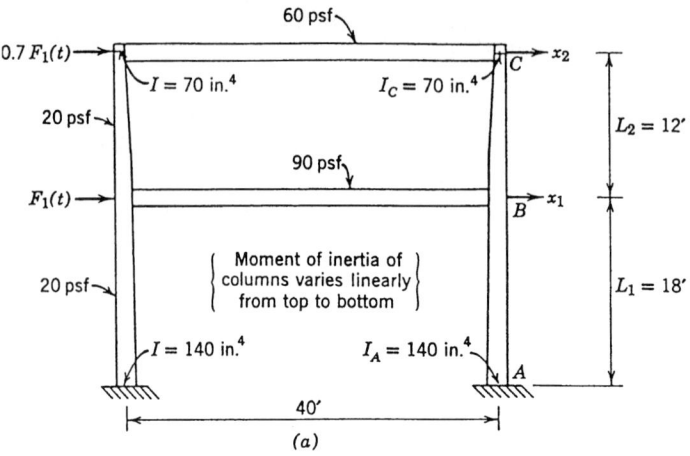

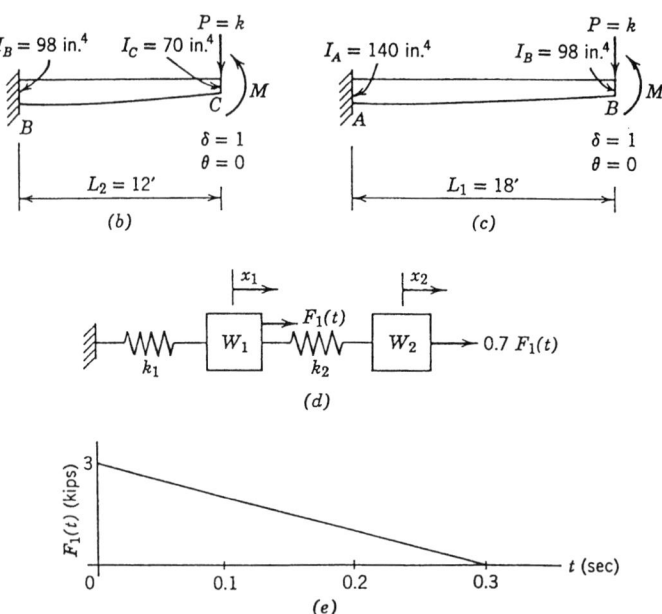

Figure 4.39. (a) Interior frame of the building. (b) Free-body diagram of column BC. (c) Free-body diagram of column AB. (d) Idealized spring-mass system of the frame. (e) Variation of the dynamically applied forces.

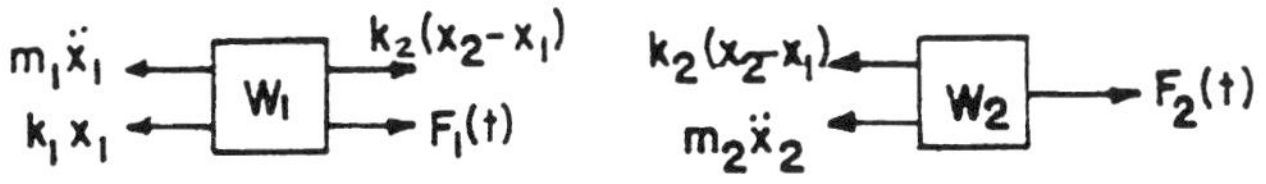

DYNAMIC RESPONSE OF RIGID FRAMES OF VARIABLE STIFFNESS 287

Figure 4.40. Free-body diagrams of W_1 and W_2.

The weight of the frame and associated wall and floor areas is lumped at the floor levers as in Chapter 3, and they are shown as lumped weights W_1 and W_2 in Fig. 4.39d. They are as follows:

$$W_1 = (90)(40)(20) + (2)(20)(15)(15)$$
$$= 84,000 \text{ lb}$$
$$= 84.0 \text{ kips}$$
$$W_2 = (60)(40)(20) + (2)(20)(6)(20)$$
$$= 52,800 \text{ lb}$$
$$= 52.8 \text{ kips}$$

The corresponding masses m_1 and m_2 are

$$m_1 = \frac{W_1}{g} = \frac{84.0}{386.0} = 0.217 \frac{\text{kip-sec}^2}{\text{in.}}$$

$$m_2 = \frac{W_2}{g} = \frac{52.8}{386.0} = 0.137 \frac{\text{kip-sec}^2}{\text{in.}}$$

By utilizing the free-body diagrams in Fig. 4.40 and applying Newton's second law of motion, the following two differential equations of motion are obtained:

$$\frac{W_1}{g}\ddot{x}_1 + k_1 x_1 - k_2(x_2 - x_1) = F_1(t) \tag{4.70}$$

$$\frac{W_2}{g}\ddot{x}_2 + k_2(x_2 - x_1) = F_2(t) \tag{4.71}$$

or

$$m_1 \ddot{x}_1 + k_1 x_1 - k_2(x_2 - x_1) = F_1(t) \tag{4.72}$$

$$m_2 \ddot{x}_2 + k_2(x_2 - x_1) = F_2(t) \tag{4.73}$$

The free undamped frequencies of variation of the idealized system in Fig. 4.39d may be determined by solving the following two homogeneous differential equations:

$$m_1\ddot{x}_1 + k_1 x_1 - k_2(x_2 - x_1) = 0 \qquad (4.74)$$

$$m_2\ddot{x}_2 + k_2(x_2 - x_1) = 0 \qquad (4.75)$$

The solution of Eqs. (4.74) and (4.75) yields the following expression for the required two frequencies ω_1 and ω_2:

$$\omega_{1,2}^2 = \frac{([(k_1+k_2)/m_1]+(k_2/m_2)) \pm \sqrt{([(k_1+k_2)/m_1]+(k_2/m_2))^2 - 4(k_1 k_2/m_1 m_2)}}{2}$$

$$(4.76)$$

By substituting the values for the k's and the m's in Eq. (4.76), we obtain

$$\omega_1 = 4.69 \text{ rps}$$
$$\omega_2 = 16.00 \text{ rps}$$

In hertz these frequencies are

$$f_1 = 0.747 \text{ Hz}$$
$$f_2 = 2.545 \text{ Hz}$$

The smallest period of vibration τ_2 is

$$\tau_2 = \frac{1}{f_2} = \frac{1}{2.545} = 0.393 \text{ sec}$$

The dynamic displacements x_1 and x_2 of the frame in Fig. 4.39a may be determined by using the idealized system in Fig. 4.39d and applying known rigorous or numerical methods of analysis. The numerical AIEM discussed in Section 3.5.2 is used here.

The time variation of the applied dynamic force $F_1(t)$ is shown in Fig. 4.39e. By using Eqs. (4.72) and (4.73) and making the required substitutions, the solution for $\ddot{x}_1$ and $\ddot{x}_2$ is

$$\ddot{x}_1 = 92.7(x_2 - x_1) - 38.3x_1 + 4.65F_1(t) \qquad (4.77)$$

$$\ddot{x}_2 = -146.5(x_2 - x_1) + 5.10F_1(t) \qquad (4.78)$$

The AIEM for the computation of the displacements x_1 and x_2 is carried out as shown in Table 4.4.

TABLE 4.4. AIEM to Determine the Displacements x_1 and x_2 at the Indicated Time Stations

(1) t (sec)	(2) $92.7(x_2-x_1)$	(3) $-38.3x_1$	(4) $4.65F_1(t)$	(5) $\ddot{x}_1$ (in./sec^2) Eq. (4.77)	(6) $\ddot{x}_1(\Delta t)^2$ (in.)	(7) x_1 (in.)	(8) $-146.5(x_2-x_1)$	(9) $5.10F_1(t)$	(10) $\ddot{x}_2$ (in./sec^2) Eq. (4.78)	(11) $\ddot{x}_2(\Delta t)^2$ (in.)	(12) x_2 (in.)
0	0	0	13.95	13.95	0.0126	0	0	15.30	15.30	0.0138	0
0.03	0.056	−0.241	12.55	12.37	0.0111	0.0063[a]	−0.088	13.75	13.66	0.0123	0.0069[b]
0.06	0.222	−0.910	11.15	10.46	0.0094	0.0237	−0.352	12.25	11.90	0.0107	0.0261
0.09	0.509	−1.930	9.75	8.33	0.0075	0.0505	−0.806	10.70	9.89	0.0089	0.0560
0.12	0.927	−3.250	8.35	6.03	0.0054	0.0848	−1.465	9.18	7.71	0.0069	0.0948
0.15	1.483	−4.770	6.95	3.66	0.0033	0.1245	−2.345	7.65	5.30	0.0048	0.1405
0.18	2.180	−6.420	5.58	1.34	0.0012	0.1675	−3.440	6.12	2.68	0.0024	0.1910
0.21	2.980	−8.300	4.18	−1.14	−0.0010	0.2117	−4.720	4.58	−0.14	−0.0001	0.2439
0.24	3.880	−9.780	2.79	−3.11	−0.0028	0.2549	−6.150	3.06	−3.09	−0.0028	0.2967
0.27	4.760	−11.300	1.40	−5.14	−0.0046	0.2953	−7.530	1.56	−5.97	−0.0054	0.3467
0.30	5.581	−12.681	0	−7.10	−0.0064	0.3311	−8.819	0	−8.82	−0.0079	0.3913
0.33	6.257	−13.807	0	−7.55	−0.0068	0.3605	−9.889	0	−9.89	−0.0089	0.4280
0.36	6.739	−14.673	0	−7.93	−0.0071	0.3831	−10.651	0	−10.651	−0.0096	0.4558
0.39	6.990	−15.266	0	−8.28	−0.0185	0.3986	−11.046	0	−11.046	−0.0099	0.4740
0.42	8.037	−15.151	0	−7.114	−0.0064	0.3956	−12.702	0	−12.702	−0.0114	0.4823
0.45						0.3862					0.4792

[a]Use Eq. (4.82) to determine this amplitude.
[b]Use Eq. (4.84) to determine this amplitude.

The time interval Δt is selected to be equal to 0.03 sec, which is in the vicinity of $\tau_2/10$. This provides ten time stations within the time duration of $F_1(t)$, which yields accurate results for practical applications even if the maximum response occurs within the time duration of $F_1(t)$. At each time station, the accelerations $\ddot{x}_1$ and $\ddot{x}_2$ are determined from Eqs. (4.77) and (4.78). For convenience, the terms of Eq. (4.77) are shown in columns 2 through 5 in Table (4.4), and the terms of Eq. (4.78) are included in columns 8 through 10.

The displacements $x_1^{(i+1)}$ and $x_2^{(i+1)}$ at any time station $(i+1)$ are obtained by using the equations

$$x_1^{(i+1)} = 2x_1^{(i)} - x_1^{(i-1)} + \ddot{x}_1^{(i)}(\Delta t)^2 \tag{4.79}$$

$$x_2^{(i+1)} = 2x_2^{(i)} - x_2^{(i-1)} + \ddot{x}_2^{(i)}(\Delta t)^2 \tag{4.80}$$

For the first time station, the values of $x_1^{(1)}$ and $x_2^{(1)}$ may be obtained from the following equations:

$$x_1^{(1)} = \tfrac{1}{6}(2\ddot{x}_1^{(0)} + \ddot{x}_1^{(1)})(\Delta t)^2 \tag{4.81}$$

$$x_1^{(1)} = \frac{\ddot{x}_1^{(0)}}{2}(\Delta t)^2 \tag{4.82}$$

$$x_2^{(1)} = \tfrac{1}{6}[2\ddot{x}_2^{(0)} + \ddot{x}_2^{(1)}](\Delta t)^2 \tag{4.83}$$

$$x_2^{(1)} = \frac{\ddot{x}_2^{(0)}}{2}(\Delta t)^2 \tag{4.84}$$

In Eqs. (4.81) and (4.83) the acceleration is assumed to be linear during the first time interval, and in Eqs. (4.82) and (4.84) the acceleration is assumed to be constant during the first time interval and equal to the acceleration at time $t = 0$. The products $\ddot{x}_1(\Delta t)^2$ and $\ddot{x}_2(\Delta t)^2$ are shown in columns 6 and 11, respectively, of Table 4.4. The displacements $x_1^{(1)}$ and $x_2^{(1)}$ at the first time station are determined by using Eqs. (4.82) and (4.84), respectively. The displacements x_1 and x_2 at any time station are shown in columns 7 and 12, respectively, of the table. The maximum displacement $x_{1\max}$ occurs at time $t = 0.39$ sec and is equal to 0.3986 in., and the maximum displacement $x_{2\max}$ is at $t = 0.42$ sec and is equal to 0.4823 in.

The total maximum shear force $V_{1\text{total}}$ at the top of the first-story columns is

$$V_{1\text{total}} = k_1 x_{1\max} = (8.32)(0.3986)$$
$$= 3.3164 \text{ kips}$$

Since both columns of the first story of the frame are of the same stiffness, the total shear force will be distributed equally to each column. Thus, each column

DYNAMIC RESPONSE OF RIGID FRAMES OF VARIABLE STIFFNESS   291

receives the value

$$V_B = \frac{3.3164}{2} = 1.6582 \text{ kips}$$

where the letter B is used to designate the top of the first-story columns.
At the top of the columns of the second story, we have

$$V_{2\text{total}} = k_2(x_2 - x_1)_{\max}$$
$$= (20.10)(0.4823 - 0.3956)$$
$$= 1.7427 \text{ kips}$$

Each of the two columns of the second story will get the same portion because they have the same stiffness. Thus,

$$V_C = \frac{1.7427}{2} = 0.8714 \text{ kips}$$

where the letter C is used to designate the top of the second story columns.

The bending moments at the ends of each column can be determined by using the appropriate problem in Figs. 4.39b and 4.39c. For example, if the moments at the ends of the second-story columns are required, Fig. 4.39b should be used. In this figure, the maximum value of P should be taken as equal to V_C, which is 0.8714 kips. Thus, with $\delta = (x_2 - x_1)_{\max}$ and $\theta = 0$, the moment M in Fig. 4.39b is the moment M_C at the top of a second-story column in Fig. 4.39a. Since M in Fig. 4.39b is derived for a deflection equal to unity, the maximum moment M_C should be equal to $M(x_2 - x_1)_{\max}$. With known M_C, the moment at end B can be determined by applying simple statics. For the first-story columns, Fig. 4.39c should be used.

At $t = 0.42$ sec, the moments M_C and M_B at the top and bottom, respectively, in each of the second-story columns, are

$$M_C = 52.95 \text{ kip-in.} \quad \text{(counterclockwise)}$$
$$M_B = 70.14 \text{ kip-in.} \quad \text{(counterclockwise)}$$

With known bending moments, the dynamic stresses produced by these moments can be determined by using known bending stress formulas. Similar procedure may be followed to determine the moments M_B and M_A at the top and bottom, respectively, in each of the first-story columns. In this case, the top of each of the first-story columns will be subjected to a bending moment of 145.48 kip-in., and at the bottom the bending moment would be 212.69 kip-in.

PROBLEMS

4.1 Determine an exact equivalent system of uniform stiffness for each of the beams loaded as shown in Fig. P4.1. In each case the modulus of elasticity and the width b are constant. Neglect the weight of the members.

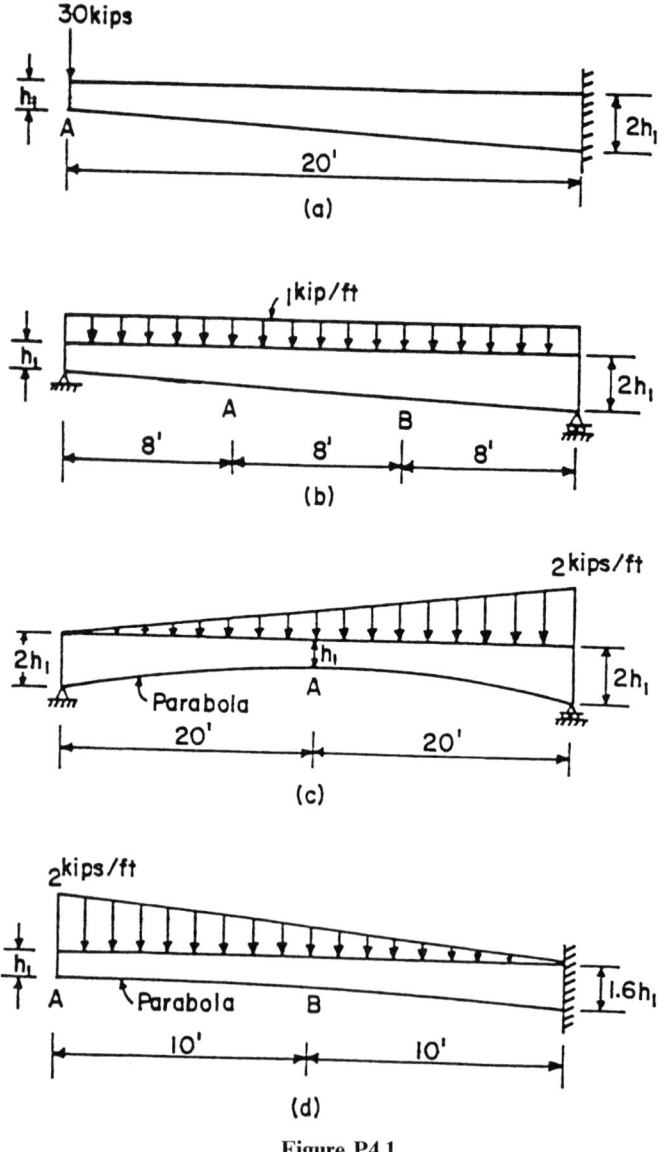

Figure P4.1.

4.2 By using the exact equivalent systems derived in Problem 4.1, determine in each case the vertical displacements and rotations at points A and (where appropriate) B.

4.3 If the variable stiffness beams in Problem 4.1 are made of steel and have a rectangular cross section of width $b = 8$ in. and depth $h_1 = 14$ in., determine an equivalent system in each case by applying the approximate method of the equivalent systems. The modulus of elasticity E is 30×10^6 psi. Neglect the weight of the members.

4.4 By using the equivalent systems derived in Problem 4.3, determine in each case the vertical displacements and rotations of points A and (where appropriate) B.

4.5 The variable stiffness steel beams loaded as shown in Fig. P4.5 are of

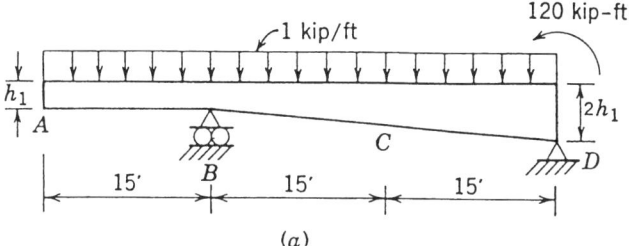

(a)

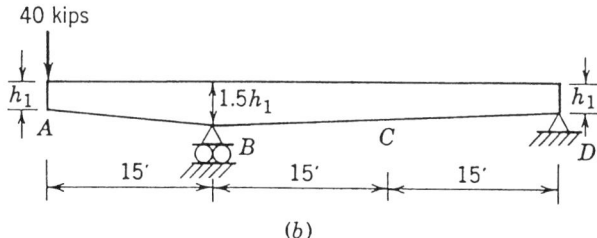

(b)

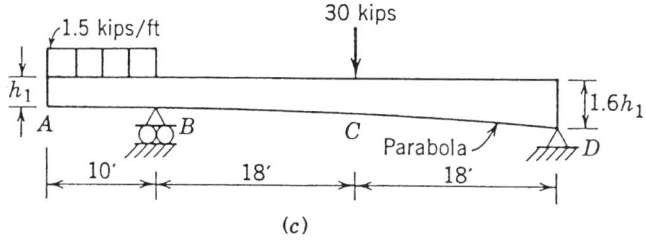

(c)

Figure P4.5.

rectangular cross section of width $b = 8$ in. and depth $h_1 = 16$ in. Determine in each case an equivalent system of uniform stiffness by applying the approximate method of the equivalent systems. The modulus of elasticity E is 30×10^6 psi. Neglect the weight of the members.

4.6 By using the equivalent systems derived in Problem 4.5, determine in each case the vertical displacements and rotations at points A and C.

4.7 The beam in Fig. P4.7 is a reinforced concrete girder of a highway bridge with a linear variation in depth. The distributed weight on the girder includes the dead weight of the girder and the associated dead weight of the bridge deck. The concentrated weights represent a standard H-S truck in accordance with AASHTO standard specifications and they are located as shown. By using the approximate method of the equivalent systems discussed in Sections 4.3 and 4.4, determine an accurate equivalent system of constant stiffness EI_1, where I_1 is the moment of inertia at the ends of the girder. The constant modulus $E = 30 \times 10^3$ ksi and $EI_1 = 460 \times 10^6$ kip-in.^{2} The depth $h_1 = 2.75$ ft.

4.8 Repeat the bridge girder problem in Example 4.3, and determine the position of the H-S truck that will yield the maximum vertical deflection in the member. Compare the results.

4.9 Repeat Problem 4.7 by placing the H-S truck loading on the second span with the first axle load located at a distance of 5 ft from the middle support.

4.10 By utilizing the equivalent system obtained in Problem 4.7, determine the vertical deflection at the center of the first span.

4.11 By utilizing the equivalent system obtained in Problem 4.9, determine the vertical deflection at the center of the first and second spans of the girder.

4.12 The variable stiffness beams in Fig. P4.12 are made of steel of rectangular cross section with constant width $b = 10$ cm and depth $h_1 = 15$ cm.

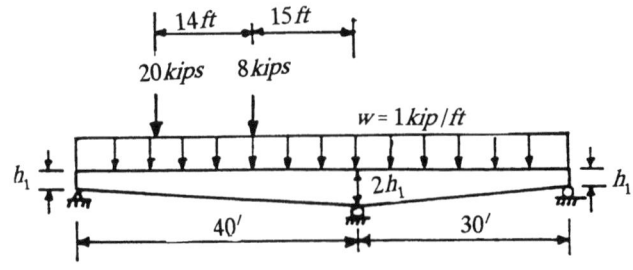

Figure P4.7.

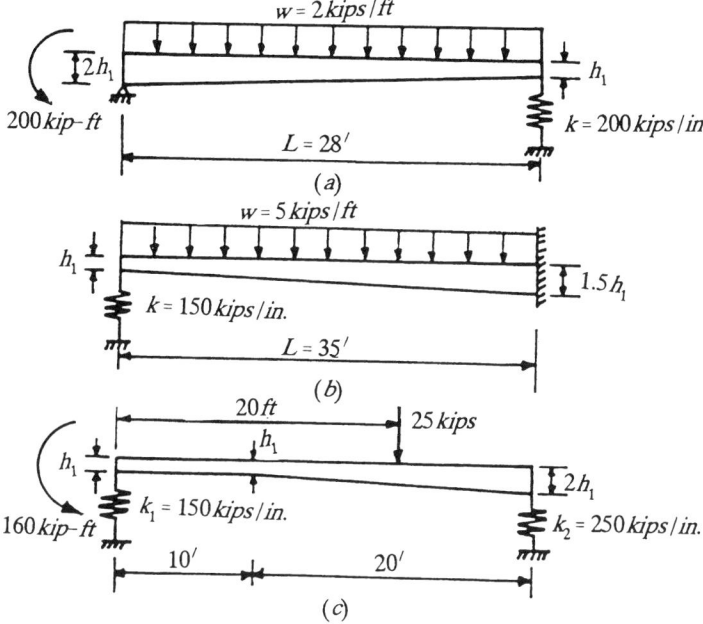

Figure P4.12.

They are loaded as shown in the figure. By using the approximate method of the equivalent systems discussed in Sections 4.3 and 4.4, determine in each case the vertical deflection at the center of the beam. The modulus of elasticity E is 207×10^6 kPa.

4.13 Repeat Problem 4.5 by assuming that the members are fixed at support D.

4.14 By using the equivalent systems obtained in Problem 4.13, determine for each case the rotations and vertical deflections at points A and C.

4.15 The beams in Fig. P4.15 are loaded as shown. By using the approximate method of the equivalent systems discussed in Section 4.4, determine in each case an equivalent system of constant stiffness EI_B, where I_B is the moment of inertia at the end B. The modulus of elasticity E is constant.

4.16 The beams in Fig. P4.16 are loaded as shown. By applying the moment-area method determine in each case the rotation at points A, B, C, and (where appropriate) D, and the vertical deflections at points C and (where appropriate) D. The stiffness EI is constant, $E = 207 \times 10^6$ kPa, and $I = 0.4 \times 10^{-3}$ m^4.

4.17 Repeat Problem 4.16 by using the conjugate-beam method.

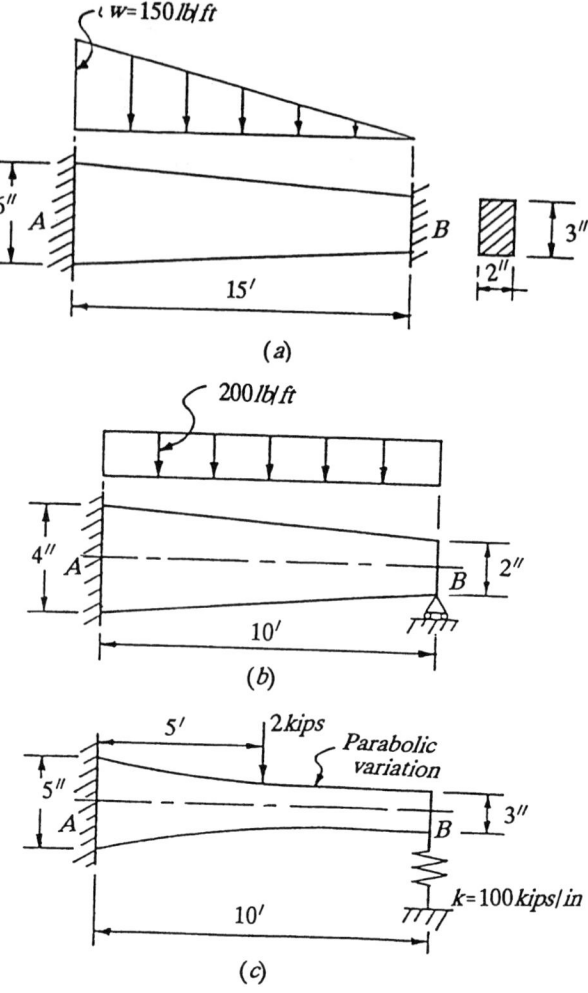

Figure P4.15.

4.18 Repeat Problem 4.16 by assuming that end A is fixed and compare the results.

4.19 Repeat Problem 4.18 by using the conjugate-beam method.

4.20 The uniform statically indeterminate beams are loaded as shown in Fig. P4.20. By applying the moment-area method and using superposition, determine in each case the rotation at the supports and the vertical deflection of point C. The constant stiffness $EI = 30 \times 10^6$ kip-in^2.

4.21 Repeat Problem 4.20 by using the conjugate-beam method.

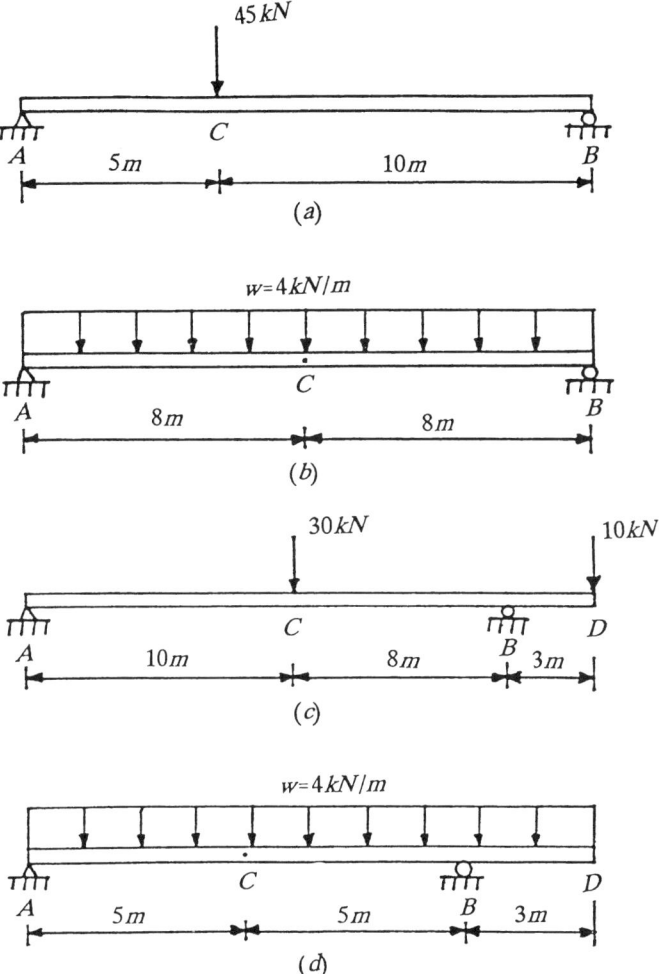

Figure P4.16.

4.22 For the beam system in Fig. P4.22, determine the force acting between the two beams at point B and the vertical deflection at points B and D, by using the moment-area method and superposition. The modulus of elasticity E is constant for both beams and the moment of inertia is $2I_0$ for beam AB and I_0 for beam BC.

4.23 Repeat Problem 4.22 by using the conjugate-beam method.

4.24 By using the moment distribution method, determine the moments at the joints of the frame loaded as shown in Fig. P4.24. The stiffness EI is constant.

298 ADVANCED METHODS AND PROBLEMS OF INFRASTRUCTURAL ANALYSIS

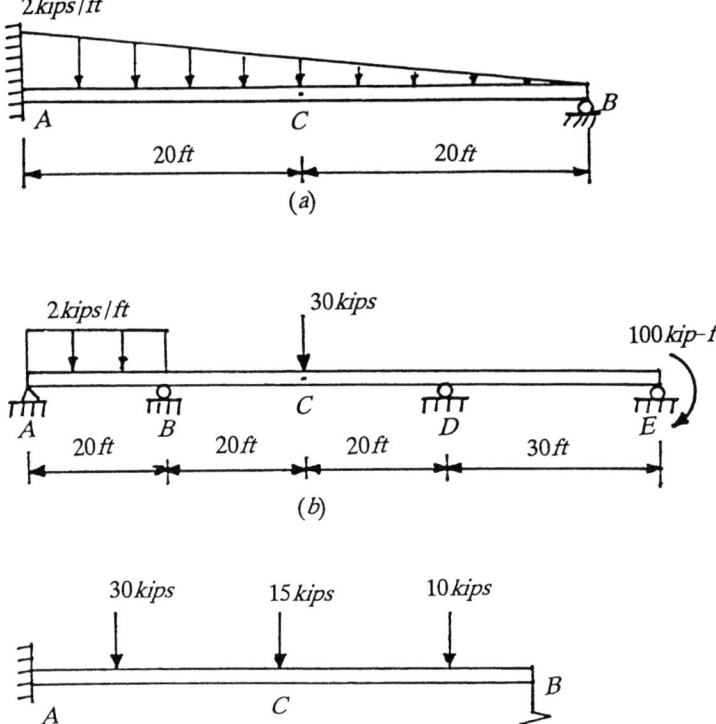

Figure P4.20.

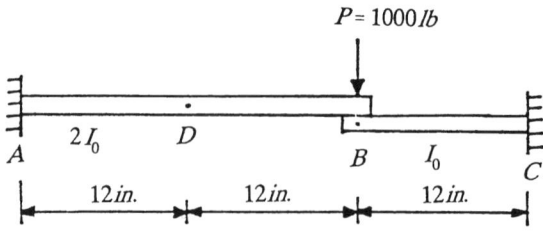

Figure P4.22.

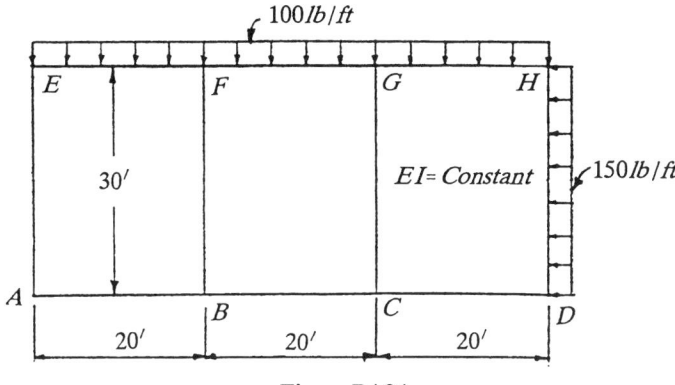

Figure P4.24.

4.25 By using the moment distribution method, determine the reactions at the supports of the symmetrical frame loaded as shown in Fig. P4.25. Assume rigid joints and that there is no sidesway.

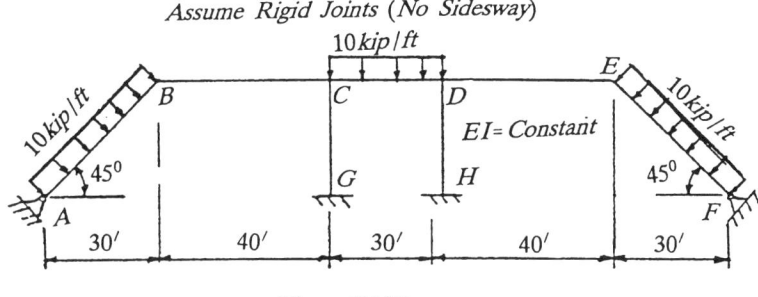

Figure P4.25.

4.26 For the problem in Fig. P4.20b, determine the internal moments and the reactions at the supports by using the moment distribution method. Also, determine the vertical deflection at point C. The constant stiffness $EI = 30 \times 10^6$ kip-in^2.

4.27 For the variable thickness bridge girders in Fig. P4.7, determine all moments at the supports and all support reactions by using the moment distribution method as discussed in Section 4.7. See also Problem 4.7 for required additional information.

4.28 The beam in Fig. P4.28 represents an interior girder of bridge $B2$ of 38-1-14 over U.S. 12, east of Jackson, Michigan. The beam is loaded with its own weight and the tributary bridge deck. By using the moment distribution method, determine the bending moments at the internal supports. The modulus of elasticity $E = 30 \times 10^6$ psi and the constant moment of inertia $I = 58 \times 10^3$ in.^{4}

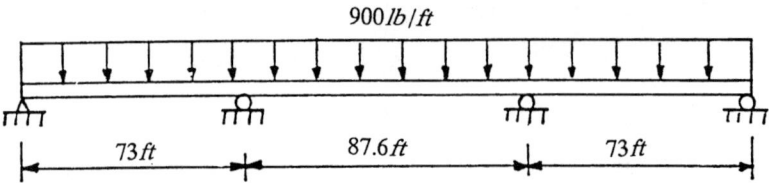

Figure P4.28.

4.29 By using the three-moment equation method, solve the problem in Fig. 4.20b, and compare the results with those already obtained.

4.30 Repeat Problem 4.28 by using the three-moment equation method. Also, determine all reactions at the supports.

4.31 The girders of the steel frames in Fig. P4.31 are assumed to be infinitely rigid compared to the columns. The dynamic force $F_1(t)$ is a suddenly applied force of 6 kips at $t = 0$ that lasts for an infinite period of time. By using the approximate method of the equivalent systems and the AIEM, determine the maximum horizontal displacements at the girder levels and the maximum bending stresses in the columns. The modulus of elasticity E is 30×10^6 psi. Neglect the weight of the columns.

4.32 Repeat Problem 4.31 for the case where the dynamic force $F_1(t)$ is a suddenly applied force of 6 kips at $t = 0$ that decreases linearly to zero at $t = 0.4$ sec.

PROBLEMS **301**

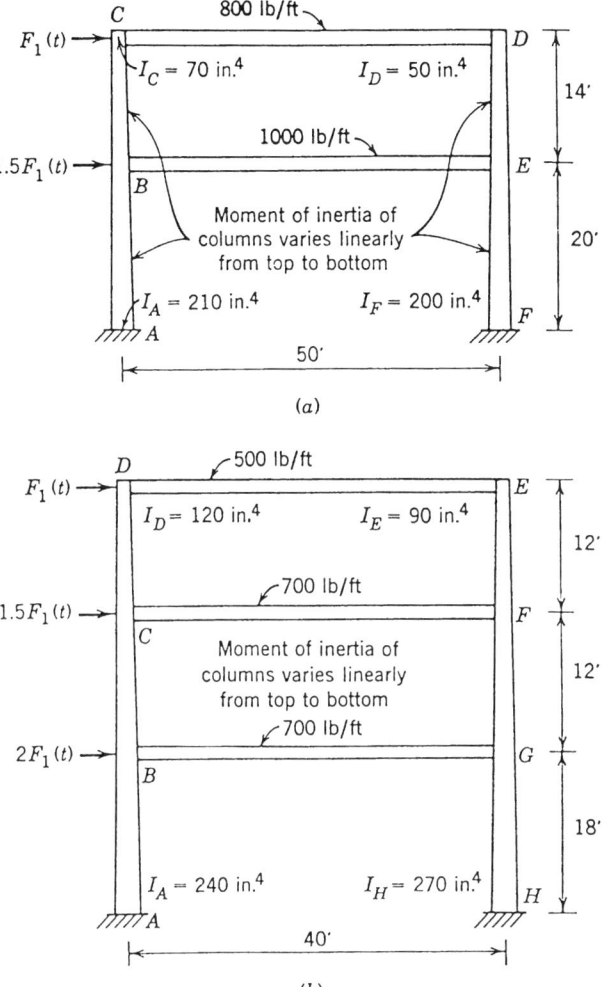

Figure P4.31.

5 Infrastructural Nonlinearities, Instabilities, and Inelastic Response

5.1 INTRODUCTION

Infrastructural elements can be nonlinear, they can experience structural instabilities when they are not appropriately designed, and the applied loads may be of magnitudes that make the structure respond inelastically. For example, several manufacturing processes require the product to be subjected to very large geometric deformations, and flexible members and aircraft structures can experience large deformations during their life span. In order for such elements to maintain their structural integrity, we need to evaluate the degree of such large deformations and their impact on the rest of the structure. The members of a truss structure that are subjected to compressive forces should be designed so that failures caused by buckling do not occur, and the effect of structural column eccentricities regarding the stability of the total structure should be carefully examined. On the other hand, when infrastructural elements such as buildings, bridges, highways, dams, water systems, and the like, are subjected to forces produced by an earthquake or an explosion of some kind, it is very possible that the stresses caused by such loads could exceed the elastic limit of their material, and the total response of the structure would be inelastic. These areas of study are increasingly important today, and design procedures have to be developed in order to safeguard the health and safety of our people.

The work in this chapter provides basic knowledge regarding such problems, and provides methods that can be used for a satisfactory and safe design. More information on these subjects may be found in the work by Fertis [3, 5, 6, 37, 39].

5.2 INFRASTRUCTURAL NONLINEAR RESPONSE

The renewed interest in nonlinear analysis and mechanics can be attributed to minimum weight criteria in the design of aircraft and aerospace vehicles, and to the larger utilization of polymer material that can undergo large deformations without becoming inelastic. Members and similar structures that are subjected to large deformations are usually called flexible members, or flexible structures. Due to the geometry of their deformation, the response of such structures is highly nonlinear, and the solution, as well as the design, of such flexible structures becomes very complex. The solution becomes immensely complex when structural components are of variable cross section along their length. Such members are often used by engineers to improve strength and weight criteria, and by architects and planners to improve the aesthetic appearance of our infrastructure.

In this section we deal with uniform and variable stiffness flexible members that are subjected to various types of loading. We present methods, developed by the author [3, 5, 6] and his collaborators, that use pseudolinear and simplified nonlinear equivalent systems to simplify the solution of such complex nonlinear problems.

5.2.1 The Elastica Theory

The exact shape of the deflection curve of a flexible member subjected to large deformations is known as the "elastica." The first known published work concerning the large deflection of flexible members was written by Euler [46] in 1744. In the appendix of his book *Des Curvis Elastics*, Euler stated that the slope of the deflection curve cannot be neglected in the expression of the curvature unless the deflections are small. In general, the development of the theory took place in the 18th century, and it should be credited to Jacob Bernoulli, his younger brother Johann Bernoulli, and Leonard Euler.

The extensively used Euler–Bernoulli law states that the bending moment M is proportional to the change in the curvature produced by the action of the load. Mathematically, we write this law as follows:

$$\frac{1}{\rho} = \frac{d\theta}{dx_0} = \frac{M}{EI} \tag{5.1}$$

where ρ is the radius of curvature and θ is the slope at any point x_0, where x_0 is measured along the length of the arc as shown in Fig. 5.1a. In this figure, the large deformation configuration of a uniform cantilever beam subjected to a concentrated load P at its free end is shown.

In rectangular x, y coordinates, Eq. (5.1) is written as

$$\frac{1}{\rho} = \frac{y''}{[1+(y')^2]^{3/2}} = -\frac{M}{EI} \tag{5.2}$$

where y is the vertical deflection at any distance x, and y' and y'' are the first and second derivatives with respect to x, respectively. Equation (5.2) is a nonlinear second order differential equation, and its exact solution is extremely difficult to obtain. This equation shows that the deflections are no longer a linear function of the bending moment, or of the load, and consequently the

INFRASTRUCTURAL NONLINEAR RESPONSE 305

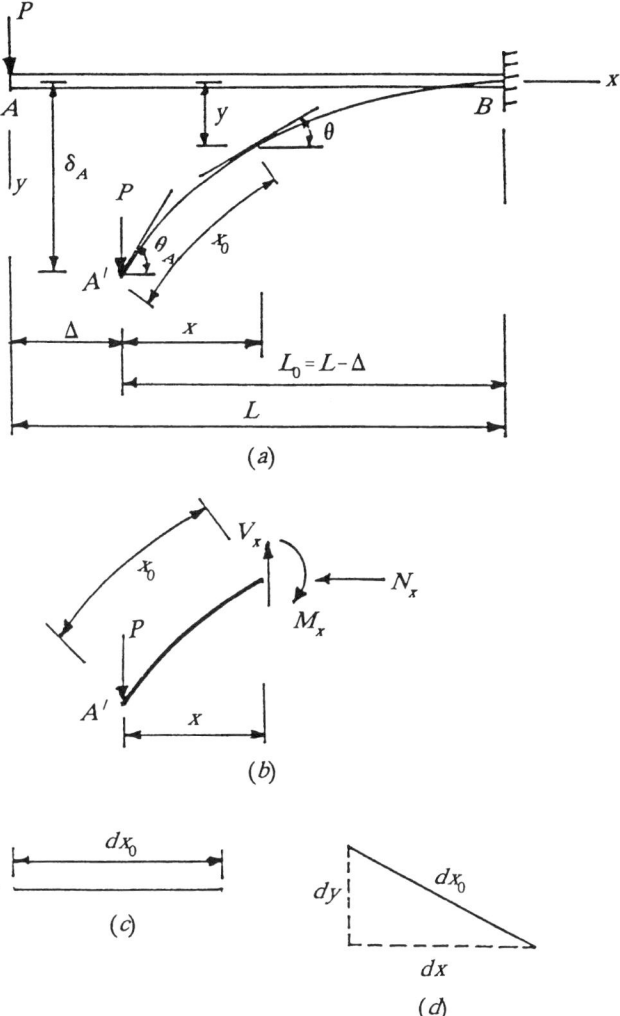

Figure 5.1. (*a*) Large deformation of a uniform flexible member. (*b*) Free-body diagram of an element of the member. (*c*) Element of the member of length dx_0. (*d*) Deformation of the element.

principles of superposition does not apply. Thus, every case that involves large deformations has to be solved separately, because combinations of load types already solved cannot be superimposed. The consequences become even greater when the stiffness EI of the member is variable.

The author and his collaborators have developed methods that simplify a great deal the solution of such complicated problems.

5.2.2 Pseudolinear Equivalent Systems of Constant Stiffness

We rewrite Eq. (5.2) as follows:

$$\frac{y''}{[1+(y')^2]^{3/2}} = -\frac{M_x}{E_x I_x} \qquad (5.3)$$

where M_x, E_x, and I_x are assumed to be variable. The left-hand side of Eq. (5.3) represents the curvature of the flexible member, and it requires that M_x, E_x, and I_x on the right-hand side of the equation be associated with the geometry of the deformation. They are, in general, nonlinear integral equations of the deformation, and contain functions of horizontal displacement. This implies that bending moment M_x, depth h_x of the flexible member, and moment of inertia I_x are all functions of both x and x_0.

To derive a pseudolinear equivalent system, we follow the procedure discussed in Section 4.2. We express the stiffness $E_x I_x$ as follows:

$$E_x I_x = E_1 I_1 \, g(x) f(x)$$

where the function $g(x)$ represents the variation of E_x with reference value E_1, and $f(x)$ represents the variation of I_x with reference value I_1.

By substituting into Eq. (5.3), we obtain

$$\frac{y''}{[1+(y')^2]^{3/2}} = -\frac{1}{E_1 I_1} \cdot \frac{M_x}{g(x) f(x)} \qquad (5.4)$$

We integrate Eq. (5.4) twice to obtain the expression of the large deformation $y(x)$, as follows:

$$y(x) = \frac{1}{E_1 I_1} \int \left\{ -\int [1+(y')^2]^{3/2} \frac{M_x}{g(x) f(x)} \, dx \right\} dx + C_1 \int dx + C_2 \qquad (5.5)$$

where C_1 and C_2 are the constants of integration.

By selecting a flexible member of constant stiffness $E_1 I_1$ that has the same length as the member represented by Eq. (5.5) and the same reference system of axes, the expression for its large deflection $y_e(x)$ is

$$y_e(x) = \frac{1}{E_1 I_1} \int \left\{ -\int [1+(y_e')^2]^{3/2} M_e \, dx \right\} dx + C_1' \int dx + C_2' \qquad (5.6)$$

where M_e is the bending moment at any x, and C_1' and C_2' are the constants of integration.

By examining Eqs. (5.5) and (5.6), we note that y and y_e will be identical if the following conditions are satisfied:

$$C_1 = C_1' \qquad C_2 = C_2' \tag{5.7}$$

$$\int \left\{ -\int [1 + (y')^2]^{3/2} \frac{M_x}{g(x) f(x)} dx \right\} dx = \int \{ -\int [1 + (y_e')^2]^{3/2} M_e dx \} dx \tag{5.8}$$

The conditions imposed by Eq. (5.7) may be satisfied if the two members have the same length and the same boundary conditions. Equation (5.8) will be satisfied if $y' = y_e'$ and

$$M_e = \frac{M_x}{g(x) f(x)} \tag{5.9}$$

Thus, Eq. (5.8) may be written as

$$[1 + (y_e')^2]^{3/2} M_e = [1 + (y')^2]^{3/2} \frac{M_x}{g(x) f(x)} \tag{5.10}$$

For small deflection theory, Eq. (5.10) reduces to Eq. (5.9), which is identical to Eq. (4.10) in Section 4.2, because y' and y_e' would be small compared with unity. By making $y' = y_e'$, Eqs. (5.10) and (5.4) suggest that the moment M_e' at any cross section x of the pseudolinear system of constant stiffness $E_1 I_1$ may be determined from the following equation:

$$M_e' = M_e [1 + (y')^2]^{3/2} = \frac{M_x z_e}{g(x) f(x)} \tag{5.11}$$

where

$$z_e = [1 + (y')^2]^{3/2} \tag{5.11a}$$

and $\theta = \tan^{-1}(y')$ is the slope of the initial system at any distance x. On this basis, Eq. (5.4) takes the form

$$y'' = -\frac{M_e'}{E_1 I_1} \tag{5.12}$$

Equation (5.12) is the differential equation of a pseudolinear system of constant stiffness $E_1 I_1$. When M_e' is known, or when we are able to determine it, then linear analysis may be used to solve the pseudolinear system. The shear

308 INFRASTRUCTURAL NONLINEARITIES

force V'_e and loading w'_e of the pseudolinear system are as follows:

$$V'_e = \frac{d}{dx}(M'_e) = \frac{d}{dx}\left(\frac{z_e}{g(x)\,f(x)}\right) M_x \tag{5.13}$$

$$w'_e = -\frac{d}{dx}(V'_e)\cos\theta = -\frac{d^2}{dx^2}\left(\frac{z_e}{g(x)\,f(x)}\right) M_x \cos\theta \tag{5.14}$$

5.2.3 Simplified Nonlinear Equivalent Systems of Constant Stiffness

Since the principle of superposition does not apply in nonlinear analysis, the solution of the problem can become very complicated, even with the utilization of pseudolinear analysis. This complexity can occur when the member is loaded with complicated loading combinations and when its moment of inertia I_x or its modulus E_x varies in some arbitrary manner along its length. In such cases, it would be advisable to derive first a simplified nonlinear equivalent system of constant stiffness $E_1 I_1$, and then solve it by using pseudolinear analysis. In many cases, the simplified nonlinear equivalent system can be derived in a way that makes it possible to use existing solutions of nonlinear analysis. All the above are possible, because in the theory of equivalent systems we use and match curvatures, and then work with the curvature dependent moment to recreate a loading that matches the curvatures. In this manner, we make it possible to work with complicated loading conditions and stiffness variations without using superposition. Such complicated situations can also be handled comparatively easily and with excellent accuracy.

The derivation of simplified nonlinear equivalent systems may be initiated by considering Eq. (5.4) and rewriting it as follows:

$$\frac{y''}{[1+(y')^2]^{3/2}} = -\frac{d}{dx}\left(\frac{M_x}{g(x)\,f(x)}\right) \tag{5.15}$$

where

$$M_e = \frac{M_x}{g(x)\,f(x)} \tag{5.16}$$

Equation (5.15) is the nonlinear differential equation of the simplified nonlinear equivalent system of constant stiffness $E_1 I_1$. Its bending moment M_e at any cross section may be determined by using Eq. (5.16), and the following equations may be used to obtain the shear force V_e and the loading w_e:

$$V_e = \frac{d}{dx}(M_e) = \frac{d}{dx}\left(\frac{M_x}{g(x)\,f(x)}\right) \tag{5.17}$$

$$w_e = -\frac{d}{dx}(V_e)\cos\theta = -\frac{d^2}{dx^2}\left(\frac{M_x}{g(x)\,f(x)}\right)\cos\theta \tag{5.18}$$

Thus, the solution of the complicated initial problem may be obtained from Eq. (5.15). Since it represents a simpler nonlinear problem, it can be solved more conveniently, using pseudolinear analysis as discussed earlier. In fact, the solution can be made much easier if we approximate the shape of the M_e diagram with straight-line segments, as was done in Section 4.2. In this manner, the equivalent nonlinear system will always be loaded with a few concentrated loads, which makes the application of the pseudolinear analysis very convenient. Similar approximation can be applied to the M'_e diagram of the pseudolinear system.

The application of the above theories is demonstrated in the following sections of this chapter. See also References [3, 5, 6] for greater detail.

5.2.4 Loading and Stiffness Dependence on the Geometry of Deformation

Before we proceed with the application of the above two theories and methods of equivalent systems, we discuss the dependence of M_x and I_x on the geometry of the deformation of flexible members. M_x and I_x are, in general, nonlinear functions of the large deformation, that is,

$$M_x = M(x, x_0) \qquad (5.19)$$

$$I_x = I_1 f(x, x_0) \qquad (5.20)$$

where x and x_0 are as shown in Fig. 5.1a, and I_1 is the reference moment of inertia.

We also observe that the bending moment M_e, or M'_e, of the equivalent system should be defined with respect to the deformed configuration of the member, where the exact solutions for M_e, V_e, and w_e are functions of the horizontal displacement $\Delta(x)$ of the member. We can reduce the complexity of the problem if we express the arc length $x_0(x)$ in terms of the horizontal displacement $\Delta(x)$ of the member, where $0 \leqslant x \leqslant (L - x)$. That is,

$$x_0(x) = x + \Delta(x) \qquad (5.21)$$

We also know that $x_0(x)$ is an integral function of the deformation of the form

$$x_0(x) = \int_0^x \{1 + [y'(x)]^2\}^{1/2} \, dx \qquad (5.22)$$

We can derive Eq. (5.21) by considering the segment of length dx_0 in Fig. 5.1c. After deformation, the segment would be deformed as shown in Fig. 5.1d. By making use of the Pythagorean theorem, we write

$$(dx_0)^2 = (dx)^2 + (dy)^2 \qquad (5.23)$$

We assume now that

$$dx_0 = dx + d\Delta(x) \qquad (5.24)$$

By substituting Eq. (5.24) into Eq. (5.23), we find

$$[dx + d\Delta(x)]^2 = (dx)^2 + (dy)^2 \qquad (5.25)$$

or

$$dx + d\Delta(x) = \{1 + [y'(x)]^2\}^{1/2}\,dx \qquad (5.26)$$

By integrating Eq. (5.26) with respect to x, we find

$$x + \Delta(x) = \int_0^x \{1 + [y'(x)]^2\}^{1/2}\,dx \qquad (5.27)$$

Equation (5.27) yields the same results as Eqs. (5.21) and (5.22).

For cases where one end of the flexible member is permitted to move in the horizontal direction, such as cantilever and simply supported beams, the following approximate expressions regarding the variation of the horizontal displacement $\Delta(x)$ are proven [3, 47] to provide accurate results:

$$\Delta(x) = \text{constant} = \Delta \qquad (5.28)$$

$$\Delta(x) = \Delta \frac{x}{L_0} \qquad (5.29)$$

$$\Delta(x) = \Delta \left(\frac{x}{L_0}\right)^{1/2} \qquad (5.30)$$

$$\Delta(x) = \Delta \sin\left(\frac{\pi x}{2L_0}\right) \qquad (5.31)$$

where Δ is the horizontal displacement of the movable end, and $L_0 = (L - \Delta)$. Equation (5.28) is an upper bound. See also Reference [3, Fig. 1.6].

For many problems [3, 47], such as cantilever beams, very accurate results may be obtained by using Eq. (5.28), which means that the variation of the bending moment M_x and, consequently, the deformation of the member, are largely dependent on the boundary condition of $\Delta(x)$ at the moving end of the member, and they are much less sensitive to the variation of $\Delta(x)$ between the ends of the member. For simply supported beams with distributed loadings, the error in deflection and rotation, as calculated using Eq. (5.28), is somewhat higher, but still less than 10 percent. In this case, any one of the expressions given by Eqs. (5.29) through (5.31) should be used if a very accurate solution of the problem is needed. For vibration analysis [3], utilization of Eq. (5.28) yields very accurate results for all beam cases that were investigated.

It should be also noted that the length L of the member may be expressed as follows:

$$L = \int_0^{L_0} [1 + (y')^2]^{1/2} \, dx \qquad (5.32)$$

where $L_0 = (L - \Delta)$. Thus, correlation between x and x_0, and between L_0 and L, can be established by using Eqs. (5.22) and (5.32), respectively. This is mandatory if an exact solution is required because, as stated earlier, the value of the moment of inertia function $f(x)$ during the solution process should be taken as the one at x_0, where $0 \leqslant x_0 \leqslant L$. The coordinate x_0 is also used for the calculation of moments when the load on the member is distributed. Once the fundamentals of the large deformation problem are well understood, the reader should be able to decide what procedure he/she wants to follow. It is recommended, however, that Eqs. (5.28) through (5.31) be used, as explained earlier, since they provide excellent results for practical applications. See Reference [3, Chapters 9, 10] for additional approaches regarding this problem.

Extensive treatment of the large deformation of flexible members for both elastic and inelastic ranges is also given in Reference [3]. The method of the equivalent systems is general, and it can be applied to members of any arbitrary stiffness variations and loading conditions, by including axial loads, cyclic loads, and so on. It is also used for both static and dynamic analysis of frames, plates, and other types of structures.

5.3 SOLUTION OF NONLINEAR PROBLEMS USING PSEUDOLINEAR EQUIVALENT SYSTEMS

The utilization of pseudolinear equivalent systems for the computation of deflections and rotations of uniform and variable stiffness members is illustrated by the following examples.

Example 5.1: Obtain a general solution for the flexible tapered cantilever beam loaded by a concentrated load P at the free end, as shown in Fig. 5.2a. Use a pseudolinear equivalent system of constant stiffness EI_A, where I_A is the moment of inertia at the free end A, and E is the constant modulus of elasticity, which implies that the modulus function $g(x) = 1$. The width b of the member is constant and n is the taper constant.

SOLUTION: The bending moment M_x at any distance $0 \leqslant x \leqslant L_0$ is

$$M_x = Px \qquad (5.33)$$

We note here that M_x is only a function of x because only a vertical concentrated load P is applied at the free end, which is assumed to stay vertical during deformation.

312　INFRASTRUCTURAL NONLINEARITIES

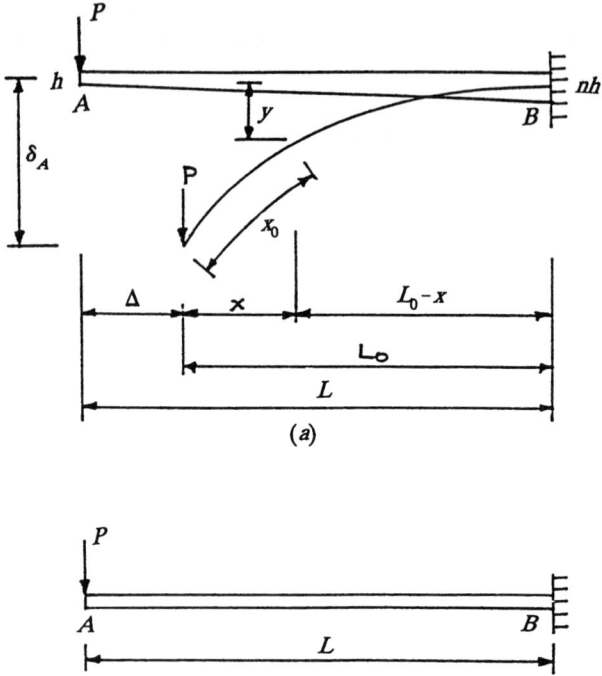

Figure 5.2. (a) Tapered flexible cantilever beam loaded as shown. (b) Uniform flexible cantilever beam loaded as shown.

The moment of inertia I_{x_0} at any $0 \leq x_0 \leq L$ is given by the expression

$$I_x = \frac{bh^3}{12} f(x_0)$$
$$= \left[1 + \frac{(n-1)}{L} x_0\right]^3 I_A = I_A f(x_0) \quad (5.34)$$

where

$$I_A = \frac{bh^3}{12} \quad (5.35)$$

$$f(x_0) = \left[1 + \frac{(n-1)}{L} x_0\right]^3 \quad (5.36)$$

$$x_0 = \int_0^x \{1 + [y'(x)]^2\}^{1/2} \, dx \quad (5.37)$$

Note that Eq. (5.37) is identical to Eq. (5.22), which relates x and x_0.

By substituting Eqs. (5.33) and (5.34) into Eq. (5.4) and assuming E constant, we obtain

$$\frac{y''}{[1+(y')^2]^{3/2}} = -\frac{P}{EI_A} \frac{x}{\left\{1 + \frac{(n-1)}{L} \int_0^x [1+(y'(x))^2]^{1/2}\, dx\right\}^3} \quad (5.38)$$

Equation (5.38) is the exact nonlinear integral differential equation representing the flexible member in Fig. 5.2a. The exact solution of this equation is very difficult, and may be impossible to obtain.

We can simplify the solution of Eq. (5.38) by expressing the variation of the depth $h(x)$ of the flexible member by the following approximate equation:

$$h(x) = \left[1 + \frac{(n-1)x}{L_0}\right] h \qquad 0 \leq x \leq L_0 \quad (5.39)$$

where x is taken as shown in Fig. 5.2a, n is the taper, h is the depth at the free end, and $L_0 = (L - \Delta)$. The error in using this equation is 3 percent or less, thus providing a reasonable accuracy for practical applications.

On the basis of Eq. (5.39), we find

$$I_x = I_A f(x) \quad (5.40)$$

where

$$I_A = \frac{bh^3}{12} \quad (5.41)$$

$$f(x) = \left[1 + \frac{(n-1)x}{L_0}\right]^3 \quad (5.42)$$

By substituting Eqs. (5.33) and (5.40) into Eq. (5.4), we find

$$\frac{y''}{[1+(y')^2]^{3/2}} = -\frac{P(L-\Delta)^3}{EI_A} \frac{x}{[(n-1)x + (L-\Delta)]^3} \quad (5.43)$$

where $(L - \Delta)$ was substituted for L_0. Equation (5.43) is much simpler to solve than Eq. (5.38).

By integrating Eq. (5.43) once and determining the constant of integration by using the boundary condition of zero rotation at $x = L_0$ (see Appendix C for detail), we find

$$y'(x) = \frac{Q(x)}{\{1 - [Q(x)]^2\}^{1/2}} \qquad \text{for } n > 1 \quad (5.44)$$

where

$$Q(x) = \frac{P(L-\Delta)^3}{EI_A}\left[\frac{2(n-1)x + (L-\Delta)}{2(n-1)^2[(n-1)x + (L-\Delta)]^2} - \frac{(2n-1)}{2(n-1)^2 n^2 (L-\Delta)}\right]$$

(5.45)

Equations (5.44) and (5.45) may be used to determine y' at any $0 \leq x \leq L_0$. In order to be able to proceed with the computation of y', the horizontal displacement Δ of the free end A of the flexible member must first be determined. This part of the work can be accomplished by using Eq. (5.32) and applying a trial-and-error procedure. This procedure may be initiated by assuming a trial value for Δ and substituting into Eq. (5.44) to determine $y'(x)$. Then we substitute $y'(x)$ into Eq. (5.32) and perform the integration to obtain the length L of the member. This procedure may be repeated until the correct length L is obtained. This procedure can be handled rather easily by using computer software. Utilization of Simpson's rule to perform the integration yields very good results.

With known $y'(x)$, the pseudolinear analysis can be easily performed in order to obtain the pseudolinear system of constant stiffness EI_A. On this basis, z_e can be determined by Eq. (5.11a), and with known z_e the moment diagram M'_e of the pseudolinear system of constant stiffness EI_A can be determined from Eq. (5.11). The loading w'_e on the pseudolinear system may be obtained by using Eq. (5.14) and performing the differentiation, or by using the approximate method of the equivalent systems, which involves the approximation of the shape of the M'_e diagram with straight-line segments, as was done in Section 4.2. The second alternative is suggested by the author because it is very convenient to do and provides excellent results for practical applications. On this basis the equivalent pseudolinear system will be loaded with a few concentrated loads, and the computation of the deflections $y(x)$ can be carried out by using linear methods of analysis, such as the moment area method, or by using available handbook formulas, such as the ones given in Appendix A.

The length of the equivalent system is L_0, and its vertical deflection y_e at the free end $x = 0$ is equal to the vertical deflection y at the free end A of the original member in Fig. 5.2a. The vertical deflection y at points $0 \leq x \leq L$ of the pseudolinear system would be equal to the vertical deflection at corresponding points $0 \leq x_0 \leq L$ of the original system. Equation (5.22) provides the correlation between points x and x_0, and provides the position x_0 when the position x is given. For the cantilever flexible beam in Fig. 5.2a, the maximum deflection and the maximum rotation occurs at the free end, and these are usually the values that are of interest to the practicing design engineer.

If the flexible beam in Fig. 5.2a is of uniform depth h with uniform width b, we have the case shown in Fig. 5.2b with taper $n = 1$, and the solution of this problem yields

$$y'(x) = \frac{G(x)}{\{1 - [G(x)]^2\}^{1/2}} \qquad (5.46)$$

where

$$G(x) = \frac{P}{EI}[x^2 - (L - \Delta)^2] \qquad (5.47)$$

and the same procedure can be used to determine the pseudolinear system. In this case, however, both the original system and the pseudolinear system have the same constant stiffness EI. Note that Eqs. (5.44) and (5.45) should be used for values of $n > 1$.

By considering the uniform flexible beam in Fig. 5.2b, the vertical deflection δ_A at the free end A was determined by using both linear and nonlinear theory. The linear theory assumed that y' is small, and it was neglected. The nonlinear theory was carried out by using Eqs. (5.46) and (5.47) and applying p-seudolinear equivalent systems. The length L of the member was assumed to be equal to 1,000 in., and the constant stiffness $EI = 180,000$ kip-in.^{2} The member was loaded by a vertical load P at the free end A, as shown in Fig. 5.2b. For values of the load P, the deflection δ_A at the free end A was determined by using both linear and nonlinear theory, and the results are plotted as shown in Fig. 5.3. For $P = 0.2$ kip, the linear theory yielded $\delta_A = 370.37$ in., and the nonlinear theory gave $\delta_A = 328.61$ in., indicating that the error in using linear theory is $+12.71$ percent. As the load P increases, the error in using linear theory becomes much higher, and it is as shown in Fig. 5.3. However, it is interesting to note that, for $P = 0.2$ kip, the deflection δ_A is 32.86 percent when it is compared with the length L of the member. This observation indicates that for many practical situations linear theory can provide reasonable results, because deflections are controlled by practical restraints. In case where large deformations are permitted, or required, as stated earlier, nonlinear theory must be used, and for the specific problem examined in Fig. 5.3, the levels of error in using linear theory are established. For various practical problems, similar studies can be performed to establish design criteria to be used by the design engineers.

Example 5.2: By using the general solution obtained in Example 5.1, derive a pseudolinear equivalent system of constant stiffness EI_A for the flexible beam in Fig. 5.2a. Assume that $L = 1,000$ in., $P = 2$ kips, $EI_A = 180,000$ kip-in.^{2}, and taper $n = 1.5$. By using the pseudolinear system, determine the deflection and rotation at the free end.

SOLUTION: For a tapered cantilever beam, the general expression for $y'(x)$ is given by Eq. (5.44) and $Q(x)$ is given by Eq. (5.45). Since $y'(x)$ is a function of the horizontal displacement Δ of the free end of the member, a trial-and-

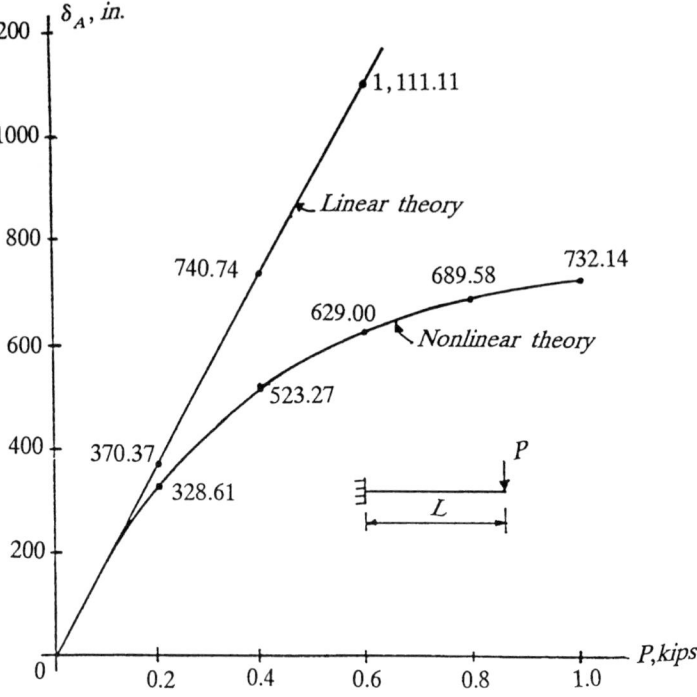

Figure 5.3. Results obtained by using linear and nonlinear theories.

error procedure using Eqs. (5.44) and (5.32) must be carried out, as explained earlier. We initiate this procedure by assuming $\Delta = 376.80$ in. By substituting for Δ, P, n, and EI_A in Eq. (5.45), we find

$$Q(x) = 2{,}689.303546 \left[\frac{x + 623.2}{0.125x^2 + 311.6x + 194{,}189.12} - 0.00285266 \right] \quad (5.48)$$

By substituting Eq. (5.44) into Eq. (5.32) and performing the required integration, we obtain

$$L = \int_0^{623.2} \left\{ 1 + \left(\frac{Q(x)}{\sqrt{1 - [Q(x)]^2}} \right)^2 \right\}^{1/2} dx \quad (5.49)$$

$$\approx 1{,}000 \text{ in.}$$

Equation (5.48) is used for $Q(x)$. In this case, the assumed value of $\Delta = 376.80$ in. is the correct one, because the correct value of the length L is obtained. Therefore, repetition of the procedure is not required. The preceding integration was carried out by using Simpson's rule.

Since the correct value of Δ was assumed, Eq. (5.48) is also correct. The values of $y'(x)$ at various locations $0 \leqslant x \leqslant 623.2$ may be determined by using Eqs. (5.48) and (5.44). For example, at $x = 0$, Eq. (5.48) yields

$$Q(0) = 2{,}689.303546 \left[\frac{623.2}{194{,}189.12} - 0.00285266 \right]$$

$$= 0.9589589$$

By substituting into Eq. (5.44), we obtain

$$y'(x = 0) = \frac{0.9589589}{\sqrt{1 - (0.9589589)^2}}$$

$$= 3.38203293$$

The rotation θ at $x = 0$ is $\theta_{x=0} = \tan^{-1}(y') = 73.528°$. This is a very large slope, as should be expected.

Other values of $y'(x)$ at selected values of x are obtained in a similar manner, and they are shown in the third column of Table 5.1. The values of $f(x)$ are computed from Eq. (5.42) with $n = 1.5$, z_e is computed from Eq. (5.11a), and M_x is computed from Eq. (5.33), while the values of the moment diagram M'_e of the equivalent pseudolinear system of constant stiffness EI_A are determined from Eq. (5.11). All these result are shown in Table 5.1. The values of M'_e are shown plotted by the solid line in Fig. 5.4a, and the shape of M'_e is approximated with five straight-line segments, as shown by the dashed lines in the same figure. The equivalent pseudolinear shear force diagram is shown in

TABLE 5.1. Calculated Values of $f(x)$, y', z_e, M_x, and M'_e for the Derivation of the Pseudolinear System

(1) x (in.)	(2) $f(x)$	(3) y' (rad)	(4) z_e	(5) M_x (kip-in.)	(6) $M'_e = \dfrac{z_e M_x}{f(x)}$ (kip-in.)
0	1.0000	3.3820	43.8654	0.00	0.00
50	1.1252	2.9218	29.4518	75.00	1,963.10
100	1.2605	2.2140	14.3376	150.00	1,706.18
200	1.5628	1.3058	4.4492	300.00	854.08
300	1.9098	0.8201	2.1631	450.00	509.68
400	2.3048	0.5032	1.4029	600.00	365.21
500	2.7508	0.2590	1.1030	750.00	300.73
600	3.2509	0.0476	1.0034	900.00	277.79
623.2	3.3750	0	1.0000	934.80	276.98

318   INFRASTRUCTURAL NONLINEARITIES

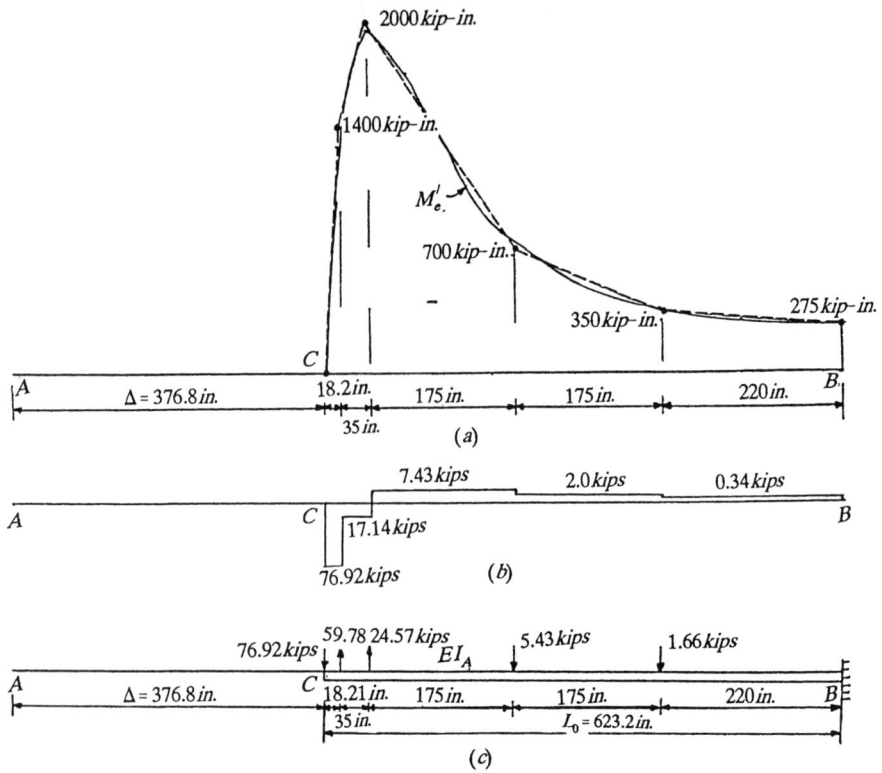

Figure 5.4. (a) M'_e diagram with its shape approximated with straight-line segments. (b) Equivalent shear force diagram. (c) Equivalent pseudolinear system of constant stiffness EI_A.

Fig. 5.4b, and the pseudolinear equivalent system of constant stiffness EI_A, loaded with five concentrated loads, is shown in Fig. 5.4c. This system is now linear, and linear methods of analysis may be used to solve it for deflections and rotations. Also, existing formulas, such as the ones given in Appendix A, may be used. The deflection y_c at the free end C of the equivalent system in Fig. 5.4c is determined here by using the moment-area method, as discussed in Section 4.5. It yields $\delta_c = 682.30$ in., which is a very large deflection. This deflection is equal to the deflection of the free end of the original variable stiffness member.

Example 5.3: Obtain a general solution for a uniform flexible simply supported beam loaded by a uniformly distributed load w as shown in Fig. 5.5. The stiffness EI is constant throughout the length of the member, which implies that $g(x) = f(x) = 1$. Use a pseudolinear equivalent system to obtain the solution.

SOLUTION OF NONLINEAR PROBLEMS   319

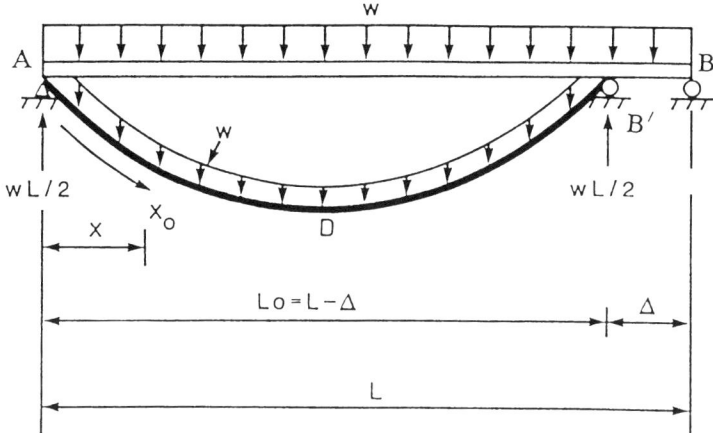

Figure 5.5. Flexible simply supported beam loaded as shown.

SOLUTION: The bending moment M_x, at any point $0 \leq x \leq L_0$, is

$$M_x = \frac{wLx}{2} - wx_0\left(\frac{x}{2}\right) \tag{5.50}$$

By substituting Eq. (5.22) into Eq. (5.50), we find

$$M_x = \frac{wLx}{2} - \frac{wx}{2}\int_0^x \{1 + [y'(x)]^2\}^{1/2}\,dx \tag{5.51}$$

In this case, the stiffness EI is constant. Thus, by substituting Eq. (5.51) into Eq. (5.3) and rearranging, we obtain

$$\frac{y''}{[1 + (y')^2]^{3/2}} = \frac{wx}{2EI}\left\{\int_0^x [1 + (y'(x))^2]^{1/2}\,dx - L\right\} \tag{5.52}$$

Equation (5.52) is the exact nonlinear integral differential equation representing the large deformation of the simply supported beam in Fig. 5.5. We can simplify again the solution of this equation by using one of the expressions given by Eqs. (5.28) through (5.31). For convenience, we use here Eq. (5.28) and we substitute into Eq. (5.50). The result is

$$M_x = \frac{wLx}{2} - \frac{wx}{2}(x + \Delta)$$

$$= \frac{w(L - \Delta)x}{2} - \frac{wx^2}{2} \tag{5.53}$$

320 INFRASTRUCTURAL NONLINEARITIES

On this basis, Eq. (5.3) yields

$$\frac{y''}{[1+(y')^2]^{3/2}} = \frac{w}{2EI}[x^2 - (L-\Delta)x] \tag{5.54}$$

Equation (5.54) is much easier to solve when it is compared with Eq. (5.52). By integrating this equation once and using the boundary condition of zero rotation at $x = L_0/2$ for the computation of the constant of integration, we find

$$y'(x) = \frac{\Phi(x)}{\sqrt{1 - [\Phi(x)]^2}} \tag{5.55}$$

where

$$\Phi(x) = \frac{w}{24EI}[6(L-\Delta)x^2 - 4x^3 - (L-\Delta)^3] \tag{5.56}$$

The unknown horizontal displacement Δ of support B may be determined by using the equation

$$L = \int_0^{L_0} [1+(y')^2]^{1/2}\, dx \tag{5.57}$$

and applying a trial-and-error procedure as discussed earlier.

With known Δ, the values of $y'(x)$ at any $0 \leq x \leq L_0$ may be determined from Eq. (5.55), and a pseudolinear system may be obtained as in Examples 5.1 and 5.2; that is, M_x can be obtained from Eq. (5.53), z_e from Eq. (5.11a), and the moment diagram M'_e of the pseudolinear equivalent system may be obtained from Eq. (5.11) with $g(x) = f(x) = 1$. By approximating the shape of M'_e with straight-line segments, the pseudolinear system can be easily obtained as in the preceding examples by using simple statics. The pseudolinear system would be loaded with concentrated loads, and linear methods of analysis, such as the moment-area method, or existing handbook formulas, may be used to determine vertical deflections and rotations. The following example illustrates the application of the methodology.

Example 5.4: By using the general solution obtained in Example 5.3, derive a pseudolinear equivalent system for the uniform flexible simply supported beam loaded as shown in Fig. 5.5. Assume that $L = 1{,}000$ in., $w = 10$ lb/in., and $EI = 75 \times 10^3$ kip-in.^{2} Use the pseudolinear system to determine the vertical deflection at the center of the beam and its rotation at support A.

SOLUTION: The general expression for $y'(x)$ is given by Eq. (5.55), and $\Phi(x)$ is given by Eq. (5.56). Since the horizontal displacement Δ of support B is not

SOLUTION OF NONLINEAR PROBLEMS 321

known, a trial-and-error procedure is applied by using Eqs. (5.55), through (5.57). We assume $\Delta = 440.53$ in., and we substitute into Eq. (5.56), yielding

$$\Phi(x) = \frac{(10)(10)^{-3}}{(24)(75)(10)^3} [6(1,000 - 440.53)x^2 - 4x^3 - (1,000 - 440.53)^3]$$

$$= 5.555556(10)^{-9}[3,356.82x^2 - 4x^3 - 175,117,847.76] \quad (5.58)$$

By substituting Eq. (5.55) into Eq. (5.57) and performing the indicated integration, we find

$$L = \int_0^{559.47} \left\{ 1 + \left(\frac{\Phi(x)}{\sqrt{1 - [\Phi(x)]^2}} \right)^2 \right\}^{1/2} dx$$

$$\approx 1,000$$

which indicates that the assumed value of Δ is the correct one and we do not need to repeat the procedure. Note that Eq. (5.56) was used to obtain $\Phi(x)$.

With correct Δ, the values of $y'(x)$ at various locations $0 \leq x \leq 559.47$ in. may be calculated by using Eqs. (5.55) and (5.56). For example, at $x = 0$, Eq. (5.56) yields

$$\Phi(x = 0) = -0.972877$$

and Eq. (5.55) yields

$$y'(x = 0) = \frac{-0.972877}{\sqrt{1 - (-0.972877)^2}} = -4.205708$$

Thus, the rotation θ_A at support A is

$$\theta_A = \tan^{-1}(y'(x = 0)) = -76.625°$$

which is considered to be a very large rotation.

Other values of $y'(x)$ at selected locations x may be obtained in a similar manner, and they are shown in the second column of Table 5.2. The values of z_e are computed by using Eq. (5.11a), those of M_x by using Eq. (5.53), and the values of the moment diagram M'_e of the pseudolinear system are computed by using Eq. (5.11). The results are shown in Table 5.2. The moment diagram M'_e is shown plotted by the solid line in Fig. 5.6a. The approximation of its shape with straight-line segments, shown by the six dashed lines in Fig. 5.6a, leads to the equivalent pseudolinear system in Fig. 5.6c. Linear methods of analysis, or handbook formulas, may be used to solve the pseudolinear system for deflections and rotations at any $0 \leq x \leq L_0$. The deflection at the center point D of the pseudolinear system is equal to the deflection at the center point of

TABLE 5.2. Calculated Values of $f(x)$, y', z_e, M_x, and M'_e for the Derivation of the Pseudolinear System

(1) x (in.)	(2) y' (rad)	(3) z_e	(4) M_x (kip-in.)	(5) M'_e (kip-in.)
0	−4.2057	80.7869	0	0
30	−3.2865	40.5399	79.42	3,219.68
50	−2.5109	19.7422	127.37	2,514.56
100	−1.3744	4.9104	229.74	1,128.12
150	−0.8076	2.1237	307.10	652.19
200	−0.4426	1.3078	357.47	470.11
300	0.1061	1.0169	389.21	395.79
400	0.7283	1.8933	318.94	603.85
450	1.2405	4.0453	246.31	996.40
500	2.2175	14.3942	148.68	2,140.13
530	3.3093	41.3773	78.10	3,231.57
559.47	4.2057	80.7869	0	0

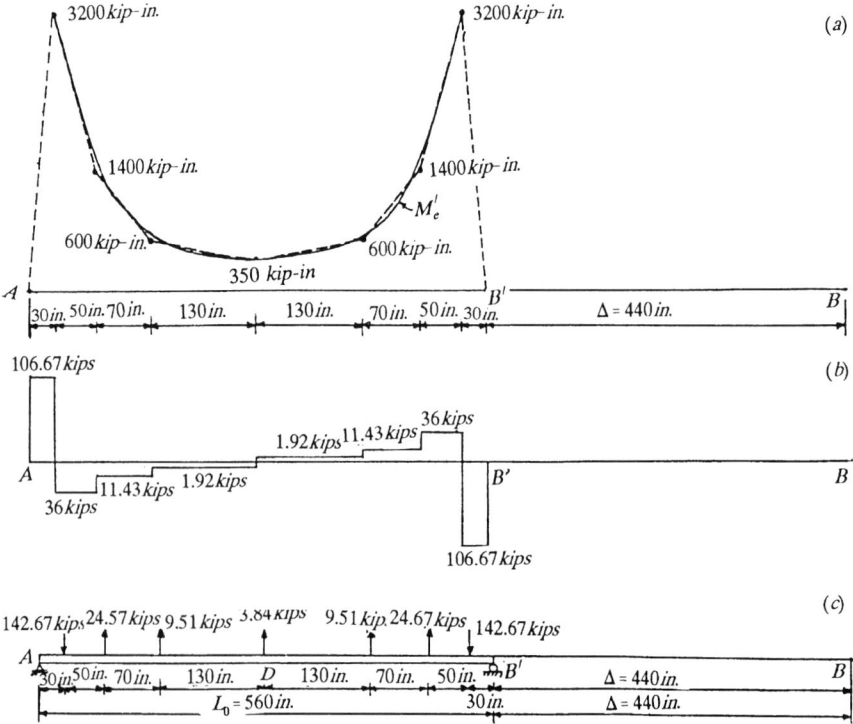

Figure 5.6. (*a*) Moment diagram M'_e of the pseudolinear system with its shape approximated with straight-line segments. (*b*) Shear force diagram. (*c*) Pseudolinear equivalent system loaded with equivalent concentrated loads.

the original system in Fig. 5.5. By using the pseudolinear system in Fig. 5.6c, the deflection at its center point D was calculated by using handbook formulas such as the ones in Appendix A. The value of this deflection is found to be equal to 367.528 in. This is practically the exact answer to this nonlinear problem.

In order to get a better insight on the large deformation problem, all the work regarding this problem, as well as the work in the preceding examples, was carried out manually, without the aid of computer software. They are fairly long problems to do by hand, but not unrealistic. The utilization of pseudolinear equivalent systems reduces a great deal the work involved and makes it possible to solve such complicated problems by hand calculations. These problems were also solved by using computer software in order to compare results. For this problem, the vertical deflection at the center of the flexible member as calculated by using the computer was found to be 367.65 in. — practically identical.

5.4 SOLUTION OF NONLINEAR PROBLEMS USING SIMPLIFIED NONLINEAR EQUIVALENT SYSTEMS

In many practical situations, the variation in stiffness along the length of a flexible member may be very arbitrary, and the applied loading may also be arbitrary. In such cases, a simplified nonlinear equivalent system of constant stiffness may first be derived, which provides us with a much simpler nonlinear system to solve by using pseudolinear analysis, or by existing solutions if such solutions are available. The following example illustrates the procedure.

Example 5.5: A flexible cantilever beam is loaded as shown in Fig. 5.7a. The width b of the beam and its modulus of elasticity E are constant, and its loading varies as shown in the figure. By using the procedure discussed in Section 5.2.3, derive one or more simplified nonlinear equivalent systems of constant stiffness EI_A that are suitable for nonlinear analysis. I_A is the moment of inertia of the member at its free end.

SOLUTION: At various cross sections, a distance $0 \leqslant x \leqslant L$ from support A, the actual moment M_x is calculated by using simple statics. At the same cross sections, the moment of inertia is determined by using the moment of inertia I_A at the free end A as the reference value. The values of the moment M_e of the simplified nonlinear equivalent system of constant stiffness EI_A, at the analogous cross sections, are calculated by using Eq. (5.16). Note that $g(x) = 1$ since E is constant. The results are shown in Table 5.3. The moment diagram M_e of the simplified nonlinear equivalent system of constant stiffness EI_A is shown plotted by the solid line in Fig. 5.8a. The approximation of the shape of M_e with three straight-line segments is shown by the dashed lines in the same figure.

324 INFRASTRUCTURAL NONLINEARITIES

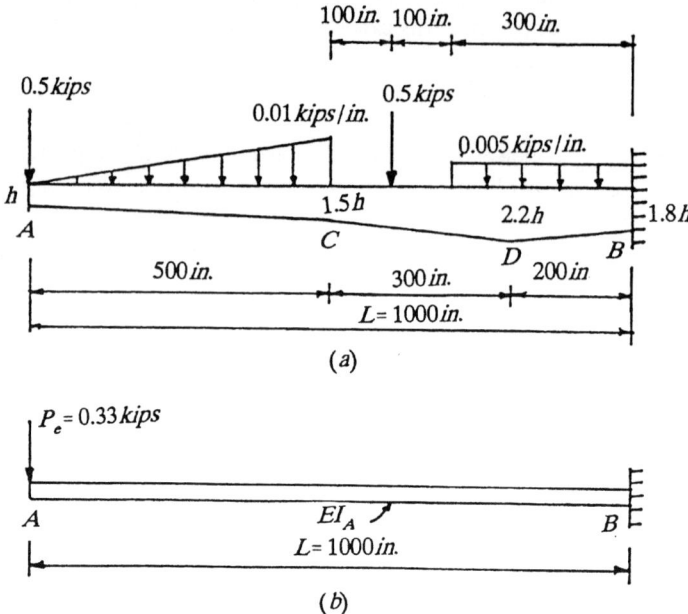

Figure 5.7. (a) Variable stiffness flexible cantilever beam loaded as shown. (b) Simplified nonlinear equivalent system of constant stiffness EI_A loaded as shown.

TABLE 5.3. Values of $f(x)$, M_x, and M_e for the Derivation of a Suitable Simplified Nonlinear Equivalent System of Constant Stiffness EI_A

(1) x (in.)	(2) $f(x)$	(3) M_x (kip-in.)	(4) M_e (kip-in.)
0	1.0000	0	0
100	1.3310	53.33	40.07
200	1.7280	126.67	73.30
300	2.1970	240.00	109.24
400	2.7440	413.33	150.63
500	3.3750	666.67	197.53
600	5.2077	966.67	185.62
700	7.6066	1,316.67	173.10
800	10.6480	1,691.67	158.87
900	8.0000	2,116.67	264.58
1,000	5.8320	2,591.67	444.39

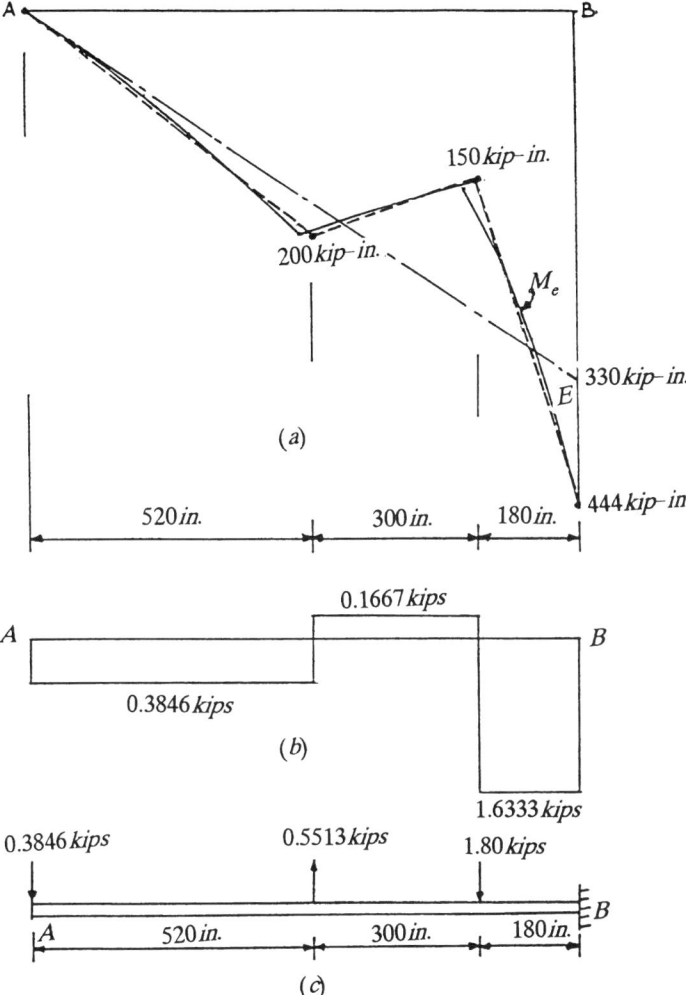

Figure 5.8. (a) M_e diagram of the simplified nonlinear equivalent system. (b) Shear force diagram. (c) Simplified nonlinear equivalent system of constant stiffness EI_A.

By proceeding in the usual way, the shear force diagram is obtained, and it is shown in Fig. 5.8b. The simplified nonlinear equivalent system of constant stiffness EI_A loaded with three equivalent concentrated loads is shown in Fig. 5.8c. The large deflection configuration of this simplified nonlinear system is closely identical to the deflection curve of the original system in Fig. 5.7a. The simplified nonlinear system may be solved for its large deflections and rotations by using pseudolinear analysis, as discussed in the preceding sections. See also the author's work in Reference [3].

For practical applications, this problem can be farther simplified by using only one straight-line segment to simplify the shape of the M_e diagram. For example, the shape of the M_e diagram in Fig 5.8a may be approximated by the straight-line segments AE, which yields the simplified nonlinear equivalent system shown in Fig. 5.7b. The straight line AE is drawn in a way that approximately balances the areas added to and subtracted from the M_e diagram. This balancing is sufficiently accurate if it is done by observation, because deflections and rotations are not very sensitive to errors introduced in the approximation of the shape of M_e, as long as the shape of M_e is approximately maintained as is done in Fig. 5.8a. Investigations by the author [3] show that the error for such one-line approximations of M_e is less than 3 percent. This would be also true if the system is linear. For example, if we solve the problems in Figs. 5.8c and 5.7b for the deflection of the free end A by using formulas of linear analysis, the difference in deflection between the two solutions would be only 1.2 percent. Of course, this would not be the expected solution to the problem, but it checks the margin of the expected error. The nonlinear analysis of the same two problems is expected to yield approximately the same error.

The above observation is extremely important, because the solution of the problem in Fig. 5.7b is much easier than that of the problem in Fig. 5.8c, and it can be carried out by using pseudolinear analysis and Eqs. (5.46) and (5.47). Also, solutions are available for problems concerning the large deformation of uniform cantilever beams loaded with a concentrated load at the free end. See Gere and Timoshenko [38, page 514]. If we assume that $EI_A = 180 \times 10^3$ kip-in.^{2}, this solution yields a deflection at the free end equal to 466 in. The error using this solution would be about 5 percent, because it utilizes elliptic integrals. The solution obtained using pseudolinear analysis, as discussed in Section 5.3, would be practically exact.

The conclusion is that simplified nonlinear equivalent systems can provide a simple and reliable answer to the practicing design engineer for very complex nonlinear problems.

5.5 INFRASTRUCTURAL INSTABILITIES

The design of an infrastructural component such as a building, a bridge, a dam, and the like, is usually completed in three stages of work.

The first stage involves the static analysis of the system. The applied loads, including the dead weight of the structure, are all considered to be static loads, and a static analysis is performed to decide on the size, cross-sectional dimensions, and material to be used for the various component elements of the structure in order to satisfy design criteria of various kinds.

Once this part of the work is completed, the particular aspects of the structure are investigated. Suppose, for example, that the structure will be subjected to dynamic loads. These might involve large trucks crossing highway bridges and pavements, or wind and gust forces affecting buildings and towers. Perhaps the structure will be subjected to the catastrophic effects of a blast or an earthquake. Whatever the dynamic criteria for a particular structure are, the design engineer has to make sure that they are satisfied by the proposed design. In many practical cases, such as tall stacks, tall buildings, and so on, these dynamic criteria are often not known in advance. There are no guidelines or codes to be followed in the design process. Hence the practicing design engineer has to figure out ways to incorporate such conditions in the design in order to come up with a safe structure. He/she often utilizes special consultants to help in the design and analysis.

The third stage of design involves stability considerations. Buildings, for example, are supported on columns, and failures in buildings can occur when a column buckles in the lateral direction. When a column of a building or a member of a truss structure is subjected to compression, there is a limit that has to be imposed on the maximum value of the compressive force in order to avoid lateral buckling of the column or the truss member. If this buckling instability is not investigated by appropriate analysis, it is possible for the column to fail by buckling even when the axial compressive stress is well below the elastic limit of the material used to design the structure.

The critical load is the limiting value of the axial compressive load that can cause lateral buckling if it is exceeded. Such load can be reached well below the elastic limit of the material if the member is slender, or it can be reached somewhere within the inelastic range. In the design of a structure, we wish to make sure that such load is not reached when the structure responds elastically. In fact, we wish to make sure that axial compressive loads are kept well below their critical value, thus providing a factor of safety for possible unforeseen conditions. On the other hand, we may design the structure in such a way that buckling is prevented in the elastic range, but can occur somewhere in the inelastic range. Although we strive to design a structure to respond elastically at all times, there are situations in which this may not be possible, or economical, to accomplish. For example, when a structure is subjected to an earthquake, or the effects of an explosion, it is possible for the structure to experience inelastic behavior. In fact, because of economic restraints, we permit the structure to behave inelastically, while still making every possible effort to retain structural integrity and prevent catastrophic failures. Thus, the practicing design engineer must decide how far in the inelastic range the structure can be permitted to go while still maintaining structural integrity. Inelastic range is a very dark engineering area, requiring special methods of analysis to secure a safe design.

The work in this and the following sections of this chapter provides fundamental concepts regarding infrastructural instabilities and inelastic behavior. More work on these subjects can be found in specialized books [3, 5, 6, 48, 49].

5.5.1 The Euler Column

We examine here the stability and instability conditions of a column hinged at both ends and loaded by an axial compressive force P, as shown in Fig. 5.9a. We assume that the column is perfectly straight and that it obeys Hooke's law. This is an ideal situation, which does not usually occur in practice, but in the majority of cases these two assumptions are considered reasonable for practical applications.

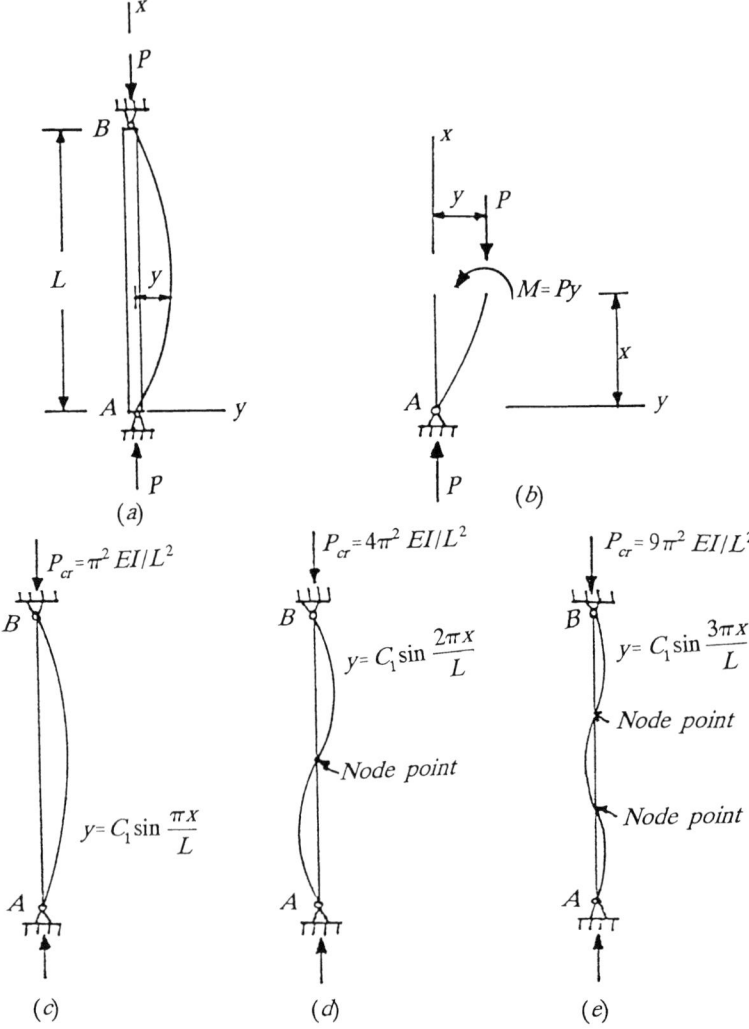

Figure 5.9. (a) Hinged-hinged column subjected to an axial compressive force. (b) Free-body diagram of an element of the column at its slightly bent configuration. (c) Buckling shape for $n = 1$. (d) Buckling shape for $n = 2$. (e) Buckling shape for $n = 3$.

When the axial compressive load P is gradually increased, the column will remain straight until a certain value of P is reached, known as the critical load P_{cr}. At the critical load, the ideal column may start to deflect in the lateral direction without any further increase in the load P. In other words, we have three stages of equilibrium. The first stage would be the one where $P < P_{cr}$ and the equilibrium of the column is stable. At this stage, if a small lateral load is applied, the column will bend, but it will return to its straight configuration if the load is removed. At $P = P_{cr}$, the column experiences what is known as neutral equilibrium. At this stage, a small lateral load will produce bending of the column that will not disappear when the lateral load is removed. Thus, static equilibrium at this stage can be maintained in both straight and slightly bent configurations. When $P > P_{cr}$, the column becomes unstable and it will collapse by bending. For an ideal column, it is possible to have equilibrium in the straight position even if $P > P_{cr}$, but such equilibrium cannot be maintained and the column will deflect sideways and fail even with an extremely small disturbance.

We can determine P_{cr} by assuming that the column is in its slightly bent configuration, as shown in Fig. 5.9a, and considering the free-body diagram in Fig. 5.9b. The bending moment M at any distance x from support A is

$$M = Py \qquad (5.59)$$

where y is the lateral deflection at any x. For small deflection theory, the Euler–Bernoulli equation is

$$\frac{d^2y}{dx^2} = -\frac{M}{EI} \qquad (5.60)$$

By substituting Eq. (5.59) into Eq. (5.60), we find

$$\frac{d^2y}{dx^2} = -\frac{Py}{EI}$$

or

$$EIy'' + Py = 0 \qquad (5.61)$$

Equation (5.61) is a second order linear homogeneous differential equation with constant coefficients, and it represents the slightly bent configuration of the column in Fig. 5.9a. We rewrite this equation as follows:

$$y'' + k^2 y = 0 \qquad (5.62)$$

where

$$k^2 = \frac{P}{EI} \qquad (5.63)$$

The general solution of Eq. (5.62) is

$$y = A \sin kx + B \cos kx \tag{5.64}$$

where A and B are constants that can be determined by using the boundary conditions of the column. In this case, we apply the following boundary conditions:

$$\text{At } x = 0 \quad y = 0 \tag{5.65}$$
$$\text{At } x = L \quad y = 0 \tag{5.66}$$

The condition given by Eq. (5.65) yields $B = 0$, and the condition given by Eq. (5.66) yields

$$A \sin kL = 0 \tag{5.67}$$

Equation (5.67) suggests that two things are possible. The first one is that $A = 0$, indicating that the deflection y is zero and the column remains straight. This is known as the trivial solution, which applies to an ideal column that can have straight stable or unstable equilibrium under the action of the compressive load P.

The second possibility suggests that

$$\sin kL = 0 \tag{5.68}$$

which is satisfied when $kL = 0, \pi, 2\pi, \ldots$. We disregard the value $kL = 0$, because it suggests that $P = 0$, which is of no interest to us. Thus, we have

$$kL = n\pi \quad n = 1, 2, 3, \ldots \tag{5.69}$$

By substituting Eq. (5.69) into Eq. (5.63) and solving for P, we obtain

$$P = P_{cr} = \frac{n^2 \pi^2 EI}{L^2} \quad n = 1, 2, 3, \ldots \tag{5.70}$$

The deflected configuration of the column may be obtained from Eq. (5.64) by substituting for k. Thus,

$$y = A \sin kx = A \sin \frac{n\pi x}{L} \quad n = 1, 2, 3, \ldots \tag{5.71}$$

The smallest value of the critical load occurs when $n = 1$, yielding

$$P_{cr} = \frac{\pi^2 EI}{L^2} \tag{5.72}$$

and the corresponding deflection shape y is

$$y = A \sin \frac{\pi x}{L} \tag{5.73}$$

Column lateral deflection configurations for $n = 1, 2,$ and 3 are shown in Figs. 5.9c, 5.9d, and 5.9e, respectively. Figure 5.9c represents the fundamental case of column buckling, and it is the one of most practical importance. The higher cases of buckling can occur only with lateral supports at node points where the deflection y is zero.

The work on the bending of a slender column and the derivation of critical loads is credited to the famous mathematician Leonard Euler (1707–1783), and the case discussed in this section is often referred to as the Euler column, or the Euler load.

5.5.2 Columns with Other Boundary Conditions

The critical loads for columns with other boundary conditions may be derived by following the procedure discussed in the preceding section and satisfying appropriate boundary conditions. Figs 5.10a, 5.10b, and 5.10c show, for each column, the fundamental case of column buckling for a cantilever column, a fixed-fixed column, and a fixed-hinged column, respectively. In all three cases the critical load P_{cr} may be also determined by using Eq. (5.72) and substituting L_e for L, where L_e is an effective length; that is,

$$P_{cr} = \frac{\pi^2 EI}{L_e^2} \tag{5.74}$$

For the cantilever column in Fig. 5.10a, we have $L_e = 2L$, for the fixed-fixed column in Fig. 5.10b, we have $L_e = L/2$, and it is $L_e = 0.699L$ for the fixed-hinged column in Fig. 5.10c.

5.5.3 Eccentrically Loaded Columns

In practical applications the compressive loads on columns are not usually perfectly axial. Such situations can develop for many reasons. One reason is that the construction of structural and mechanical components is not perfect. Construction imperfections can produce column eccentricities that were never intended to exist; these can result in axial loads that are no longer axial.

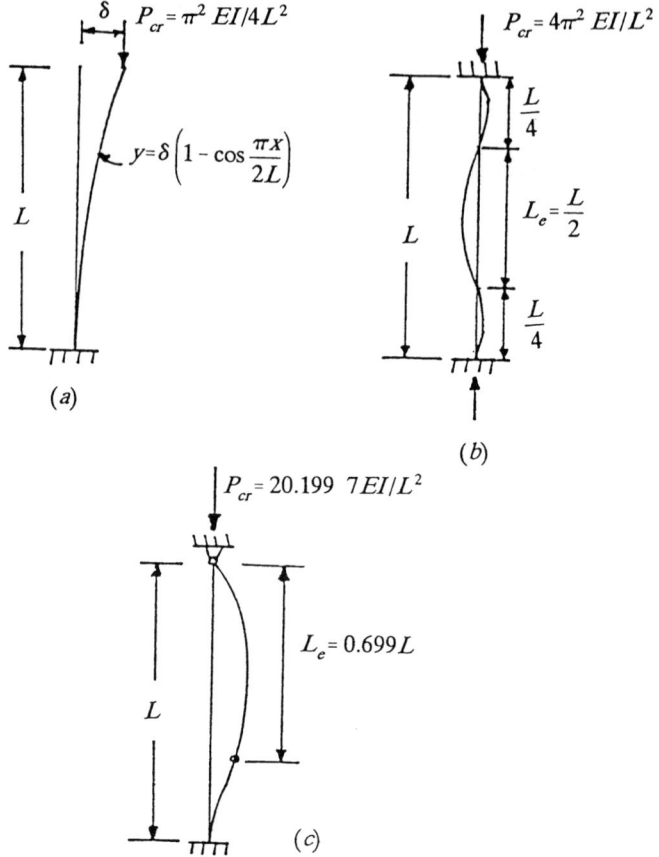

Figure 5.10. (*a*) Cantilever column. (*b*) Fixed-fixed column. (*c*) Fixed-hinged column.

Construction imperfections can also result in a column that is no longer straight, and the assumption that requires the column to be straight is violated. In other situations, the design of the total structure is performed in a way that permits compressive loads not to be axial. With such eccentricities, the column will bend even with very small compressive loads, and it will not be initially straight as was assumed in the preceding sections on ideal columns. As the compressive load increases, bending of the column will increase, and the critical load for such columns will be different than the one obtained for ideal columns.

The preceding discussion can be illustrated graphically by observing the load-deflection curves shown in Fig. 5.11. For an ideal column, at $P = P_{cr}$, the column may be subjected to any lateral small deflection y that is represented by curve 1 in the figure. The deflection must be small, because small deflection theory is used to determine P_{cr}. If deflections are permitted to be large and large deflection theory is used to determine P_{cr}, then the ideal column will

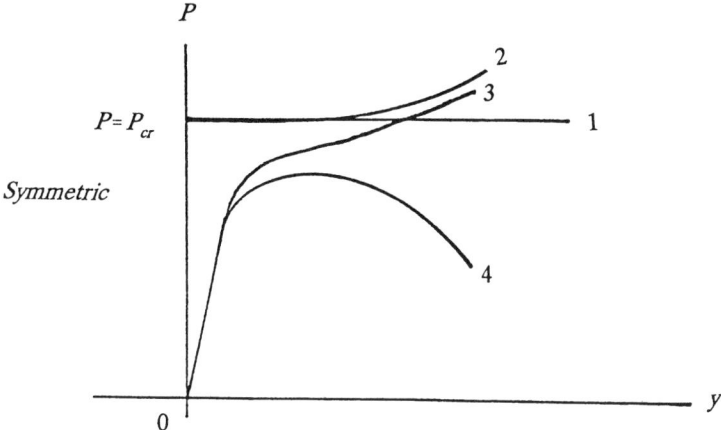

Figure 5.11. Load-deflection curves for buckled columns.

follow curve 2, as shown in Fig. 5.11, indicating that when an elastic column starts to buckle it requires increasingly larger compressive loads to cause an increase in deflection.

If the column is not constructed perfectly, as discussed above, the load deflection curve follows the shape shown by curve 3 in Fig. 5.11. For small deflections, it approaches curve 1, but when the deflections become large, it approaches curve 2. However, the degree of closeness to either curve 1 or curve 2 depends on the degree of the imperfections. When the imperfections are very small, curve 3 approaches curve 1, and curve 3 will move further to the right if the imperfections are larger.

If the compressive stresses are permitted to exceed the proportional limit of the material of the column and Hooke's law does not apply, then the load-deflection curve would be as shown by curve 4 in Fig 5.11. Initially curves 3 and 4 coincide, since the response is elastic, but curve 4 departs from curve 3 as the column response becomes inelastic. It continues upward, reaches a maximum value for the compressive load P, and then turns downward. Stockier columns fall into the inelastic response category and follow curve 4, while only very slender columns may remain elastic up to $P = P_{cr}$. The disadvantage with curve 4 is that after the maximum compressive load is reached, larger deflections can be produced with smaller P. On the other hand, the critical load increases as the deflections become large, and the maximum compressive load is smaller than the one for the other three curves. Thus, if a structure is permitted to respond inelastically, different design criteria may have to be followed by the practicing design engineer regarding allowable limits for compressive loads.

For eccentrically loaded columns, we consider the column shown in Fig. 5.12a in order to determine its response in terms of critical load and lateral deflection. By using the free-body diagram in Fig. 5.12b and applying statics,

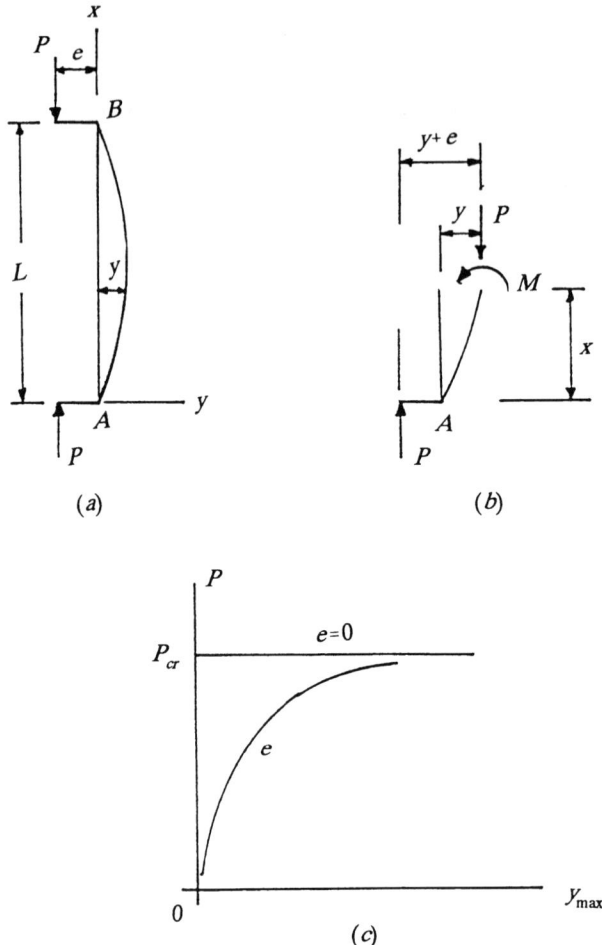

Figure 5.12. (*a*) Eccentrically loaded column. (*b*) Free-body diagram. (*c*) Plot of *P* versus y_{max} for values of $e = e$ and $e = 0$.

the bending moment *M*, at any distance *x* from end *A*, is

$$M = P(e + y) \tag{5.75}$$

By substituting Eq. (5.75) into Eq. (5.60), we find

$$EIy'' = -P(e + y) \tag{5.76}$$

or

$$y'' + k^2 y = -k^2 e \tag{5.77}$$

where

$$k^2 = \frac{P}{EI} \tag{5.78}$$

Equation (5.77) is a nonhomogeneous differential equation in y, and its solution y is

$$y = y_c(x) + y_p(x) \tag{5.79}$$

where the complementary solution $y_c(x)$ is the solution of the homogeneous equation $y'' + k^2 y = 0$, and it is given by Eq. (5.64). The particular solution $y_P(x)$ may be assumed to be equal to a constant C, since $k^2 e$ in Eq. (5.77) is constant. By substituting into Eq. (5.77), we obtain $k^2 C = -k^2 e$, or $C = -e$. Thus, $y_P = -e$. On this basis, Eq. (5.79) yields

$$y = A \sin kx + B \cos kx - e \tag{5.80}$$

The constants A and B can be determined by using the boundary conditions that the deflection y is zero at $x = 0$ and $x = L$, yielding

$$B = 0 \tag{5.81}$$

$$A = \frac{e(1 - \cos kL)}{\sin kL} = \frac{e \tan kL}{2} \tag{5.82}$$

By substituting Eqs. (5.81) and (5.82) into Eq. (5.80), the general solution for y is

$$y = e\left(\tan \frac{kL}{2} \sin kx + \cos kx - 1\right) \tag{5.83}$$

Equation (5.83) may be used to determine the deflection y at any distance x from support A for a given value of the eccentricity e. The maximum deflection y_{max} occurs at $x = L/2$, and it is as follows:

$$y_{max} = e\left(\sec \frac{kL}{2} - 1\right) \tag{5.84}$$

Equation (5.84) shows that $y_{max} = 0$ when $e = 0$, or when $P = 0$.

We note here that the behavior of eccentricity loaded columns is not similar to the response of ideal columns. Ideal columns are initially straight, and they stay straight until the critical value of P is reached. At $P = P_{cr}$, the static equilibrium becomes neutral, and the lateral deflection changes from zero to undefined. On the other hand, for a given value of P, eccentrically loaded

336   INFRASTRUCTURAL NONLINEARITIES

columns are subjected immediately to definite values of the deflection y. Also, if we select a value for the eccentricity e and plot P versus y_{max}, we note that this curve is nonlinear and, consequently, the superposition principle cannot be used to calculate deflection when the loads are more than one.

The plot of P versus y_{max} for a given value e is shown in Fig. 5.12c. This graph shows that as P approaches the value $P = P_{cr} = \pi^2 EI/L^2$, the secant term in Eq. (5.84) approaches infinity, which indicates that y_{max} increases without limit as $P \to P_{cr}$, and the horizontal line representing $e = 0$ becomes an asymptote for any curve e. Thus, the ideal column is the limiting case of an eccentrically loaded column. Note, however, that the differential equation used in the derivation applies only to small deflections, and large nonlinear deflection theory must be used for large deflections.

5.6 INFLUENCE OF THE AXIAL FORCE ON THE FREE VIBRATION AND FLEXIBILITY OF COLUMNS OR BEAM-COLUMNS

Consider the hinged-hinged column in Fig. 5.13. The free undamped frequencies of vibration of this column may be obtained using the following equation (for details see the author's work in Reference [5]):

$$\omega_n = \frac{\pi^2}{L^2} \sqrt{\frac{EI}{\rho A}} \left[\sqrt{n^2 \left(n^2 - \frac{P}{P_{cr}} \right)} \right] \qquad n = 1, 2, 3, \ldots \qquad (5.85)$$

where

$$P_{cr} = \frac{\pi^2 EI}{L^2} \qquad (5.86)$$

ρ is the mass density per unit volume, A is the cross-sectional area of the column, and EI is its stiffness. In Eq. (5.85), positive P indicates compression and tensile P is considered to be negative. For $n = 1$, we obtain the free undamped fundamental frequency of vibration of the column, which is as follows:

$$\omega_1 = \frac{\pi^2}{L^2} \sqrt{\frac{EI}{\rho A}} \left[\sqrt{\left(1 - \frac{P}{P_{cr}}\right)} \right] \qquad (5.87)$$

Examining Eq. (5.87), we note that, when the compressive force P is zero, the fundamental frequency of the column becomes

$$\omega_1(P = 0) = \frac{\pi^2}{L^2} \sqrt{\frac{EI}{\rho A}} \qquad (5.88)$$

INFLUENCE OF THE AXIAL FORCE ON FREE VIBRATION 337

Figure 5.13. Hinged-hinged column subjected to an axial tensile or compressive force P.

which is identical to the well-known expression for a simply supported beam such as the one shown in Fig. 5.14 with $P = 0$. On the other hand, as the compressive force P increases, the frequency ω_1 decreases. For example, if $P = 0.5P_{cr}$, Eq. (5.87) yields

$$\omega_1(P = 0.5P_{cr}) = 0.707\omega_1(P = 0) \qquad (5.89)$$

and when $P = P_{cr}$, we have

$$\omega_1(P = P_{cr}) = 0 \qquad (5.90)$$

Since the rigidity of a structure is directly dependent on its free frequency of vibration, lower frequencies of vibration are associated with more flexible structures, and the rigidity of a structure increases as its free frequency of vibration increases. Therefore, we can conclude that when a column is subjected to an axial force P, it becomes more flexible, and it loses completely

Figure 5.14. Beam-column.

its rigidity when $P = P_{cr}$. The same conclusions can be drawn for the beam-column in Fig. 5.14.

If, on the other hand, the axial force is tensile, for $P = 0$ we again obtain Eq. (5.88). However, as P increases, the frequency ω_1 increases and the column, or beam-column, becomes stiffer. For example, when $P = 0.5P_{cr}$, we have

$$\omega_1(P = 0.5P_{cr}) = 1.225\omega_1(P = 0) \tag{5.91}$$

Columns, however, are intended to be subjected to compressive loads, but beam-columns may be subjected to both tensile and compressive loads.

From the preceding discussion, we can conclude that the design of slender columns, or beam-columns, subjected to axial compressive loads, should take into consideration the reduction in rigidity caused by the presence of an axial compressive force. This becomes even more important when the structure is subjected to dynamic loads.

5.7 INFLUENCE OF AXIAL COMPRESSIVE FORCE ON THE STABILITY AND FREE VIBRATION OF ELASTICALLY SUPPORTED BEAM-COLUMNS

Another type of instability that was recently investigated by the author [3, 5] is related to members that are elastically supported at the ends. The cases examined involved: (a) uniform members elastically supported at the ends by vertical springs and subjected to an axial compressive force; (b) uniform members supported at the ends by both vertical and horizontal springs and subjected to an axial compressive force; and (c) uniform members elastically supported at the ends with vertical and horizontal springs and with initial vertical and horizontal restraints, and also subjected to an axial compressive force. Such initial restraints could involve construction imperfections or, possibly, imposed design condition. Both static and flutter instabilities were examined in detail.

In this section we are only concerned with the instabilities that are associated with a uniform beam subjected to an axial compressive force P at both ends, and supported elastically at both ends by vertical springs, as shown in Fig. 5.15a. For more details on these problems, consult the author's work in the indicated references.

The differential equation of motion for this problem is derived by the author in Reference [3, Chapter 12] and it includes the effects of shear and rotatory inertia. It is as follows:

$$\frac{EI(K'AG)}{K'AG + P}y'''' - \frac{EI\rho}{K'AG + P}\left(A + \frac{K'AG}{E}\right)\ddot{y}'' + m\ddot{y}$$
$$+ Py'' + \frac{I\rho^2 A}{K'AG + P}\ddddot{y} = 0 \tag{5.92}$$

INFLUENCE OF AXIAL COMPRESSIVE FORCE ON BEAM-COLUMNS 339

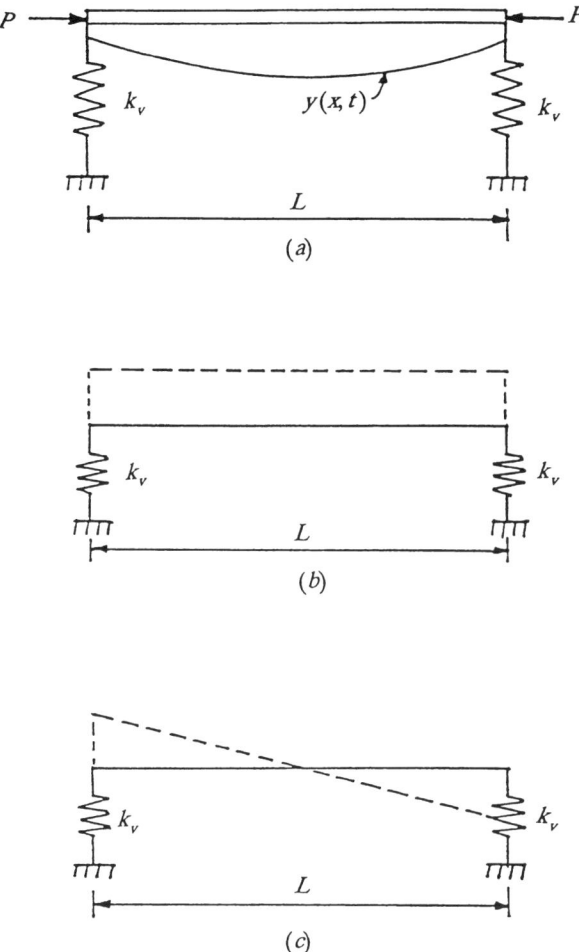

Figure 5.15. (a) Uniform elastically supported member subjected to axial compressive forces at both ends. (b) Vertical translational motion. (c) Rotational motion.

where A is the cross-sectional area of the member, ρ is the material density, P is the axial compressive force, m is the mass per unit length of the member, E is the modulus of elasticity, G is the shear modulus, K' is the shear coefficient, and I is the cross-sectional moment of inertia.

The solution [3, 5] of Eq. (5.92) and the analysis of the results obtained by many problem combinations, reveal the following:

1. For a given length L and a spring constant k_v, there exists a value of the axial compressive force P that makes the first free frequency ω_1 and the second free frequency ω_2 coincide ($\omega_1 = \omega_2$). If this value of P is denoted

340 INFRASTRUCTURAL NONLINEARITIES

as P_{flutter}, it means that at $P = P_{\text{flutter}}$ the elastically supported member experiences a state of flutter instability. Since ω_1 and ω_2 represent the translational and rotational motions of the elastically supported member shown in Figs. 5.15b and 5.15c, respectively, we can deduce that the member becomes unstable when the translational and rotational motions coincide.

2. We also note that higher values of the spring constant k_v require higher values of the axial compressive force P for ω_1 and ω_2 to become equal. This observation suggests that there is a value of k_v that establishes the limit for such flutter instability taking place. If we denote this value as $k_{v_{cr}}$, then we can say that flutter instability occurs when $k_v < k_{v_{cr}}$, which means that for every $k_v < k_{v_{cr}}$ there exists an axial load P_{flutter} that makes the frequencies ω_1 and ω_2 equal.

3. When $k_v = k_{v_{cr}}$, the compressive critical axial load P_{cr} is reached, and P_{cr} remains constant with increasing k_v. Thus, for $k_v > k_{v_{cr}}$, static instability governs with $\omega_1 = 0$. On this basis, we can say that we have two definite regions of instability: (a) the region of flutter instability with $k_v < k_{v_{cr}}$; and (b) the region of static instability with $k_v > k_{v_{cr}}$ and $\omega_1 = 0$.

4. It was also observed that, for a given value of the axial load P smaller than P_{cr} and a given value of k_v, there exists a value of the length L of the elastically supported member that makes $\omega_1 = \omega_2$, thus establishing a value $L = L_{\text{flutter}}$ where flutter instability occurs. We also note that there is a critical value L_{cr} of the length L where $\omega_1 = 0$, and static instability occurs.

5. The results also show that at the higher values of k_v the existence of flutter instability may not be recognized if the shear and rotatory effects are not included in the analysis.

The preceding results are very important to the design engineers, since such flutter and static instabilities do exist, and the integrity of a structure subjected to such conditions must be maintained. For more information on this subject see the author [3, 5, 6].

5.8 SUSPENSION BRIDGES AND THE COLLAPSE OF THE TACOMA NARROWS BRIDGE

The collapse of the first Tacoma Narrows suspension bridge triggered the great concern regarding the dynamic performance and response of suspension bridges. The first Tacoma Narrows suspension bridge was built near Seattle, Washington, between November 23, 1938 and July 1, 1940, at a cost of approximately $6.4 million at that time. The total length of the bridge was 5,000 ft, the length of the center span was 2,800 ft, the width of the bridge was 39 ft, accommodating two lanes of traffic, the height of the side girders was 8 ft, and the two end spans were each 1,100 ft long.

Problems with this bridge were apparent even during construction, as large vertical oscillations were exhibited at that time. After completion, the deck of both the side spans and the main span of the bridge oscillated in a twisting fashion under the action of relatively light winds. Finally, on November 7, 1940, the suspension bridge collapsed. It is now generally assumed that the collapse of the bridge was due to a combination of torsional and vertical oscillations of the deck, and that this mode of failure was not taken into consideration by the designer.

It is interesting to note the various different explanations that were given by investigators regarding the failure of Tacoma Narrows bridge. Six such explanations are given below:

1. Board of Investigation, Tacoma Narrows Bridge, L. J. Svendrup, Chairman, June 26, 1941. Primary cause lies in the general proportions of the bridge and the type of stiffening girders and floors. Ratio of width to length of main span much smaller, and vertical stiffness much less compared to previously constructed bridges.
2. A Report to the Honorable John M. Carmody, Administrator, Federal Works Agency, Washington, D.C., March 28, 1941. The vortex formation and frequency is determined by the oscillation of the structure rather than the oscillatory motion is induced by the vortex formation.
3. *Bridges and Their Builders*, D. Steinman and S. Watson, Putnam's Sons, New York, 1941. Small undulation, once started, the resultant effect of a wind tends to cause a building up of vertical undulations with a tendency to change to a twisting motion, until the torsional oscillations reach destructive proportions.
4. University of Washington Engineering Experiment Station Bulletin No. 116, 1952. The motions were a result of vortex shedding.
5. *Bridges and Men*, J. Gies, Doubleday and Co., New York, 1963. Long narrow, shallow, and therefore very flexible structure standing in a wind ridden valley. Its stiffening support was a solid girder, which, combined with a solid floor, produced a cross section peculiarly vulnerable to aerodynamic effects.
6. *Wind Forces on Buildings and Structures*, E. Houghton and N. Carruthers, John Wiley & Sons, New York, 1976. Aerodynamic instability was responsible for the failure of the Tacoma Narrows Bridge in 1940. The oscillations are caused by the periodic shedding of vortices on the leeward side of the structure, a vortex being shed first from the upper section and then the lower section.

In this section, fundamental aspects associated with the dynamic behavior of suspension bridges are discussed. Additional work on this subject may be found in the literature [3, 53–58]. The section also includes additional discussion on the collapse of the Tacoma Narrows suspension bridge, as well as other structural failures.

5.8.1 Brief Discussion on Fundamental Aspects of Suspension Bridges

The following list is intended to pinpoint some of the problems, concerns, suggestions, and experimental investigations that need to be taken into consideration regarding the design of suspension bridges.

1. *Drag Forces:* Suspension bridges must be designed to withstand drag forces that are produced by severe winds.
2. *Aeroelastic Effects:* Suspension bridges must be capable of resisting aeroelastic effects, which involve vortex-induced oscillation, flutter, torsional divergence or lateral buckling, galloping, and buffeting caused by the presence of self-excited forces. Tunnel tests can provide the required information to perform such studies.
3. *Wind Action:* The action of the wind must be taken into consideration both for the completed bridge and for its state during the construction stage.
4. *Precautions During Construction:* Temporary ties and damping mechanisms may be needed for partially completed suspension bridges, and when possible, construction of suspension bridges should take place at times and seasons where the probability of severe storms is low.
5. *Other Parts of the Structure:* Aeroelastic studies must be also performed for suspension bridge towers, hangers, and cables, in addition to the ones performed for the bridge deck, because these bridge elements can be affected by the various aeroelastic phenomena mentioned in item 2 above.
6. *Wind Tunnel Tests:* These tests are used extensively by experimental researchers to obtain the necessary data for calculating the aerodynamic response of suspended-span bridges. The following test models are currently used for such studies:
    a. *Full-Bridge Models:* Scales in the order 1/300 are used to prepare models of the full bridge. Such scales are geometrically similar to the full bridge and satisfy similarity requirements with regard to mass distribution, reduced frequency, mechanical damping, and mode shapes. Such full-scale models are very costly and very elaborate to construct.
    b. *Taut Strip Models:* These models are prepared in such a way that they will respond in a manner similar to the center span of the suspension bridge when they are placed in the laboratory wind flow. Two wires are stretched across the wind tunnel and are internally clad to represent the geometry of a given bridge, and also to permit the duplication to model frequency scale of the fundamental bending and torsional frequencies of the bridge.
    c. *Section Models:* Section models are usually inexpensive, and they can be constructed to scales of the order of 1/50 to 1/25 in order to

reduce the discrepancies in Reynolds number between the full-scale bridge and the scale model. They consist of representative spanwise sections of the deck that are supported at the ends by springs that allow both torsional and vertical motion. They are usually enclosed between end plates in order to reduce the end aerodynamic effects. Based on simple tests, they are used to make initial assessments regarding the aeroelastic stability of the shape of the bridge deck, and also to make measurement of fundamental aerodynamic characteristics of the bridge deck that are required for comprehensive analytical studies. These aerodynamic characteristics involve measurements of the steady-state drag, lift and moment coefficients, as well as motional aerodynamic coefficients that characterize the self-excited forces acting on the oscillating bridge, and the Strouhal number that describes the vortex-shedding phenomenon.

7. *Torsional Divergence or Lateral Buckling:* When the deck of a suspension bridge is subjected to a slight twist, the drag load and the self-excited aerodynamic moment will cause a torsional divergence instability. When the wind acts on a bridge, a drag force, a lift force, and a twisting moment will be developed, and these must be resisted by the bridge. The twisting moment increases with increasing wind velocity and increasing angle of attack of the wind relative to the structure, thus demanding additional reactive moment from the structure. A velocity, however, may be reached where the magnitude of the wind-induced moment, together with the tendency of the twist to require additional structural reaction, develops an unstable condition that causes the structure to twist to destruction. This can be thought of as a problem of instability analogous to the one associated with column buckling. Torsional divergence will occur at some critical divergence velocity of the wind, while column buckling will occur when a critical value of the compressive column load is reached. Such conditions do not depend upon ultimate structural strength, but they do depend upon the structural flexibility and the way the aerodynamic moments develop with twist.

    In the case of a full bridge, the analysis can be carried out by using the experimentally measured moment coefficients and the torsional flexibility matrix of the bridge deck. An iteration procedure may be used such that, for any chosen velocity less than the critical divergence velocity, the procedure will converge. The iteration procedure will approach the critical divergence velocity in an asymptotic manner. This instability, for wind speeds attainable in practice, is found to occur only for torsionally weak bridges.

8. *Aeroelastic Stability:* The aeroelastic stability of a suspension bridge is controlled by three important factors, namely, geometry of the bridge deck, frequencies of vibration of the bridge, and mechanical damping of

the bridge. Regarding the first factor, unstable shapes are considered to be solid girders of deck form, very bluff cross section, and open-truss deck sections with closed or unvented roadways. Streamlined forms and open-truss sections with vents or grills through the roadway surface enhance the stability of the bridge deck. Regarding the second factor, stability is enhanced with high torsional frequencies and high ratio of torsional-to-bending frequency. The aerodynamic stability of a suspension bridge is also enhanced if its mechanical damping ratios are high.

9. *Galloping:* The influence of galloping on suspension bridge decks may be evaluated by examining the graphs of the lift and drag coefficients that are plotted as functions of the angle α between the horizontal plane and the plane of the bridge deck. Bridge deck shapes with regions of strongly negative slopes of the lift curve should be avoided.

10. *Buffeting:* Buffeting is defined as the unsteady loading of a structure caused by velocity fluctuations in the oncoming flow. Wide bridge decks generally exhibit higher buffeting response. Also, bridges with deck sections that have high drag, or that have steep slopes in their lift and moment curves tend to be more susceptible to buffeting. Reduction of buffeting response may be obtained by increased mechanical damping.

5.8.2 More on the Disaster of the Tacoma Narrows Suspension Bridge

As stated earlier, the construction of the first Tacoma Narrows bridge was finished on July 1, 1940, and the construction of a second bridge at the same location was finished on October 14, 1950. On November 7, 1940, at 10 a.m., winds had reached 42 mph, at 10:30 collections of tolls was discontinued, and at 11:08 a.m. the first Tacoma Narrows suspension bridge collapsed. Figures 5.16a and 5.16b show photographs of the failing suspension bridge. These two figures show the predominant influence of twisting during failure.

To this date, we do not really have a complete explanation regarding the exact factors that caused the collapse of the Tacoma Narrows suspension bridge. It is, however, proven that this bridge was significantly different in many ways from other suspension bridges built at that time. The cross section of the deck consisted of two very narrow I beams, 39 ft apart and 8 ft deep. The deck was braced laterally to resist static wind loads. It was unusually narrow and less deep compared to previous suspension bridges. The deck was stiffened by girders (flanges) rather than trusses, and such girders do not provide good aeroelastic stability. After work on the bridge had started, the deck itself was redesigned in order to save on the structural steel reinforcement of the concrete roadway. These factors must have contributed significantly to the torsional stiffness reduction of the bridge deck.

Another factor that is believed to have contributed significantly to the failure of this suspension bridge is the relative closeness of the normal vertical

THE COLLAPSE OF THE TACOMA NARROWS BRIDGE 345

(a)

(b)

Figure 5.16. (a) Fatal torsional oscillation mode of Tacoma Narrows Bridge. (b) Collapsing of the Tacoma Narrows Bridge. Photos by Farguharson, negative nos. 4 & 12, Special Collections Division University of Washington Libraries.

TABLE 5.4. Vertical f_v and Torsional f_t Oscillations of Suspension Bridges

Bridge	Span Length (ft)	f_v (cpm)	f_t (cpm)	f_t/f_v
Verrazano	4,260	6.2	11.9	1.92
Golden Gate	4,200	5.6	11.0	1.96
Sevem	3,240	7.7	30.6	3.97
First Tacoma Narrows	2,800	8.0	10.0	1.25

and torsional frequencies. The normal vertical frequency represents the motion in which the amplitudes of oscillation of the two sides are equal in magnitude and direction. The torsional frequency represents the motion in which the amplitudes of the two sides are equal in magnitude but opposite in direction. It was reported [60] that the calculated value of the torsional frequency was 10 cycles per minute (cpm), while that of the normal vertical frequency was 8 cpm, giving a torsional-to-vertical frequency ratio of 1.25. This is very low compared to general practices for other existing suspension bridges, Table 5.4, indicating that the first Tacoma Narrows Bridge was torsionally flexible. Since the calculated values of these two frequencies are only approximate, the probability of resonant vibration and the creation of divergence instability is rather high. See also the author's work in Reference [3, Chapter 12].

The collapse of the Tacoma Narrows bridge, however, created a great deal of interest in research and discussion of the aerodynamic and general dynamic behavior of suspension bridges. See, for example, References [3, 53–59]. A great deal of this research concentrated on suspension bridge aerodynamics, identifying the aerodynamic behavior of bridge sections by experimentally determining motional aerodynamic coefficients of freely oscillating models. Studies in turbulent flow and wake effects were performed by using bridge section models, and full-scale investigations were used to improve the design of suspension bridges. Extensive work was also performed in obtaining motional aerodynamic coefficients to be used in analytical studies and nonlinear aerodynamic investigations.

The controversy regarding the causes that contributed to the collapse of the first Tacoma Narrows suspension bridge still continues today, because no satisfactory answer to these questions has yet been obtained. An interesting discussion on this subject is given by Berreby [59], who presented the wide difference in opinion between the mathematicians Joe McKenna and A. C. Lazer and Professor of Johns Hopkins University Robert H. Scanlan.

Joe McKenna, a math professor at the University of Connecticut at Storrs, and his Florida colleague, A. C. Laser of the University of Miami, say that their equations and computer models suggest that bridges do not always behave as their designers assume they will. McKenna and Laser believe that their model, often associated with chaos theory, provides a more accurate representation of the way a suspension bridge vibrates. McKenna believes that their model confirms the old approach in most everyday instances, but that it is superior in accounting for certain special cases. Any structure that must endure the buffeting of wind, ground tremors, or passing traffic will vibrate in response to the forces assaulting it. McKenna says that the engineers' math is insufficiently sophisticated, and that their models are inadequate.

In an interview with a reporter from the *San Francisco Examiner*, June 7, 1990, McKenna explained his views as to how a severe earthquake could destroy the Golden Gate Bridge. This interview sparked an ongoing debate between McKenna and Robert H. Scanlan regarding the design efficiency of suspension bridges. The engineers, led by Scanlan, say that McKenna and Lazer have yet to build a physical model and aren't likely to do so soon. Instead, they explore equations that describe simple oscillating systems like a pendulum or a bar suspended from springs.

When the mathematicians run their equations on a computer, they find that a single large disturbance—such as might come from a high wind or an earthquake—can lead to predictable, wild oscillations that sometimes get bigger and bigger rather than slowly dying out, as smaller vibrations generally do. "Basically, we discovered a mathematical phenomenon," McKenna says. "And then we said, if this is happening in the math, then it's probably happening in the real world." The problem, according to McKenna, is that engineers rely too heavily on linear equations to describe bridge behavior. According to McKenna and Lazer, once bridge behavior has entered a nonlinear mode, the effects of the forces that hit it are not proportional, or even predictable. In fact, exactly the same amount of force can produce very little vibration or a huge amount—depending on slight variations in other variables, such as the angle at which the wind strikes the bridge.

Many other points have been raised by both sides regarding the collapse of the Tacoma Narrows bridge. The details of the discussions may be found in Reference [59].

My purpose in discussing these differences in technical views between McKenna, Lazer, and Scanlan is to illustrate that complete agreement on important technical problems may not always be available. The positive aspects of such arguments, however, is that they raise important questions about technical problems, questions that demand solutions. Such solutions may finally be obtained by the individuals participating in the argument, or by researchers who happened to learn about the argument and whose own curiosity then took its course.

As the author of this book, I tend to disagree with McKenna's suggestion that engineers rely too much on linear equations to describe bridge behavior, because a great deal of both experimental and analytical work on nonlinear behavior has been done on many subjects. See, for example, the author's work in References [3, 5, 6, 47, 50, 52]. I do, however, agree with McKenna and Lazer on what they call a "mathematical phenomenon," where large oscillations may be produced in a suspension bridge. I am not familiar with their mathematical model and I cannot say if their nonlinear analysis is applicable to suspension bridges.

The author has done extensive work regarding the solution of many nonlinear problems; see for example, References [3, 5, 6], and accurate methods of analysis have been developed for the solution of many engineering nonlinear problems. In particular, I wish to single out the work in Reference [3, Chapter 12], the concept of the dynamic hinge in References [5, 37], and the discussion in Section 5.7 of this text. Utilization of the concept of the dynamic hinge, as developed by the author and his collaborators, makes it possible to isolate a portion of a span of a bridge and to develop a dynamically equivalent system by satisfying appropriately, using springs, its end boundary conditions. The vibration response of the dynamically equivalent system would be closely identical to the response of the complete system. In Chapter 12 Reference [3], and in Section 5.7, methods are provided that could be modified and used for the stability analysis of suspension bridges, and may be able to determine the real cause of failure of the Tacoma Narrows bridge.

The analytical results obtained from the dynamically equivalent system could be checked by experimental results, which can be obtained by testing a scale model of the dynamically equivalent system. The key factor concerning the accuracy of such analytical and experimental results would lie in the derivation of the dynamically equivalent system using the dynamic hinge concept. This concept and theory has worked very well for problems worked out by the author, and it was proved mathematically that flutter divergence instabilities do exist for elastically supported members subjected to axial restraints. No experimental investigations, however, were performed by the author to verify experimentally the existence of such instabilities. Details regarding this work may be found in the indicated references.

The preceding idea could be used by McKenna and Lazer to develop a representative mathematical model of a suspension bridge, preferably the one representing the first Tacoma Narrows bridge, and examine its dynamic behavior. McKenna's claim that "if the simple model shows nonlinear behavior, the more complex one will be even more nonlinear," is not necessarily valid for actual engineering applications. The model, no matter how simple or how complex it is, must represent to a reasonable degree the actual engineering structure in order for the solution to be valid for structural applications.

5.8.3 Other Failures and What We Learn from Them

The Tacoma Narrows bridge is not the only suspension bridge that has failed. Table 5.5 lists 10 additional suspension bridges that failed before the first Tacoma Narrows bridge. Note, however, that the main span of the Tacoma Narrows bridge is much longer than the span lengths of all other bridges that have failed.

TABLE 5.5. Collapsed Suspension Bridges

Bridge	Designer	Span Length (ft)	Failure Date
Dryburgh (Scotland)	J. and H. Smith	260	1818
Union (England)	Sir Samuel Brown	449	1821
Nassau (Germany)	L and W (Lossen and Wolf)	245	1834
Brighton (England)	Sir Samuel Brown	255	1836
Montrose (Scotland)	Sir Samuel Brown	432	1838
Menai (Wales)	T. Telford	580	1839
Roche (France)	Le Blanc	641	1852
Wheeling (USA)	C. Ellet	1,010	1854
Niagara (USA)	E. Serrell	1,041	1864
Niagara (USA)	S. Keefer	1,260	1889
Tacoma Narrows	L. Moisseff	2,800	1940

The failure of the first Tacoma Narrows suspension bridge, as well as many other structural failures, has taught us that a structure may be considered adequately designed and may provide problem-free service for decades, while still having a serious design flaw present, a flaw that will eventually result in a sudden collapse. For example, the elevated walkways of the Kansas City Hyatt Regency Hotel were considered safe to carry pedestrian crowds until they collapsed on the afternoon of July 17, 1981. The Mianus River bridge near Greenwich, Connecticut, was considered to be adequately designed to carry the heavy traffic of Interstate 95, but it suddenly collapsed early in the morning of June 28, 1983.

Failure analysis, coupled with comprehensive knowledge of the causes of failure, can help to eliminate design errors and prevent the repetition of such catastrophic failures. We, the engineers, as Lev Zetlin [61] has said, "have to be somewhat paranoiac during the design stage of a structure." We should consider and imagine that the impossible could happen. We should not feel secure that the structure will be safe and sound merely because all the requirements of the design handbooks and manuals have been satisfied. If we do not remember the past and we do not learn from past mistakes, then we will be condemned to repeat them.

John Augustus Roebling, a well-known suspension bridge design engineer, realized the economical suspension bridge principle and he did not abandon it because of past failures. Instead, he used the concept of failure analysis to learn what mistakes should be avoided and eliminate in his new designs the weaknesses that had caused the collapse of past suspension bridges. He used this knowledge in his successful design of the Niagara Gorge Bridge, between western New York State and Canada, and the Brooklyn Bridge. In fact the span of the Brooklyn Bridge was twice as long as that of the Niagara Gorge Bridge.

Civil engineering structures are not the only ones that have failed. Other engineering fields, such as electrical, mechanical, computer, manufacturing, aeronautical, aerospace, and so on, have also recorded catastrophic failures. Such examples, to name just a few, are the 1979 loss of coolant at the Three Mile Island Nuclear Plant near Harrisburg, Pennsylvania, and the 1986 explosion and fire at the Chernobyl Nuclear Plant near Kiev, Ukraine, where radioactive material was spread well beyond its point of release. Another devastating failure was the 1984 release of toxic gas from a Union Carbide insecticide plant at Bhopal, India, which killed more than 2,000 people and harmed about 150,000 more.

The decisions regarding the design of engineering structures are made by people, and people are responsible when errors are made and the structure fails. Therefore, in order to develop successful designs that protect the safety and welfare of people, we must learn from past mistakes and also develop sound new technologies that help to improve the functioning and service of our infrastructure systems. This is the lesson we learn from past failures.

5.9 INELASTIC ANALYSIS AND DESIGN OF INFRASTRUCTURAL ELEMENTS

Today, many structures and mechanical systems are subjected to stresses beyond the elastic limit of their material, thus creating a loading condition of great significance to the design engineer. Such a condition is of particular importance for aircraft and aerospace structures, and for any structure subjected to the effects of an earthquake or blast. Since both uniform and variable stiffness members are used as component elements of such structures, their analysis can become very complex, and in many cases unrealistic, for a reasonable solution of such problems. When the material of a member is permitted to be stressed beyond its elastic limit, the modulus of elasticity E_x along the length of the member becomes variable, and special analysis is required to determined the variation of E_x across the length of the member. On this basis, the stiffness $E_x I_x$ of the member will vary along its length regardless of whether the moment of inertia I_x is constant or variable.

The inelastic analysis of such members may be carried out conveniently and reliably by using the author's [3, 6, 39] method of the equivalent systems. For small deformations, the fundamentals of this method with several applications are discussed in Chapter 4. In the earlier sections of this chapter more challenging cases were considered by including large deformations. In all these cases, however, the modulus of elasticity of the material was assumed to be constant. The analysis in this section is based on the idea that both E_x and I_x can vary arbitrarily along the length of a member. The variation of E_x is obtained by using Timoshenko's concept of reduced modulus E_r [3, 5, 6, 51].

5.9.1 Theory of Equivalent Systems for Inelastic Analysis and Reduced Modulus E_r

In Section 4.2 we proved that the analysis of members with variable stiffness $E_x I_x$ along their length can be made convenient by using equivalent systems of constant stiffness $E_1 I_1$, where E_1 and I_1 are reference values for E_x and I_x. If $g(x)$ and $f(x)$ represent the variations of E_x and I_x, respectively, with respective reference values E_1 and I_1, the moment diagram M_e of the equivalent system of constant stiffness $E_1 I_1$ may be determined by using Eq. (4.10). This equation is as follows:

$$M_e = \frac{M_x}{g(x) f(x)} \quad (5.93)$$

where M_x is the bending moment at cross sections along the length of the original variable stiffness member.

The application of Eq. (5.93) for inelastic analysis requires the evaluation of the modulus function $g(x)$, which represents the variation of E_x along the length of the member when its material is stretched beyond the elastic limit. Such modulus variation and, consequently, $g(x)$, may be determined by using Timoshenko's [3, 5, 6, 51] concept of reduced modulus E_r. In this section, the discussion is centered on statically determinate members of rectangular cross section where both E_x and I_x can vary in any arbitrary manner along their length. The deformation of the member is assumed to be small, so that it would be reasonable to use small deflection theory. It should be stated, however, that the methodology can be used for members subjected to large deformations, and also for members under the influence of axial restraints. See the author's work in Reference [3]. The reduced modulus E_r may be also determined for nonrectangular across-section beams [51].

In order to determine E_r, we need to know the stress-strain curve of the material that is used to construct the member. Such stress-strain curves are usually available from the manufacturers, but they can also be determined experimentally by using standard procedures. We start the discussion by using a general shape of stress-strain curve, such as the one shown in Fig. 5.17a. Actual stress-strain curves of known materials will be used later on to illustrate the methodology.

A portion of a member that is subjected to bending is shown in Fig. 5.17b, where r is the radius of curvature, M is the bending moment, h_1 is the distance from the neutral axis to the bottom surface of the member, and h_2 is the distance from the neutral axis to the top surface of the beam.

In Fig. 5.17a, the stains ε_1 and ε_2 represent the unit elongations of the extreme fibers of the member, and strain ε represents the unit elongation of a fiber at a distance y from the neutral axis, as shown in Fig. 5.17b. The symbol Δ is used to denote the sum of the absolute values of the strains ε_1 and ε_2.

352 INFRASTRUCTURAL NONLINEARITIES

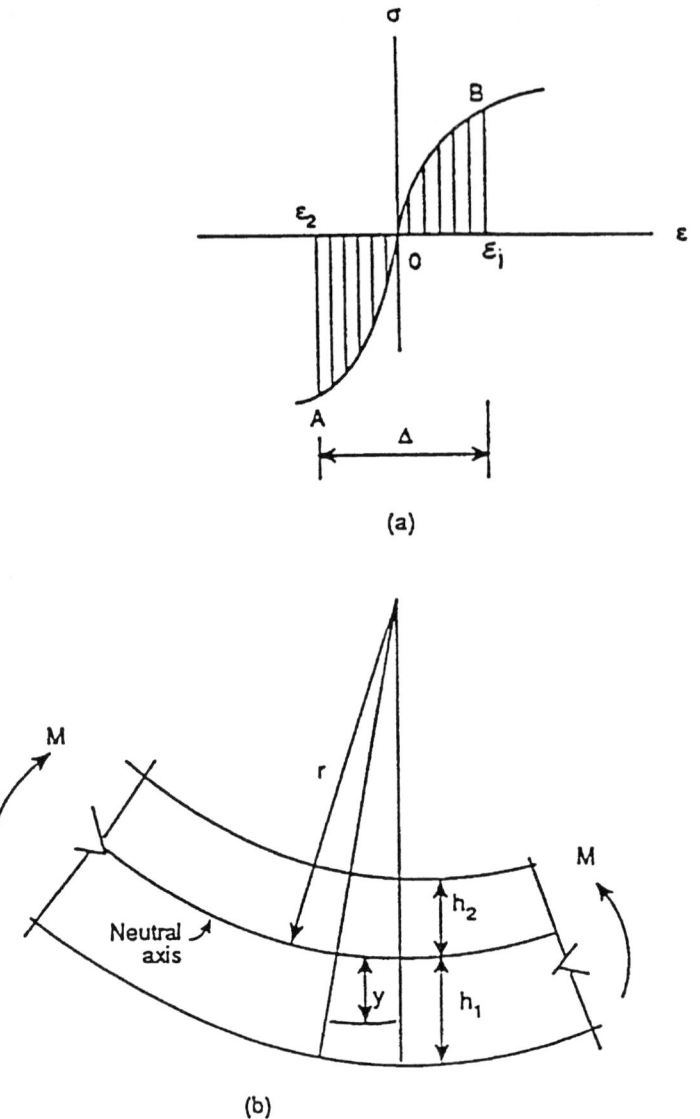

Figure 5.17. (a) General shape of a stress-strain curve. (b) Portion of a member subjected to bending.

Figure 5.17a is also assumed to represent the cross-sectional bending sress distribution if the depth h of the member is substituted for Δ. This is a reasonable assumption for practical applications. As stated earlier, the member is free of axial restraints and, consequently, the neutral and centroidal axes coincide.

INELASTIC ANALYSIS AND DESIGN OF INFRASTRUCTURAL ELEMENTS   353

From basic mechanics we have the following relations:

$$\varepsilon_1 = \frac{h_1}{r} \tag{5.94}$$

$$\varepsilon_2 = \frac{h_2}{r} \tag{5.95}$$

$$\varepsilon = \frac{y}{r} \tag{5.96}$$

$$\frac{h_1}{h_2} = \frac{|\varepsilon_1|}{|\varepsilon_2|} \tag{5.97}$$

where r is the radius of curvature. We can determine the position of the neutral axis by using the equations

$$b \int_{-h_2}^{h_1} \sigma \, dy = 0 \tag{5.98}$$

$$b \int_{-h_2}^{h_1} \sigma y \, dy = M \tag{5.99}$$

where σ is the bending stress and b is the width of the member.

From Eq. (5.96) we find

$$y = \varepsilon r \tag{5.100}$$

Thus,

$$dy = r \, d\varepsilon \tag{5.101}$$

By substituting Eq. (5.101) into Eq. (5.98), we obtain

$$br \int_{\varepsilon_2}^{\varepsilon_1} \sigma \, d\varepsilon = 0 \tag{5.102}$$

Since b and r are different from zero, Eq. (5.102) shows that the position of the neutral axis should be the one that satisfies the equation

$$\int_{\varepsilon_2}^{\varepsilon_1} \sigma \, d\varepsilon = 0 \tag{5.103}$$

By substituting Eqs. (5.100) and (5.101) into Eq. (5.99), we find

$$br^2 \int_{\varepsilon_2}^{\varepsilon_1} \sigma \varepsilon \, d\varepsilon = M \tag{5.104}$$

354 INFRASTRUCTURAL NONLINEARITIES

From Fig. 5.17a, we note that

$$\Delta = |\varepsilon_1| + |\varepsilon_2|$$
$$= \frac{h_1}{r} + \frac{h_2}{r} \qquad (5.105)$$
$$= \frac{h}{r}$$

Thus,

$$r = \frac{h}{\Delta} \qquad (5.106)$$

By substituting Eq. (5.106) into Eq. (5.104), we find

$$\frac{bh^2}{\Delta^2} \int_{\varepsilon_2}^{\varepsilon_1} \sigma\varepsilon\, d\varepsilon = M \qquad (5.107)$$

Equation (5.107) may be easily rearranged so that it can be written as follows:

$$\frac{I}{r}\left(\frac{12}{\Delta^3}\right) \int_{\varepsilon_2}^{\varepsilon_1} \sigma\varepsilon\, d\varepsilon = M \qquad (5.108)$$

In Eq. (5.108), the symbol I denotes the member's moment of inertia at the rectangular cross section under consideration, and it is given by the well-known expression

$$I = \frac{bh^3}{12} \qquad (5.109)$$

For the elastic range, the curvature $1/r$ that is produced by bending of the member can be obtained by using the equation

$$M = \frac{EI}{r} \qquad (5.110)$$

where M is the elastic bending moment and E is the elastic modulus. If the elastic limit of the material is exceeded, the curvature produced by the bending moment M may be determined by using Eq. (5.108), which can be rewritten as follows:

$$M = \frac{E_r I}{r} \qquad (5.111)$$

where

$$E_r = \frac{12}{\Delta^3} \int_{\varepsilon_2}^{\varepsilon_1} \sigma\varepsilon\, d\varepsilon \qquad (5.112)$$

INELASTIC ANALYSIS AND DESIGN OF INFRASTRUCTURAL ELEMENTS   355

At cross sections along the length of the member, the reduced modulus E_r may be determined from Eq. (5.112) when the member is stressed beyond the elastic limit of its material. The integral in Eq. (5.112) represents the first moment of the shaded area in Fig. 5.17a with respect to the vertical axis through the origin 0. The units of E_r are force per unit area, since the abscissas in Fig. 5.17a represent strain and the ordinates represent stress. Since at cross sections along the length of the member, the stresses and strains will vary, the reduced modulus E_r will also vary. At cross sections where the stresses are elastic, the reduced modulus E_r would be equal to the elastic modulus E.

Since Δ is unknown, the reduced modulus E_r at a cross section of depth h and moment of inertia I may be determined by using a trial-and-error procedure. We can initiate the trial-and-error procedure by assuming values of Δ and using in each case the stress-strain curve in Fig. 5.17a to determine the extreme elongations ε_1 and ε_2 for each value of Δ. For the assumed Δ, we determine E_r by using Eq. (5.112), and we apply Eq. (5.111) to determine the required moment M_{req}. If the problem is statically determinate, the procedure may be repeated until M_{req} is equal to the actual moment M_x that is obtained at the considered cross section of the member by applying statics. The procedure is explained in more detail in the numerical example. Equation (5.106) provides the expression to be used for the computation of the radius of curvature r when Δ is known.

With known E_r, we use the expression

$$E_r = Eg(x) \tag{5.113}$$

to determine the modulus function $g(x)$ as follows:

$$g(x) = \frac{E_r}{E} \tag{5.114}$$

In Eqs. (5.113) and (5.114), the elastic modulus E is used as the reference value.

When both functions $g(x)$ and $f(x)$ are known along the length of a member, then the method of the equivalent systems, as discussed in the early sections of Chapter 4, may be used to determine inelastic deflections and rotations of members. This analysis may be also used to determine ultimate load capabilities of members that can be used for the design of structural and mechanical systems. The following example illustrates the application of the method to a practical infrastructural problem.

Example 5.6: The tapered cantilever beam in Fig. 5.18a is made out of aluminum alloy 7075-T651, which is used for very high strength structural aircraft parts. The distributed load $w = 2,500\,\text{lb/in.}$, the depth h at the free end B is 6 in., the constant width $b = 2$ in., the length $L = 60$ in., and the taper parameter $n = 1.75$. Perform an inelastic analysis of the aluminum member by using the method of the equivalent systems and the reduced modulus concept.

356 INFRASTRUCTURAL NONLINEARITIES

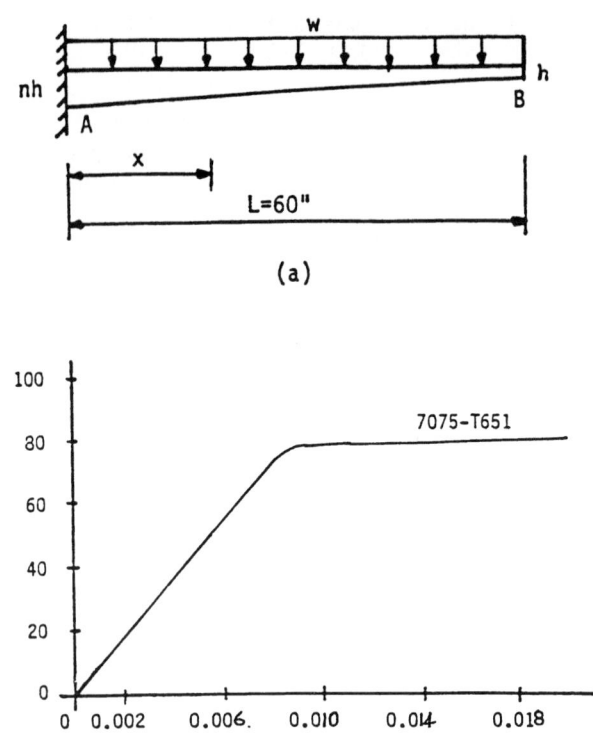

Figure 5.18. (a) Tapered aluminum cantilever beam loaded as shown. (b) Experimental stress-strain curve of the 7075-T651 aluminum alloy.

SOLUTION: The experimental stress-strain curve of the 7075-T651 aluminum alloy is shown in Fig. 5.18b. The curve is based on a 0.505-in. diameter test sample, resulting in a yield strength of 79,920 psi, a tensile strength of 86,913 psi, and a modulus of elasticity $E = 9.50 \times 10^6$ psi. See also Reference [52] by Fertis and Schubert. The applied load $w = 2,500$ lb/in. will stress the member well beyond the elastic limit of its material and, consequently, the modulus of elasticity E will not be constant along its length.

The moment of inertia I_x of the tapered member is variable, and at any distance x from its fixed end may be obtained from the following expression:

$$I_x = \frac{bh^3}{12}\left[\frac{(n-1)(L-x)+L}{L}\right]^3$$

$$= I_B f(x)$$

(5.115)

where

$$I_B = \frac{bh^3}{12} \tag{5.116}$$

is the moment of inertia at the free end B of the member, which is taken as the reference value, and

$$f(x) = \left[\frac{(n-1)(L-x) + L}{L}\right]^3 \tag{5.117}$$

is the moment of inertia function that expresses its variation along the length of the member. For $n = 1.75$, we have

$$f(x) = \left[\frac{1.75L - 0.75x}{L}\right]^3 \tag{5.118}$$

The evaluation of the modulus function $g(x)$ is based on the determination of a reduced modulus E_r, as discussed earlier in this section. A very accurate approximate solution for the evaluation of E_r may be obtained by approximating the shape of this stress-strain curve with two, three, or six straight-line segments, depending on the desired accuracy and the practical purpose of the analysis. For example, if it is required to determine maximum deflections, the two-line approximation should yield reasonable results for practical applications. If, however, we wish to know the ultimate load-carrying capacity of the member, the three-line or the four-line approximation might be more desirable. The use of the computer program in Appendix D makes the computation of E_r and the application of the equivalent systems for inelastic analysis very convenient.

The two-line, three-line, and six-line approximations of the stress-strain curve of the 7075-T651 aluminum alloy yield the values of E shown in Table 5.6, and the corresponding stresses σ are shown in Table 5.7. In order to

TABLE 5.6. Values of E for the Two-Line, Three-Line, and Six-Line Approximations of the Stress-Strain Curve of the 7075-T651 Aluminum Alloy

(1) Modulus E (psi)	(2) Two-Line Approximation	(3) Three-Line Approximation	(4) Six-Line Approximation
E_1	9.50×10^6	9.51×10^6	9.49×10^6
E_2	2.00×10^5	8.63×10^6	8.65×10^6
E_3	—	1.90×10^5	4.99×10^6
E_4	—	—	1.24×10^6
E_5	—	—	5.00×10^5
E_6	—	—	1.99×10^5

TABLE 5.7. Values of σ for the Two-Line, Three-Line, and Six-Line Approximations of the Stress-Strain Curve of the 7075-T651 Aluminum Alloy

(1) Stress σ (psi)	(2) Two-Line Approximation	(3) Three-Line Approximation	(4) Six-Line Approximation
σ_1	79.79×10^3	59.94×10^3	56.94×10^3
σ_2	—	79.67×10^3	72.93×10^3
σ_3	—	—	77.92×10^3
σ_4	—	—	79.42×10^3
σ_5	—	—	79.92×10^3

simplify the illustration of the methodology, the two-line (or bilinear) approximation is used here, which yields $E_1 = 9.50 \times 10^6$ psi, $E_2 = 2 \times 10^5$ psi, and yield stress $\sigma_y = 79.795 \times 10^3$ psi. The results, however, are compared by using the six-line approximation of the stress-strain curve.

A schematic representation of the bilinear stress-strain curve of the 7075-T651 aluminum alloy is shown in Fig. 5.19a. The compressive yield strength is taken to be equal in magnitude to the tensile yield strength. This is not mandatory for the application of this method, but it follows typical engineering practice for metals. The curve AOB in Fig. 5.19a represents the bending stress distribution along the depth h at a cross section of the member, if h is substituted for Δ.

An iteration procedure is used here to calculate E_r for any specific cross section of the member that has a depth h and moment of inertia I. The iteration procedure may be initiated by assuming values for Δ. Since in this case $\varepsilon_1 = \varepsilon_2$, we have $\varepsilon_1 = \varepsilon_2 = \Delta/2$. For each assumed value of Δ, we use Eq. (5.112) to calculate the corresponding value of E_r. Note that the integral in Eq. (5.112) represents the first moment of the stress-strain area AOB about the center 0 shown in Fig. 5.19a. Also, we use Eq. (5.106) to determine the radius of curvature r corresponding to the assumed Δ. The calculated values of E_r, r, and moment of inertia I of the considered cross section are substituted into Eq. (5.111) to determine the required moment $M = M_{req}$ that produces the strains ε_1 and ε_2. Since the member is statically determinate, M_{req} should be equal to the static moment M_x at the section that is produced by the application of the external distributed load w. The procedure is repeated for as many values of Δ as are required to make M_{req} sufficiently equal to M_x.

Table 5.8 provides a summary of the calculated values of E_r and M_{req} at the indicated distances x from the fixed end of the cantilever beam. Note that, at locations $x \geq 21$ in., the bending stress σ is lower than the yield strength of the material and, consequently, E_r is equal to the elastic E at these locations.

In order to illustrate the procedure, we consider the cross section at the location $x = 6$ in., and assume that $\Delta = 30.6 \times 10^{-3}$ which makes $\varepsilon_1 = \varepsilon_2 = \Delta/$

INELASTIC ANALYSIS AND DESIGN OF INFRASTRUCTURAL ELEMENTS 359

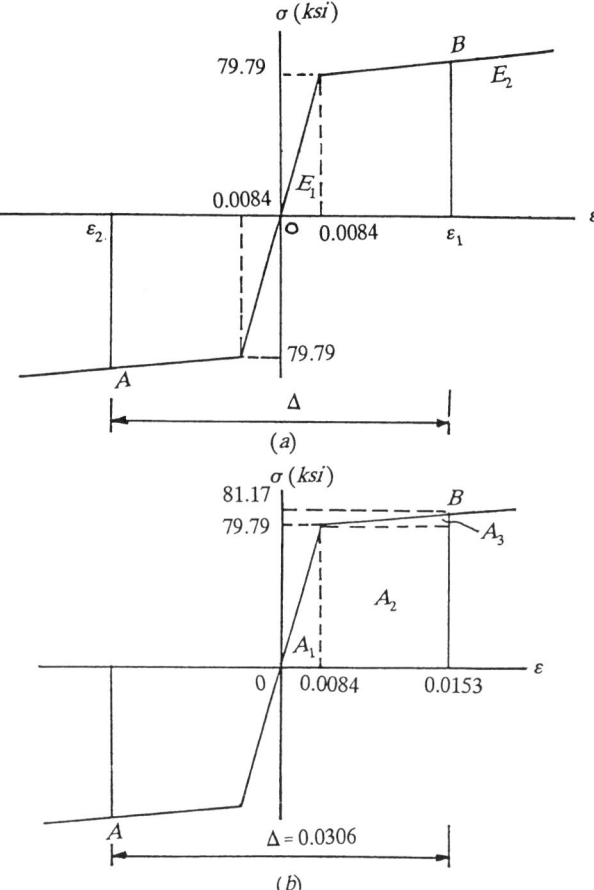

Figure 5.19. (a) Schematic representation of the bilinear approximation of the stress-strain curve of the 7075-T651 aluminum alloy. (b) Bilinear stress-strain approximation with $\Delta = 0.0306$.

$2 = 15.3 \times 10^{-3}$. These values of ε_1, ε_2, and Δ are shown in Fig. 5.19b. The first moment of the area under the curve $A0B$ about point 0 is

Area $A0B = \frac{1}{2}(79.79)(0.0084)(\frac{2}{3})(0.0084)(2)$

$\qquad + (79.79)(0.0153 - 0.0084)\left(0.0084 + \dfrac{0.0153 - 0.0084}{2}\right)(2)$

$\qquad + \frac{1}{2}(81.17 - 79.79)(0.0153 - 0.0084)[0.0084$

$\qquad + \frac{2}{3}(0.0153 - 0.0084)](2)$

$\qquad = 0.016925 \text{ ksi}$

TABLE 5.8. Values of Δ, r, E_r, and M_{req} at the Indicated Cross Sections of the Cantilever Beam Made of 7075-T651 Aluminum Alloy, Using the Two-Line Approximation of the Stress-Strain Curve

(1) x (in.)	(2) h (in.)	(3) I_x (in.⁴)	(4) $\Delta \times 10^{-3}$ (in./in.)	(5) r (in.)	(6) $E_r \times 10^6$ (psi)	(7) $M_{req} = M_x \times 10^6$ (lb-in.)
0	10.500	192.93	76.0	138.16	3.23	4.50
3	10.275	180.80	43.4	236.75	5.32	4.06
6	10.050	169.18	30.6	328.43	7.09	3.65
9	9.825	158.07	24.6	399.40	8.24	3.25
12	9.600	147.45	20.8	461.54	9.00	2.88
15	9.375	137.33	18.4	509.51	9.38	2.53
18	9.150	126.63	16.8	544.64	9.49	2.21
21	8.925	118.48	15.1	592.24	9.50	1.90

From Eq. (5.106) we obtain

$$r = \frac{h}{\Delta} = \frac{10.05}{0.0306} = 328.43 \text{ in.}$$

On this basis, Eq. (5.112) yields

$$E_r = \frac{12}{\Delta^3} \int_{\varepsilon_2}^{\varepsilon_1} \sigma\varepsilon \, d\varepsilon$$

$$= \frac{12}{(0.0306)^3} (0.016925) = 7.088 \times 10^3 \text{ ksi}$$

Thus, from Eq. (5.111), we find

$$M_{req} = \frac{E_r I}{r} = \frac{(7.088)(10)^3(169.18)}{328.43}$$

$$= 3.65 \times 10^3 \text{ kip-in.}$$

$$= 3.65 \times 10^6 \text{ lb-in.}$$

The actual moment M_x at $x = 6$ in. is

$$M_x = w(L-x)\left(\frac{L-x}{2}\right)$$

$$= (2,500)(54)(27)$$

$$= 3.645 \times 10^6 \text{ lb-in.}$$

INELASTIC ANALYSIS AND DESIGN OF INFRASTRUCTURAL ELEMENTS   361

TABLE 5.9. Values of $f(x)$, E_r, $g(x)$, M_x, and M_e for the Inelastic Analysis of the Nonprismatic Cantilever Beam Using the Bilinear Approximation of the Stress-Strain Curve of the 7075-T651 Aluminum Alloy

(1) x (in.)	(2) $f(x)$	(3) $E_r \times 10^6$ (psi)	(4) $g(x)$	(5) $M_x \times 10^6$ (lb-in.)	(6) $M_e = M_x/f(x)\,g(x)(10^6)$ (lb-in.)
0	5.36	3.23	0.34	4.50	2.469
3	5.02	5.32	0.56	4.06	1.444
6	4.70	7.08	0.745	3.65	1.042
9	4.39	8.24	0.867	3.25	0.853
12	4.09	9.00	0.947	2.88	0.743
15	3.81	9.38	0.987	2.53	0.672
18	3.52	9.49	0.998	2.21	0.629
21	3.29	9.50	1.00	1.90	0.577
24	3.05	9.50	1.00	1.62	0.531
27	2.82	9.50	1.00	1.36	0.482
30	2.60	9.50	1.00	1.13	0.434
33	2.39	9.50	1.00	0.911	0.381
36	2.20	9.50	1.00	0.720	0.327
39	2.01	9.50	1.00	0.551	0.274
42	1.84	9.50	1.00	0.405	0.220
45	1.67	9.50	1.00	0.281	0.168
48	1.52	9.50	1.00	0.180	0.118
51	1.38	9.50	1.00	0.101	0.073
54	1.24	9.50	1.00	0.045	0.036
57	1.11	9.50	1.00	0.011	0.010
60	1.00	9.50	1.00	0	0

This shows that the assumed value of $\Delta = 30.6$ yields a value of M_{req} that, for practical purposes, can be assumed closely equal to the actual moment M_x. Thus, repetition of the iteration procedure is not required.

With known E_r and using the elastic modulus $E = 9.5 \times 16^6$ psi as a reference value, the modulus function $g(x)$ at the indicated locations x can be determined from Eq. (5.114). The moment diagram M_e of the equivalent system of constant stiffness EI_B, where I_B is the moment of inertia at the free end of the member, may be determined from Eq. (5.93). Table 5.9 gives the values of $f(x)$, E_r, $g(x)$, M_x, and M_e at the indicated locations x.

The moment diagram M_e of the equivalent system of constant stiffness EI_B is shown plotted by the solid line in Fig. 5.20a. The approximation of its shape with four straight-line segments is shown by the dashed line in the same figure. This approximation leads to the constant stiffness equivalent system shown in Fig. 5.20b, which is loaded with four equivalent concentrated loads located as shown. The deflections and rotations at any point along the length of the equivalent system may be determined by using standard handbook formulas,

362   INFRASTRUCTURAL NONLINEARITIES

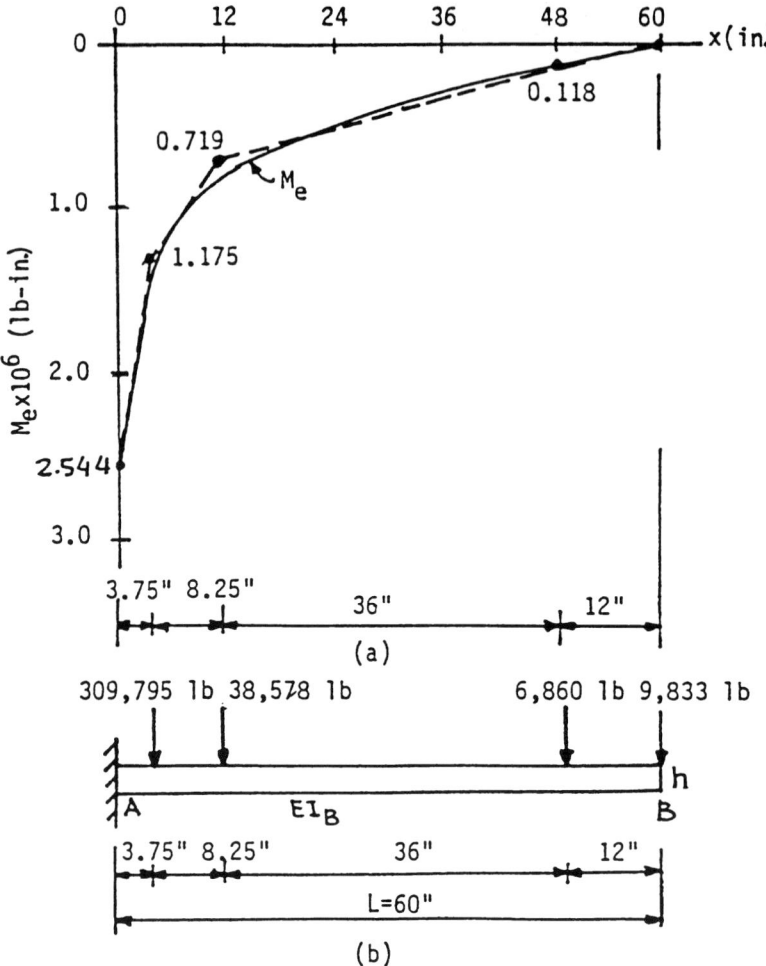

Figure 5.20. (*a*) Moment diagram M_e of the equivalent system of constant stiffness EI_B approximated with four straight-line segments. (*b*) Equivalent system of constant stiffness EI_B.

or by applying linear methods of elementary mechanics such as the moment-area and conjugate-beam methods. These deflections and rotations will be closely identical to the corresponding deflections and rotations of the initial variable stiffness member in Fig. 5.18*a*. For example, by using the equivalent system in Fig. 5.20*b* and applying the moment-area method as discussed in Section 4.5.1, we find that the deflection y_B at the free end *b* of the member is 4.142 in. Note that the juncture points of the four straight-line segments in Fig. 5.20*a* can lie above or below the solid-line curve so that the areas added to or subtracted from the M_e diagram are approximately balanced. This is why the

TABLE 5.10. Values of Δ, r, E_r, and M_{req} at the Indicated Cross Sections of the Cantilever Beam Made of 7075-T651 Aluminum Alloy, Using the Six-Line Approximation of the Stress-Strain Curve

(1) x (in.)	(2) h (in.)	(3) I_x (in.^{4})	(4) $\Delta \times 10^{-3}$ (in./in.)	(5) r (in.)	(6) $E_r \times 10^6$ (psi)	(7) $M_{req} = M_x \times 10^6$ (lb-in.)
0	10.500	192.93	78.0	134.61	3.14	4.50
3	10.275	180.80	45.4	228.33	5.13	4.06
6	10.050	169.18	31.80	316.03	6.82	3.65
9	9.825	158.07	25.50	387.25	7.98	3.25
12	9.600	147.45	21.36	449.43	8.79	2.88
15	9.375	137.33	18.80	498.67	9.21	2.53
18	9.150	126.63	17.00	538.23	9.40	2.21
21	8.925	118.48	15.14	589.50	9.48	1.90
23	8.700	109.75	13.60	639.70	9.50	1.62

amount of 2.544×10^6 lb-in. in Fig. 5.20a is somewhat higher than the actual value of 2.469×10^6 lb-in. shown in Table 5.9.

The solution of this problem is repeated by using the six-line approximation of the stress-strain curve of the 7075-T651 aluminum alloy. The results are shown in Tables 5.10 and 5.11. From the point of view of the practicing design engineer, reasonable agreement is obtained when they are compared with the analogous results in Tables 5.8 and 5.9, which are obtained by using the two-line approximation of the stress-strain curve. The solution of the constant stiffness equivalent system that is obtained by using the six-line approximation

TABLE 5.11. Values of $f(x)$, E_r, $g(x)$, M_x, and M_e for the Inelastic Analysis of the Nonprismatic Cantilever Beam Using the Six-Line Approximation of the Stress-Strain Curve of the 7075-T651 Aluminum Alloy

(1) x (in.)	(2) $f(x)$	(3) $E_r \times 10^6$ (psi)	(4) $g(x)$	(5) $M_x \times 10^6$ (lb-in.)	(6) $M_e = M_x/f(x)\, g(x)(10^6)$ (lb-in.)
0	5.36	3.14	0.330	4.50	2.544
3	5.02	5.13	0.540	4.06	1.498
6	4.70	6.82	0.717	3.65	1.083
9	4.39	7.98	0.840	3.25	0.881
12	4.09	8.79	0.925	2.88	0.761
15	3.81	9.21	0.969	2.53	0.685
18	3.52	9.40	0.989	2.21	0.634
21	3.29	9.48	0.997	1.90	0.579
24	3.05	9.50	1.000	1.62	0.531

of the stress-strain curve yields, at the free end B, the maximum deflection $y_B = 3.820$ in., which is 8.4 percent lower than the one obtained by using the two-line approximation. The value of y_B obtained by using the six-line approximation is considered to be more accurate. The exact value is not yet known, because we do not have an analytical method that can be used to obtain the exact solution. The methodologies used in this section are, however, reliable, and they provide a very reasonable estimate regarding these quantities. The reason is that the method of the equivalent systems is mathematically exact, since it is based on exact mathematical analogies. The simplification of the M_e diagram with straight-line segments, which is introduced for convenience, has proved to yield very accurate results, and its accuracy is thoroughly investigated. Almost exact solutions may be obtained. The next approximation is in the bending stress distribution at a cross section of the inelastic member, which is assumed to have the shape of the stress-strain curve of the material that is used to construct the member. The extensive research on this subject by the author of this text and his collaborators indicated that this is a reasonable assumption. Experimental investigations may be used to investigate this assumption in greater detail. For the practicing design engineer, however, the above methodology has the capability of providing him/her with the answers she/he needs to carry out an efficient design.

5.10 ULTIMATE DESIGN LOADS BASED ON INELASTIC RESPONSE

The analysis in this section deals with the determination of ultimate loads for simply supported and cantilever beams of both uniform and variable depth along their length. The loading w on the beam is assumed to be distributed uniformly along its entire length, and its cross section is rectangular. The methodology, however, is general, and it can be applied to other types of loadings and cross-sectional shapes, as well as to other beam types. Three types of materials are considered here, namely, monel, mild steel, and aluminum. Ultimate loads for each beam case are determined by using all three material cases. The method of the equivalent systems and the reduced modulus E_r, as discussed in Section 5.9, are used here to carry out the analytical work. The computer program in Appendix D is used to facilitate the computational work.

In the computation of the reduced modulus E_r by using Eq. (5.112), the shape of the stress-strain curve for each material may be approximated with two, three, or more straight-line segments, as discussed in Section 5.9. In the work of this section, the cases shown schematically in Fig. 5.21 are used. Figure 5.21a represents the commonly used elastic-perfectly-plastic case, and the additional cases shown in the same figures are improvements. The results obtained for each of the cases shown in Fig. 5.21 are compared, and useful conclusions are drawn. See also Fertis [3] and Fertis and Lu [50].

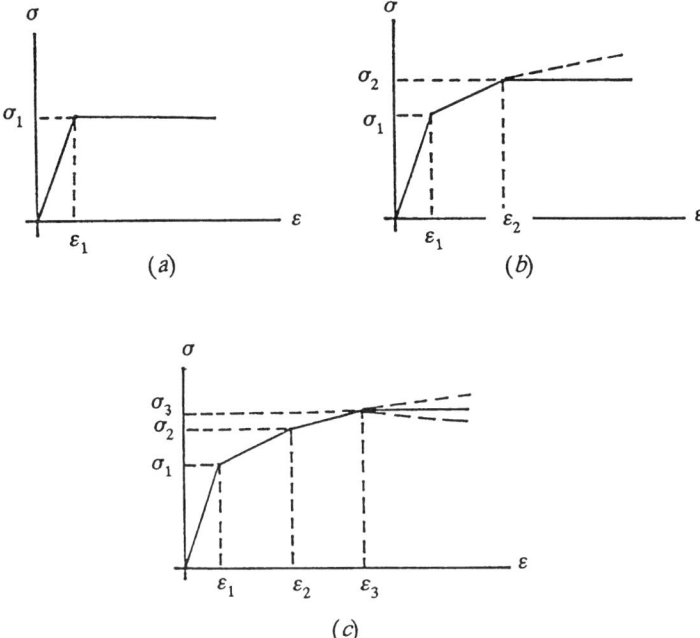

Figure 5.21. (a) Two-line approximation of the stress-strain curve. (b) Three-line approximation of the stress-strain curve. (c) Four-line approximation of the stress-strain curve.

5.10.1 Ultimate Loads for Beams of Monel Material

Prismatic and nonprismatic simply supported and cantilever beams made out of monel material are examined here. The stress-strain curve of monel is shown in Fig. 5.22a, and Figs. 5.22b and 5.22c illustrate a cantilever beam and a simply supported beam, respectively, loaded with a uniformly distributed load w. The constant width b of each member is 6 in., and the depth varies linearly along the length. The variation of the cross-sectional moment of inertia I_x is given by the equation

$$I_x = I_B f(x) \tag{5.119}$$

where

$$I_B = \frac{bh_2^3}{12} \tag{5.120}$$

$$f(x) = \left[\frac{(n-1)(L-x) + L}{L}\right]^3 \tag{5.121}$$

Note that the moment of inertia I_B at end B of each member is taken as the reference value, and $f(x)$ is the function representing the variation of I_x.

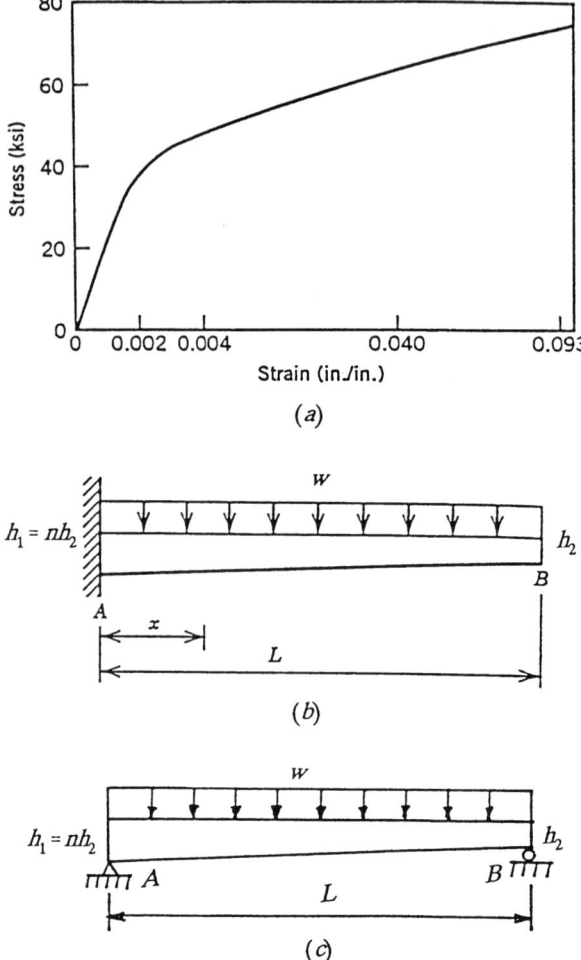

Figure 5.22. (a) Stress-strain curve of monel. (b) Tapered cantilever beam loaded with a uniformly distributed load w. (c) Tapered simply supported beam loaded with a uniformly distributed load w.

For monel, the stresses and strains corresponding to the stress-strain curve approximations shown in Fig. 5.21 are $\sigma_1 = 48$ ksi, $\sigma_2 = 59$ ksi, $\sigma_3 = 75$ ksi, $\varepsilon_1 = 2.1818 \times 10^{-3}$, $\varepsilon_2 = 24.007 \times 10^{-3}$, and $\varepsilon_3 = 152.007 \times 10^{-3}$. For the cantilever beam, the ultimate load w_u for each case is determined by using the vertical deflection y at its free end B and plotting the load w versus the deflection y curve. At the ultimate load w_u, the y versus w cure becomes flat and no further increase of w is possible, indicating that the ultimate load capacity of the member is reached. The ultimate load capacity for the simply supported beam is established by using the vertical deflection at the center of the member. For uniform beams, the maximum deflection occurs at the center. For tapered beams, the deflection at the center would be slightly less than the maximum one, but reasonable results are obtained for practical design applications. However, if desired, the location of the maximum deflection can easily be taken into account. The method of the equivalent systems can easily provide the location and value of such maximum deflections.

By considering a uniform simply supported beam of width $b = 6$ in. and depth $h = 8$ in., an inelastic analysis was performed by using the four-line approximation of the stress-strain curve of monel, such as the one shown by the solid line in Fig. 5.21c. The values of σ_1, σ_2, σ_3, ε_1, ε_2, and ε_3 are given earlier in this section. The graphs of y versus w for lengths $L = 60$, 70, 80, 90, 100, 110, 120, 130, 140, 160, 180, 200, 220, 240, 280, and 320 in. are shown in Fig. 5.23. Note that y is the vertical deflection at the center of the member. For each curve, there corresponds a maximum value w_u of the distributed load w, which represents the ultimate load capacity of the beam. At this load, each curve in Fig. 5.23 ends with a very distinct upward direction, indicating that

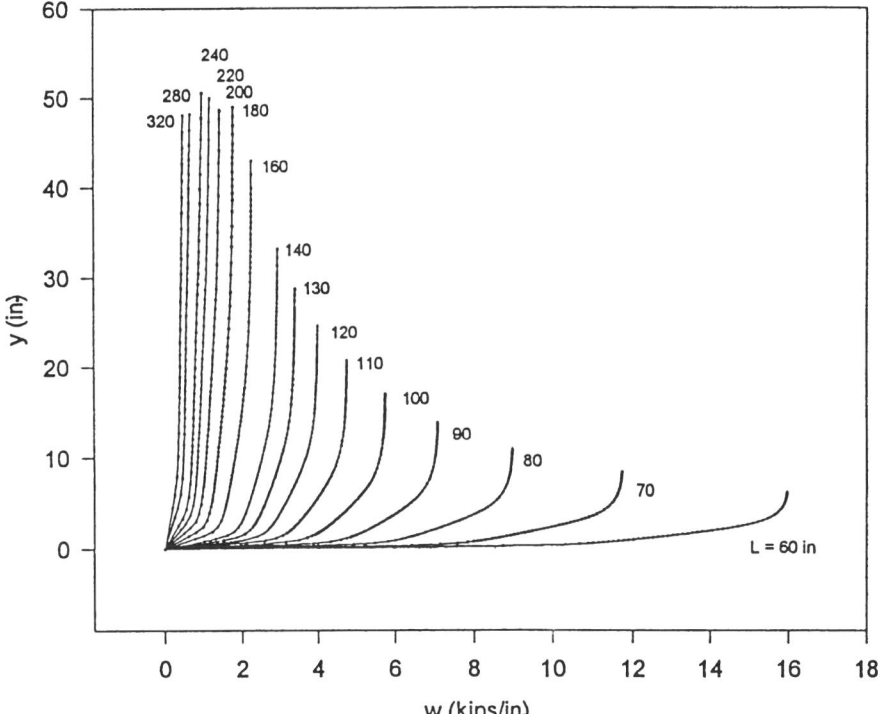

Figure 5.23. Curves of y versus w for a uniform simply supported beam of various lengths L, using the four-line approximation of the stress-strain curve of monel represented by the solid lines in Fig. 5.21c.

TABLE 5.12. Ultimate Loads w_u for a Uniform Simply Supported Beam of Various Lengths L, Using the Straight-Line Approximation of Monel Shown by the Solid Lines in Fig. 5.21c

(1) Length L (in.)	(2) Width b (in.)	(3) Depth h (in.)	(4) Ultimate Load w_u (kips/in.)
60	6	8	15.96
70	6	8	11.73
80	6	8	8.98
90	6	8	7.09
100	6	8	5.75
110	6	8	4.75
120	6	8	3.99
130	6	8	3.40
140	6	8	2.93
160	6	8	2.24
180	6	8	1.77
200	6	8	1.42
220	6	8	1.16
240	6	8	0.96
280	6	8	0.66
320	6	8	0.47

the ultimate load capacity of the member is reached. The fourth column of Table 5.12 provides the values of the ultimate loads w_u corresponding to the considered lengths L of the simply supported member. It starts with $w_u = 15.96$ kips/in. for $L = 60$ in., and ends with $w_u = 0.47$ kips/in. for $L = 320$ in.

Figures 5.24a and 5.24b show the results obtained for the same uniform simply supported beam with lengths $L = 80$ in. and 120 in., respectively, using in each case the three approximations of the stress-strain curve of monel shown in Fig. 5.21. The 48-ksi curve in Fig. 5.24a represents the approximation in Fig. 5.21a with $\sigma_1 = 48$ ksi, yielding $w_u = 5.76$ kips/in.; the 59-ksi curve represents the solid-line case in Fig. 5.21b with $\sigma_1 = 48$ ksi and $\sigma_2 = 59$ ksi, yielding $w_u = 7.06$ kips/in.; and the 75-ksi curve represents the solid-line case in Fig. 5.21c with $\sigma_1 = 48$ ksi, $\sigma_2 = 59$ ksi, and $\sigma_3 = 75$ ksi, yielding $w_u = 8.98$ kips/in. The portion of the curve marked unlimited is obtained by assuming that the stress σ increases without bounds beyond σ_3, as shown by the upward dashed line in Fig. 5.21c. Note that the ultimate load w_u corresponding to the 75-ksi curve is much larger and, consequently, more accurate, compared to the ultimate loads obtained by the 48-ksi and 59-ksi curves. Similar observations can be made by examining the results in Fig. 5.24b, which are obtained for beam length $L = 120$ in. This trend is typical for all cases examined, and

ULTIMATE DESIGN LOADS BASED ON INELASTIC RESPONSE 369

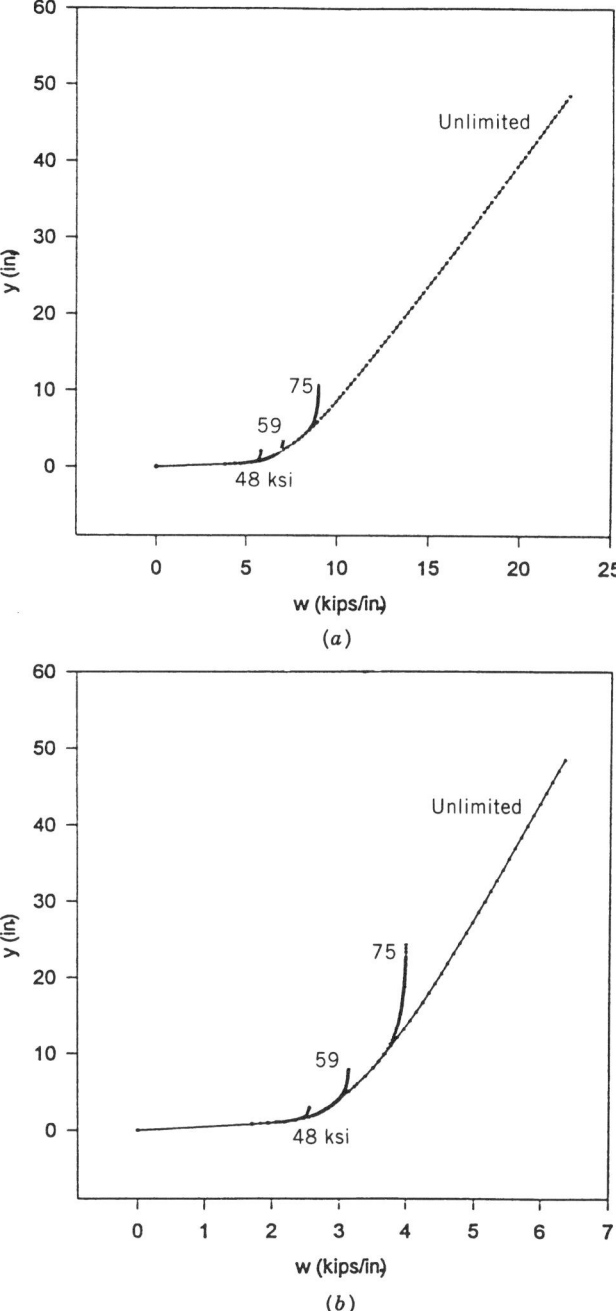

Figure 5.24. (a) Ultimate loads for a simply supported beam of length $L = 80$ in., using the monel stress-strain approximations in Fig. 5.21. (b) Analogous results for a simply supported beam of length $L = 120$ in.

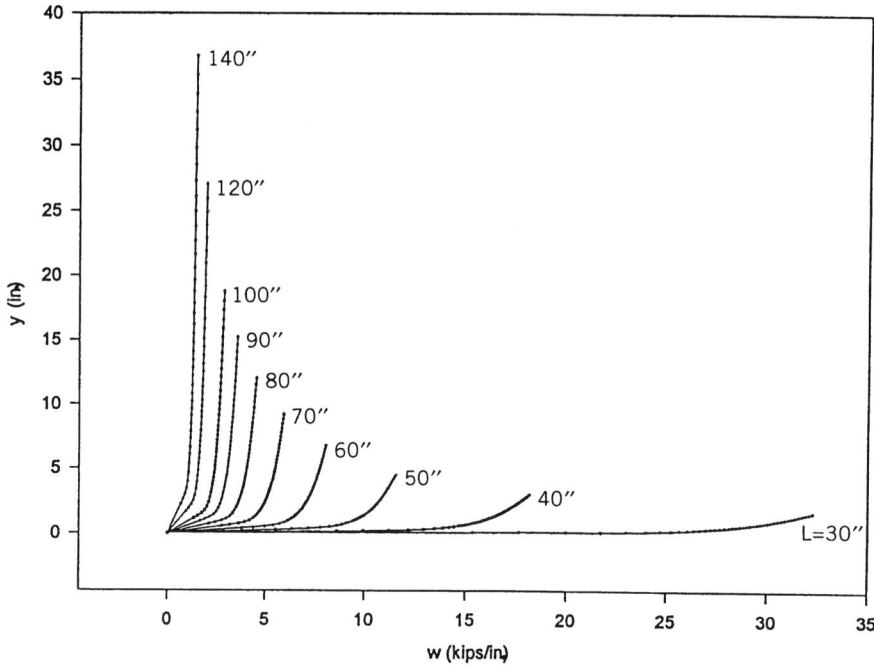

Figure 5.25. Curves of y versus w for a tapered cantilever beam of various lengths L, using the four-line approximation of the stress-strain curve of monel represented by the solid lines in Fig. 5.21c.

provides very useful information regarding the design of structural and mechanical components.

We consider now a tapered cantilever beam of width $b = 6$ in., depth $h_1 = 12$ in., and depth $h_2 = 8$ in. By using the same four-line approximation of the stress-strain curve of monel as the one used for the simply supported beam, the inelastic analysis was carried out for lengths $L = $ 30, 40, 50, 60, 70, 80, 90, 100, 120, and 140 in. The y versus w curves are shown in Fig. 5.25, where y is the deflection at the free end and w is the uniformly distributed load. The calculated ultimate load w_u for each length case is shown in the fifth column of Table 5.13. It starts with 34.97 kips/in. for $L = 30$ in., and reduces to 1.51 kips/in. for $L = 140$ in.

Figures 5.26a and 5.26b show the results of the cantilever beam for lengths $L = 50$ in. and 90 in., respectively. They are obtained by using, in each case, the three approximations of the stress-strain curve of monel shown in Fig. 5.21. The 48-ksi curve in Fig. 5.26a, which represents the approximation in Fig. 5.21a, yields $w_u = 8.23$ kips/in. The 59-ksi curve yields $w_u = 10.08$ kips/in., and the 75-ksi curve yields $w_u = 12.59$ kips/in. The portion of the curve marked unlimited is obtained by assuming that σ increases without bounds beyond σ_3,

ULTIMATE DESIGN LOADS BASED ON INELASTIC RESPONSE 371

TABLE 5.13. Ultimate Loads w_u for Tapered Cantilever Beam of Various Lengths L Using the Straight-Line Approximation of Monel Shown by the Solid Lines in Fig. 5.21c

(1) Length L (in.)	(2) Width b (in.)	(3) h_1 (in.)	(4) h_2 (in.)	(5) Ultimate Load w_u (kips/in.)
30	6	12	8	34.97
40	6	12	8	19.67
50	6	12	8	12.59
60	6	12	8	8.74
70	6	12	8	6.42
80	6	12	8	4.91
90	6	12	8	3.87
100	6	12	8	3.11
120	6	12	8	2.10
140	6	12	8	1.51

as shown by the upward dashed line in Fig. 5.21c. We note here that the ultimate load changes from 8.23 kips/in. to 12.59 kips/in., when four lines are used to approximate the shape of the stress-strain curve of monel. Similar observations can be made by examining the results in Fig. 5.26b, which are obtained for length $L = 90$ in. The analogous values in this case are 2.54, 3.11, and 3.87 kips/in.

5.10.2 Ultimate Loads for Mild Steel Beams

The inelastic analysis of the preceding section is repeated here by considering a simply supported beam and a cantilever beam that are made out of mild steel. In this case we have $\sigma_1 = 33.5$ ksi, $\sigma_2 = 54.1$ ksi, $\sigma_3 = 60.0$ ksi, $\varepsilon_1 = 3.55 \times 10^{-3}$, $\varepsilon_2 = 76.356 \times 10^{-3}$, and $\varepsilon_3 = 234.877 \times 10^{-3}$. The actual stress-strain curve of the mild steel is shown in Fig. 5.27. The applied distributed load w is uniform throughout the length of the member under consideration.

For a tapered simply supported beam of width $b = 6$ in., depth $h_1 = 12$ in., and depth $h_2 = 8$ in., the deflection y at its center versus the distributed load w produced the graphs shown in Fig. 5.28. The values of lengths L used in the inelastic analysis are 80, 100, 120, 160, 200, 240, 280, and 320 in. The three-line approximation of the stress-strain curve of the mild steel shown by the solid lines in Fig. 5.21b is used in this analysis. The fifth column in Table 5.14 shows the value of the ultimate load w_u for each beam-length case. It varies from $w_u = 9.66$ kips/in. for $L = 80$ in. to $w_u = 0.50$ kips/in. for $L = 320$ in.

Figure 5.29a shows the results of the same tapered cantilever beam for length $L = 120$ in., results obtained by using the two-line, three-line, and four-line approximations of the stress-strain curve of the mild steel, as shown in Fig. 5.21. The two-line approximation in Fig. 5.21a, with $\sigma_1 = 33.5$ ksi, yields

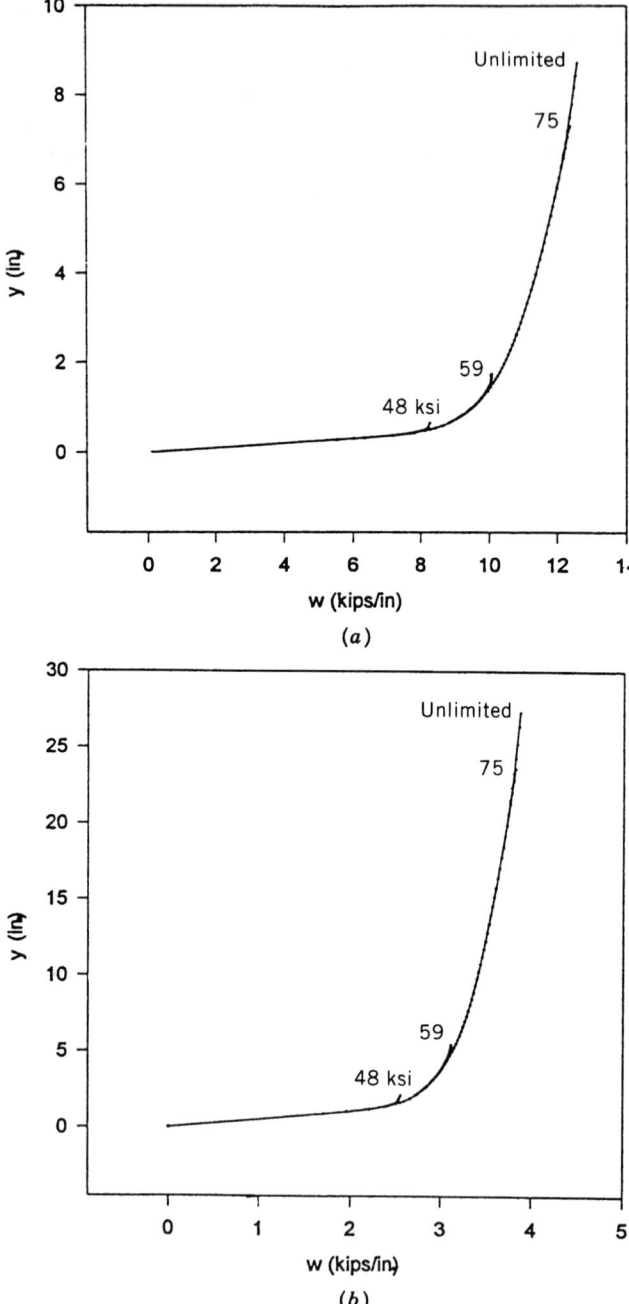

Figure 5.26. (a) Ultimate loads for a tapered cantilever beam of length $L = 50$ in., using the monel stress-strain curve approximations in Fig. 5.21. (b) Similar results for a tapered cantilever beam of length $L = 90$ in.

ULTIMATE DESIGN LOADS BASED ON INELASTIC RESPONSE 373

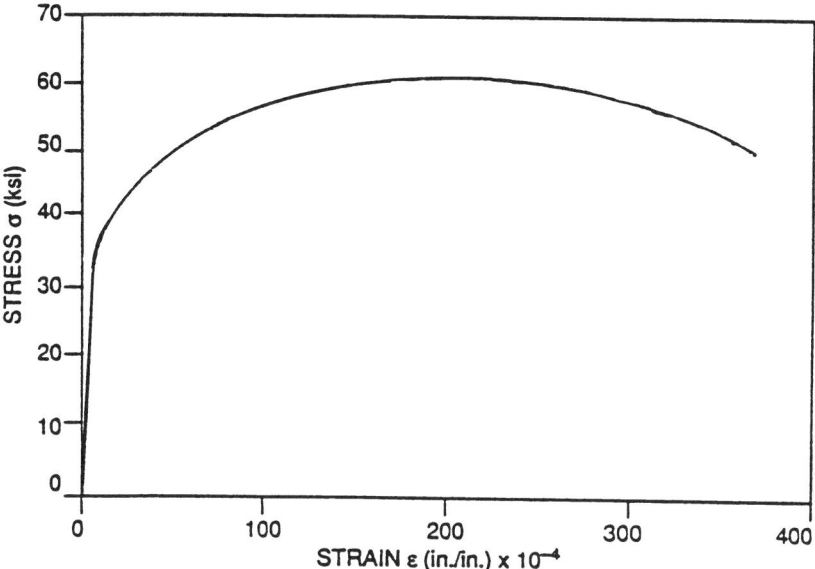

Figure 5.27. Stress-strain curve of the mild steel.

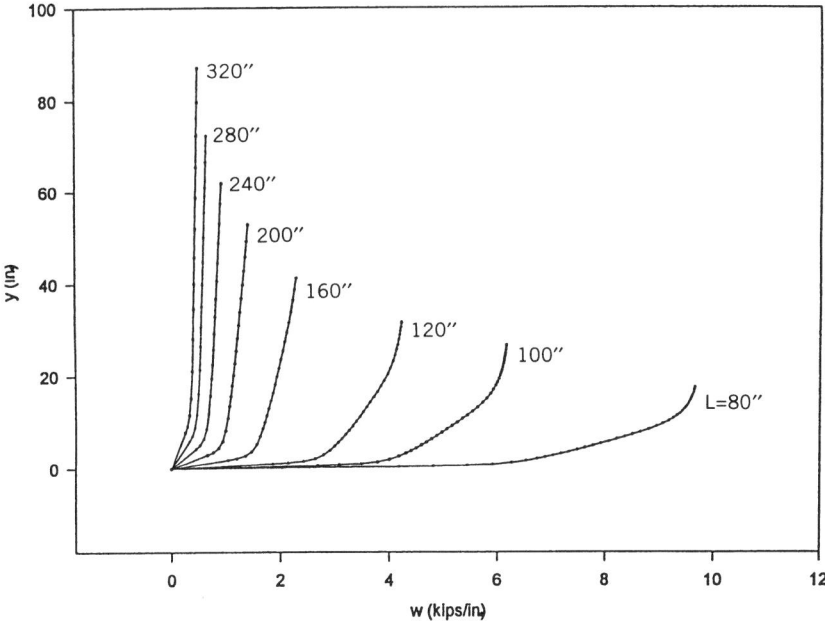

Figure 5.28. Curves of y versus w for a tapered simply supported beam of various lengths L, using the three-line approximation of the stress-strain curve of mild steel represented by the solid lines in Fig. 5.21b.

TABLE 5.14. Ultimate Loads w_u for Mild Steel Simply Supported Beam of Various Lengths L, Using the Straight-Line Approximation Shown by the Solid Lines in Fig. 5.21b

(1) Length L (in.)	(2) Width b (in.)	(3) h_1 (in.)	(4) h_2 (in.)	(5) Ultimate Load w_u (kips/in.)
80	6	12	8	9.66
100	6	12	8	6.18
120	6	12	8	4.24
160	6	12	8	2.30
200	6	12	8	1.41
240	6	12	8	0.93
280	6	12	8	0.66
320	6	12	8	0.50

$w_u = 2.68$ kips/in., the three-solid-line approximation in Fig. 5.21b, with $\sigma_1 = 33.5$ ksi and $\sigma_2 = 54.1$ ksi, yields $w_u = 4.243$ kips/in., and the four-solid-line approximation in Fig. 5.21c, with $\sigma_1 = 33.5$ ksi, $\sigma_2 = 54.1$ ksi, and $\sigma_3 = 60.0$ ksi, yields $w_u = 4.334$ kips/in. The case marked unlimited, which is represented by Fig. 5.21c with the fourth line downward as shown by the dashed line in the same figure, yields results that are very close to the results obtained by using the solid-line curve in Fig. 5.21c. This is easily observed by examining Fig. 5.29a, where these two curves are shown to be the same. However, a large difference in ultimate load w_u is obtained by using the two-line and the three-line approximations of the stress-strain curve of the mild steel.

Figure 5.29b illustrates the results that are obtained from the inelastic analysis of a uniform simply supported beam that is made out of mild steel. The width $b = 6$ in., depth $h = 8$ in, length $L = 100$ in., and the applied distributed load w is uniform throughout its length. In this case, the two-line approximation yields $w_u = 2.573$ kips/in., the three-solid-line approximation yields $w_u = 4.033$ kips/in., and the four-solid-line approximation yields $w_u = 4.088$ kips/in. The results for the case marked unlimited are practically identical to the results obtained from the four-solid-line approximation of the mild steel. The large difference in w_u exists only between the two-line and the three-line approximations of the stress-strain curve of the mild steel. This indicates that a three-solid-line approximation of the stress-strain curve of mild steel provides accurate results for practical design purposes regarding the ultimate load w_u.

Here we consider a cantilever beam that is made out of mild steel of width $b = 6$ in., depth $h_1 = 12$ in., depth $h_2 = 8$ in., and of lengths $L = 30, 40, 50, 60, 70, 80, 90, 100, 120,$ and 140 in. The inelastic analysis is carried out by using the three-line approximation of the mild steel that is represented by the solid lines in Fig. 5.21b. The corresponding curves of the deflection y at the free end

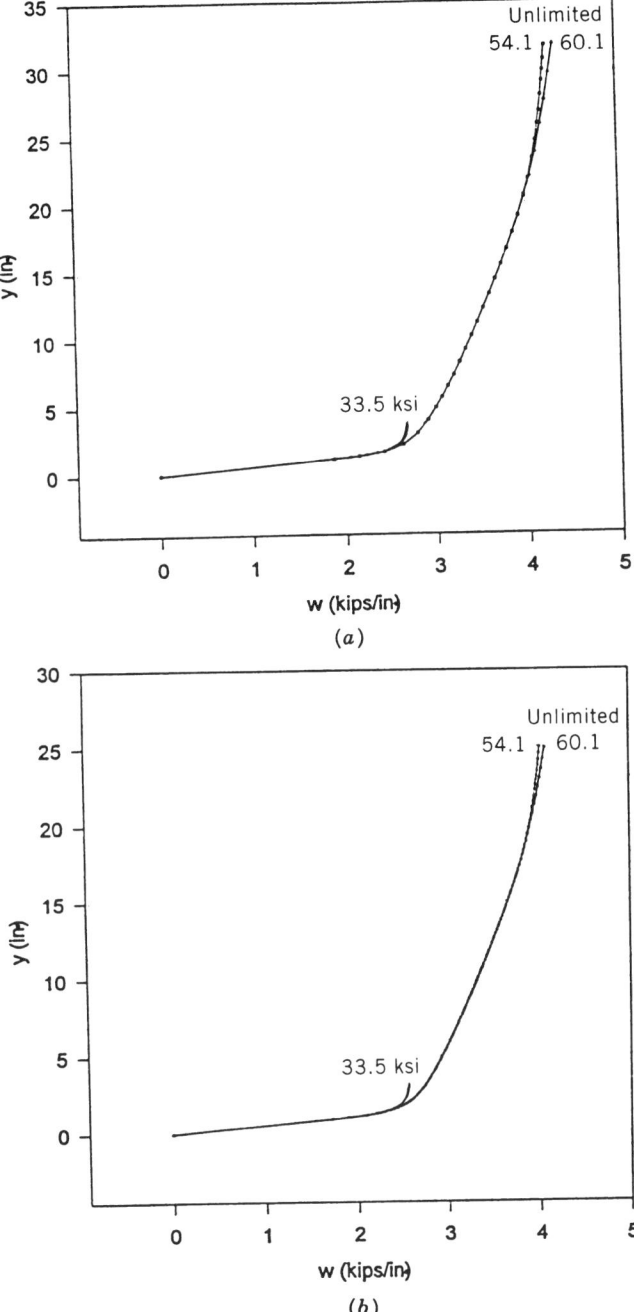

Figure 5.29. (*a*) Ultimate loads for a tapered simply supported beam of length $L = 120$ in., using the mild steel stress-strain approximations in Fig. 5.21. (*b*) Analogous results for a uniform simply supported beam of length $L = 100$ in.

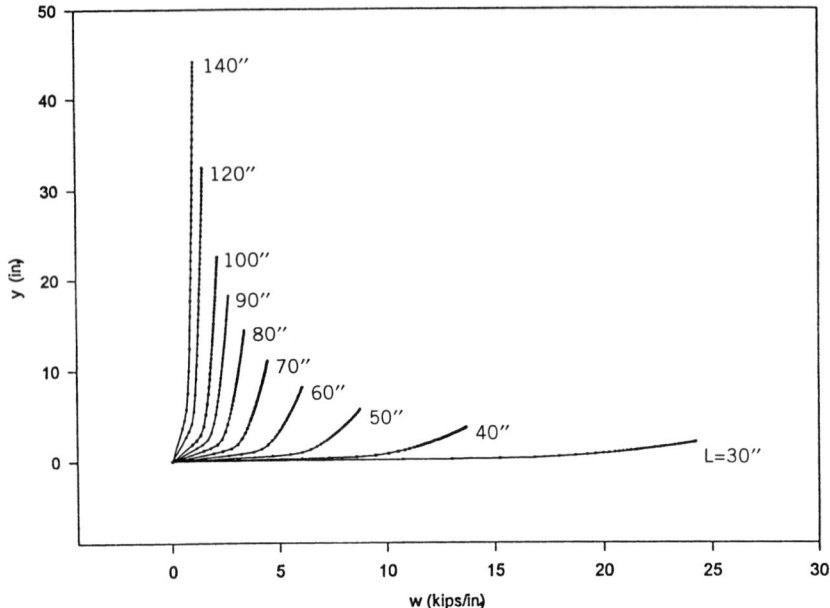

Figure 5.30. Curve of y versus w for a tapered cantilever beam of various lengths L, using the three-line approximation of the stress-strain curve of mild steel represented by the solid lines in Fig. 5.21b.

versus the distributed load w are shown in Fig. 5.30, and the corresponding ultimate loads w_u are listed in the fifth column of Table 5.15. They vary from $w_u = 25.53$ kips/in. for $L = 30$ in. to $w_u = 1.16$ kips/in. for $L = 140$ in.

Figure 5.31a shows the inelastic analysis results of the same tapered cantilever beam of length $L = 60$ in. We note that the two-line approximation yields $w_u = 3.989$ kips/in., the three-line approximation yields $w_u = 6.832$ kips/in., and the four-line approximation yields $w_u = 6.845$ kips/in. The case marked unlimited produces results very similar to those obtained by using the solid-line four-line approximation of the stress-strain curve. Again, we see here that a large difference in w_u exists between the two-line and the three-line approximations of the mild steel stress-strain curve.

Figure 5.31b illustrates the ultimate load results for a uniform cantilever beam of width $b = 6$ in., depth $h = 8$ in., and length $L = 60$ in., which is made out of mild steel. The two-line approximation yields $w_u = 1.773$ kips/in., the three-line approximation gives $w_u = 2.829$ kips/in., and the four-line approximation yields $w_u = 3.04$ kips/in. For all practical purposes, the curve marked unlimited is identical to the one obtained by using the four-solid-line-approximation of the stress-strain curve of the mild steel. We also observe here that the large difference in w_u exists between the two-line and the three-line approximations of the stress-strain curve of the mild steel.

ULTIMATE DESIGN LOADS BASED ON INELASTIC RESPONSE   377

TABLE 5.15. Ultimate Loads w_u for Mild Steel Tapered Cantilever Beam of Various Lengths L, Using the Straight-Line Approximation of Monel Shown by the Solid Lines in Fig. 5.21b

(1) Length L (in.)	(2) Width b (in.)	(3) h_1 (in.)	(4) h_2 (in.)	(5) Ultimate Load w_u (kips/in.)
30	6	12	8	25.53
40	6	12	8	14.36
50	6	12	8	9.19
60	6	12	8	6.38
70	6	12	8	4.69
80	6	12	8	3.59
90	6	12	8	2.84
100	6	12	8	2.30
120	6	12	8	1.60
140	6	12	8	1.16

5.10.3 Ultimate Loads for Aluminum Alloy Beams

In this section, we perform an inelastic analysis for a simply supported beam and for a cantilever beam, both made of aluminum alloy. The stress-strain curve of the material is shown in Fig. 5.18b, and the inelastic analysis is carried out by using the stress-strain curve approximations shown in Figs. 5.21a and 5.21b, where $\sigma_1 = 59.94$ ksi, $\sigma_2 = 79.67$ ksi, $\varepsilon_1 = 6.3284 \times 10^{-3}$, and $\varepsilon_2 = 8.589 \times 10^{-3}$. The applied distributed load w is uniform throughout the length of the member.

We consider first a tapered simply supported beam of width $b = 6$ in., depth $h_1 = 12$ in., depth $h_2 = 8$ in., and lengths $L = 80, 100, 120, 160, 200, 240, 280,$ and 320 in. The inelastic analysis using the two-line approximation of the stress-strain curve with $\sigma_1 = 59.94$ ksi yields the y versus w curves shown in Fig. 5.32, where y is the vertical deflection at the center of the member. The ultimate load w_u for each beam length is shown in the fifth column of Table 5.16, and it varies from $w_u = 10.79$ kips/in. for $L = 80$ in. to $w_u = 0.68$ kips/in. for $L = 320$ in.

Figure 5.33a illustrates the results for a tapered simply supported beam of length $L = 120$ in., $b = 6$ in., $h_1 = 12$ in., and $h_2 = 8$ in.; these results were obtained by using the stress-strain curve approximations shown in Figs. 5.21a and 5.21b. The two-line approximation yields $w_u = 4.796$ kips/in., and the three-line approximation yields $w_u = 6.373$ kips/in. The case marked unlimited is also shown in the same figure, and it represents the approximation in Fig. 5.21b with the third dashed line directed upwards. Note the large difference in w_u between the two-line and the three-line approximations of the stress-strain curve of the aluminum.

Figure 5.33b illustrates the results for a uniform simply supported beam of length $L = 120$ in., $b = 6$ in., and depth $h = 8$ in. The two-line approximation

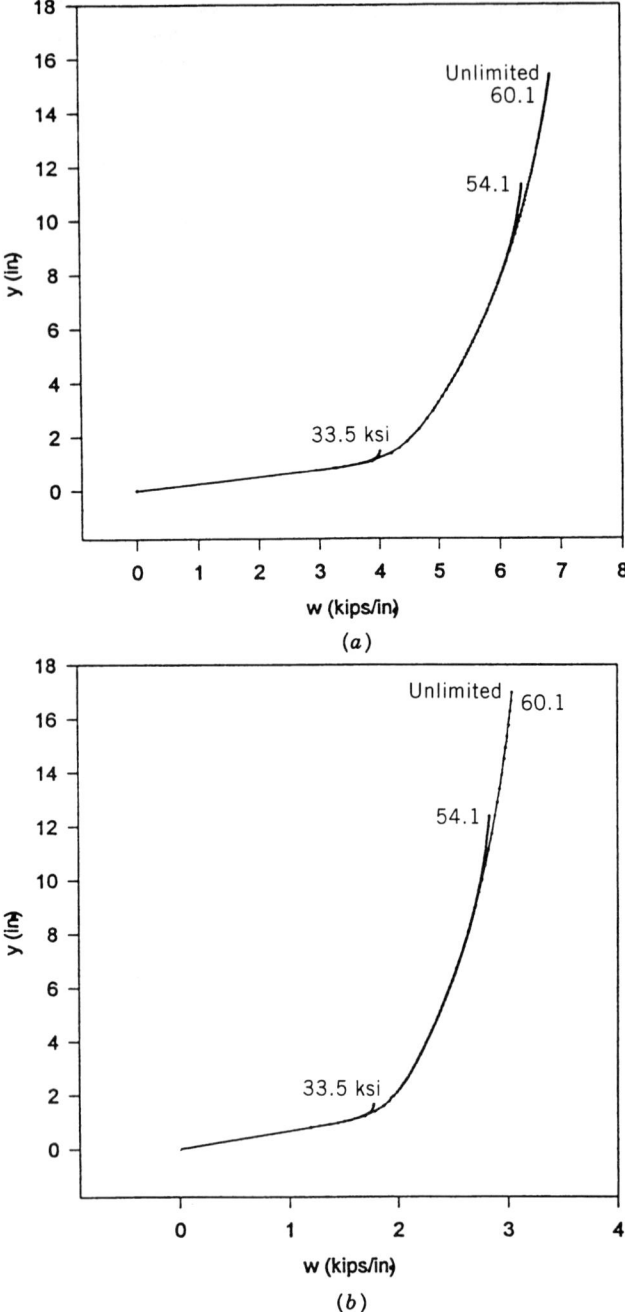

Figure 5.31. (*a*) Ultimate loads for a tapered mild steel cantilever beam of length $L = 60$ in., using the stress-strain curve approximations in Fig. 5.21. (*b*) Analogous results for a uniform cantilever beam of length $L = 60$ in.

ULTIMATE DESIGN LOADS BASED ON INELASTIC RESPONSE

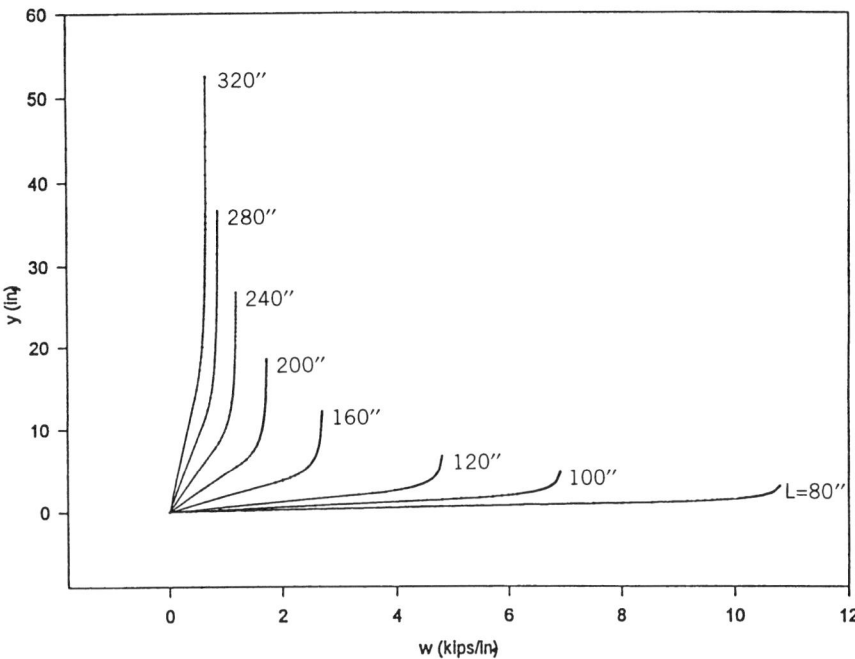

Figure 5.32. Curves of y versus w for a tapered simply supported beam of various lengths L, using the two-line approximation of the stress-strain curve of the aluminum alloy.

TABLE 5.16. Ultimate Loads w_u for Tapered Aluminum Alloy Simply Supported Beam of Various Lengths L, Using the Straight-Line Approximation Shown by the Solid Lines in Fig. 5.21b

(1) Length L (in.)	(2) Width b (in.)	(3) h_1 (in.)	(4) h_2 (in.)	(5) Ultimate Load w_u (kips/in.)
80	6	12	8	10.79
100	6	12	8	6.91
120	6	12	8	4.80
160	6	12	8	2.70
200	6	12	8	1.73
240	6	12	8	1.20
280	6	12	8	0.88
320	6	12	8	0.68

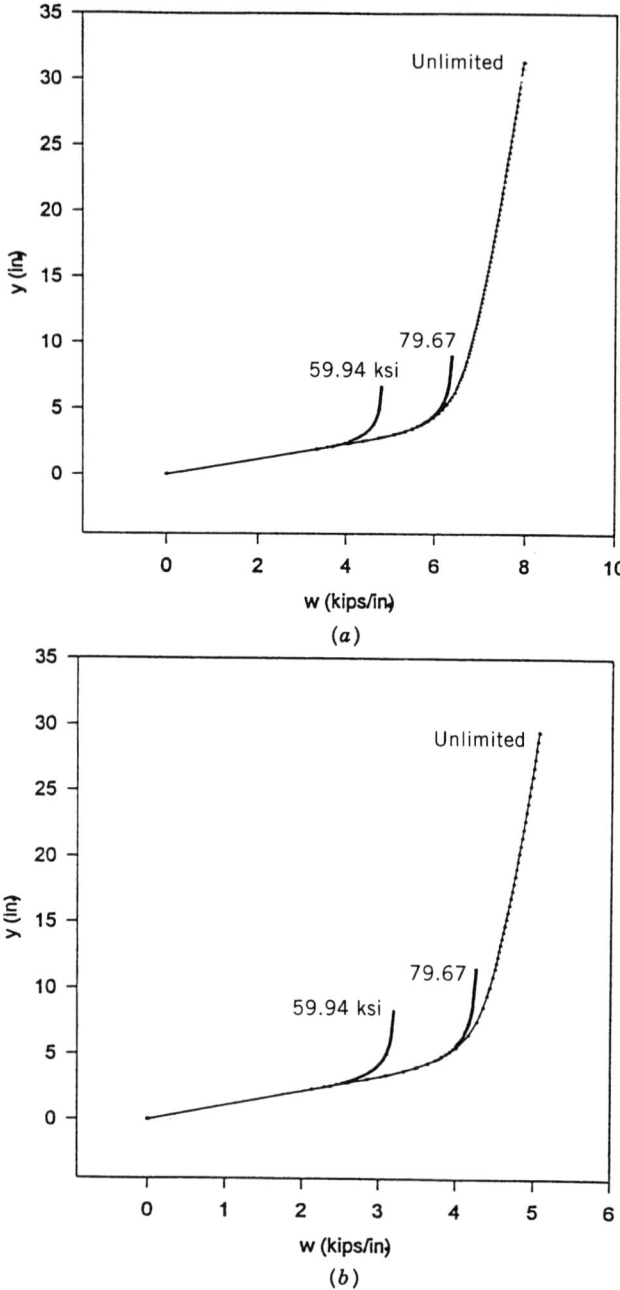

Figure 5.33. (*a*) Ultimate loads for a tapered simply supported beam of length $L = 120$ in., using the aluminum stress-strain curve approximations shown in Figs. 5.21*a* and 5.21*b*. (*b*) Analogous results for a uniform simply supported beam of length $L = 120$ in.

ULTIMATE DESIGN LOADS BASED ON INELASTIC RESPONSE 381

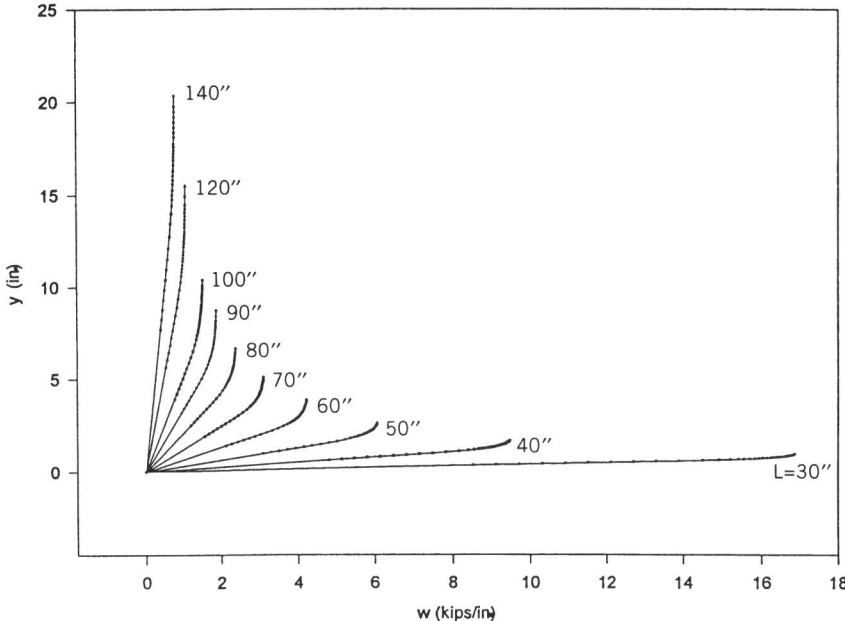

Figure 5.34. Curves of y versus w for a uniform aluminum cantilever beam of various lengths L, using the three-line approximation of the stress-strain curve.

yields $w_u = 3.197$ kips/in., and the three-line approximation yields $w_u = 4.251$ kips/in. The case named unlimited is also shown in the same figure. Note again the large difference in ultimate load w_u for the two-line and the three-line approximations of the aluminum stress-strain curve. Such results are typical for all cases that were investigated in detail.

We consider now a uniform aluminum cantilever beam of width 6 in., and depth $h = 8$ in. The inelastic analysis was carried out for lengths $L = 30, 40, 50, 60, 70, 80, 90, 100, 120$, and 140 in. The curves of y versus w corresponding to these lengths are shown in Fig. 5.34. These results are obtained by using the three-line approximation of the stress-strain curve of the aluminum shown by the solid line in Fig. 5.21b. The corresponding ultimate loads w_u are shown in the fourth column of Table 5.17. They vary from $w_u = 16.87$ kips/in. for $L = 30$ in., to $w_u = 0.77$ kips/in. for $L = 140$ in.

Figure 5.35a shows the inelastic analysis results for the uniform aluminum cantilever beam of length $L = 100$ in. The two-line approximation of the aluminum stress-strain curve yields $w_u = 1.141$ kips/in., and the three-line approximation yields $w_u = 1.516$ kips/in. The results of the case marked unlimited with the third line inclined upwards are also shown in the same figure. Again, we note a large difference in w_u between the two-line and the three-line approximations of the aluminum stress-strain curve.

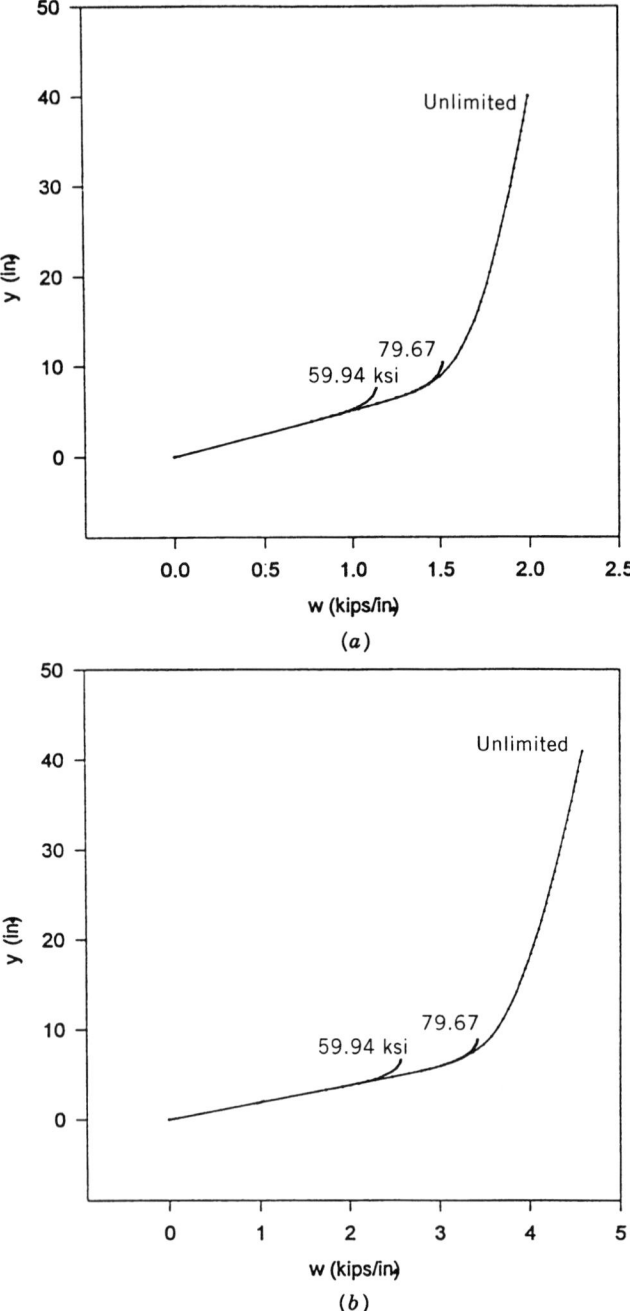

Figure 5.35. (*a*) Ultimate loads for uniform aluminum cantilever beam of length $L = 100$ in., using the straight-line approximations in Figs. 5.21*a* and 5.21*b*. (*b*) Analogous results for a tapered aluminum cantilever beam of length $L = 100$ in.

TABLE 5.17. Ultimate Loads w_u for Uniform Aluminum Cantilever Beam of Various Lengths L Using the Straight-Line Approximation of the Stress-Strain Curve Shown by the Solid Lines in Fig. 5.21b

(1) Length L (in.)	(2) Width b (in.)	(3) Depth h (in.)	(4) Ultimate Load w_u (kips/in.)
30	6	8	16.87
40	6	8	9.48
50	6	8	6.06
60	6	8	4.22
70	6	8	3.09
80	6	8	2.37
90	6	8	1.87
100	6	8	1.52
120	6	8	1.05
140	6	8	0.77

Figure 5.35b shows the results of the inelastic analysis for a tapered aluminum cantilever beam of length $L = 100$ in., width $b = 6$ in., depth $h_1 = 12$ in., and depth $h_2 = 8$ in. The two-line approximation of the stress-strain curve of the aluminum yields $w_u = 2.57$ kips/in., and the three-line approximation yields $w_u = 3.416$ kips/in. The case marked unlimited is also shown in the same figure. Note again the typical large difference in w_u between the two-line and the three-line approximations of the aluminum stress-strain curve.

PROBLEMS

5.1 For the uniform cantilever beam in Fig. 5.1a, we have $P = 0.8$ kips, $L = 1,000$ in., and $EI = 180 \times 10^3$ kip-in.² By using pseudolinear analysis, determine the moment diagram M'_e of the pseudolinear system. The horizontal displacement Δ_A of the free end A is 356.7 in.

5.2 By using the M'_e diagram obtained in Problem 5.1 and approximating its shape with a small number of straight-line segments, determine a pseudolinear equivalent system. By using the pseudolinear system and applying the moment-area method, determine the vertical deflection δ_A and rotation θ_A at the free end A.

5.3 For the tapered cantilever beam in Fig. 5.2a, we have $L = 1,000$ in., $P = 3$ kips, $n = 1.5$, and $EI_A = 180 \times 10^3$ kip-in.², where I_A is the cross-sectional moment of inertia at the free end A of the member. By using pseudolinear analysis, determine a pseudolinear equivalent system of constant stiffness EI_A. The horizontal displacement Δ_A at the free end A of the member is 483.11 in.

5.4 By using the pseudolinear system obtained in Problem 5.3 and applying the moment-area method, determine the vertical deflection δ_A and rotation θ_A at the free end A of the member.

5.5 Determine the horizontal displacement Δ_A of the free end A of the cantilever beam in Problem 5.1 when $P = 2.5$ kips.

5.6 Determine the horizontal displacement Δ_A of the free end A of the cantilever beam in Problem 5.3 when $P = 4.0$ kips.

5.7 Obtain a general solution for the uniform flexible simply supported beam of Example 5.3 by using Eq. (5.21) and assuming that $\Delta(x) = \Delta x/L_0$, which is represented by Eq. (5.29). Compare the results.

5.8 Obtain a general solution for a flexible uniform cantilever beam loaded with a uniformly distributed load w, as shown in Fig. P5.8. Assume that $\Delta(x)$ is constant and equal to the horizontal displacement Δ at the free end of the beam.

5.9 Repeat Problem 5.8 by assuming that $\Delta(x) = \Delta x/L_0$ and compare the results.

5.10 By using the general solution obtained in Problem 5.8 and applying pseudolinear analysis, determine the vertical deflection δ_A, rotation θ_A, and horizontal displacement Δ_A at the free end A of the member. Assume that $L = 1,000$ in., $w = 1$ lb/in., and $EI = 180 \times 10^3$ kip-in.^{2}

5.11 By using the general solution obtained in Problem 5.9 and applying pseudolinear analysis, determine the vertical deflection δ_A, rotation θ_A, and horizontal displacement Δ_A at the free end A of the member. Assume that $L = 1,000$ in., $w = 1$ lb/in., and $EI = 180 \times 10^3$ kip-in.^{2}

5.12 Repeat Problem 5.8 by assuming that $\Delta(x) = \Delta(x/L_0)^{1/2}$, $L = 1,000$ in., $w = 1$ lb/in., and $EI = 180 \times 10^3$ kip-in.^{2} By using an equivalent pseudolinear system, determine the vertical deflection δ_A, rotation θ_A, and horizontal displacement Δ_A at the free end A of the member.

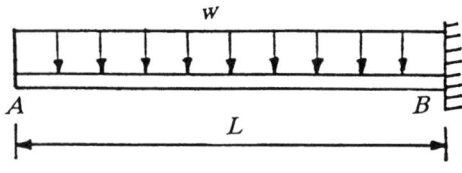

Figure P5.8.

5.13 Repeat Problem 5.12 by assuming that $\Delta(x) = \sin(\pi x/2L_0)$. All other conditions are the same.

5.14 Repeat the problem in Examples 5.3 and 5.4 by assuming that $\Delta(x) = \Delta(x/L_0)^{1/2}$ and retaining all other conditions. By using a pseudolinear equivalent system, determine the vertical deflection δ_D at the center point D, and the rotation θ_B at the moving end B. Compare the results with the results obtained in Example 5.4.

5.15 By using the simplified nonlinear equivalent system in Fig. 5.7b and applying pseudolinear analysis, determine the vertical deflection δ_A, rotation θ_A, and horizontal displacement Δ_A at the free end A of the member. Assume that $EI_A = 180 \times 10^3$ kip-in.^{2}, where I_A is the constant moment of inertia.

5.16 For each flexible beam case shown in Fig. P5.16, determine a simplified nonlinear equivalent system of constant stiffness $EI_A = 180 \times 10^3$ kip-in.^{2}, where EI_A is the stiffness at the end A of each member. Approximate the shape of the M_e diagram so that the simplified equivalent system is loaded with only one equivalent concentrated load. Assume that $P_1 = 2$ kips and $w = 5$ lb/in.

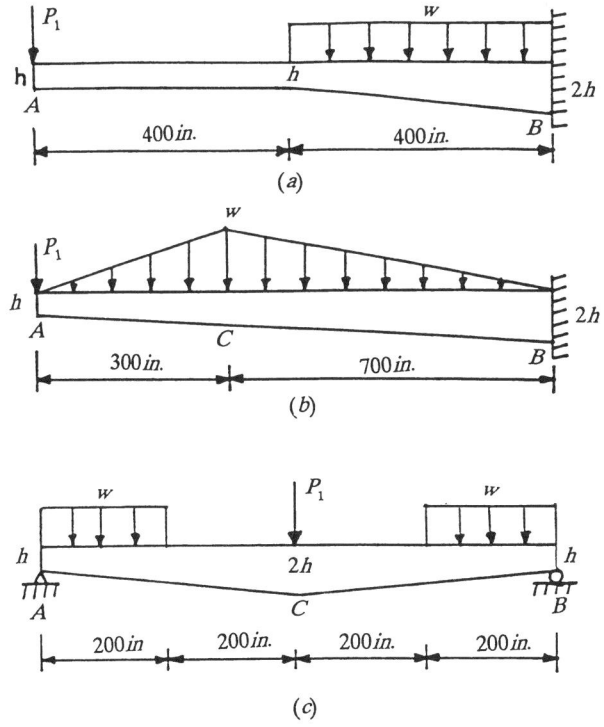

Figure P5.16.

5.17 By using the simpified nonlinear equivalent system obtained in Problem 5.16, determine in each case the vertical deflection and rotation under the equivalent concentrated load.

5.18 The trusses in Fig. P5.18 are made out of steel pipes of the same cross-sectional diameter. Certain members of these trusses are subjected to compression by the application of a vertical load P, located as shown in the figures. If the members that are subjected to compression are treated as hinged-hinged columns, determine for each truss case the magnitude of the applied load P so that the axial compressive force

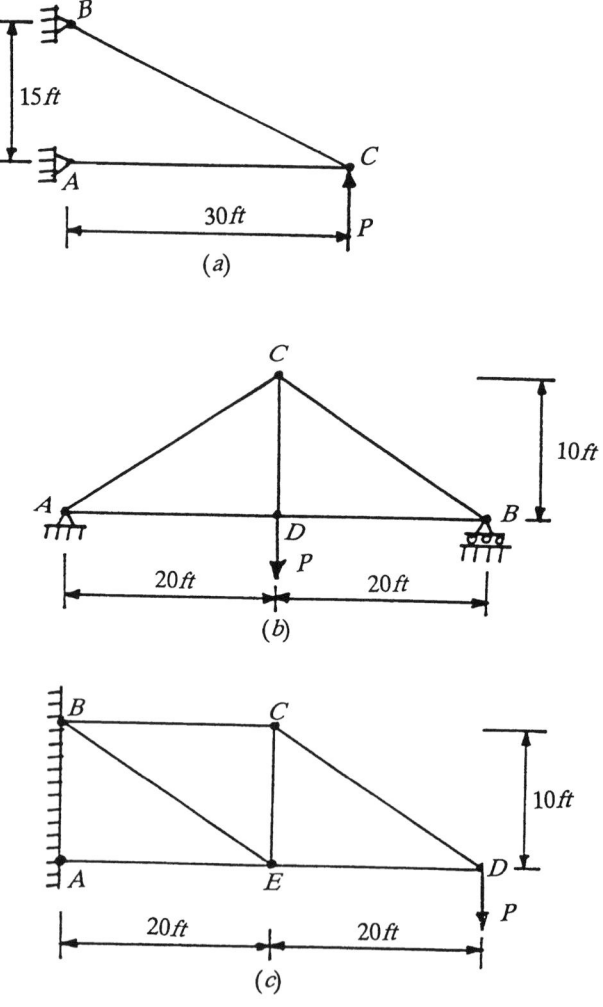

Figure P5.18.

attains its critical value. The modulus $E = 30 \times 10^6$ psi, $I = 28.10$ in.^{4}, and the radius of gyration $r = 2.25$ in. Also, determine the axial stresses in all members of each truss that are produced by the load P. The cross-sectional area of each pipe is 5.58 in.^{2}

5.19 If the applied load P in each truss in Fig. P5.18 is 65,000 lb, determine the cross-sectional area of each truss member so that the magnitude of the axial stress does not exceed 20,000 psi. Check also if stability criteria in terms of buckling are satisfied.

5.20 The horizontal bar ABC in Fig. P5.20 is hinged at end A and carries a vertical load P at end C, as shown. At point B, the bar is supported by a column BD of length $L = 10$ ft. In Fig. P5.20a, the column is pinned at

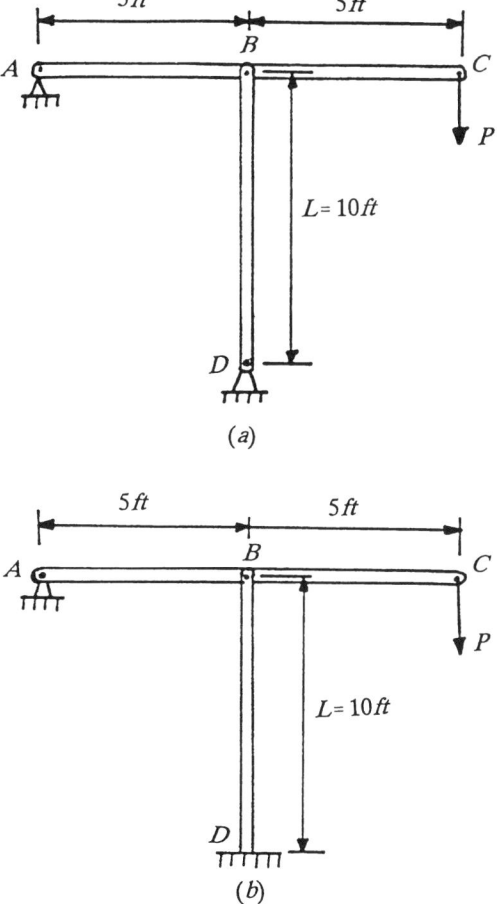

Figure P5.20.

both ends, and in Fig. P5.20b, the column is pinned at B and fixed at end D. If the column is a steel pipe of moment of inertia $I = 15.2 \text{ in.}^4$ and cross-sectional area $A = 4.30 \text{ in.}^2$, determine in each case the value of the load P that will cause the system to collapse by buckling. Also, determine for each case the axial stress in the column. The modulus $E = 30 \times 10^6$ psi.

5.21 The diameter of a solid circular steel column that is pinned at both ends is 4 in., and its length is 30 ft. If the allowable axial stress is 20,000 psi, determine the maximum axial load that can be applied to the column in order to satisfy both stress and buckling requirements. Buckling is not permitted to occur for axial stresses less than 36,000 psi.

5.22 Select a reasonable wide-flange steel cross section from your steel manual, in order to support an axial load of 150,000 lb by a 180-in. long column. The allowable stress is 20,000 psi and buckling is not permitted to occur for stresses equal or smaller than 20,000 psi. The modulus $E = 30 \times 10^6$ psi.

5.23 The steel frame $ABCD$ in Fig. P5.23 has a square cross section of sides equal to 6 in., and it is loaded by a force P as shown in the same figure. Determine the value of the load P so that the maximum deflection of member BC does not exceed 1.5 in. The modulus $E = 30 \times 10^6$ psi.

5.24 The beam-column in Fig. 5.14 is subjected to a compressive axial force $P = 0.4P_{cr}$. The cross section of the column is a W16 × 67 and its length is 18 ft. Determine its first three free frequencies of vibration. Neglect damping.

5.25 Repeat Problem 5.24 by assuming that P is an axial tensile force and compare the results.

5.26 Repeat Problem 5.24 for values of $P = 0, 0.1P_{cr}, 0.2P_{cr}, \ldots, P_{cr}$, and prepare for each mode of free vibration a graph showing the variation of ω with respect to P/P_{cr}. Discuss the results.

5.27 Repeat Problem 5.26 when P is an axial tensile load and compare the results.

5.28 Investigate in detail the collapse of the first Tacoma Narrows suspension bridge and produce a report of your findings. Based on your findings, provide an opinion as to what you think was the reason(s) of its collapse.

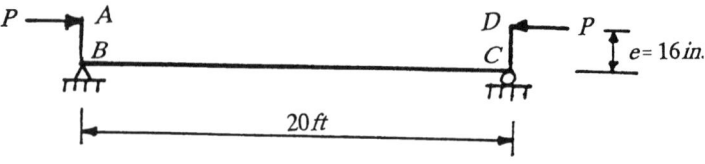

Figure P5.23.

PROBLEMS

5.29 Investigate in detail the second Tacoma Narrows suspension bridge and provide a report in which you state the changes that were made in order to obtain a safe bridge design.

5.30 In your opinion, what is the lesson we learn from the collapse of the first Tacoma Narrows suspension bridge? Support your thoughts with factual data.

5.31 Choose a well-known failure where sufficient data are available in the literature, and investigate it in detail. Provide a report of your findings and conclude with the lesson we learn from this failure.

5.32 For the tapered cantilever beam in Example 5.6, determine the values of the reduced modulus E_r and required moment M_{req} at the fixed end A by using the two-line approximation of the stress-strain curve of the 7075-T651 aluminum alloy. Also, at the same end, determine the value of the modulus function $g(x)$ and the value of the moment M_e of the equivalent system of constant stiffness EI_B, where the elastic modulus $E = 9.50 \times 10^6$ psi, and I_B is the moment of inertia at the free end of the member. Show your calculations and procedures in detail.

5.33 Repeat Problem 5.32 by using the six-line approximation of the stress-strain curve of the 7075-T651 aluminum alloy. Compare the results. Show all calculations and procedures.

5.34 Approximate with two straight lines the exact shape of the M_e diagram shown by the solid line in Fig. 5.20a, and obtain an equivalent system of constant stiffness EI_B that is loaded with two equivalent concentrated loads. By using the equivalent system and applying the moment-area method, determine the vertical deflection at the free end B of the member. Compare the result with the value obtained in Example 5.6 and make appropriate remarks.

5.35 Table 5.12 gives the values of the ultimate load w_u for a uniform simply supported beam that is made out of monel material and loaded with a uniformly distributed load w throughout its length. Various beam lengths are considered. By using the computer program in Appendix D, verify the value $w_u = 3.99$ kips/in., which corresponds to the length $L = 120$ in.

5.36 Table 5.14 gives the values of the ultimate load w_u for a uniform simply supported beam that is made out of mild steel material and loaded with a uniformly distributed load w throughout its length. Various lengths are considered. By using the computer program in Appendix D, verify the value $w_u = 2.30$ kips/in., which corresponds to the length $L = 160$ in.

5.37 Table 5.16 gives the values of the ultimate load w_u for a tapered aluminum alloy simply supported beam that is loaded with a uniformly distributed load w throughout its length. Various lengths are considered. By using the computer program in Appendix D, verify the value $w_u = 1.20$ kips/in., which corresponds to the length $L = 240$ in.

6 Structures Subjected to Earthquake Excitations and Blast Loadings

6.1 INTRODUCTION

A great deal of emphasis during the past four decades has been placed on the problems associated with blast and earthquake, because both problems create devastating effects on our infrastructure. The earthquake is rather old, most probably as old as the earth itself, but most knowledge on the subject has been accumulated during the past four or five decades. The blast problem is not very old, and most of the work in this field is made available through publications of the U.S. Army Corps of Engineers, Department of Defense, U.S. Air Force, and other governmental offices and public institutions. In general, the field of structural dynamics may be considered to have been born during the 1950s and 1960s in an effort to land a person on the moon by the year 1970. This mission was accomplished in 1969, a year earlier.

Despite the recent progress, the complexities in these fields of dynamics are enormous, and require sound judgment and exceptional experience on the part of the design engineer. In this chapter, fundamental knowledge regarding the design and analysis of structures to resist the effects of blast and earthquake are provided. Additional work on these subjects may be found in specialized references, which are listed as the discussion in the various sections of this chapter progresses.

6.2 LAGRANGE'S EQUATION

In Chapter 3, the differential equations of motion for structures under the action of dynamic loads have been derived by using the Newton's second law of motion. For many structural problems, the derivation of such equations is rather straightforward, and application of Newton's second law is usually sufficient. In certain cases, however, the analysis of structures for dynamic response becomes more convenient if we use a very powerful tool known as Lagrange's equation. This equation, developed by Joseph-Louis Lagrange, was published in his book *Mechanics Analytique* in 1788. Since its inception, the equation has enjoyed a wide popularity in the solution of many classical and modern dynamics problems. Lagrange's equation makes use of energy relations for the system, as well as generalized coordinates, and it is extensively used in this chapter for the solution of idealized systems subjected to dynamic loadings.

The generalized coordinates for a multiple degree of freedom system are designated by $q_1, q_2, \ldots, q_i, \ldots, q_n$. The number of such coordinates required to define a system's configuration during motion should be equal to the number of degrees of freedom of the system. The coordinates are independent of each other, and there is one generalized coordinate associated with each degree of freedom.

Our derivation of Lagrange's equation starts here by considering the beam in Fig. 6.1a, which supports the weights $W_1, W_2, \ldots, W_j, \ldots, W_n$ at points $1, 2, \ldots, j, \ldots, n$, respectively. In addition, the member is acted upon by the dynamic forces $F_1, F_2, \ldots, F_k, \ldots, F_m$. It is further assumed that the beam is of negligible weight, and thus the number of weights supported by this member defines its degrees of freedom. Since there are n weights, the system has n degrees of freedom, and n generalized coordinates are needed to define its configuration during motion.

For a deflection configuration such as the one in Fig. 6.1b, the n generalized coordinates are $q_1, q_2, \ldots, q_j, \ldots, q_n$, and they are taken at the positions shown in the figure. Other locations could be selected, because the only requirement

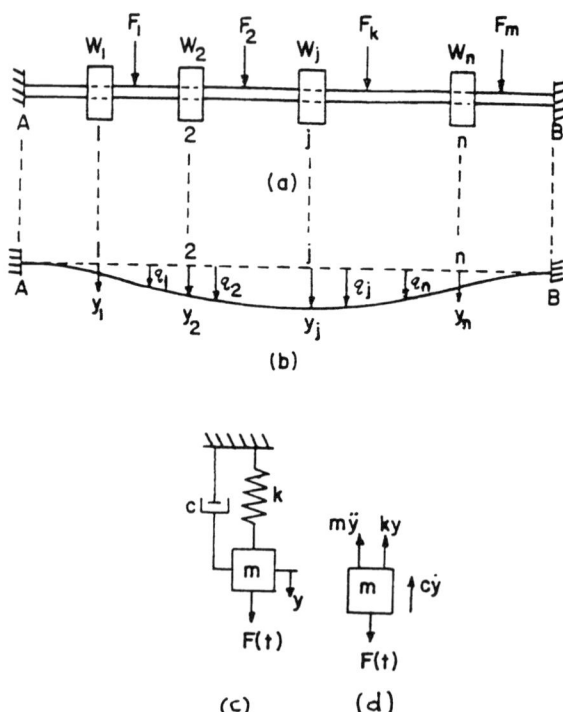

Figure 6.1. (a) Fixed-fixed beam supporting n weights and subjected to m concentrated dynamic loads. (b) Deformation configuration of the member. (c) General single-degree of freedom spring-mass system. (d) Free-body diagram of mass m.

in their choice is that they should be equal in number to the number of degrees of freedom of the system. In addition, the deflections $y_1, y_2, \ldots, y_j, \ldots, y_n$, under the weight concentration points of the beam, should be defined by the generalized coordinates. In other words, each deflection y should be a function of the coordinates $q_1, q_2, \ldots, q_j, \ldots, q_n$. Mathematically, such relationships are written as:

$$y_1 = f_1(q_1, q_2, \ldots, q_n) \qquad (6.1)$$

$$y_2 = f_2(q_1, q_2, \ldots, q_n) \qquad (6.2)$$

$$y_3 = f_3(q_1, q_2, \ldots, q_n) \qquad (6.3)$$

$$\ldots$$

$$y_n = f_n(q_1, q_2, \ldots, q_n) \qquad (6.4)$$

By considering the generalized coordinate q_i and introducing a virtual displacement δq_i, the work δW that is produced by the external forces during the virtual displacement is equal to the change δU in the internal strain energy U of the system; that is

$$\delta U = \delta W \qquad (6.5)$$

The two terms in the above equation can be written as

$$\delta U = \frac{\partial U}{\partial q_i} \delta q_1 \qquad (6.6)$$

$$\delta W = \frac{\partial W}{\partial q_i} \delta q_1 \qquad (6.7)$$

By substituting Eqs. (6.6) and (6.7) into Eq. (6.5), we find

$$\frac{\partial U}{\partial q_i} \delta q_i = \frac{\partial W}{\partial q_i} \delta q_1 \qquad (6.8)$$

In a dynamic system, forces producing work δW are usually the external forces $F_1, F_2, \ldots, F_m$, the inertial forces $m_j \ddot{y}_j = W_j \ddot{y}_j / g$ acting on the m_j masses ($j = 1, 2, \ldots, n$), and the forces due to damping. If δW_e, δW_d, and δW_{in} represent the virtual work done by the external, damping, and inertia forces, respectively, then Eq. (6.8) may be written as follows:

$$\frac{\partial U}{\partial q_i} \delta q_i = \frac{\partial W_e}{\partial q_i} \delta q_1 + \frac{\partial W_d}{\partial q_i} \delta q_i - \sum_{j=1}^{n} (m_j \ddot{y}_j) \frac{\partial y_j}{\partial q_i} \delta q_i \qquad (6.9)$$

In the right-hand side of Eq. (6.9), the amplitude y_j in the last term is the total displacement y_j of the mass m_j.

394 EARTHQUAKE EXCITATIONS AND BLAST LOADINGS

The kinetic energy T of the system, by definition, is given by the expression

$$T = \sum_{j=1}^{n} \tfrac{1}{2} m_j \dot{y}_j^2 \tag{6.10}$$

Thus,

$$\frac{\partial T}{\partial \dot{q}_i} = \sum_{j=1}^{n} m_j \dot{y}_j \frac{\partial \dot{y}_j}{\partial \dot{q}_i} \tag{6.11}$$

The preceding equation may be written as

$$\frac{\partial T}{\partial \dot{q}_i} = \sum_{j=1}^{n} m_j \dot{y}_j \frac{\partial y_j}{\partial q_i} \tag{6.12}$$

because, from the expressions given by Eqs. (6.1) through (6.4), we find that

$$\dot{y}_j = \frac{\partial y_j}{\partial q_i} \dot{q}_i \tag{6.13}$$

and

$$\frac{\partial \dot{y}_j}{\partial \dot{q}_i} = \frac{\partial y_j}{\partial q_i} \tag{6.14}$$

The last term in the right-hand side of Eq. (6.9) may be written as follows:

$$-\sum_{j=1}^{n} (m_j \ddot{y}_j) \frac{\partial y_j}{\partial q_i} \delta q_i = -\frac{d}{dt} \sum_{j=1}^{n} (m_j \dot{y}_j) \frac{\partial y_j}{\partial q_i} \delta q_i \\ + \sum_{j=1}^{n} (m_j \dot{y}_j) \frac{\partial \dot{y}_j}{\partial q_i} \delta q_i \tag{6.15}$$

From Eq. (6.10), by differentiation, we find

$$\frac{\partial T}{\partial q_i} = \sum_{j=1}^{n} m_j \dot{y}_j \frac{\partial \dot{y}_j}{\partial q_i} \tag{6.16}$$

Thus, by substituting Eqs. (6.11) and (6.16) into Eq. (6.15), the following equation is obtained:

$$-\sum_{j=1}^{n} (m_j \ddot{y}_j) \frac{\partial y_j}{\partial q_i} \delta q_i = -\frac{d}{dt} \frac{\partial T}{\partial \dot{q}_i} \delta q_i + \frac{\partial T}{\partial q_i} \delta q_i \tag{6.17}$$

Equation (6.17) may be substituted into Eq. (6.9) to obtain the following result:

$$\frac{\partial U}{\partial q_i}\delta q_i = \frac{\partial W_e}{\partial q_1}\delta q_i + \frac{\partial W_d}{\partial q_i}\delta q_i - \frac{d}{dt}\frac{\partial T}{\partial \dot{q}_i}\delta q_i + \frac{\partial T}{\partial q_i}\delta q_i$$

By rearranging terms and simplifying, we obtain

$$\frac{d}{dt}\frac{\partial T}{\partial \dot{q}_i} - \frac{\partial T}{\partial q_i} + \frac{\partial U}{\partial q_i} - \frac{\partial W_d}{\partial q_i} = \frac{\partial W_e}{\partial q_i} \qquad i = 1, 2, 3, \ldots, n \qquad (6.18)$$

Equation (6.18) is known as the Lagrange's equation. Its application to actual practical problems will yield as many differential equations of motion as there are degrees of freedom. The quantities required for the application of Lagrange's equation are the expressions for T, U, W_d, and W_e in terms of the generalized coordinates $q_1, q_2, \ldots, q_n$, and there are as many Lagrange's equations as there are generalized coordinates.

The following examples illustrate the application of Lagrange's equation.

Example 6.1: By using Lagrange's equation, derive the differential equation of motion for the one-degree spring-mass system shown in Fig. 6.1c. The mass m is restricted to move only in the vertical direction.

SOLUTION: The free-body diagram of mass m is shown in Fig. 6.1d. The displacement y of mass m is taken as the generalized coordinate q_i. Only one generalized coordinate is needed here because the spring-mass system has only one degree of freedom. The expressions for the energies required for the application of Eq. (6.18) are as follows:

$$T = \tfrac{1}{2}m\dot{y}^2 \qquad (6.19)$$

$$W_d = (-c\dot{y})y \qquad (6.20)$$

$$U = \tfrac{1}{2}ky^2 \qquad (6.21)$$

$$W_e = F(t)y \qquad (6.22)$$

where c in Eq. (6.20) is the viscous damping factor. By substituting Eqs. (6.19) through (6.22) into Eq. (6.18) and performing the required mathematical operations, we find

$$m\ddot{y} + ky + c\dot{y} = F(t) \qquad (6.23)$$

Equation (6.23) is the familiar general differential equation of motion for the one-degree system in Fig. 6.1c.

Example 6.2: For the spring-mass system shown in Fig. 6.2a, determine the differential equations of its motion by using Lagrange's equation. The masses m_1, m_2, and m_3 are restricted to move in the vertical direction only.

SOLUTION: The free-body diagrams of masses m_1, m_2, and m_3 are shown in Fig. 6.2b. The displacements y_1, y_2, and y_3 of the masses m_1, m_2, and m_3, respectively, are considered to be the generalized coordinates q_1, q_2, and q_3 of the spring-mass system. We have here three generalized coordinates, because the spring-mass system has three degrees of freedom.

The required energy expressions for T, U, W_d, and W_e are as follows:

$$T = \tfrac{1}{2}m_1\dot{y}_1^2 + \tfrac{1}{2}m_2\dot{y}_2^2 + \tfrac{1}{2}m_3\dot{y}_3^2 \tag{6.24}$$

$$U = \tfrac{1}{2}k_1 y_1^2 + \tfrac{1}{2}k_2(y_2 - y_1)^2 + \tfrac{1}{2}k_3(y_3 - y_2)^2 \tag{6.25}$$

$$W_d = (-c\dot{y}_1)y_1 + [-c(\dot{y}_2 - \dot{y}_1)](y_2 - y_1) + [-c(\dot{y}_3 - \dot{y}_2)](y_3 - y_2) \tag{6.26}$$

$$W_e = F_1(t)y_1 + F_2(t)y_2 + F_3(t)y_3 \tag{6.27}$$

The required derivatives with respect to the generalized coordinates y_1, y_2, and

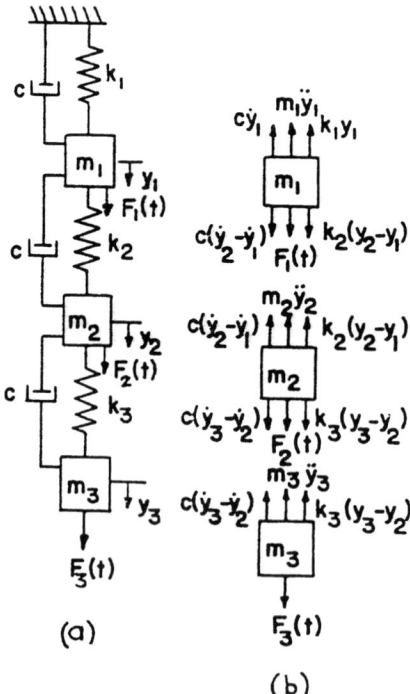

Figure 6.2. (a) Spring-mass system with three degrees of freedom. (b) Free-body diagrams of masses m_1, m_2, and m_3.

y_3 are as follows:

$$\frac{\partial T}{\partial \dot{y}_1} = m_1 \dot{y}_1 \tag{6.28}$$

$$\frac{\partial T}{\partial \dot{y}_2} = m_2 \dot{y}_2 \tag{6.29}$$

$$\frac{\partial T}{\partial \dot{y}_3} = m_3 \dot{y}_3 \tag{6.30}$$

$$\frac{\partial T}{\partial y_1} = \frac{\partial T}{\partial y_2} = \frac{\partial T}{\partial y_3} = 0 \tag{6.31}$$

$$\frac{\partial U}{\partial y_1} = k_1 y_1 - k_2(y_2 - y_1) \tag{6.32}$$

$$\frac{\partial U}{\partial y_2} = k_2(y_2 - y_1) - k_3(y_3 - y_2) \tag{6.33}$$

$$\frac{\partial U}{\partial y_3} = k_3(y_3 - y_2) \tag{6.34}$$

$$\frac{\partial W_d}{\partial y_1} = -c\dot{y}_1 + c(\dot{y}_2 - \dot{y}_1) \tag{6.35}$$

$$\frac{\partial W_d}{\partial y_2} = -c(\dot{y}_2 - \dot{y}_1) + c(\dot{y}_3 - \dot{y}_2) \tag{6.36}$$

$$\frac{\partial W_d}{\partial y_3} = -c(\dot{y}_3 - \dot{y}_2) \tag{6.37}$$

$$\frac{\partial W_e}{\partial y_1} = F_1(t) \tag{6.38}$$

$$\frac{\partial W_e}{\partial y_2} = F_2(t) \tag{6.39}$$

$$\frac{\partial W_e}{\partial y_3} = F_3(t) \tag{6.40}$$

Since we have three degrees of freedom and three generalized coordinates, Eq. (6.18) would have to be applied three times for $i = 1$, 2, and 3. By doing so and collecting appropriate expressions from the results given by Eqs. (6.28) through (6.40), the following three differential equations of motion representing the three-degree system in Fig. 6.2a are obtained:

$$m_1 \ddot{y}_1 + k_1 y_1 - k_2(y_2 - y_1) + c\dot{y}_1 - c(\dot{y}_2 - \dot{y}_1) = F_1(t) \tag{6.41}$$

$$m_2 \ddot{y}_2 + k_2(y_2 - y_1) - k_3(y_3 - y_2) + c(\dot{y}_2 - \dot{y}_1) - c(\dot{y}_3 - \dot{y}_2) = F_2(t) \tag{6.42}$$

$$m_3 \ddot{y}_3 + k_3(y_3 - y_2) + c(\dot{y}_3 - \dot{y}_2) = F_3(t) \tag{6.43}$$

6.3 MODAL EQUATIONS AND ANALYSIS OF STRUCTURAL SYSTEMS THAT ARE IDEALIZED AS SPRING-MASS SYSTEMS

The analysis for the dynamic response of complicated multiple degree of freedom infrastructural problems can become rather convenient by using what is known as the method of modal analysis. The required modal equations for the application of this method to structural dynamics problems may be derived by using Lagrange's equation, which is given by Eq. (6.18). Theoretically, the applicability of this method is supposed to be limited to linearly elastic systems and to cases where the dynamic forces acting on a structure have the same time variation. If these restrictions, however, cannot be removed, then numerical methods of analysis, such as the acceleration impulse extrapolation method, could be used.

In modal analysis, the response of a multidegree system is computed individually for each mode of vibration, and the total response is obtained by superimposing the responses of the individual modes. The modal equations that are derived in this section permit each mode to be represented by an equivalent one degree of freedom system, whose dynamic response is obtained independently by using methods discussed in earlier chapters.

The development of the modal equations is initiated by considering a general system consisting of r masses, j springs, and M normal modes. If $\dot{a}_{ip}$ is the velocity component of mass i in the p mode, then the total kinetic energy T of the system at a time t is

$$T = \sum_{i=1}^{r} \tfrac{1}{2} m_i \left(\sum_{P=1}^{M} \dot{a}_{ip} \right)^2 \qquad (6.44)$$

The total strain energy U stored in the springs is

$$U = \sum_{n=1}^{j} \tfrac{1}{2} k_n \left(\sum_{P=1}^{M} \delta_{np} \right)^2 \qquad (6.45)$$

where δ_{np} is the relative displacement of the ends of the spring n in the p mode, and k_n is the spring constant of the nth spring.

Equations (6.44) and (6.45) can be also written as follows:

$$T = \sum_{i=1}^{r} \tfrac{1}{2} m_i \sum_{P=1}^{M} \dot{a}_{ip}^2 \qquad (6.46)$$

$$U = \sum_{n=1}^{j} \tfrac{1}{2} k_n \sum_{P=1}^{M} \delta_{np}^2 \qquad (6.47)$$

because, according to the orthogonality properties of normal modes, the cross products in the squared series of Eqs. (6.44) and (6.45) are zero.

MODAL EQUATIONS AND ANALYSIS OF STRUCTURAL SYSTEMS   399

The work W_e of the external forces $F_i(t)$ acting on the r masses is

$$W_e = \sum_{i=1}^{r} F_i(t) \sum_{P=1}^{M} a_{ip} \tag{6.48}$$

If we select an arbitrary modal displacement Y_P, preferably the displacement of one of the masses, then all the displacements a_{ip} of the masses could be expressed in proportion of the modal displacement Y_P. On this basis, the quantities a_{ip}, $\dot{a}_{ip}$, and δ_{np} may be expressed as follows:

$$a_{ip} = Y_p \beta_{ip} \tag{6.49}$$

$$\dot{a}_{ip} = \dot{Y}_p \beta_{ip} \tag{6.50}$$

$$\delta_{np} = Y_p \beta_{\delta np} \tag{6.51}$$

where β_{ip} and $\beta_{\delta np}$ define the characteristic shape of the mode and they are given by the expressions

$$\beta_{ip} = \frac{a_{ip}}{Y_p} = \frac{\dot{a}_{ip}}{\dot{Y}_p} \tag{6.52}$$

$$\beta_{\delta np} = \frac{\delta_{np}}{Y_p} \tag{6.53}$$

Equations (6.46) through (6.48) are now written as follows:

$$T = \sum_{i=1}^{r} \tfrac{1}{2} m_i \sum_{P=1}^{M} \dot{Y}_p^2 \beta_{ip}^2 \tag{6.54}$$

$$U = \sum_{n=1}^{j} \tfrac{1}{2} k_n \sum_{P=1}^{M} Y_p^2 \beta_{\delta np}^2 \tag{6.55}$$

$$W_e = \sum_{i=1}^{r} F_i(t) \sum_{P=1}^{M} Y_p \beta_{ip} \tag{6.56}$$

By selecting the modal displacement Y_p as the generalized coordinate q_i, we write the following expressions:

$$\frac{d}{dt}\left(\frac{\partial T}{\partial \dot{Y}_p}\right) = \ddot{Y}_p \sum_{i=1}^{r} m_i \beta_{ip}^2 \tag{6.57}$$

$$\frac{\partial U}{\partial Y_p} = Y_p \sum_{n=1}^{j} k_n \beta_{\delta np}^2 \tag{6.58}$$

$$\frac{\partial W_e}{\partial Y_p} = \sum_{i=1}^{r} F_i(t) \beta_{ip} \tag{6.59}$$

By using Eqs. (6.57) through (6.59) and Lagrange's equation, given by Eq. (6.18), the following result is obtained:

$$\ddot{Y}_p \sum_{i=1}^{r} m_i \beta_{ip}^2 + Y_p \sum_{n=1}^{j} k_n \beta_{\delta np}^2 = \sum_{i=1}^{r} F_i(t) \beta_{ip} \quad (6.60)$$

Equation (6.60) is known as the modal equation of motion for any p mode. It could be also written in a more compact form by making the following substitutions:

$$m_e = \sum_{i=1}^{r} m_i \beta_{ip}^2 \quad (6.61)$$

$$k_e = \sum_{n=1}^{j} k_n \beta_{\delta np}^2 \quad (6.62)$$

$$F_e(t) = \sum_{i=1}^{r} F_i(t) \beta_{ip} \quad (6.63)$$

where m_e, k_e, and $F_e(t)$ are considered to be the equivalent mass, equivalent spring constant, and equivalent force, respectively, of an equivalent one-degree spring-mass system. With this in mind, Eq. (6.60) is now written as follows:

$$m_e \ddot{Y}_p + k_e Y_p = F_e(t) \quad (6.64)$$

By examining Eq. (6.64), it becomes obvious that this is the equation of an equivalent one-degree spring-mass system corresponding to the p mode of vibration of a multiple-degree of freedom system. The kinetic energy, internal strain energy, and the work done by all external forces in the equivalent one-degree system are equal to the same quantities of the complete system while it is vibrating in the p mode alone. Thus, each normal mode can be analyzed independently as a one-degree spring-mass system, and the dynamic response corresponding to a mode can be determined by using the methods discussed in Chapter 3.

Equation (6.64) can also be written as

$$\ddot{Y}_p + \omega_p^2 Y_p = \frac{F_e(t)}{m_e} \quad (6.65)$$

or, by using the expressions given by Eqs. (6.61) through (6.63), we find

$$\ddot{Y}_p + \omega_p^2 Y_p = g(t) \frac{\sum_{i=1}^{r} F_i \beta_{ip}}{\sum_{i=1}^{r} m_i \beta_{ip}^2} \quad (6.66)$$

where $g(t)$ in Eq. (6.66) is the time variation of the force $F_i(t) = g(t) F_i$.

The two summations in Eq. (6.66) are constant and they can be computed for a given mode and loading. In other words, for a given multiple-degree of freedom system, the first step in applying the modal equations would be to determine the required natural frequencies ω and the corresponding modes of vibration of the system. Methods for such computations are already discussed in Chapter 3, and additional ones may be found in References [5, 37, 39]. When these quantities are determined, then the dynamic response of the system for each mode is computed by using Eq. (6.66). It should be noted, however, that the time function $g(t)$ should be the same for all applied forces $F_i(t)$. This limitation, however, may be removed if the modal equations are solved by using numerical methods.

When each normal mode is considered as an independent one-degree spring-mass system, as discussed earlier, the modal static deflection Y_{pst} may be determined from the equation

$$Y_{pst} = \frac{F_e}{k_e} = \frac{F_e}{\omega_p^2 m_e} \qquad (6.67)$$

By substituting Eqs. (6.61) and (6.63) into Eq. (6.67), we find

$$Y_{pst} = \frac{\sum_{i=1}^{r} F_i \beta_{ip}}{\omega_p^2 \sum_{i=1}^{r} m_i \beta_{ip}^2} \qquad (6.68)$$

The dynamic modal displacement $Y_p(t)$ may be written as

$$Y_p(t) = Y_{pst} \Gamma_p \qquad (6.69)$$

and the maximum modal displacement $Y_{p\max}$ is

$$Y_{p\max} = Y_{pst} \Gamma_{p\max} \qquad (6.70)$$

where Γ_p is the magnification factor for the p mode, and $\Gamma_{p\max}$ is its maximum value.

In Eq. (6.69), the magnification factor Γ_p depends only on the time function $g(t)$ and the natural frequency ω_p. Thus, the solutions presented in Chapter 3 can be used for the analysis of multiple-degree of freedom systems. By superimposing the responses of all modes, the total deflection $y_i(t)$ of mass m_i may be obtained from the equation

$$y_i(t) = \sum_{P=1}^{M} Y_{pst} \beta_{ip} \Gamma_p \qquad (6.71)$$

For many practical applications the response of the first mode predominates, and it can be as high as 98%, or higher. This is usually the case when the values of the free frequencies of vibration of the multidegree spring-mass system are well apart, and the applied dynamic loads are acting in the same direction. Buildings that are idealized as spring-mass systems, as discussed in Chapter 3, would fall into this category. In such cases, numerical superposition of the responses of the first few modes would produce very accurate results. In many practical situations, the contributions of the higher modes are all together neglected.

If, on the other hand, the contributions of the higher modes are appreciable, but not predominant, it would be more reasonable to consider the response of the fundamental mode, plus the square root of the sum of the squares of the higher modes. For random inputs, or for cases where the modal components are random variables, a total response may be obtained by taking the square root of the sum of the squares of the modal maximum responses.

The following example illustrates the application of the modal analysis method to structures that can be idealized as spring-mass systems.

Example 6.3: For the steel frame shown in Fig. 6.3a, determine the maximum horizontal displacements at the girder levels, and the maximum bending stresses in the columns by using the method of modal analysis. The time variation of the dynamic force $F(t)$ is shown in Fig. 6.3b. Assume that the girders are rigid as compared to the columns and that there is no damping. The modulus $E = 30 \times 10^6$ psi.

SOLUTION: The idealized two-degree spring-mass system for the frame is shown in Fig. 6.3c. The constants k_1 and k_2 represent the stiffness at points 1 and 2, respectively, of the frame. Since the girders are assumed to be rigid, k_1 represents the sum of the stiffnesses at the tops of the first-story columns, and k_2 is the sum of the stiffnesses at the tops of the second-story columns. On this basis, we have

$$k_1 = \frac{3EI}{L^3} + \frac{12EI}{L^3} = \frac{(3)(30)(10)^6 \ (394)}{(15)^3 \ (12)^3} + \frac{(12)(30)(10)^6 \ (394)}{(15)^3 \ (12)^3}$$

$$= 30{,}401 \ \text{lb/in.}$$

$$= 30.40 \ \text{kips/in.}$$

$$k_2 = \frac{12EI}{L^3} + \frac{12EI}{L^3} = \frac{(24)(30)(10)^6 \ (133.2)}{(12)^3 \ (12)^3}$$

$$= 32{,}118 \ \text{lb/in.}$$

$$= 32.12 \ \text{kips/in.}$$

The masses m_1 and m_2 of the idealized spring-mass system represent the lumped weights of the frame at the first- and second-story levels, respectively, divided by the acceleration of gravity $g = 386$ in./sec². Thus, m_1 would include

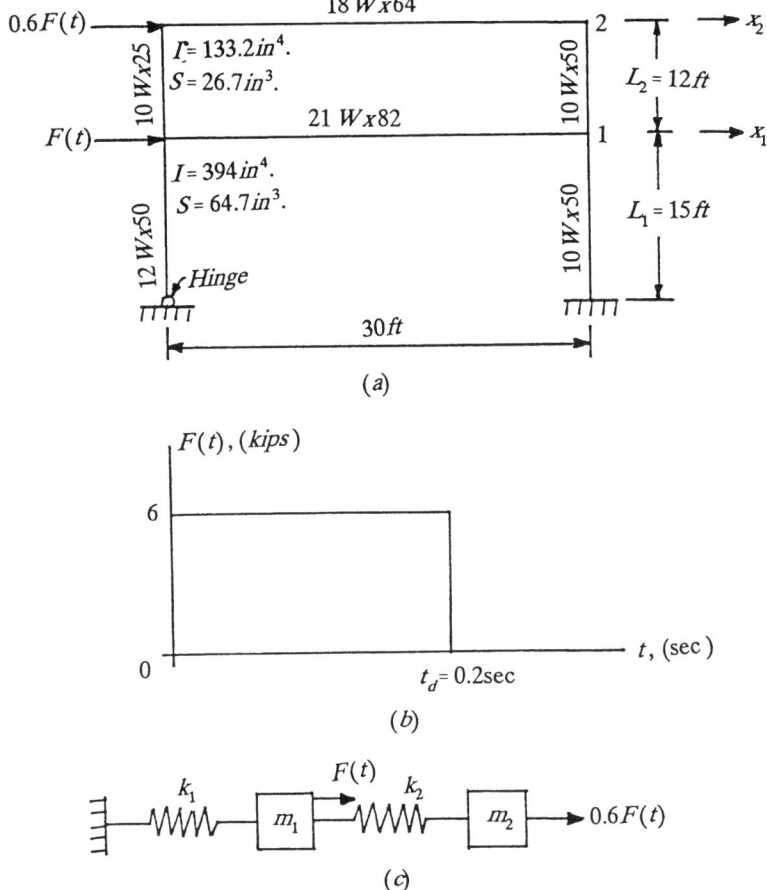

Figure 6.3. (a) Two-story steel frame loaded as shown. (b) Time variation of the dynamic force $F(t)$. (c) Idealized two-degree spring-mass system.

the mass of the first-story girder plus half the mass of the first- and second-story columns. Mass m_2 would include the mass of the second-story girder plus half the mass of the second-story columns. On this basis, we have

$$m_1 = \frac{1}{386}[(30)(82) + (2)(7.5)(50) + (2)(6)(25)]$$

$$= 0.0091 \text{ kip} \cdot \sec^2/\text{in}.$$

$$m_2 = \frac{1}{386}[(30)(64) + (2)(6)(25)]$$

$$= 0.0058 \text{ kip} \cdot \sec^2/\text{in}.$$

404   EARTHQUAKE EXCITATIONS AND BLAST LOADINGS

The frequencies of vibration ω_1 and ω_2 of the idealized spring-mass system may be obtained from the equation

$$\omega_{1,2}^2 = \frac{1}{2}\left(\frac{k_1+k_2}{m_1}+\frac{k_2}{m_2}\right) \pm \frac{1}{2}\left[\left(\frac{k_1+k_2}{m_1}+\frac{k_2}{m_2}\right)^2 - 4\frac{k_1 k_2}{m_1 m_2}\right]^{1/2} \quad (6.72)$$

By substituting the values of the m's and k's into Eq. (6.72), we find

$$\omega_{1,2}^2 = 6{,}201.31 \pm 4{,}462.49$$

Thus,

$$\omega_1 = 41.63 \text{ rps}$$
$$\omega_2 = 103.29 \text{ rps}$$

The periods of vibration τ_1 and τ_2 are

$$\tau_1 = \frac{2\pi}{\omega_1} = \frac{2\pi}{41.63} = 0.151 \text{ sec}$$

$$\tau_2 = \frac{2\pi}{\omega_2} = \frac{2\pi}{103.29} = 0.061 \text{ sec}$$

and the ratios t_d/τ, where t_d is the time duration of the force $F(t)$, are

$$\frac{t_d}{\tau_1} = \frac{0.20}{0.151} = 1.325 \quad (6.73)$$

$$\frac{t_d}{\tau_2} = \frac{0.20}{0.061} = 3.279 \quad (6.74)$$

If the mode amplitudes of the first mode are denoted as $X_1^{(1)}$ and $X_2^{(1)}$, they are related to each other by the following equation:

$$X_2^{(1)} = C_1 X_1^{(1)} \quad (6.75)$$

where

$$C_1 = \frac{k_2}{k_2 - m_2\omega_1^2} = \frac{32.12}{32.12 - (0.0058)(41.63)^2}$$
$$= 1.4555$$

Thus, by assuming $X_1^{(1)} = 1.00$, we find from Eq. (6.75) that $X_2^{(1)} = 1.456$.
For the second mode we have

$$X_2^{(2)} = C_2 X_1^{(2)} \quad (6.76)$$

MODAL EQUATIONS AND ANALYSIS OF STRUCTURAL SYSTEMS   405

where
$$C_2 = \frac{k_2}{k_2 - m_2\omega_2^2} = \frac{32.12}{32.12 - (0.0058)(103.29)^2}$$
$$= -1.078$$

Therefore, by assuming $X_1^{(2)} = 1.00$, Eq. (6.76) yields $X_2 = -1.078$. In summary, we have the following mode amplitudes:

First Mode	Second Mode
$X_1^{(1)} = 1.000$	$X_1^{(2)} = 1.000$
$X_2^{(1)} = 1.456$	$X_2^{(2)} = -1.078$

If Y_{pst} is replaced by X_{pst}, the modal static displacements X_{1st} and X_{2st} for the first and second mode, respectively, may be determined by using Eq. (6.68). Thus, for $p = 1$, we have

$$X_{1st} = \frac{\sum_{i=1}^{2} F_i \beta_{i1}}{\omega_1^2 \sum_{i=1}^{2} m_i \beta_{i1}^2}$$
$$= \frac{(6)(1.000) + (3.6)(1.456)}{(41.63)^2[(0.0091)(1.000)^2 + (0.0058)(1.456)^2]}$$
$$= 0.3031 \text{ in.}$$

$$X_{2st} = \frac{\sum_{i=1}^{2} F_i \beta_{i2}}{\omega_2^2 \sum_{i=1}^{2} m_i \beta_{i2}^2}$$
$$= \frac{(6)(1.000) + (3.6)(-1.078)}{(103.29)^2[(0.0091)(1.000)^2 + (0.0058)(-1.078)^2]}$$
$$= 0.0126 \text{ in.}$$

For certain force functions, Appendix H provides graphs that can be used to determine maximum magnification factors, and times at which maximum response occurs for undamped one-degree spring-mass systems. These graphs are plotted for a wide range of ratios t_d/τ, and they can be used to determine the maximum magnification factors for our problem. By consulting Fig. H.2 of Appendix H, we find that, for both ratios given by Eqs. (6.73) and (6.74), the maximum magnification factor is 2; that is, $\Gamma_{1max} = \Gamma_{2max} = 2$.

If x_{21} and x_{22} denote the contributions of the first and second mode, respectively, to the horizontal deflection at point 2 of the frame in Fig. 6.3a, then numerical addition of these two responses should yield the total horizontal displacement x_{2max} at point 2. Thus, by using Eq. (6.71), we find

$$x_{2max} = x_{21} + x_{22} = X_{1st}\beta_{21}\Gamma_{1max} + X_{2st}\beta_{22}\Gamma_{2max}$$
$$= (0.3031)(1.456)(2) + |(0.0126)(-1.078)(2)|$$
$$= 0.8826 + 0.0272 = 0.9098 \text{ in.}$$

For the maximum horizontal displacement, $x_{1\max}$, at point 1 in Fig. 6.3a, we have

$$x_{1\max} = x_{11} + x_{12} = X_{1st}\beta_{11}\Gamma_{1\max} + X_{2st}\beta_{12}\Gamma_{2\max}$$
$$= (0.3031)(1.00)(2) + (0.0126)(1.00)(2)$$
$$= 0.6062 + 0.0252 = 0.6314 \text{ in.}$$

Note that the contribution of the second mode response is 3 percent of the total for point 2, and 4 percent of the total for point 1.

With known $x_{1\max}$ and $x_{2\max}$, the maximum shear force, bending moment, and bending stress in each column of the frame can be determined. At the top of the first-story columns the total shear force $V_{1\max}$ is

$$V_{1\max} = k_1 x_{1\max} = (0.6314)(30.4) = 19.1946 \text{ kips}$$

This shear force is distributed among the two first-story columns in proportion to their stiffness. For the first column, the shear force V_{11} is

$$V_{11} = 19.1946 \left(\frac{3}{15}\right) = 3.8389 \text{ kips}$$

For the second first-story column, the shear force V_{12} is

$$V_{12} = 19.1946 \left(\frac{12}{15}\right) = 15.3557 \text{ kips}$$

At the top of the second-story columns, the total maximum shear force $V_{2\max}$ is

$$V_{2\max} = k_2(x_{2\max} - x_{1\max})$$
$$= 32.12(0.9098 - 0.6314) = 8.9366 \text{ kips}$$

This shear force is distributed equally among the two columns of the second story because their stiffnesses are equal. Thus, we have

$$V_{21} = V_{22} = \frac{8.9366}{2} = 4.4683 \text{ kips}$$

The maximum moments M_{11} and M_{12} of the first and second columns of

MODAL EQUATIONS AND ANALYSIS OF STRUCTURAL SYSTEMS   407

the first story are

$$M_{11} = V_{11}L_1 = (3.8383)(15)(12) = 691.002 \text{ kip-in.}$$

$$M_{12} = V_{12}\frac{L_1}{2} = (15.3557)\left(\frac{15}{2}\right)(12) = 1{,}382.0130 \text{ kip-in.}$$

For the second-story columns, the bending moments M_{21} and M_{22} for the first and second column, respectively, are

$$M_{21} = M_{22} = \frac{(8.9366)(12)(12)}{2} = 643.4352 \text{ kip-in.}$$

Thus, for the first and second columns of the first story, the maximum bending stresses σ_{11} and σ_{12}, respectively, are

$$\sigma_{11} = \frac{M_{11}}{S} = \frac{691.002}{64.70} = 10.68 \text{ ksi}$$

$$\sigma_{12} = \frac{M_{12}}{S} = \frac{1{,}382.0130}{64.70} = 21.36 \text{ ksi}$$

For the second-story columns, we have

$$\sigma_{21} = \sigma_{22} = \frac{M_{21}}{S} = \frac{643.4352}{26.70} = 24.10 \text{ ksi}$$

The above values of the column bending stresses include only the effects of the applied dynamic forces $F(t)$ and $0.6\ F(t)$. If stresses are caused by other types of loadings being applied to the frame, they can be superimposed.

Numerical Solution of the Modal Equations: Numerical methods of analysis may be used to solve the modal equation given by Eq. (6.66). This would be particularly useful for forcing functions where the maximum magnification factor Γ_p is not known, or is very difficult to determine. For illustration purposes, we determine here the first mode response by using the numerical acceleration impulse extrapolation method (AIEM) to solve the modal equation.

From Eq. (6.66), for $p = 1$, we find

$$g(t)\frac{\sum_{i=1}^{2} F_i \beta_{i1}}{\sum_{i=1}^{2} m_i \beta_{i1}^2} = \frac{(6)(1.00) + (3.6)(1.456)}{(0.0091)(1.000)^2 + (0.0058)(1.456)^2} g(t)$$

$$= 525.2895 g(t)$$

Thus, from Eq. (6.66), the modal equation for $p = 1$ is

$$\ddot{X}_1 + (41.63)^2 X_1 = 525.2895g(t)$$

or

$$\ddot{X}_1 + 1{,}733.0569 X_1 = 525.2895g(t) \qquad (6.77)$$

The modal acceleration $\ddot{X}_1$ is

$$\ddot{X}_1 = 525.2895g(t) - 1{,}733.0569 X_1 \qquad (6.78)$$

In Eq. (6.78) the time function $g(t)$ represents the time variation of the forcing function $F(t)$. Since $F(t)$ is constant for $0 \leq t \leq 0.2$ sec, $g(t)$ would be constant and equal to 1 for $0 \leq t \leq 0.2$ sec. For $t > 0.2$ sec, $g(t)$ would be equal to zero. By selecting a time interval $\Delta t = 0.015$ sec, we can determine the modal displacement $X_1^{(1)}$ at the first time station by using the following equation:

$$X_1^{(1)} = \frac{\ddot{X}_1^{(0)}}{2} (\Delta t)^2 \qquad (6.79)$$

This equation yields

$$X_1^{(1)} = \frac{525.2895}{2} (0.015)^2 = 0.0591 \text{ in.}$$

At all other time stations, the following equation may be used:

$$X_1^{(i+1)} = 2X_1^{(i)} - X_1^{(i-1)} + \ddot{X}_1^{(i)}(\Delta t)^2 \qquad (6.80)$$

Table 6.1 illustrates the AIEM for the computation of the dynamic modal displacement $X_1(t)$. Its maximum value occurs at $t = 0.075$ sec, and it is equal

TABLE 6.1. Variation of Modal Displacement X_1 with Time t for First Mode Response

t (sec)	525.2895 (in./sec^2)	$-1{,}733.0569 X_1$ (in./sec^2)	$\ddot{X}_1$ (in./sec^2)	$\ddot{X}_1(\Delta t)^2$ (in.)	X_1 (in.)
0	525.2895	0	525.2895	0.1182	0
0.015	525.2895	−102.4151	422.8744	0.0951	0.0591
0.030	525.2895	−369.6610	155.6285	0.0350	0.2133
0.045	525.2895	−697.5554	−172.2659	−0.0388	0.4025
0.060	525.2895	−958.2072	−432.9177	−0.0974	0.5529
0.075	525.2895	−1{,}050.0592	−524.7697	−0.1181	0.6059
0.090					0.5408

to 0.6059 in. This value agrees very well with the value obtained earlier, where

$$X_1(t) = X_{1_{st}}\Gamma_{1_{max}} = (0.3031)(2) = 0.6062 \text{ in.}$$

In a similar manner, Eq. (6.66) may be solved for $p = 2$ in order to determine the maximum value of the dynamic modal displacement $X_2(t)$. With known maximum modal displacements, the maximum horizontal displacements at points 1 and 2 of the frame in Fig. 6.3a may be determined as discussed earlier in this problem. The same procedure may be used to determine shear forces, bending moments, and bending stresses in the columns. Since this procedure is already explained earlier in this problem, it will not be repeated here.

6.4 USING MODAL ANALYSIS FOR EARTHQUAKE RESPONSE OF IDEALIZED FRAMES AND BUILDINGS

The response of structural systems, such as frames and buildings, that are subjected to earthquake accelerations and are idealized as spring-mass systems, can be determined by using modal analysis. The development of the appropriate modal equations for earthquake response may be initiated by first considering the one-story frame in Fig. 6.4a, which is subjected to an earthquake ground motion. This motion is represented by a support displacement u_s, as shown in the figure. At the top of the frame, the displacement relative to the ground is denoted as u. Thus, at the top of the frame, the total horizontal displacement x is

$$x = u_s + u \tag{6.81}$$

We assume here that the girder of the frame is infinitely stiff when it is compared with the stiffness of the columns and, on this basis, we derive the idealized one-degree spring-mass system shown in Fig. 6.4b. From the free-body diagram of the mass m in Fig. 6.4c, we write the following differential equation of motion:

$$m\ddot{x} + c\dot{u} + ku = 0 \tag{6.82}$$

By substituting Eq. (6.81) into Eq. (6.82) and rearranging, we find

$$m\ddot{u} + c\dot{u} + ku = -m\ddot{u}_s \tag{6.83}$$

Equation (6.83) may be written as

$$\ddot{u} + 2\mu\dot{u} + \omega^2 u = -\ddot{u}_s \tag{6.84}$$

410 EARTHQUAKE EXCITATIONS AND BLAST LOADINGS

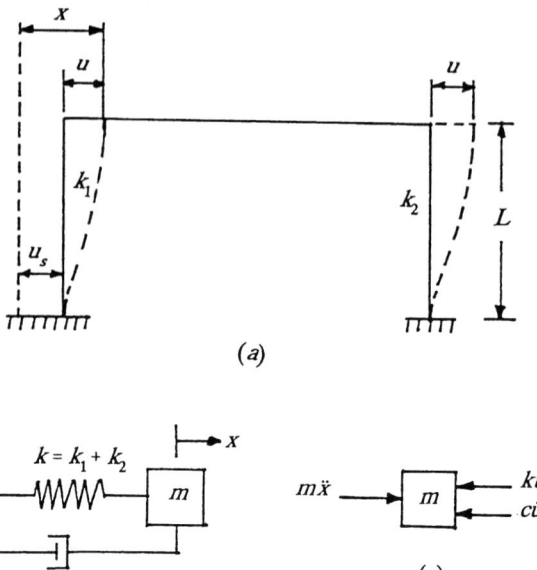

Figure 6.4. (a) One-story frame subjected to an earthquake motion as shown. (b) Idealized one-degree spring-mass system representing the frame. (c) Free-body diagram of mass m.

because $k = m\omega^2$, $c/m = 2\zeta\omega = 2\mu$, and $\mu = \zeta\omega$. Note that ζ is the viscous damping constant.

The solution of Eq. (6.84) is similar to the one obtained in Section 3.5, Eq. (3.105a), and it can be written as follows:

$$u(t) = \frac{1}{\omega} \int_0^t (-\ddot{u}_s) \exp[-\mu(t-T)] \sin \omega(t-T) \, dT \qquad (6.85)$$

where $\ddot{u}_s$ is the applied ground or support acceleration, and $u(t)$ is the displacement of the top of the frame relative to the ground.

We can express the support acceleration $\ddot{u}_s$ as

$$\ddot{u}_s = \ddot{u}_{s0} f_s(t) \qquad (6.86)$$

where $\ddot{u}_{s0}$ is the maximum support acceleration, and $f_s(t)$ is the time variation. By substituting Eq. (6.86) into Eq. (6.85), we obtain

$$u(t) = \frac{\ddot{u}_{s0}}{\omega} \int_0^t f_s(T) \exp[-\mu(t-T)] \sin \omega(t-T) \, dT \qquad (6.87)$$

USING MODAL ANALYSIS FOR EARTHQUAKE RESPONSE   411

If we assume that there is no damping in the system, we have $\mu = 0$, and Eq. (6.87) yields

$$u(t) = -\frac{\ddot{u}_{s0}}{\omega} \int_0^t f_s(T) \sin \omega(t - T) \, dT \qquad (6.88)$$

If we consider Eq. (6.87) and we multiply and divide its right-hand side by $m\omega$ and substitute k for $m\omega^2$, we find

$$u(t) = \frac{m\ddot{u}_{s0}\omega}{k} \int_0^t f_s(T) \exp[-\mu(t - T)] \sin \omega(t - T) \, dT \qquad (6.89)$$

If we write the magnification factor Γ as

$$\Gamma = \omega \int_0^t f_s(T) \exp[-\mu(t - T)] \sin \omega(t - T) \, dT \qquad (6.90)$$

and the static displacement u_{st} as

$$u_{st} = \frac{m\ddot{u}_{s0}}{k} = \frac{\ddot{u}_{s0}}{\omega^2} \qquad (6.91)$$

then Eq. (6.89) may be written as

$$u(t) = -u_{st}\Gamma = -\frac{\ddot{u}_{s0}}{\omega^2} \Gamma \qquad (6.92)$$

If damping is neglected, Eq. (6.90) yields

$$\Gamma = \omega \int_0^t f_s(T) \sin \omega(t - T) \, dT \qquad (6.93)$$

From the preceding derivation, we can easily observe that support motions produce responses that are equivalent to the ones produced by the applied forces $-m\ddot{u}_s$, where $\ddot{u}_s$ may be expressed as shown in Eq. (6.86). For spring-mass systems with many degrees of freedom, the applied force on the mass m_i is $-m_i\ddot{u}_s$. On this basis, the modal equation for a p mode for structures that can be idealized as spring-mass systems would be similar to the one shown by Eq. (6.66). Thus, by replacing $F_i g(t)$ in Eq. (6.66) with $-m_i \ddot{u}_s$, and also by using Eq. (6.86), we have the following modal equation, which can be used for earthquake response:

$$\ddot{Y}_p + \omega_p^2 Y_p = -f_s(t)\ddot{u}_{s0} \frac{\sum_{i=1}^r m_i \beta_{ip}}{\sum_{i=1}^r m_i \beta_{ip}^2} \qquad (6.94)$$

where Y_p is the modal displacement. If there is viscous damping in the system, it can be included in Eq. (6.94) by the addition of a third term in its left-hand side. That is,

$$\ddot{Y}_p + \omega_p^2 Y_p + 2\mu \dot{Y}_p = -f_s(t)\ddot{u}_{s0} \frac{\sum_{i=1}^r m_i \beta_{ip}}{\sum_{i=1}^r m_i \beta_{ip}^2} \quad (6.95)$$

In Eqs. (6.94) and (6.95) we note that the summations on the right-hand side of the equations are constant. In practice, the factor Λ_p, known as the modal participation factor, is introduced, which is defined as

$$\Lambda_p = \frac{\sum_{i=1}^r m_i \beta_{ip}}{\sum_{i=1}^r m_i \beta_{ip}^2} \quad (6.96)$$

On this basis, the modal displacement $Y_p(t)$ relative to the support may be determined from the equation

$$Y_p(t) = \Lambda_p u_p(t) = -\Lambda_p \frac{\ddot{u}_{s0}}{\omega_p^2} \Gamma_p \quad (6.97)$$

where $u_p(t)$ is given by Eq. (6.92). For a given frequency ω_p and the corresponding mode, the magnification factor may be determined from Eq. (6.90) when viscous damping is considered, and from Eq. (6.93) when there is no damping.

For the mass m_i, its displacement u_{ip} in the p mode may be determined from the equation

$$u_{ip} = -\Lambda_p \Gamma_p \beta_{ip} \frac{\ddot{u}_{s0}}{\omega_p^2} \quad (6.98)$$

If the response contributions of all modes are numerically superimposed, the total displacement $u_i(t)$ of the mass m_i is

$$u_i(t) = -\sum_{p=1}^N \Lambda_p \Gamma_p \beta_{ip} \frac{\ddot{u}_{s0}}{\omega_p^2} \quad (6.99)$$

where N in the summation sign represents the total number of modes. Usually the contribution of the fundamental mode predominates.

The preceding methodology is illustrated by the following example.

Example 6.4: The two-story rigid frame building shown in Fig. 6.5a is subjected to an artificial earthquake of ground acceleration $\ddot{u}_s = 0.05g \sin 4\pi t$, where g is the acceleration of gravity and t is time. The duration of the

USING MODAL ANALYSIS FOR EARTHQUAKE RESPONSE 413

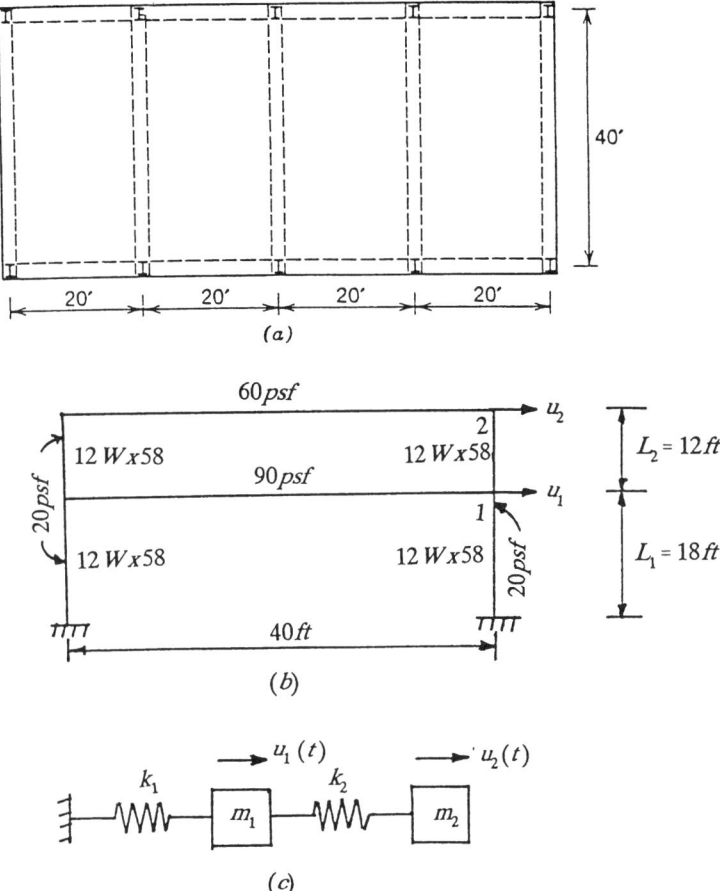

Figure 6.5. (a) Top view of a two-story building. (b) An interior frame of the building. (c) Idealized two-degree spring-mass system for the frame.

acceleration $\ddot{u}_s$ is 1 sec. The steel sections of the girders and columns are W24 × 76 and W12 × 58, respectively. By considering an interior frame such as the one in Fig. 6.5b and applying the method of modal analysis, determine the maximum bending stresses in the columns of the frame. Assume that the girders are infinitely stiff compared to the stiffness of the column. The modulus of elasticity $E = 30 \times 10^6$ psi. Neglect damping. Note that modal analysis assumes completely elastic response.

SOLUTION: Since the elastic properties of the building in Fig. 6.5a are uniform throughout its length, the dynamic analysis of an interior frame, such as the one in Fig. 6.5b, together with the associated wall and floor areas, would be sufficient for the analysis of the building. On this basis,

$$m_1 = \frac{1}{386.4}\left[(90)(20)(40) + (76)(40) + (2)(20)\left(\frac{18+12}{2}\right)(20)\right]$$

$$= 0.22526 \frac{\text{kip-sec}^2}{\text{in.}}$$

$$m_2 = \frac{1}{386.4}\left[(60)(20)(40) + (76)(40) + (2)(20)\left(\frac{12}{2}\right)(20)\right]$$

$$= 0.14451 \frac{\text{kip-sec}^2}{\text{in.}}$$

By considering the two columns of the first story, the total stiffness k_1 is

$$k_1 = 2\frac{12EI}{L^3} = \frac{(2)(12)(30)(10)^3\ (476)}{(18)^3\ (12)^3}$$

$$= 34.01 \text{ kips/in.}$$

For the second-story columns, the total stiffness k_2 is

$$k_2 = 2\frac{12EI}{L^3} = \frac{(2)(12)(30)(10)^3\ (476)}{(12)^3\ (12)^3}$$

$$= 114.78 \text{ kips/in.}$$

By applying Eq. (6.72), we find that the undamped free frequencies ω_1 and ω_2 of the idealized spring-mass system are

$$\omega_1 = 9.366 \text{ rps}$$
$$\omega_2 = 36.974 \text{ rps}$$

The undamped periods of vibration τ_1 and τ_2 are

$$\tau_1 = \frac{2\pi}{\omega_1} = 0.67085 \text{ sec}$$

$$\tau_2 = \frac{2\pi}{\omega_2} = 0.16994 \text{ sec}$$

By applying Eqs. (6.75) and (6.76) and assuming $X_1 = 1$, the first and second mode amplitudes are found to be as follows:

First Mode	Second Mode
$X_1^{(1)} = \beta_{11} = 1.000$	$X_1^{(2)} = \beta_{12} = 1.000$
$X_2^{(1)} = \beta_{21} = 1.124$	$X_2^{(2)} = \beta_{22} = -1.387$

By applying Eq. (6.96) for $p = 1$ and 2, the corresponding modal participation factors Λ_1 and Λ_2 are

$$\Lambda_1 = \frac{\sum_{i=1}^{2} m_i \beta_{i1}}{\sum_{i=1}^{2} m_i \beta_{i1}^2} = \frac{(0.22526)(1.000) + (0.14451)(1.124)}{(0.22526)(1.000)^2 + (0.14451)(1.124)^2}$$

$$= 0.9506$$

$$\Lambda_2 = \frac{\sum_{i=1}^{2} m_i \beta_{i2}}{\sum_{i=1}^{2} m_i \beta_{i2}^2} = \frac{(0.22526)(1.000) + (0.14451)(-1.387)}{(0.22526)(1.000)^2 + (0.14451)(-1.387)^2}$$

$$= 0.0493$$

From Reference [5, Appendix A], the magnification factors Γ_1 and Γ_2 for first and second mode response, respectively, may be determined from the following equation:

$$\Gamma_p = \frac{1}{(1 - (\omega_s^2/\omega^2))} \left(\sin \omega_s t - \frac{\omega_s}{\omega} \sin \omega t \right) \qquad t \leqslant 1.0 \text{ sec} \qquad (6.100)$$

Thus, for $\omega = \omega_1 = 9.366$ rps and $\omega = \omega_2 = 36.974$ rps, Eq. (6.100) yields

$$\Gamma_1 = -1.2497 \sin 4\pi t + 1.6768 \sin 9.366t \qquad (6.101)$$

$$\Gamma_2 = 1.1306 \sin 4\pi t - 0.38426 \sin 36.974t \qquad (6.102)$$

The relative story displacements $u_1(t)$ and $u_2(t)$ of the masses m_1 and m_2 in Fig. 6.5c, respectively, may be determined by using Eq. (6.99). They are

$$u_1(t) = 0.260852 \sin 4\pi t - 0.35106 \sin 9.366t \qquad (6.103)$$
$$+ 0.000268 \sin 36.974t$$

$$u_2(t) = 0.29517 \sin 4\pi t - 0.39459 \sin 9.366t \qquad (6.104)$$
$$- 0.000372 \sin 36.974t$$

The displacement $u_1(t)$ gives the horizontal movement of the tops of the first-story columns relative to their lower end, which in this case is fixed. The horizontal displacement $u_{r2}(t)$ of the tops of the second-story columns, relative to the column tops of the first story columns, is

$$u_{r2}(t) = u_2(t) - u_1(t)$$
$$= 0.034318 \sin 4\pi t - 0.04353 \sin 9.366t \quad (6.105)$$
$$- 0.00064 \sin 36.974t$$

Since the time duration t_d of the earthquake ground acceleration $\ddot{u}_s$ is 1 sec, we have to examine maximum conditions for the displacements $u_1(t)$ and $u_{r2}(t)$ for times $t \leq t_d$ and $t > t_d$. This problem is worked out for both time intervals, and it is found that maximum conditions occur at $t = 1.165$ sec, yielding the following results:

$$u_{1\max} = 0.6990 \text{ in.} \quad \text{at } t = 1.165 \text{ sec}$$
$$u_{r2\max} = 0.0872 \text{ in.} \quad \text{at } t = 1.165 \text{ sec}$$

With known displacements, the maximum total shear forces $V_{1\max}$ and $V_{2\max}$ at the tops of the first- and second-story columns, respectively, are

$$V_{1\max} = k_1 u_{1\max} = (34.01)(0.699)$$
$$= 23.77 \text{ kips}$$
$$V_{2\max} = k_2 u_{r2\max} = (114.78)(0.0872)$$
$$= 10.01 \text{ kips}$$

Since in each story the columns are of the same stiffness, the shear forces $V_{1\max}$ and $V_{2\max}$ are distributed equally among the columns of each story level.

For the first-story columns, the maximum moment $M_{1\max}$ for each first-story column is

$$M_{1\max} = \frac{V_{1\max}}{2} \cdot \frac{L}{2} = \frac{23.77}{2} \cdot \frac{(18)(12)}{2}$$
$$= 1{,}284 \text{ kip-in.}$$

For the second-story columns, we have

$$M_{2\max} = \frac{V_{2\max}}{2} \cdot \frac{L}{2} = \frac{10.01}{2} \cdot \frac{(12)(12)}{2}$$
$$= 360 \text{ kip-in.}$$

Thus, the maximum bending stresses $\sigma_{1\max}$ and $\sigma_{2\max}$ for the first- and second-story columns, respectively, are

$$\sigma_{1\max} = \frac{M_{1\max}}{S} = \frac{1284}{78.1}$$

$$= 16.44 \text{ ksi}$$

$$\sigma_{2\max} = \frac{M_{2\max}}{S} = \frac{360}{78.1}$$

$$= 4.61 \text{ ksi}$$

6.5 ELASTOPLASTIC ANALYSIS FOR EARTHQUAKE RESPONSE

In the preceding section, the analysis was based on elastic response, and it would be appropriate for moderate to low intensity earthquakes. For strong earthquakes, on the other hand, an elastoplastic analysis would be more appropriate, because such earthquakes are experienced by a structure only a few times, if at all, in its life span, and some damage to the structure could be tolerated. It becomes very uneconomical to design a structure to behave entirely within the elastic range when the dynamic loads are very severe, and when such loads are expected to occur only a few times, if any, in the structure's life span.

Many ductile materials are characterized by a large yielding range that permits them to absorb large amounts of energy before complete failure. For such special cases, the important decision to be made by a designer is the amount of energy that a structure should be permitted to absorb for a still-safe design.

6.5.1 Resistance R of a Structure

The resistance R of a structure, or of a structural component such as a beam, a plate, and so on, is defined here as the internal force that tends to restore the structure, or a structural component, to its unloaded static position. It may have a variety of forms. For example, if a structural element is made up of brittle material, the resistance R will be represented by curve 1 in Fig. 6.6a; on the other hand, R will have the form given by curve 2 if the element is made up of a ductile material. Curve 3 would represent the resistance R when the material is plain concrete.

The resistance functions are, in most cases, idealized in order to simplify the analysis of the problem at hand. For static response, advanced methods of analysis dealing with the inelastic response of structural elements, such as beams and plates, are discussed in Chapter 5 of this text and in References [3, 5, 6, 50, 52], as well as in many journal papers published by the author and his collaborators. For many practical dynamics problems, it is often reasonable to use the bilinear resistance function form shown in Fig. 6.6b.

418 EARTHQUAKE EXCITATIONS AND BLAST LOADINGS

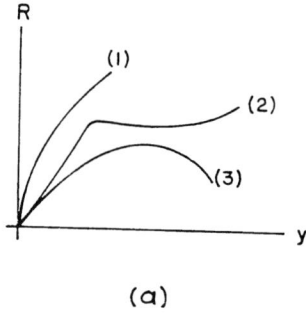

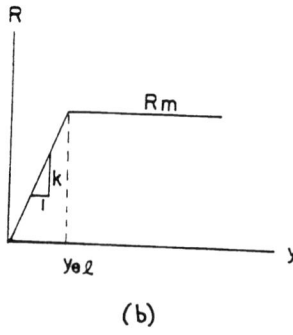

Figure 6.6. (*a*) Resistance functions for various materials. (*b*) Bilinear form of the resistance *R*.

For structures that can be idealized as a one-degree spring-mass system, as discussed earlier in this text (see also Fig. 6.7*a*), the resistance R may be assumed to have the bilinear form in Fig. 6.7*b*. In this case, R increases linearly until the elastic limit displacement y_{el} is reached, and the slope is equal to the spring constant k, as shown in the figure. At y_{el}, the resistance function attains its maximum value R_m, and it will remain constant and equal to R_m with increasing y. The extent of this yielding range is dependent on the ductility limit of the material.

If y reaches its maximum value y_m before the ductility limit is attained, while R_m remains constant, the structural system is said to rebound, and during rebound the resistance R is assumed to decrease linearly along a line parallel to its initial elastic line, as shown in Fig. 6.7*b*, and the decrease will continue until R becomes equal to $-R_m$. The other possibility is for y to continue to increase while R_m remains constant until the ductility limit is attained. In this case, the resistance function can be assumed to be composed of two lines, as shown in Fig. 6.6*b*.

For the spring-mass system in Fig. 6.7*a*, the resistance function R is represented by the spring force ky. By using the free-body diagram in Fig. 6.7*c*

ELASTOPLASTIC ANALYSIS FOR EARTHQUAKE RESPONSE

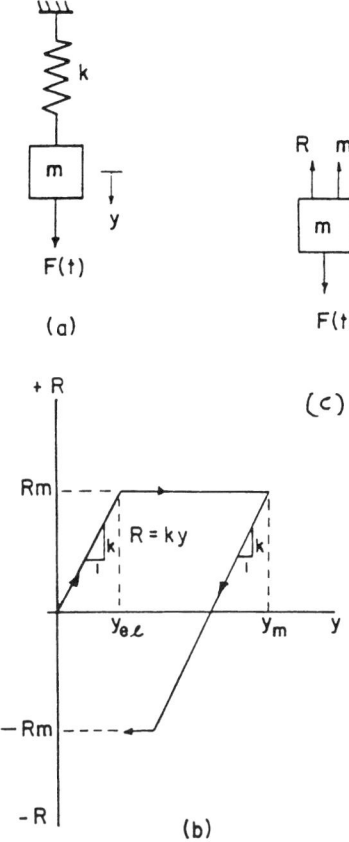

Figure 6.7. (*a*) Spring-mass system with one degree of freedom. (*b*) Resistance function representation for the one-degree spring-mass system. (*c*) Free-body diagram of mass *m*.

and applying the second law of motion, we find the following differential equation representing the spring-mass system:

$$m\ddot{y} + R - F(t) = 0 \tag{6.106}$$

By using the resistance function shape shown in Fig. 6.7*b*, R is equal to ky for the elastic range, and Eq. (6.106) yields

$$m\ddot{y} + ky - F(t) = 0 \qquad 0 \leqslant y \leqslant y_{el} \tag{6.107}$$

For values of y between y_{el} and y_m, R_m remains constant, and Eq. (6.106) yields

$$m\ddot{y} + R_m - F(t) = 0 \qquad y_{el} \leqslant y \leqslant y_m \tag{6.108}$$

When R_m starts to decrease linearly until it becomes $-R_m$, y will attain values between y_m and $y_m - 2y_{el}$, and Eq. (6.106) takes the following form:

$$m\ddot{y} + R_m - k(y_m - y) - F(t) = 0 \qquad (y_m - 2y_{el}) \leqslant y \leqslant y_m \qquad (6.109)$$

Additional equations, however, may be obtained for the negative plastic range in a similar manner, but this is not usually required, since this range is not of interest to the practicing engineer.

Equations (6.107) through (6.109) may be solved by using a rigorous solution, or by using a numerical method such as the AIEM discussed earlier. If a rigorous solution is used, the initial conditions of Eq. (6.107) are the initial conditions of the problem under consideration. The initial conditions of Eq. (6.108) are the final displacement and final velocity obtained from the solution of Eq. (6.107), and the initial conditions of Eq. (6.109) are the final displacement and final velocity obtained from the solution of Eq. (6.108). This procedure would be sufficient if $F(t)$ is a continuous function of time and it does not reduce to zero before the analysis for y is completed. If $F(t)$ is of finite duration, or has discontinuities, additional stages of the equations, one for each discontinuity, should be included in the solution. See References [5, 37, 39].

In practical applications, the ductility ratio η is introduced, which is the ratio of y_m and y_{el}; that is,

$$\eta = \frac{y_m}{y_{el}} \qquad (6.110)$$

The ductility ratio η provides a measure that can be used to decide how far into the inelastic range a member can be permitted to be exposed. In practice, ductility ratios of 5, and ever higher, are used for structures that are subjected to blast, earthquake, and dynamic loading conditions that occur only a few times during their life span.

The following example illustrates the application of the preceding methodology and equations. Also, Appendix G provides useful charts of elastoplastic response for one-degree spring-mass systems; these results are very convenient for the solution of practical engineering problems.

Example 6.5: The uniform simply supported steel beam in Fig. 6.8a supports the heavy weight $W = 20$ kips as shown, and also carries the dynamic force $F(t)$ whose time variation is as shown in Fig. 6.8b. The cross section of the beam is a W24 × 76 wide flange with moment of inertia $I = 2{,}100.0 \text{ in.}^4$, section modulus $S = 196.3 \text{ in.}^3$, modulus of elasticity $E = 30 \times 10^6$ ksi, and plastic modulus $Z_p = 200 \text{ in.}^3$ By following elastoplastic analysis, determine the maximum vertical displacement of the beam that is produced by $F(t)$. Use an appropriate idealized one-degree spring-mass system. Assume that the beam is weightless, and neglect damping.

ELASTOPLASTIC ANALYSIS FOR EARTHQUAKE RESPONSE

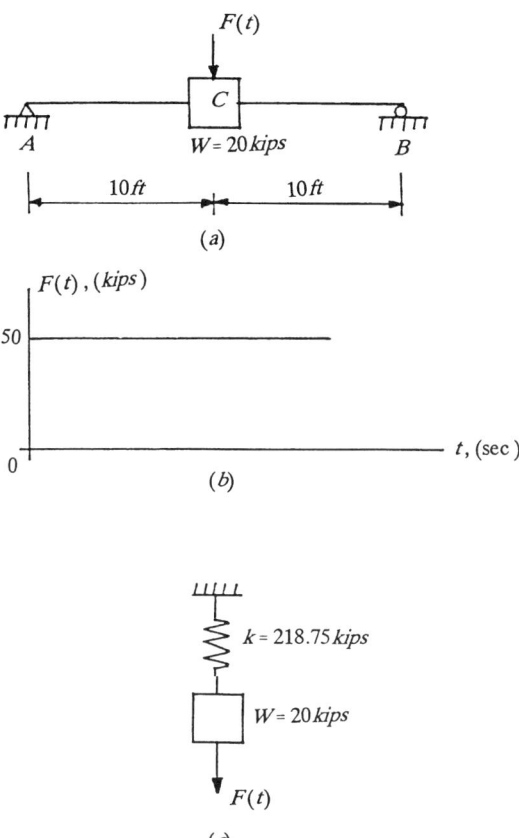

Figure 6.8. (a) Uniform simply supported beam supporting a weight W and a dynamic force $F(t)$. (b) Time variation of $F(t)$. (c) Idealized one-degree spring-mass system.

SOLUTION: Since the beam and loading conditions are symmetrical, the maximum vertical displacement will occur at its center C. For point C, the idealized one-degree spring-mass system is as shown in Fig. 6.8c. The spring constant k represents the stiffness of the beam at its center C, which is the required vertical force at C for a vertical deflection equal to unity. Thus,

$$y_c = \frac{PL^3}{48EI} = 1$$

or

$$P = k = \frac{48EI}{L^3} = \frac{(48)(30)(10)^3(2{,}100)}{(20)^3(12)^3}$$

$$= 218.75 \text{ kips/in.}$$

The plastic moment M_p, based on a yield-point stress of 30 ksi, is

$$M_p = (30)(200) = 6,000.0 \text{ kip-in.}$$

The maximum resistance R_m is equal to the load at the center C of the beam, which produces the ultimate moment M_p. For this beam case and loading, the plastic hinge develops at the center C. Thus,

$$M_p = \frac{R_m L}{4}$$

or

$$R_m = \frac{4M_p}{L} = \frac{(4)(6,000)}{(20)(12)}$$
$$= 100.0 \text{ kips}$$

Since the beam carries also the weight $W = 20$ kips, the maximum resistance that is available to resist the dynamic load is

$$R_m = 100.0 - 20.0 = 80.0 \text{ kips}$$

The resistance function R is assumed to have the form shown in Fig. 6.7b. The elastic deflection y_{el} of the idealized one-degree system in Fig. 6.8c, is

$$y_{el} = \frac{R_m}{k} = \frac{80.0}{218.75} = 0.3657 \text{ in.}$$

By using Eq. (6.108) and substituting appropriate values, we have

$$0.05176\ddot{y} + R_m - F(t) = 0 \tag{6.111}$$

or

$$\ddot{y} = 19.3199F(t) - 19.3199R_m \tag{6.112}$$

For the various intervals of the vertical deflection y, we have the following three equations:

$$\ddot{y} = 965.995 - 4{,}226.228y \qquad 0 \leq y \leq 0.3657 \tag{6.113}$$
$$\ddot{y} = -579.597 \qquad 0.3657 \leq y \leq y_m \tag{6.114}$$
$$\ddot{y} = -579.597 + 4{,}226.228(y_m - y) \qquad (y_m - 0.7314) \leq y \leq y_m \tag{6.115}$$

The numerical AIEM will be used here to carry out the analysis.

ELASTOPLASTIC ANALYSIS FOR EARTHQUAKE RESPONSE

The period of vibration τ is

$$\tau = 2\pi \sqrt{\frac{m}{k}} = 2\pi \sqrt{\frac{0.05176}{218.75}}$$

$$= 0.0967 \text{ sec}$$

A time interval $\Delta t = 0.01$ sec would be reasonable to use for the numerical method. Application of the AIEM yields the results shown in Table 6.2. The maximum deflection y_m occurs at time $t = 0.05$ sec, and it is equal to 0.4811 in. Note that Eq. (6.113) was used to determine $\ddot{y}$ up the $t = 0.04$ sec, Eq. (6.114) was used for $0.04 \leqslant t \leqslant 0.06$, and Eq. (6.115) was used for the remaining computations shown in the table. The computations, however, could go on indefinitely until other peaks of the displacement y are obtained.

The ductility ratio η in this case is

$$\eta = \frac{y_m}{y_{el}} = \frac{0.4811}{0.3657} = 1.316$$

A larger dynamic force $F(t)$ would produce larger η. A larger η would be also obtained if a lighter wide-flange section is used to support W and $F(t)$. The design engineer can decide at this point how far into the inelastic range the beam should be permitted to go. This would largely depend upon the type of $F(t)$, and upon how many times in the lifespan of the beam the dynamic force is applied to the beam.

TABLE 6.2. Variation of the Displacement y with Time

t (sec)	$\ddot{y}$ (in./sec^2)	$\ddot{y}(\Delta t)^2$ (in.)	y (in.)
0	965.995	0.0966	0
0.01	761.868	0.0762	0.0483
0.02	235.703	0.0236	0.1728
0.03	−390.202	−0.0390	0.3209
0.04	−579.597	−0.0580	0.4300
0.05	−579.597	−0.0580	0.4811
0.06	−550.436	−0.0550	0.4742
0.07	−288.833	−0.0289	0.4123
0.08	94.909	0.0095	0.3215
0.09	435.543	0.0436	0.2402
0.10	597.830	0.0598	0.2025
0.11	504.430	0.0504	0.2246
0.12			0.2971

The solution of this problem is based on elastoplastic response, and certain amount of permanent deformation y_p will take place. In this solution, the amount of y_p is

$$y_p = y_m - y_{el} = 0.4811 - 0.3657$$
$$= 0.1154 \text{ in.}$$

When the elastoplastic system attains its maximum deflection y_m in Fig. 6.7b and starts to rebound, the application of Eq. (6.115) shows that it will vibrate harmonically for an indefinite period of time since there is no damping to die out the motion. This residual vibration is elastic and it takes place about an equilibrium position y, which can be determined from the following equation:

$$y = y_m - \frac{R_m - F}{k} \tag{6.116}$$

that is,

$$y = 0.4811 - \frac{80.0 - 50.0}{218.75}$$
$$= 0.344 \text{ in.}$$

The peak amplitude Y of the residual vibration is

$$Y = \frac{R_m - F}{k} = \frac{80.0 - 50}{218.75}$$
$$= 0.1371 \text{ in.}$$

It may be also stated that the residual vibration takes place about a position y that is equal to $y_p + y_{st}$, where $y_{st} = F/k$ is the static equilibrium position about which the system would vibrate if its response were considered to be elastic at all times. Figure 6.9 illustrates the motion of the system.

6.5.2 Inelastic Earthquake Response of Multistory Structures

The preceding analysis and methodology will be extended here to apply for the dynamic inelastic response of multistory structures that are exposed to the effects of rather strong earthquakes. For such an inelastic response, it would be reasonable to neglect the flexibility of the girders by assuming that their stiffness is infinite when it is compared to the stiffness of the columns. For such a condition, yielding will usually take place in whichever story is, in comparison, the weakest in transmitting the magnitudes of the shear forces. Practical experience and research [36, 70, 72] shows that yielding in many cases occurs near the base of the structure. On this basis, the magnitudes of the shear forces

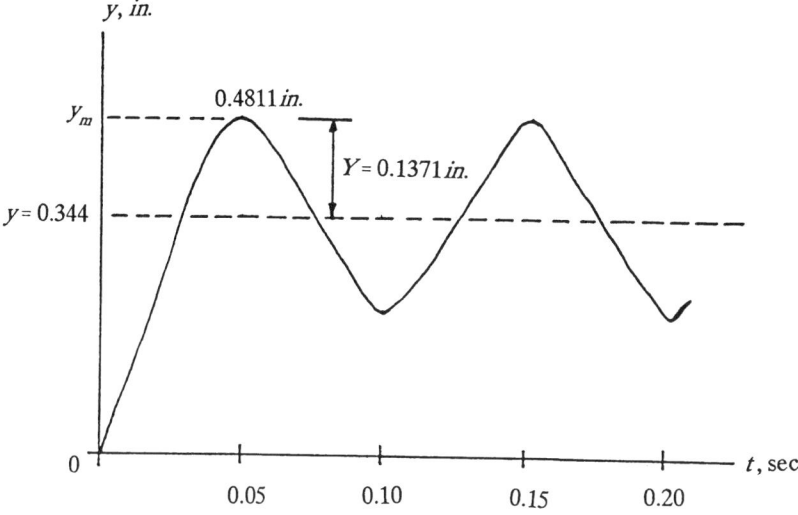

Figure 6.9. Variation of the displacement y with respect to time t.

in the upper part of the structure are reduced when they are compared with the values obtained by an elastic analysis for the same base earthquake motion. Thus, if the design of a structure is based on some fraction of the maximum value of the critical shear obtained by an elastic analysis, yielding will occur in the weakest story and the shear forces in the remaining parts of the structure will have appropriately revised values.

We start the development of the required equations by considering the two-story frame in Fig. 6.10a, and assuming that its girders are infinitely stiff in comparison with its columns. The idealized two-degree spring-mass system for the frame is shown in Fig. 6.10b, where u_s is the earthquake support motion, x_1 and x_2 are the first and second story horizontal displacements, respectively, and k_1 and k_2 are the spring constants. The resistance functions R_1 and R_2 are characterized by the total shear forces V_1 and V_2, respectively, and they are shown in Fig. 6.10c. On this basis we have

$$R_1 = k_1(x_1 - u_s) \qquad (6.117)$$
$$R_2 = k_2(x_2 - x_1) \qquad (6.118)$$

By considering the free-body diagrams of masses m_1 and m_2 in Fig. 6.10c and applying the second law of motion, the following two differential equations are obtained:

$$m_1\ddot{x}_1 + R_1 - R_2 = 0 \qquad (6.119)$$
$$m_2\ddot{x}_2 + R_2 = 0 \qquad (6.120)$$

426 EARTHQUAKE EXCITATIONS AND BLAST LOADINGS

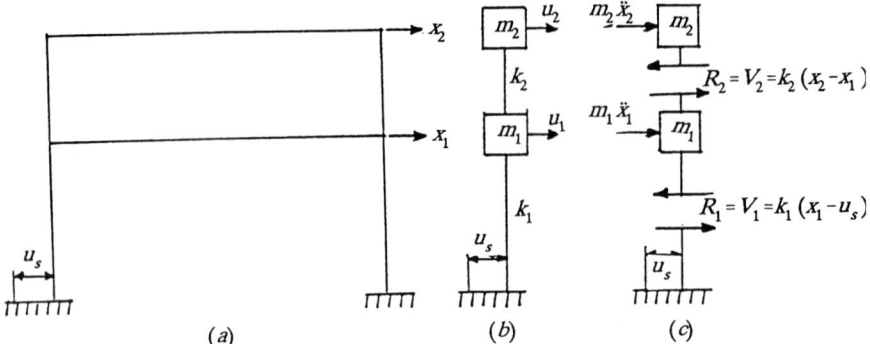

Figure 6.10. (*a*) Two-story frame subjected to an earthquake support motion u_s. (*b*) Idealized two-degree spring-mass system for the frame. (*c*) Resistance functions R_1 and R_2.

The resistance functions R_1 and R_2 in Eqs. (6.119) and (6.120) are given by Eqs. (6.117) and (6.118), respectively.

The dynamic elastoplastic analysis of the frame may be initiated by assuming that R_1 and R_2 are bilinear, and assuming maximum plastic story resistances R_{1_m} and R_{2_m} for R_1 and R_2, respectively. If an elastic analysis is performed, then the maximum plastic story resistances R_{1_m} and R_{2_m} may be taken as equal to one-half the maximum shears obtained from the elastic analysis. This is equivalent to selecting elastic limits for the story shears. With this in mind, the dynamic response of the system may be determined easily by using the AIEM to solve Eqs. (6.119) and (6.120). In the solution, R_1 and R_2 should be assumed to remain constant after the maximum values R_{1_m} and R_{2_m} are attained. The analysis presupposes that some plastic distortions of the system are permitted, and that their extent can be controlled by the ductility ratio η, as stated earlier.

If an elastic analysis is not carried out, a trial-and-error procedure may be used until a satisfactory design is obtained. The following example illustrates the application of the methodology.

Example 6.6: The two-story building in Fig. 6.5*a* is subjected to an earthquake motion with ground displacement versus time diagram as shown in Fig. 6.11. By considering the frame in Fig. 6.5*b*, perform an elastoplastic analysis to arrive at a suitable column size so that the ductility ratio η is somewhere between 2.5 and 3.

SOLUTION: The graph in Fig. 6.11 is, in reality, the ground displacement versus time graph in Fig. 1.27*c*, representing the Kern County, California, earthquake. In Fig. 6.11, however, the horizontal time scale is one tenth compared to the one shown in Fig. 1.27*c*. Table 6.3 shows the values of the ground displacements u_s, in inches, at time increments $\Delta t = 0.02$ sec.

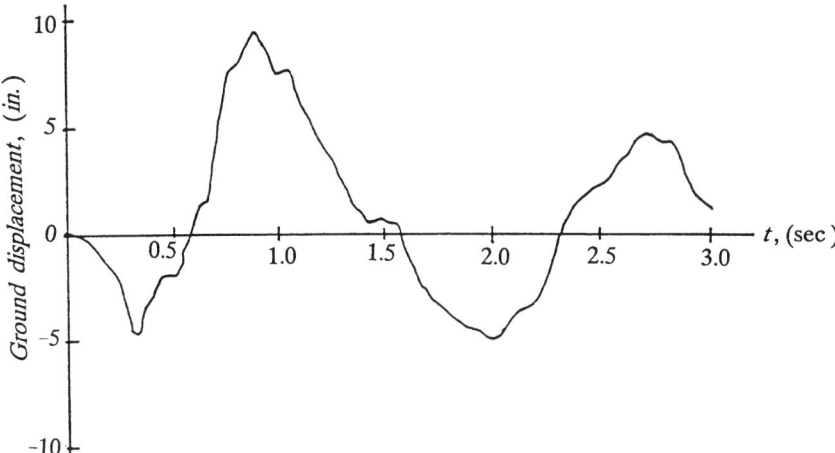

Figure 6.11. Earthquake ground displacement curve.

We start the analysis by assuming that all column sizes in Fig. 6.5b are W12 × 58 wide-flange sections. In addition, we assume that the girders are infinitely stiff when they are compared with the stiffness of the columns. On this basis, the idealized two-degree spring-mass system is the same as the one shown in Fig. 6.5c. From Example 6.4, we have the following values for m_1, m_2, k_1, and k_2:

$$m_1 = 0.22526 \frac{\text{kip-sec}^2}{\text{in.}}$$

$$m_2 = 0.14451 \frac{\text{kip-sec}^2}{\text{in.}}$$

$$k_1 = 34.005 \text{ kips/in.}$$

$$k_2 = 114.768 \text{ kips/in.}$$

For the W12 × 58 columns, we have moment of inertia $I_x = 476$ in.^{4}, section modulus $S_x = 78.1$ in.^{3}, and plastic section modulus $Z_p = 86.5$ in.^{3} We calculate the maximum elastic moment M_e in the columns by using the equation

$$M_e = S_x f_y = (78.1)(36) = 2{,}811.6 \text{ kip-in.}$$

where $f_y = 36$ ksi is the yield stress.

For a column that is fixed at both ends, but where one end can be displaced by an horizontal displacement x, we have the following well-known expression

428 EARTHQUAKE EXCITATIONS AND BLAST LOADINGS

TABLE 6.3. Values of Earthquake Ground Displacements Obtained from the Graph in Fig. 6.11

t	u_s	t	u_s	t	u_s	t	u_s	t	u_s
0.02	−0.02	0.56	−0.40	1.10	6.20	1.64	−1.90	2.18	−3.55
0.04	−0.05	0.58	+0.30	1.12	5.80	1.66	−2.40	2.20	−3.40
0.06	−0.15	0.60	0.80	1.14	5.60	1.68	−2.60	2.22	−3.00
0.08	−0.29	0.62	1.40	1.16	4.80	1.70	−2.80	2.24	−2.40
0.10	−0.41	0.64	1.45	1.18	4.60	1.72	−3.10	2.26	−2.00
0.12	−0.70	0.66	2.00	1.20	4.10	1.74	−3.30	2.28	−1.40
0.14	−0.90	0.68	3.40	1.22	3.70	1.76	−3.50	2.30	−0.80
0.16	−1.15	0.70	4.40	1.24	3.40	1.78	−3.60	2.32	0.00
0.18	−1.40	0.72	5.80	1.26	3.00	1.80	−3.80	2.34	+0.60
0.20	−1.70	0.74	6.80	1.28	2.60	1.82	−4.00	2.36	1.00
0.22	−2.00	0.76	7.60	1.30	2.20	1.84	−4.20	2.38	1.20
0.24	−2.50	0.78	7.85	1.32	1.90	1.86	−4.40	2.40	1.40
0.26	−3.20	0.80	7.90	1.34	1.60	1.88	−4.50	2.42	1.60
0.28	−4.00	0.82	8.20	1.36	1.20	1.90	−4.55	2.44	1.80
0.30	−4.20	0.84	8.60	1.38	0.80	1.92	−4.59	2.46	1.95
0.32	−4.70	0.86	9.20	1.40	0.50	1.94	−4.70	2.48	2.05
0.34	−4.00	0.88	9.45	1.42	0.40	1.96	−4.90	2.50	2.15
0.36	−3.30	0.90	9.20	1.44	0.50	1.98	−5.00	2.52	2.40
0.38	−3.00	0.92	9.00	1.46	0.65	2.00	−4.90	2.54	2.60
0.40	−2.60	0.94	8.40	1.48	0.65	2.02	−4.80	2.56	3.00
0.42	−2.20	0.96	7.90	1.50	0.60	2.04	−4.75	2.58	3.20
0.44	−1.90	0.98	7.50	1.52	0.50	2.06	−4.60	2.60	3.40
0.46	−1.88	1.00	7.40	1.54	0.49	2.08	−4.30	2.62	3.60
0.48	−2.00	1.02	7.50	1.56	0.35	2.10	−4.00	2.64	3.90
0.50	−1.80	1.04	7.60	1.58	−0.40	2.12	−3.80	2.66	4.35
0.52	−1.60	1.06	7.40	1.60	−1.00	2.14	−3.70	2.68	4.50
0.54	−1.00	1.08	6.60	1.62	−1.40	2.16	−3.65	2.70	4.60

relating M and x:

$$M = \frac{6EIx}{L^2} \tag{6.121}$$

where L is the length of the column. On this basis, by using Eq. (6.121), the maximum elastic displacement $x_{1_{el}}$ for the first-story columns is

$$x_{1_{el}} = \frac{(2,811.6)(18)^2(12)^2}{(6)(30)(10)^3(476)} = 1.531 \text{ in.}$$

For the second-story columns, we have

$$(x_2 - x_1)_{el} = \frac{(2,811.6)(12)^2(12)^2}{(6)(30)(10)^3(476)} = 0.6805 \text{ in.}$$

ELASTOPLASTIC ANALYSIS FOR EARTHQUAKE RESPONSE 429

The second thing to do here is to establish the maximum resistances R_{1_m} and R_{2_m} for the first and second stories, respectively, of the frame. The plastic moment M_p of the columns is

$$M_p = Z_p f_y = (86.5)(36) = 3{,}114 \text{ kip-in.}$$

Thus,

$$R_{1_m} = V_{1_m} = \frac{2M_p}{L} = \frac{(2)(3{,}114)}{(18)(12)}$$
$$= 28.83 \text{ kips}$$

$$R_{2_m} = V_{2\max} = \frac{2M_p}{L} = \frac{(2)(3{,}114)}{(12)(12)}$$
$$= 43.25 \text{ kips}$$

The preceding values of R_{1_m} and R_{2_m} are assigned here as the maximum plastic story resistances. For a more conservative design, lower values for R_{1_m} and R_{2_m} may be assigned. This can be also controlled by the ductility ratio η. The value of 3 for η is rather conservative.

The numerical AIEM is used here for the computation of the maximum story displacements x_{1_m} and x_{2_m}. We select $\Delta t = 0.02$ sec. At the first time station, the displacements $x_1^{(1)}$ and $x_2^{(1)}$ are as follows:

$$x_1^{(1)} = \tfrac{1}{6}[2\ddot{x}_1^{(0)} + \ddot{x}_1^{(1)}](\Delta t)^2 = \tfrac{1}{6}[\ddot{x}_1^{(1)}](\Delta t)^2 \qquad (6.122)$$

$$x_2^{(1)} = \tfrac{1}{6}[2\ddot{x}_2^{(0)} + \ddot{x}_2^{(1)}](\Delta t)^2 = \tfrac{1}{6}[\ddot{x}_2^{(1)}](\Delta t)^2 \qquad (6.123)$$

By using Eqs. (6.117) through (6.120), we find

$$\ddot{x}_1^{(1)} = \frac{R_2 - R_1}{m_1} = \frac{k_2(x_2^{(1)} - x_1^{(1)}) - k_1(x_1^{(1)} - u_s^{(1)})}{m_1} \qquad (6.124)$$

$$\ddot{x}_2^{(1)} = -\frac{R_2}{m_2} = -\frac{k_2}{m_2}(x_2^{(1)} - x_1^{(1)}) \qquad (6.125)$$

Thus,

$$x_1^{(1)} = \left[\frac{k_2}{6m_1}(x_2^{(1)} - x_1^{(1)}) - \frac{k_1}{6m_1}(x_1^{(1)} - u_s^{(1)})\right](\Delta t)^2 \qquad (6.126)$$

$$x_2^{(1)} = \frac{k_2}{6m_1}(x_2^{(1)} - x_1^{(1)})(\Delta t)^2 \qquad (6.127)$$

From Eq. (6.127), we obtain

$$x_2^{(1)} = \frac{(k_2/6m_2)(\Delta t)^2}{[1 + (k_2/6m_2)(\Delta t)^2]} x_1^{(1)} \qquad (6.128)$$
$$= C x_1^{(1)}$$

where

$$C = \frac{(k_2/6m_2)(\Delta t)^2}{[1 + (k_2/6m_2)(\Delta t)^2]} \qquad (6.129)$$

By substituting Eq. (6.128) into Eq. (6.126), we find

$$x_1^{(1)} = \frac{[k_1 u_s^{(1)}(\Delta t)^2]/6m_1}{1 + \{[k_2(\Delta t)^2]/Cm_1\}(1 + C) + \{[k_1(\Delta t)^2]/6m_1\}} \qquad (6.130)$$

where C is given by Eq. (6.129). Thus, Eqs. (6.130) and (6.128) may be used to determine the displacements $x_1^{(1)}$ and $x_2^{(1)}$, respectively, at the first time station.

The computer program in Appendix I was written in order to perform the AIEM. For given values of I_x, S_x, and Z_p, it performs the following:

1. It calculates k_1 and k_2 values from the equation

$$k = \frac{(2)(12) EI_x}{L^3}$$

2. It calculates ω_1, ω_2, τ_1, and τ_2 from the equations

$$\omega_{1,2}^2 = \frac{1}{2}\left(\frac{k_1 + k_2}{m_1} + \frac{k_2}{m_2}\right) \pm \frac{1}{2}\left[\left(\frac{k_1 + k_2}{m_1} + \frac{k_2}{m_2}\right)^2 - \frac{4k_1 k_2}{m_1 m_2}\right]^{1/2}$$

$$\tau_1 = \frac{2\pi}{\omega_1} \qquad \tau_2 = \frac{2\pi}{\omega_2}$$

3. It calculates maximum elastic deflection values for each story column from the equation

$$y = \frac{M_e L^2}{6EI_x} = \frac{S_x(36)L^2}{6EI_x}$$

4. It calculates maximum plastic shears for each story by using the equation

$$R_m = \frac{2M_p}{L} = \frac{2Z_p(36)}{L}$$

ELASTOPLASTIC ANALYSIS FOR EARTHQUAKE RESPONSE   431

For the given value of time increment Δt and the displacement $u_s^{(1)}$ at the first time station, the program performs the following calculations:

1. It calculates $x_1^{(1)}$ and $x_2^{(1)}$ using Eqs. (6.130) and (6.128).
2. It calculates $x_1^{(1)} - u_s^{(1)}$.
3. It calculates $R_1 = k_1(x_1^{(1)} - u_s^{(1)})$.
4. It calculates $x_2^{(1)} - x_1^{(1)}$.
5. It calculates $R_2 = k_2(x_2^{(1)} - x_1^{(1)})$.
6. It calculates $\ddot{x}_1 = (R_2 - R_1)/m_1$ and $\ddot{x}_1(\Delta t)^2$.
7. It calculates $\ddot{x}_2 = -R_2/m_2$ and $\ddot{x}_2(\Delta t)^2$.
8. It calculates

$$\eta_1 = \left|\frac{x_1}{y_{1el}}\right|$$

$$\eta_2 = \left|\frac{x_2 - x_1}{y_{2el}}\right|$$

9. By using the equation

$$x^{(i+1)} = 2x^{(i)} - x^{(i-1)} + \ddot{x}^{(i)}(\Delta t)^2$$

it calculates $x_1^{(2)}$ and $x_2^{(2)}$ at the second time station.

Then, by imputing the ground displacement $u_s^{(2)}$, it performs the calculations in steps 2 through 9 again, and so on. Note that whenever $|R_1| > R_{1_m}$ and/or $|R_2| > R_{2_m}$, the program sets $R_1 = R_{1_m}$ if $R_1 > 0$ and/or $R_2 = R_{2_m}$ if $R_2 > 0$, $R_1 = -R_{1_m}$ if $R_1 < 0$ and $R_2 = -R_{2_m}$ if $R_2 < 0$.

The first design, which incorporates W12 × 58 wide flange section for the columns, yields $x_{1_m} = -5.9501$ in. at time $t = 0.68$ sec, and

$$\eta_1 = \left|\frac{x_{1_m}}{x_{1el}}\right| = 3.89 \tag{6.131}$$

For the second story we have $(x_2 - x_1)_m = -0.3145$ in. at time $t = 0.46$ sec, and

$$\eta_2 = \left|\frac{(x_2 - x_1)_m}{x_{2el}}\right| = 0.46 \tag{6.132}$$

The preceding results show that for the first peak of maximum response, the first story undergoes plastic response and deformation for the indicated column sizes, and the response of the second-story columns is elastic. Since the structure did not yet experience the larger positive peak of the ground displacement, which comes at $t = 0.88$ sec, the value of η may go even higher. Since in this problem we require $2.5 \leqslant \eta \leqslant 3.0$, the procedure will be repeated by using a different column size.

We use for this design trial a W10 × 25 wide-flange section. In this case, the AIEM yields $x_1 = 4.9793$ in. occurring at time $t = 2.00$ sec, and

$$\eta_1 = \left|\frac{x_{1m}}{x_{1el}}\right| = 2.68 \tag{6.133}$$

For the second story, we have $(x_2 - x_1)_m = 0.2777$ in. at time $t = 1.52$ sec, and

$$\eta_2 = \left|\frac{(x_2 - x_1)_m}{x_{2el}}\right| = 0.34 \tag{6.134}$$

The results of this design trial indicate that the first story undergoes plastic response and deformation for its columns, and that the maximum deflections occur well beyond the large positive peak of the ground earthquake displacement. Since the maximum value of the ductility ratio η is between 2.5 and 3.0, as desired, the W10 × 25 column sizes for the frame are acceptable. Note that the lighter columns produced a more satisfactory design for the given earthquake ground motion, indicating that other structural characteristics, such as the structure's periods of vibration, play a key role in its ability to resist the effects of an earthquake.

If a computer and/or a computer program is not available, the AIEM may be carried out manually, as was done in earlier sections of this text. For buildings having three or more stories, utilization of computer software would be advisable.

6.6 INTRODUCTORY ASPECTS OF NUCLEAR AND CONVENTIONAL EXPLOSIONS

6.6.1 Nuclear Explosions

The principal categories of damage resulting from a nuclear explosion, that is, a nuclear weapon, are three: radioactivity, thermal and blast effects, and electromagnetic effects. Although damages in all three categories can occur simultaneously, radioactivity is considered to be the dominant concern at short distances from the center of the explosion, while electromagnetic effects pose the largest risk at great distances. The thermal and blast effects are considered to be the most important for structures that are subjected to the explosion at intermediate ranges.

Radioactivity produced from the explosion of a nuclear weapon consists of gamma rays, neutrons, and beta and alpha particles emitted in the nuclear reaction. There are also secondary nuclear processes caused by energetic particles of the reaction or by the fission product decay of radioactive materials. These secondary nuclear processes cause radiation but take place long after the original nuclear explosion.

For convenience, nuclear radiations are classified as initial or residual. Initial nuclear radiations are the ones that occur within the first minute of a nuclear explosion. Residual radiations occur after the first minute of the explosion and are the result of the decay of fission products or radioactive weapon materials. In fission weapons, residual radiations are also caused by the reaction of high speed neutrons with weapon materials or with the surrounding atmosphere.

Residual radiations can be further divided into early fallout, which reaches the ground within 24 hours, and delayed fallout, which arrives in small quantities over several weeks or months. Because of radioactive decay, the radiation levels of delayed fallout are too low to present an immediate danger to health. Early fallout can produce hazardous radiations, consisting mainly of relatively large particles, but the radioactivity decreases rapidly with time after the explosion. In a nuclear warfare environment, residual radiations are assumed to represent only a minor threat, since they can be washed away by ships equipped with washdown systems.

Initial nuclear radiations, however, present an immediate danger to people who are within a relatively short distance of a nuclear weapon explosion. The alpha and beta particles have short ranges, and from the practical point of view are not considered to be harmful. However, the gamma rays and neutrons can travel considerable distances through air, and both can have harmful effects on living organisms. They constitute about 3 percent of the total explosive energy or weapon yield.

Thermal radiation and air blast from nuclear explosions pose a threat to people over larger areas, or ranges, than the ones affected by initial radiation. Thermal radiation results from very high temperatures, tens of millions of degrees, caused by the rapid release of energy in the nuclear reaction. This high temperature at the site of the reaction causes electromagnetic radiation of short wavelength (X-rays) to be emitted. This electromagnetic radiation in turn heats the air surrounding the blast site to high temperatures, several thousand degrees. The radiation is absorbed readily, within a few feet, by the surrounding air, thus creating a fireball, which in turn emits relatively long wavelength radiation in the ultraviolet, visible, and near infrared spectral region, which is not readily absorbed by the atmosphere. The effects of this radiation begin immediately after the formation of the fireball, because it travels at the speed of light (3×10^8 m/sec).

Even relatively small amounts of thermal radiation can be very damaging. For example, 5 cal/cm^2 from a short thermal pulse (low weapon yield) can cause disabling skin burns on exposed personnel. Clothing material may ignite when exposed to short pulses yielding 10–12 cal/cm^2 (dark colors), and rubber or plastic material may burn or melt in the 10–20 cal/cm^2 range. Critically important elements of radar antennas may be damaged at relatively low thermal irradiance levels, and higher radiant exposure levels can weaken, or even melt, aluminum structures.

The thermal radiation pulse travels ahead of the blast wave and heats any exposed structure or system in its path. The resulting temperature rise can cause serious problems in temperture-sensitive materials. In aluminum material, for example, the temperature rise produces a drop in mechanical strength, which can cause failure of load-carrying structural elements even before the blast wave hits, or it can create a weakness that will cause the element to fail at blast arrival. In addition, thermal stresses may develop, stresses that can also be large enough to cause failure before the arrival of the blast wave. However, such thermal stresses will develop only in structures that have relatively large temperature gradients from the front to rear surfaces; thus, it takes a thick structure to develop a large enough thermal gradient to cause a large thermal stress. Thermal stresses differ in this way from blast stresses, which develop the largest magnitudes when the elements are thin.

In a nuclear explosion, the sudden release of energy initiates a pressure wave, known as shock, or blast, wave, in the surrounding medium. This shock, or blast, wave results in a sudden increase of pressure at the front, followed by a gradual decrease, as shown in Fig. 6.12a. The destructive action of a nuclear weapon is much more severe than that of a conventional weapon, and it is due mainly to this blast or shock wave. In a typical air burst at an altitude below 100,000 ft, an approximate distribution of energy would consist of 50 percent blast and shock, 35 percent thermal radiation, 10 percent residual nuclear radiation, and 5 percent initial nuclear radiation. For more information on this subject, the reader may consult References [37, 39, 40, 73].

The phenomena associated with nuclear explosions vary with the location of the point of burst in relation to the surface of the earth. Bursts are usually classified as air bursts, high-altitude bursts, underwater bursts, underground bursts, and surface bursts. In the case of air and surface bursts, when an explosion takes place the expansion of hot gases produces a pressure wave in the surrounding air. As this wave moves away from the center of the explosion, the inner part moves through the region that was previously compressed, and is now heated by the leading part of the wave. As the pressure wave moves with the velocity of sound, the temperature and pressure of the air cause this velocity to increase. The inner part of the wave starts to move faster, and gradually overtakes the leading part of the wave. After a short period of time, the pressure wave front becomes abrupt, thus forming a shock front somewhat similar to the one illustrated in Fig. 6.12b. The maximum overpressure occurs at the shock front and is called the peak overpressure. Behind the shock front, the overpressure drops very rapidly to about one-half the peak overpressure, and remains almost uniform in the central region of the explosion.

As the expansion proceeds, the overpressure in the shock front decreases steadily, and the pressure behind the front falls off in a regular manner. After a short time, at a certain distance from the center of the explosion, the pressure

ASPECTS OF NUCLEAR AND CONVENTIONAL EXPLOSIONS   435

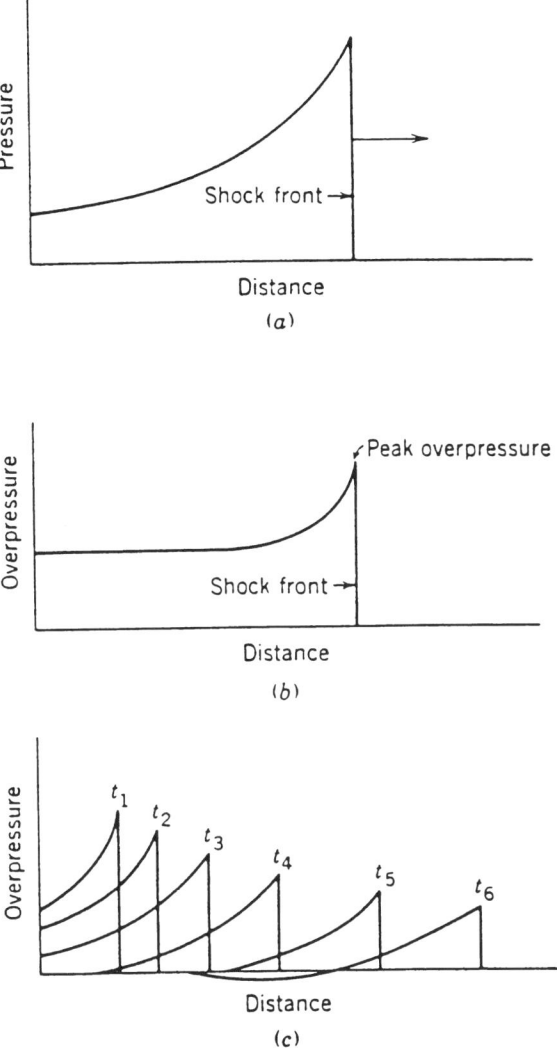

Figure 6.12. (*a*) Variation of pressure with distance in a shock wave. (*b*) Formation of the shock front. (*c*) Variation of overpressure with distance from the center of explosion at the indicated times.

behind the shock front becomes smaller than that of the surrounding atmosphere, and the so-called negative phase, or suction, develops. The front of the blast weakens as it progresses outward, and its velocity drops toward the velocity of sound in the undisturbed atmosphere. The sequence of events is illustrated in Fig. 6.12c, showing the overpressures at times t_1 through t_6. In the curves marked t_1 through t_5, the pressure in the blast wave does not fall below that of the atmosphere. In the curve marked t_6, at some distance behind the shock front, the overpressure becomes negative. For greater clarity this last curve is shown again, by itself, in Fig. 6.13a. In this figure, U_o is the velocity of the shock front, P_{so} is the peak overpressure, P_s is the overpressure behind the shock front, and the arrows under the curve indicate the direction of the air mass movement. The same blast wave is also shown in Fig. 6.13b, where t_o denotes the time duration of the positive phase and the time at the end of the positive phase.

An additional quantity of equivalent importance is the pressure caused by the strong winds accompanying the blast wave, known as the dynamic

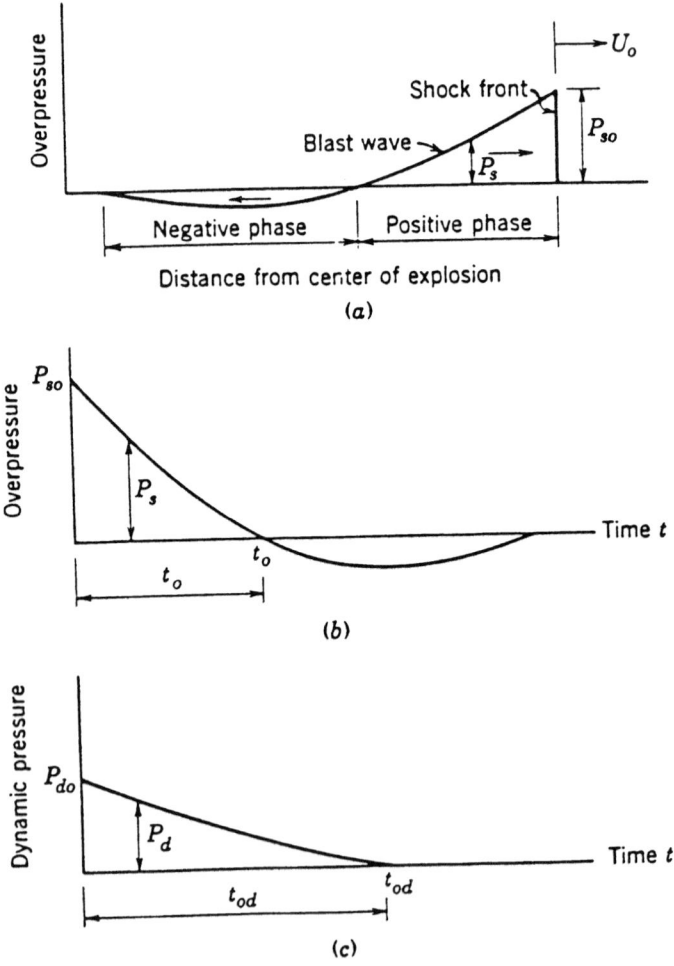

Figure 6.13. (a) Blast wave showing its positive and negative phases. (b) Blast wave. (c) Dynamic pressure variation.

pressure. At a given distance from the center of the explosion, its variation is shown in Fig. 6.13c, where P_{do} is the dynamic peak pressure, P_d is the dynamic pressure at any time t, and t_{od} is as shown in the figure. The peak dynamic pressure decreases with increasing distance from the center of the explosion, but its rate of decrease is different than the one shown for the peak overpressure. In Fig. 6.13c, we note that the dynamic pressure P_d behind the shock front is somewhat similar in shape to that of the overpressure P_s, but with different rate of decrease. For practical design purposes, the negative phase of the overpressure in Fig. 6.13a, or Fig. 6.13b, is of no interest to the practicing engineer, and it is usually neglected.

In Fig. 6.13b, the variation of P_s in the positive phase only, for peak overpressures of about 10 psi or less, is given by the following equation:

$$P_s = P_{so}\left(1 - \frac{t}{t_o}\right)e^{-t/t_o} \qquad (6.135)$$

where e is the base of natural system of logarithms and is equal to 2.7182. Equation (6.135) shows that the rate of the decay of the overpressure P_s is a function of the peak overpressure P_{so}.

The corresponding expression for the variation of the dynamic pressure P_d in Fig. 6.13c is

$$P_d = P_{do}\left(1 - \frac{t}{t_{od}}\right)e^{-2t/t_{od}} \qquad (6.136)$$

where t_{od} is the duration of the dynamic pressure.

The velocity U_o of the shock front is a function of the peak overpressure, and it can be obtained from the equation

$$U_o = U_s\left(1 + \frac{6P_{so}}{7P_o}\right)^{1/2} \qquad (6.137)$$

where P_o is the atmospheric pressure, and U_s is the velocity of sound. At sea level and normal atmospheric conditions, we have

$$U_o = 1{,}117\left[1 + \frac{6P_{so}}{(7)(14.7)}\right]^{1/2} \qquad (6.138)$$

A great deal of information regarding nuclear explosions may be found in References [37, 39, 40, 73, 74].

6.6.2 Conventional Explosions

The similarity between nuclear and the more conventional types of explosions lies in the fact that their destructive action is due mainly to blast or shock. In the case of conventional explosions, the energy comes from chemical reactions that involve a rearrangement among the atoms, while in a nuclear explosion, the energy arises from the formation of different atomic nuclei. Other important differences are that nuclear explosions can be thousands and millions of times more powerful than the largest conventional detonations. Also, a fairly large proportion of the energy in a nuclear explosion is thermal radiation, which is also accompanied by initial and residual nuclear radiations, as discussed earlier in this section.

Conventional explosions are of great concern to the practicing design engineer, because methods of design are needed for protective construction of facilities to be used for the manufacture, maintenance, modification, inspection, and storage of explosive materials. The design engineer must be familiar with design procedures and construction techniques that can prevent propagation of explosions from one building, or part of a building, to another, that can avoid mass detonations, and that can provide protection for personnel and valuable equipment.

These design procedures must also: (a) establish blast load parameters, which are required for the design of protective structures; (b) provide methods for calculating the dynamic response of reinforced concrete and other materials; (c) establish construction details necessary to afford the required strength to resist the applied blast loads; and (d) establish guidelines for siting explosive facilities in order to obtain maximum cost effectiveness in both site planning and structural arrangement, to provide closures, and to prevent damage to interior portions of structures due to structural motion and shock.

A great deal of information and convenient methodologies regarding the design of structures to resist the effects of accidental explosions may be found in Reference [75]. Other manuals [40, 73] are also available that can be used for the design of structures protected against the effects of nuclear detonations, as discussed earlier in this section. These procedures are also applicable to the design of protective structures at considerable distance from very large quantities of explosives. The primary concern in the manual of Reference [75] is the design of structural elements, or structures, located close to a potential high-explosive hazard.

In order to provide the required level of protection against the hazards of accidental explosions, we need to know the donor system that produces the damaging output, the receiver system that requires a level of protection, and the protective structure, or structural elements, that need to be protected from the hazardous effects. The donor system includes the type and amount of the donor explosive and its location relative to the components of the protective facility. The output of the donor explosive includes blast overpressures, and also primary fragments for cased explosives. It may also include ground shock, fire, heat, dust, and so on, but for quantities of explosives up to 25,000 lb, blast pressures are usually the primary parameter governing the design of protective structures. There are situations, however, where primary fragments from cased explosives may assume equal importance in the planning of the protective system. Each situation has to be examined individually in order to determine what effects, other than blast overpressures, need to be considered in the design.

The chemical and physical properties of the donor explosive determine the magnitude of the blast pressure, while the distribution of the pressure patterns is mainly a function of the distance of the donor explosive relative to the components of the protective facility. The explosive properties of the explosive determine the limitation of the detonation process, which may result in either high- or low-order detonation. For protection against primary fragments, sufficient structure thickness must be provided to prevent full penetration.

Another component of the problem is provision of protection for the receiver, which in this case would be personnel, equipment, or explosives that require protection. Full protection is usually required for personnel and equipment, but the degree of protection for explosives ranges from full protection to allowable partial or total collapse of the protective structure.

Protective structures may consist of shelters and barriers. Shelters are fully enclosed structures used to protect personnel from injury, to prevent damage to valuable equipment, and also to prevent detonation of sensitive explosives. Barriers are structures used as shields between two or more potentially detonating explosives. Entrance to shelters must be sealed with blast doors and may also require blast locks.

Design pressure ranges may be classified as high, intermediate, and low. The high-pressure design range is associated with extremely high initial pressures that are further amplified by their reflection from the structure, and with time durations of applied loads that are short compared to the time it takes for the structure to reach maximum deflection. In such cases the applied load may be considered as an impulse, if preferred, which includes the area under the pressure-time curve. The intermediate-pressure design range is associated with smaller intensity pressures, and the response of structures designed for this range involves the combined effects of both the pressure and impulse associated with the blast output. In the low-pressure design range, the duration of the blast loads acting on protective structures is very long in comparison to the other two ranges. In this case, the structure responds primarily to the peak pressure in a manner similar to structures that are designed to resist the effects of nuclear detonations.

6.7 DYNAMIC LOADING ON CLOSED RECTANGULAR STRUCTURES SUBJECTED TO NUCLEAR BLAST WAVES

We consider here the closed rectangular structure shown in Fig. 6.14a, which is subjected to the blast waves of a nuclear explosion. We assume that the front face of the rectangular structure is normal to the direction of travel of the shock front; in this case the angle of incidence $\alpha = 0$.

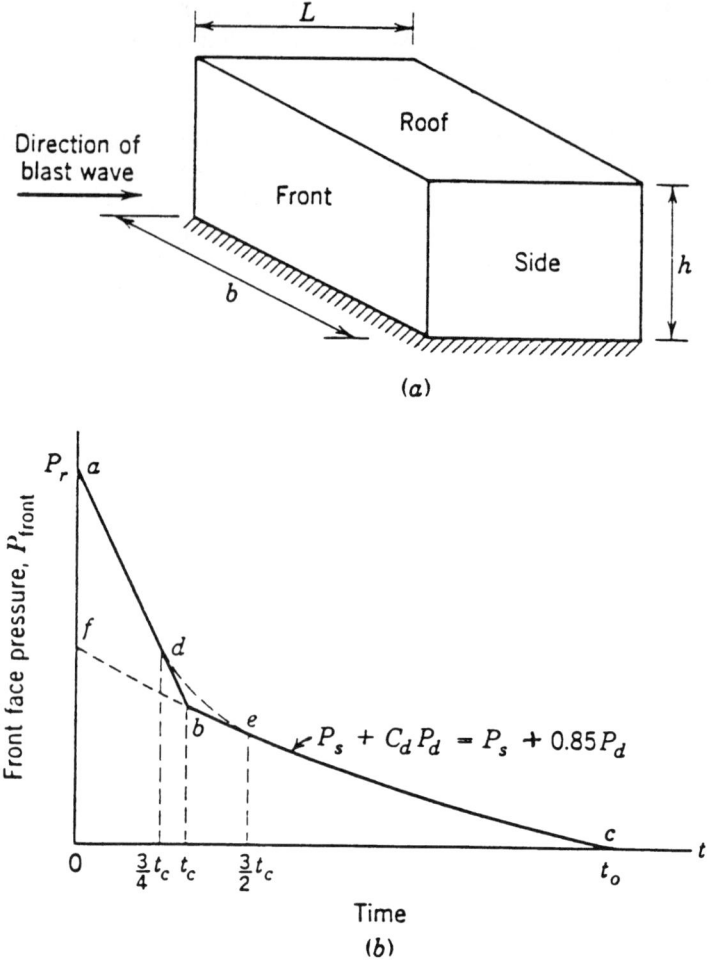

Figure 6.14. (a) Closed rectangular structure. (b) Average front-face pressure versus time [37, 73].

6.7.1 Average Loading on Front Face

The average loading P_{front} on the front face of the rectangular structure consists of the reflected pressure P_r up to the clearing time t_c, Fig. 6.14b, and the summation

$$P_c = P_s + C_d P_d \tag{6.139}$$

from time t_c to time t_o. In Eq. (6.139), P_s is the overpressure of the blast wave, and the product $C_d P_d$ is the drag pressure, where P_d is the dynamic pressure and C_d is the drag coefficient.

DYNAMIC LOADING ON CLOSED RECTANGULAR STRUCTURES 441

The reflected overpressure P_r can be determined from the equation

$$P_r = 2P_{so}\left(\frac{7P_o + 4P_{so}}{7P_o + P_{so}}\right) \quad (6.140)$$

where P_o is the atmospheric pressure and P_{so} is the peak overpressure. At sea level, we have

$$P_r = 2P_{so}\left(\frac{103 + 4P_{so}}{103 + P_{so}}\right) \quad (6.141)$$

The value of P_r approaches $8P_{so}$ for strong shocks and moves toward $2P_{so}$ for weak shocks.

For solid flat surfaces, such as walls of a building, the clearing time t_c may be determined from the equation

$$t_c = \frac{3S}{U_r} \quad (6.142)$$

where S is either the height of the reflecting surface or one-half its width, whichever is smaller, and U_r is the velocity of sound in the region of the reflected overpressure. For surface bursts, such as the one under consideration, U_r can be taken equal to U_o, and it can be determined from Eq. (6.137) or Eq. (6.138). The velocity U_r can also be determined from the following equation:

$$U_r = 422\left(\frac{1.0088P_{so}^2 + 70P_{so} + 720}{102.9 + 6P_{so}}\right)^{1/2} \quad (6.143)$$

The decay P_s of the overpressure and the decay P_d of the dynamic pressure can be determined from Eqs. (6.135) and (6.136), respectively, when the peak overpressure does not exceed 10 psi. For large values of P_{so}, the graphs in Reference [39, Chapter 11] may be used. The drag coefficient C_d in Eq. (6.139) is usually taken as equal to 0.85 by many practicing engineers.

On this basis, the average pressure on the front face of the rectangular structure would have the distribution shown by the curve *abc* in Fig. 6.14b. The incompatible discontinuity at point *b* may be smoothed out by the fairing curve *de*, as shown in the same figure.

6.7.2 Average Loading P_{back} on Back Face

The shock front arrives at the back face of the rectangular structure at a time $t_d = L/U_o$, where L is the length of the structure, and the velocity U_o of the shock front can be determined from Eq. (6.138). When the shock front reaches the back wall, it requires an additional time t_b for the average pressure to build up to its maximum value $(P_{back})_{max}$. The time t_b may be determined from the equation

$$t_d = \frac{4S}{U_o} \quad (6.144)$$

where S is either the full height h of the back wall or one-half its width $(b/2)$, whichever is smaller.

After the buildup is completed, the maximum value $(P_{back})_{max}$ can be determined from the following expression:

$$(P_{back})_{max} = \frac{P_{sb}}{2}[1 + (1-\beta)e^{-\beta}] \quad (6.145)$$

In Eq. (6.145), P_{sb} is the blast wave overpressure at time $t = t_d + t_b$, $\beta = 0.5P_{so}/14.7$, and e is the base of natural logarithms. The peak value $(P_{back})_{max}$ of the average pressure occurs at time $t = t_d + t_b$.

The variation of P_{back}, for times larger than $t = t_d + t_b$, is given by the following equation:

$$\frac{P_{back}}{P_s} = \frac{(P_{back})_{max}}{P_{sb}} + \left[1 - \frac{(P_{back})_{max}}{P_{sb}}\right]\left[\frac{t - (t_d + t_b)}{t_o - t_b}\right]^2 \quad (6.146)$$

where t_o is the time duration of the positive phase of the overpressure P_s. Figure 6.15 illustrates the variation with time of the average back-face pressure. Note that between times t_d and $t_d + t_b$ the variation is assumed to be linear. It is also assumed that the peak overpressure P_{so} does not reduce in strength in

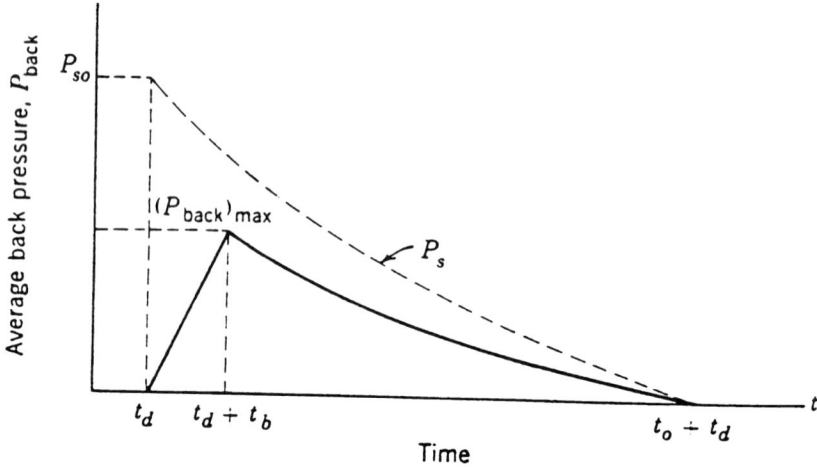

Figure 6.15. Time variation of the average back-face pressure [37, 73].

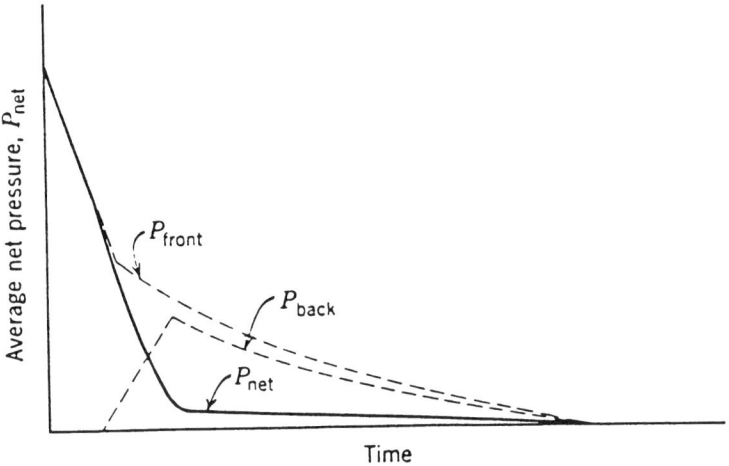

Figure 6.16. Time variation of the average net horizontal pressure [37, 73].

the time interval t_d, which is the time where the blast wave passes over the rectangular structure.

The required overpressure P_{sb} in Eq. (6.145) may be determined from Eq. (6.135) by making $t = t_b$, provided that the peak overpressure P_{so} is 10 psi or less. For larger values of P_{so}, consult References [37, 39]. The values of P_{back}/P_s in Eq. (6.146) can be determined for a sequence of times in excess of $t = t_d + t_b$. For the corresponding times, P_s is computed from Eq. (6.135).

The average net horizontal loading P_{net} acting on a rectangular structure is equal to the front-face loading P_{front} minus the back-face loading P_{back}. Thus,

$$P_{net} = P_{front} - P_{back} \qquad (6.147)$$

The algebraic sum of these two loading is illustrated in Fig. 6.16.

6.7.3 Average Pressure on Roof and Sides

On the roof and sides, the variation of the average pressure for the closed rectangular structure is approximated as shown in Fig. 6.17. The sides and roof are not fully loaded until the blast wave has traveled the length L of the building, that is at time $t = L/U_o$. At this time, the average pressure reaches its maximum value P_m, and is equal to the algebraic sum of the overpressure P_s and the drag pressure $C_d P_d$ at the distance $L/2$ from the front face of the structure. On this basis, at $t = L/U_o$, we have

$$P_m = P_s\left(\frac{L}{2U_o}\right) + C_d P_d\left(\frac{L}{2U_o}\right) \qquad (6.148)$$

where U_o can be obtained from Eq. (6.137) or Eq. (6.138).

444 EARTHQUAKE EXCITATIONS AND BLAST LOADINGS

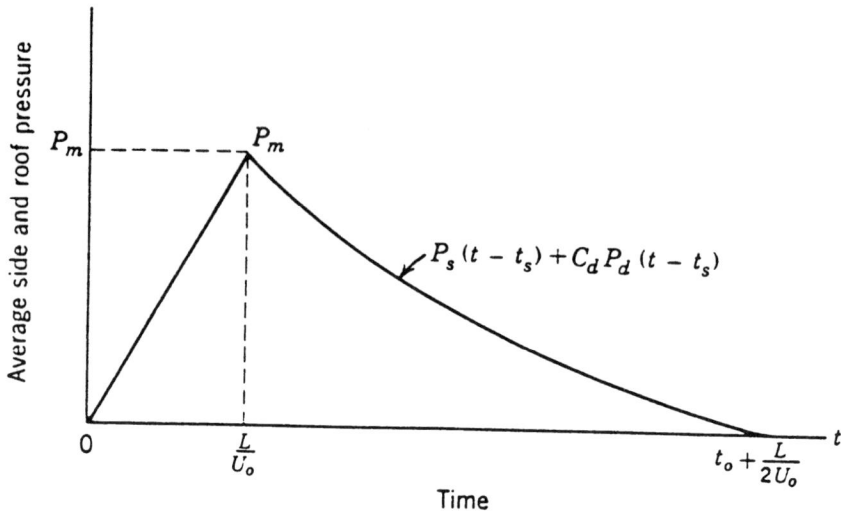

Figure 6.17. Time variation of the average roof and side pressure [37, 73].

At times $L/U_o \leq t \leq t_o + L/2U_o$, the average pressure P_a may be determined from the following expression:

$$P_a = P_s(t - t_s) + C_d P_d(t - t_s) \qquad (6.149)$$

where $t_s = L/2U_o$, P_s is the overpressure at time $t - t_s$, and P_d is the dynamic pressure at time $t - t_s$. For roof and sides, the drag coefficient $C_d = -0.4$ when the peak dynamic pressure is 25 psi or less. It decreases to -0.3 for peak dynamic pressures between 25 and 50 psi, and is equal to -0.2 for peak dynamic pressures between 50 and 130 psi.

A rectangular structure may be assumed to be a closed one, if the front and back faces have opening or windows of area less than 30 percent. If the open area is more than 30 percent, some corrections to the methodology should be applied. See Reference [39].

The following example illustrates the preceding methodology.

Example 6.7: The building in Fig. 6.18 is located at a distance of 3,000 ft from the center of an 18-kiloton (18-KT) typical surface burst. Determine the net horizontal pressure P_{net} that acts on the rectangular aboveground building. Also, determine the average side and roof pressures. The length of the building parallel to the shock front is 200 ft.

SOLUTION: From Fig. 6.19, for an 18-KT surface burst, we find that the peak overpressure $P_{so} = 10$ psi. From Eq. (6.138), the shock front

DYNAMIC LOADING ON CLOSED RECTANGULAR STRUCTURES   445

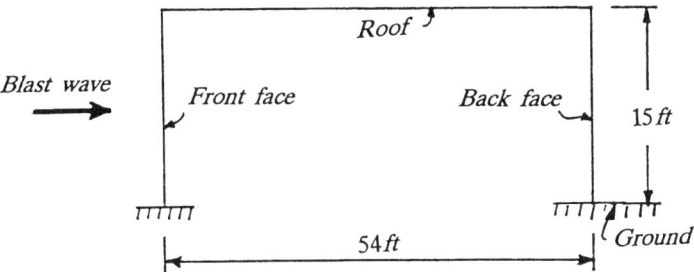

Figure 6.18. Aboveground rectangular building subjected to an 18-KT surface burst.

velocity U_o is

$$U_o = 1{,}117\left[1 + \frac{(6)(10)}{(7)(14.7)}\right]^{1/2} = 1{,}405\,\text{fps}$$

From Fig. 6.20, the positive phase duration t_{o1}, of the overpressure P_s for 1-KT surface burst is 0.26 sec. For a burst size $Z = 18$ KT, the positive phase

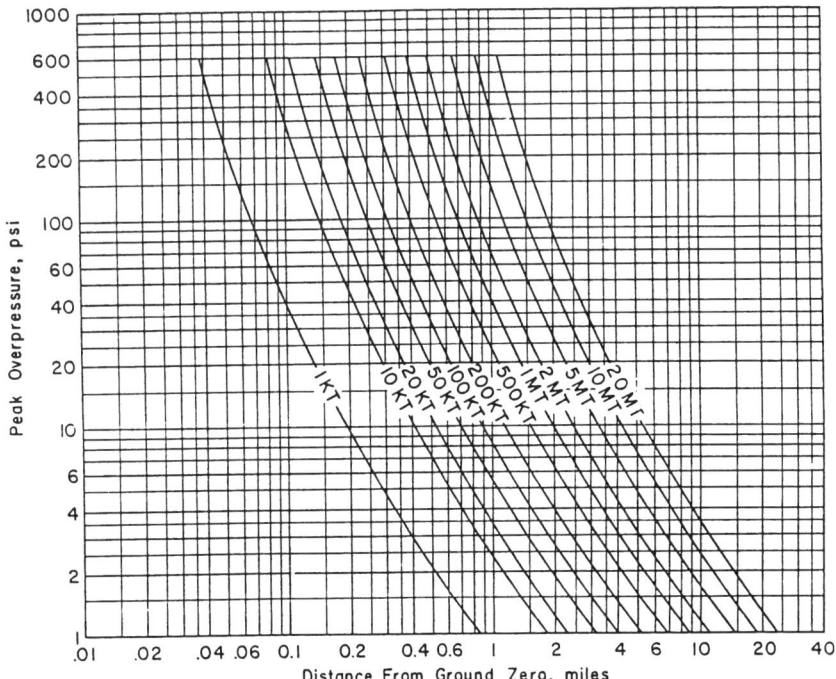

Figure 6.19. Peak overpressures for surface bursts [40].

446 EARTHQUAKE EXCITATIONS AND BLAST LOADINGS

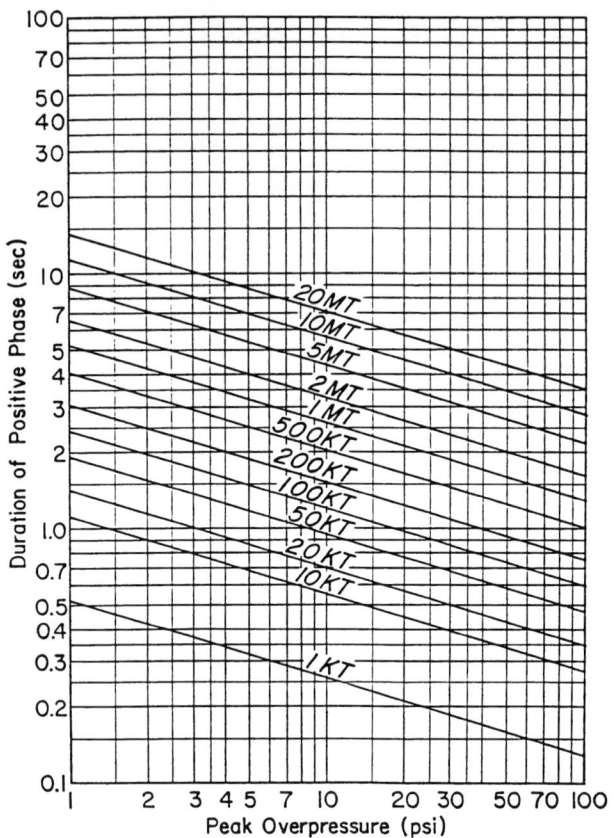

Figure 6.20. Duration of positive phase for surface bursts [40].

duration t_{o2} may be determined from the following expression:

$$t_{o2} = t_{o1} \sqrt[3]{Z} \qquad (6.150)$$

Thus,

$$t_{o2} = (0.26) \sqrt[3]{18} = 0.68 \text{ sec}$$

From Fig. 6.21 and $P_{so} = 10$ psi, the range d_1, for a 1-KT burst, is 1,000 ft, and from the same figure, for the same range, the peak dynamic pressure $P_{do} = 2.23$ psi. For $Z = 18$ KT, the range d can be determined from the equation

$$d = d_1 \sqrt[3]{Z} \qquad (6.151)$$

Thus,

$$d = 1,000 \sqrt[3]{18} = 2,621 \text{ ft}$$

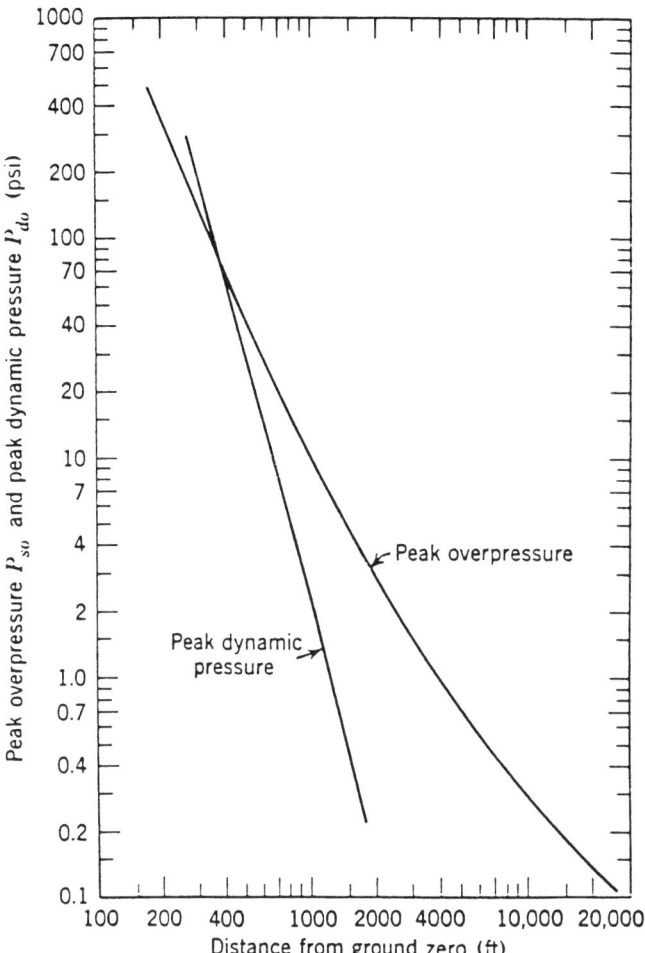

Figure 6.21. Peak overpressure and peak dynamic pressure for 1-KT surface burst [73].

We use Fig. 6.22 to find the positive phase duration t_{od} of the dynamic pressure P_d for the 1-KT burst. Thus, for $d_1 = 1,000$ ft, Fig. 6.22 yields $t_{od_1} = 0.347$ sec. Thus, by using Eq. (6.150), we find that the positive phase duration t_{od}, for $Z = 18$ KT, is

$$t_{od} = (0.347)\sqrt[3]{18} = 0.909 \text{ sec}$$

The peak reflected overpressure P_r may be determined from Eq. (6.140), and it is as follows:

$$P_r = (2)(10)\left[\frac{(7)(14.7) + (4)(10)}{(7)(14.7) + 10}\right] = 25.31 \text{ psi}$$

448   EARTHQUAKE EXCITATIONS AND BLAST LOADINGS

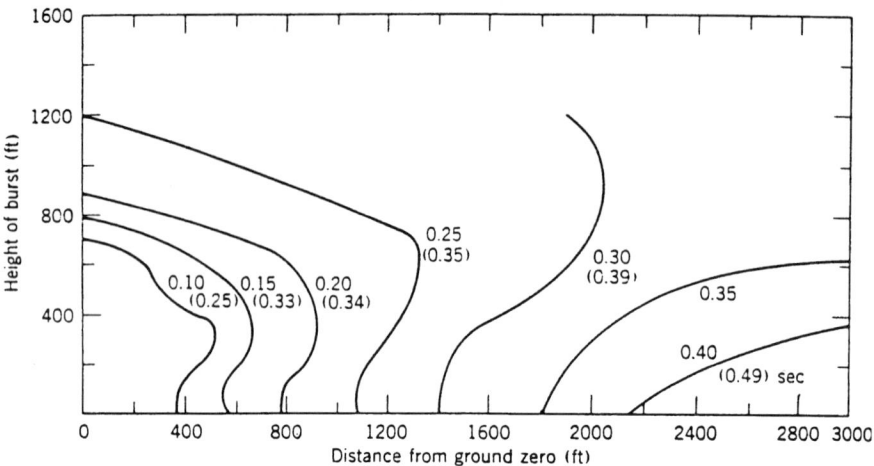

Figure 6.22. Positive phase duration of overpressure and dynamic pressure (in parentheses) for 1-KT burst [73].

The velocity U_r in the region of the reflected overpressure may be determined from Eq. (6.143). Thus,

$$U_r = 422 \left[\frac{(1.0088)(10)^2 + (70)(10) + 720}{102.9 + (6)(10)} \right]^{1/2}$$

$$= 1{,}289 \text{ fps}$$

From Eq. (6.142), the clearing time t_C of the reflected overpressure is

$$t_C = \frac{(3)(15)}{1{,}289} = 0.0349 \text{ sec}$$

Average Pressure on Front Face: At $t = 0$, the pressure P_{front} is equal to P_r, which is equal to 25.31 psi. At $t = t_C = 0.0349$ sec, the values of P_s and P_d can be determined from Eqs. (6.135) and (6.136), respectively. Thus, at $t = t_C$, we have

$$P_s = (10)\left(1 - \frac{0.0349}{0.68}\right)(2.7182)^{-0.0349/0.68}$$

$$= 9.01 \text{ psi}$$

$$P_d = 2.23\left(1 - \frac{0.0349}{0.909}\right)(2.7182)^{-2(0.0349)/0.909}$$

$$= 1.91 \text{ psi}$$

On this basis, we have

$$P_{\text{front}} = P_s + 0.85P_d$$
$$= 9.01 + (0.85)(1.94)$$
$$= 10.66 \text{ psi}$$

The values of P_{front} for the time interval between $t = t_C$ and $t = 0.909$ sec may be obtained from the summation $P_s + 0.85P_d$, which is the sum of P_s and $0.85P_d$ for various times $t_C \leqslant t \leqslant 0.909$ sec. The procedure is similar to the one used to determine P_{front}, at $t = t_C$. The results are shown plotted in Fig. 6.23. The time variation of the pressure at $0 \leqslant t \leqslant t_C$ is assumed to be linear. The incompatible discontinuity at $t = t_C$, if preferred, can be smoothed out by the fairing curve de shown by the dashed line.

Average Pressure on Back Face: The shock front arrives at the back face of the structure at time t_d. Thus,

$$t_d = \frac{L}{U_o} = \frac{54}{1{,}405} = 0.0384 \text{ sec}$$

The time t_b that is required for the back pressure P_{back} to build up to its maximum value $(P_{\text{back}})_{\max}$ is

$$t_b = \frac{4S}{U_o} = \frac{(4)(15)}{1{,}405} = 0.0427 \text{ sec}$$

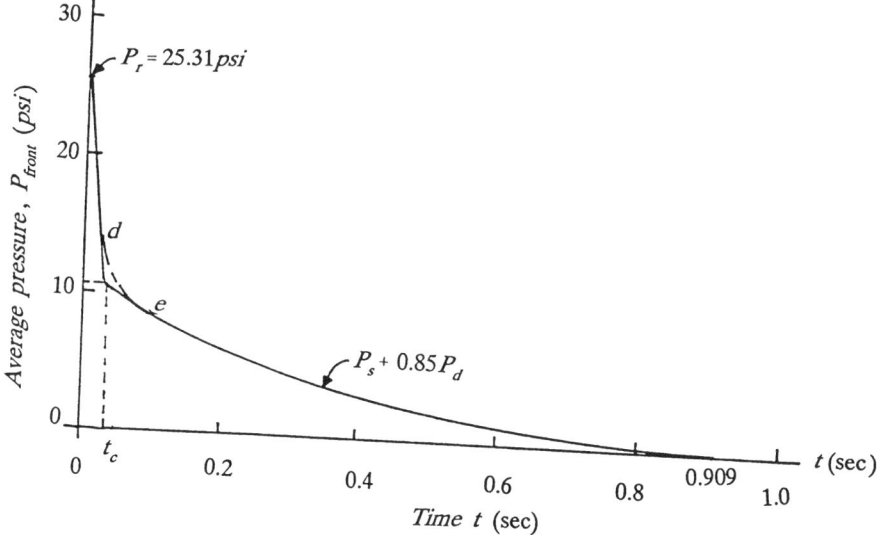

Figure 6.23. Time variation of the average front-face pressure of the building.

450 EARTHQUAKE EXCITATIONS AND BLAST LOADINGS

By using Eq. (6.135) and the times $t = t_b = 0.0427$ sec and $t_o = t_{o2} = 0.68$ sec, the value P_{sb} of the overpressure P_s is

$$P_{sb} = (10)\left(1 - \frac{0.0427}{0.68}\right)(2.7182)^{-0.0427/0.68}$$

$$= 8.80 \text{ psi}$$

In Eq. (6.145), the quantity β is

$$\beta = \frac{0.5 P_{so}}{14.7} = \frac{(0.5)(10)}{14.7} = 0.340$$

On this basis, Eq. (6.145) yields

$$(P_{back})_{max} = \frac{8.80}{2}[1 + (1 - 0.340)e^{-0.34}]$$

$$= 6.467 \text{ psi}$$

The variation of P_{back} during the time interval $t_d \leqslant t \leqslant (t_d + t_b)$ is assumed to be linear. For the time interval $(t_d + t_b) < t \leqslant (t_{o2} + t_d)$, the variation of P_{back} can be determined from Eq. (6.146). For example, at $t = 0.381$ sec, Eq. (6.146) yields

$$\frac{P_{back}}{P_s} = \frac{6.467}{8.80} + \left[1 - \frac{6.467}{8.80}\right]\left[\frac{0.381 - (0.0384 + 0.0427)}{0.68 - 0.0427}\right]^2$$

$$= (0.7349) + (0.2651)\left(\frac{0.2999}{0.6373}\right)^2$$

$$= 0.7936$$

Thus,

$$P_{back} = 0.7936 P_s$$

At $t = 0.381 - t_d = 0.3426$ sec, Eq. (6.135) yields

$$P_s = 10\left(1 - \frac{0.3426}{0.68}\right)e^{-0.3426/0.68}$$

$$= 10(1 - 0.5038)e^{-0.5038}$$

$$= 3.00 \text{ psi}$$

Therefore,

$$P_{back} = 0.7936 P_s = (0.7936)(3.00)$$

$$= 2.38 \text{ psi}$$

In a similar manner, the pressure P_{back} at other times $(t_d + t_b) < t \leq (t_{o2} + t_d)$ can be obtained. The complete results are shown plotted in Fig. 6.24.

Average Pressure on Roof and Sides: For the roof and sides, we have

$$\frac{L}{U_o} = \frac{54}{1,405} = 0.0384 \text{ sec}$$

$$\frac{L}{2U_o} = \frac{54}{(2)(1,405)} = 0.0192 \text{ sec}$$

From Eq. (6.148), we find

$$P_m = P_s(t = 0.0192) + C_d P_d(t = 0.0192)$$

$$= (10)\left(1 - \frac{0.0192}{0.68}\right)\exp\frac{-0.0192}{0.68}$$

$$+ (-0.4)(2.23)\left(1 - \frac{0.0192}{0.909}\right)\exp\frac{-(2)(0.0192)}{0.909}$$

$$= (10)(1 - 0.0282)e^{-0.0282} + (-0.4)(2.23)(1 - 0.0211)e^{-0.0422}$$

$$= 9.45 - 0.84$$

$$= 8.61 \text{ psi}$$

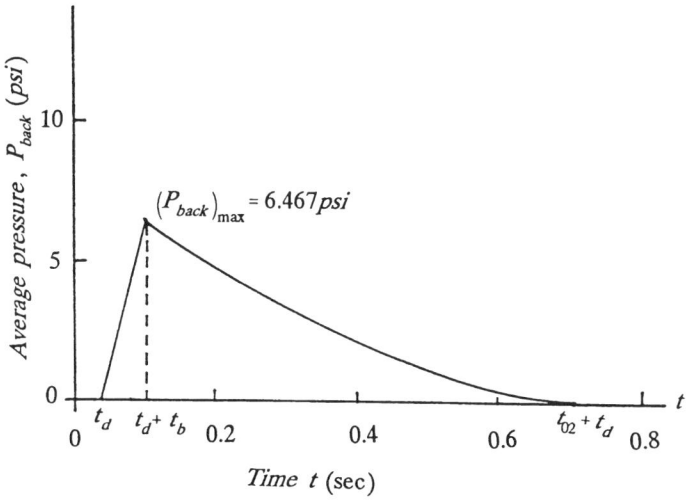

Figure 6.24. Time variation of the average back pressure of the building.

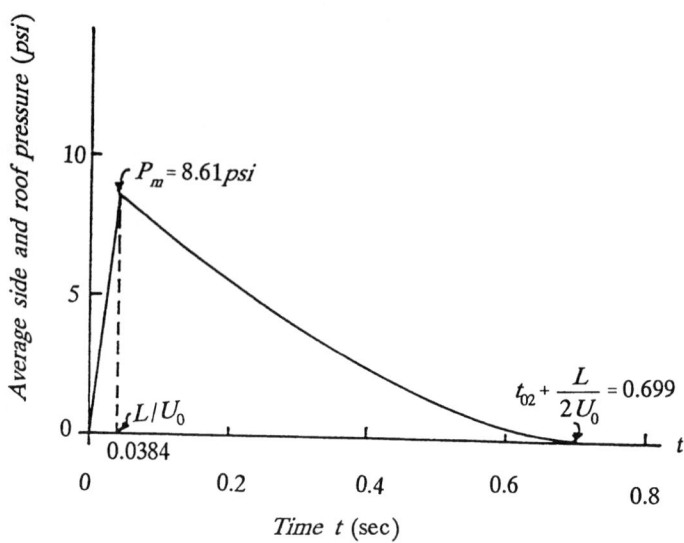

Figure 6.25. Time variation of the average pressure on roof and sides.

where P_s and P_d are computed by using Eqs. (6.135) and (6.136), respectively.

The average pressure P_a at times exceeding $L/U_o = 0.0384$ sec can be obtained from Eq. (6.149). The time variation of the pressure on roof and sides is shown in Fig. 6.25.

Average Net Horizontal Pressure: If the values of the back-face pressure in Fig. 6.24 are subtracted from the corresponding values of the front-face pressure in Fig. 6.23, the variation of the average net horizontal pressure P_{net} is obtained, and it is shown in Fig. 6.26.

PROBLEMS

6.1 For the spring-mass systems shown in Fig. P6.1, determine in each case the differential equation(s) of motion by using Lagrange's equation.

6.2 Repeat the problem in Example 6.3 by assuming that $F(t)$ is a suddenly applied force of 6 kips that decreases linearly to zero at time $t_d = 0.2$ sec. Compare the results.

6.3 The two-story frame in Fig. P6.3a is acted upon by the horizontal forces $F(t)$ and $0.6F(t)$ at the indicated locations. The time variation of $F(t)$ is shown in Fig. P6.3b. By idealizing the frame as a two-degree spring-mass system, determine the maximum horizontal displacements at points 1 and 2 of the frame and the maximum bending stresses in the columns by using the method of modal analysis. Neglect the weight of the columns

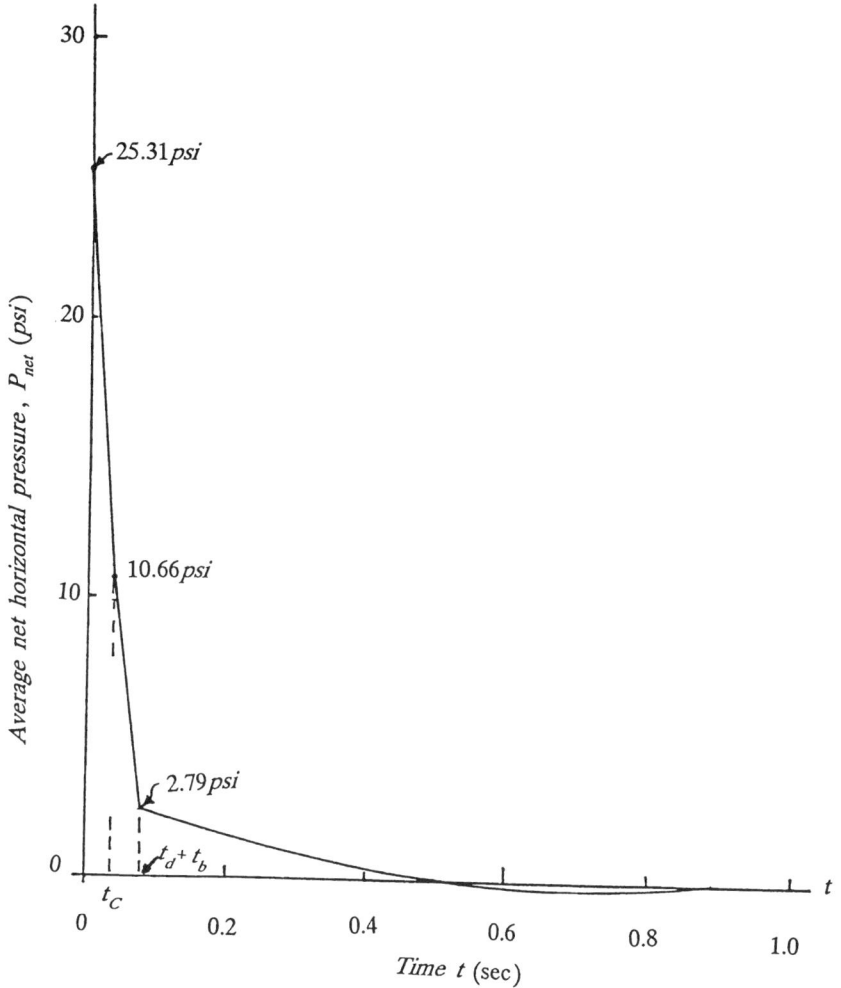

Figure 6.26. Average net horizontal pressure versus time.

and assume that the girders are infinitely stiff compared with the stiffness of the columns.

6.4 Repeat Problem 6.3 by assuming that the time variation of $F(t)$ is as shown in Fig. P6.3c. Compare the results.

6.5 Repeat the problem in Example 6.3 by assuming that the dynamic force $F(t)$ is a suddenly applied force of 6 kips and of infinite duration. Compare the results.

6.6 The two-story frame in Fig. P6.6a represents the interior bay (20 ft wide)

454 EARTHQUAKE EXCITATIONS AND BLAST LOADINGS

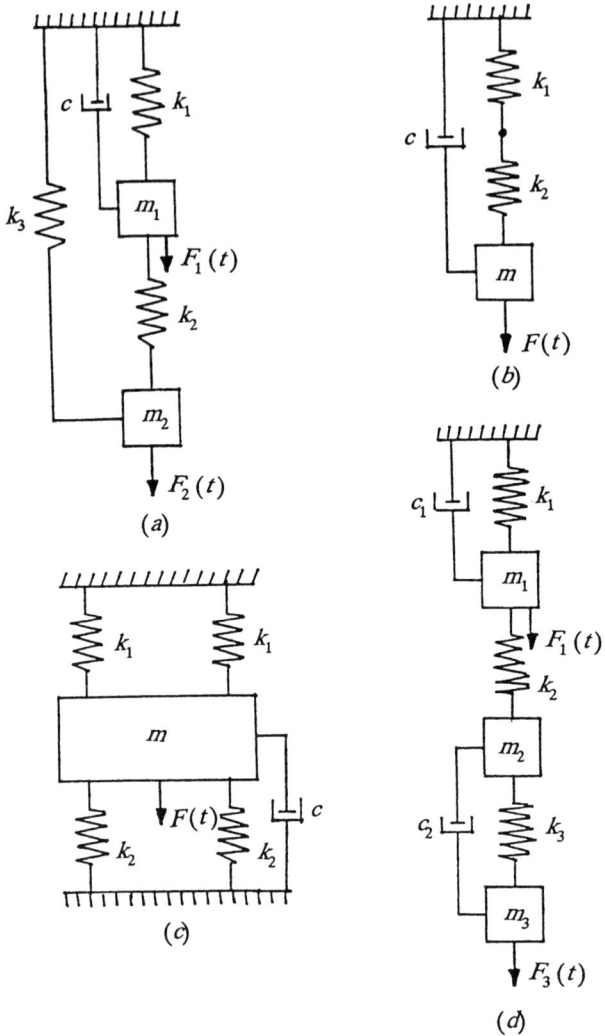

Figure P6.1.

of a two-story building having the top view shown in Fig. 6.5a. The moment of inertia of the W10 × 25 columns is 133.2 in.^{4}, their section modulus $S = 26.4$ in.^{3}, and the modulus of elasticity $E = 30 \times 10^6$ psi. The applied dynamic force $F(t)$ has the time variation shown in Fig. P6.6b. By applying the method of modal analysis, determine the maximum horizontal displacements at the girder levels, and the maximum bending stresses in the columns. Assume that the girders are infinitely stiff.

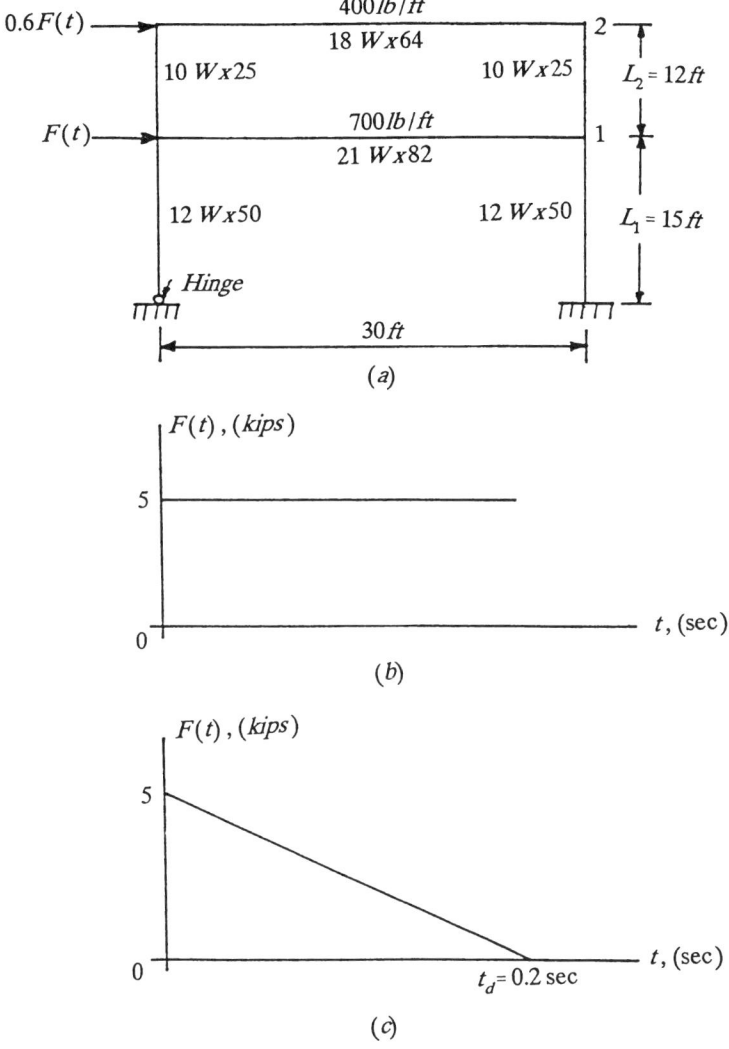

Figure P6.3.

6.7 Repeat Problem 6.6 by assuming that the dynamic force $F(t)$ is a suddenly applied force of 5 kips that lasts 0.3 sec. Compare the results.

6.8 The two-story frame in Fig. P6.8 represents the interior bay of a two-story building that is subjected to an artificial earthquake of ground acceleration $\ddot{u}_s = 0.05g \sin 4\pi t$, where g is the acceleration of gravity and t is time. The time duration of the acceleration $\ddot{u}_s$ is 1 sec. All steel columns of the frame are W12 × 58 sections with moment of inertia

456 EARTHQUAKE EXCITATIONS AND BLAST LOADINGS

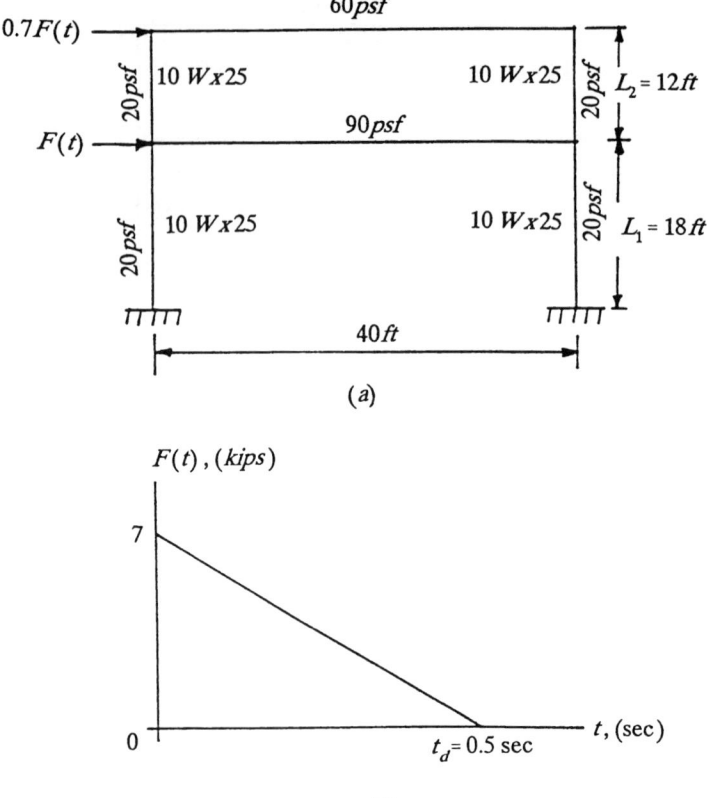

Figure P6.6.

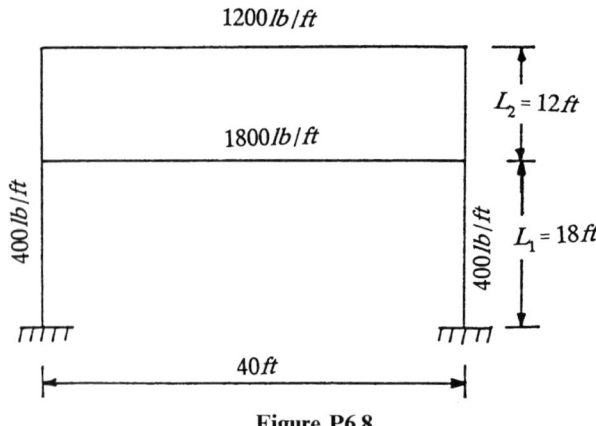

Figure P6.8.

$I = 475 \text{ in.}^4$ and section modulus $S = 78 \text{ in.}^3$ By applying the method of modal analysis, determine the maximum displacements $u_{1\text{max}}$ and $u_{2\text{max}}$ at the first- and second-story levels, respectively, and the maximum bending stresses in the columns. Assume that the girders of the frame are infinitely stiff compared to the column, and the modulus of elasticity $E = 30 \times 10^6$ psi.

6.9 The two-story frame in Fig. P6.9 represents an interior bay, 25 ft wide, of a two-story building, that is subjected to an artificial earthquake of ground displacement $u_s = 0.5 \sin 4\pi t$ in., where t is time. By applying the method of modal analysis, determine the modal participation factors Λ_1 and Λ_2, and the maximum story shears that determine the magnitude of column bending. Assume that the girders are infinitely stiff compared to the columns and that the modulus $E = 30 \times 10^6$ psi.

6.10 Repeat the problem in Example 6.4 by assuming that $\ddot{u}_s = 0.08g \sin 4\pi t$ and compare the results. Assume that the time duration of the acceleration $\ddot{u}_s$ is 2 sec.

6.11 Repeat Problem 6.8 by assuming that $\ddot{u}_s = 0.10g \sin 4\pi t$ and compare the results. The time duration of the acceleration $\ddot{u}_s$ is 1.5 sec.

6.12 Repeat the problem in Example 6.5 by assuming that $F(t)$ is a dynamic force of 60 kips that is suddenly applied to the beam at point C and that lasts indefinitely. Compare the results.

6.13 Rework the problem in Example 6.5 and determine the magnitude of the dynamic load $F(t)$ so that the ductility ratio η is equal to about 4.

6.14 The fixed-fixed uniform beam in Fig. P6.14a supports the weight $W = 40$ kips at its center. The beam is also subjected to the dynamic force $F(t)$, located as shown, and has the time variation shown in Fig.

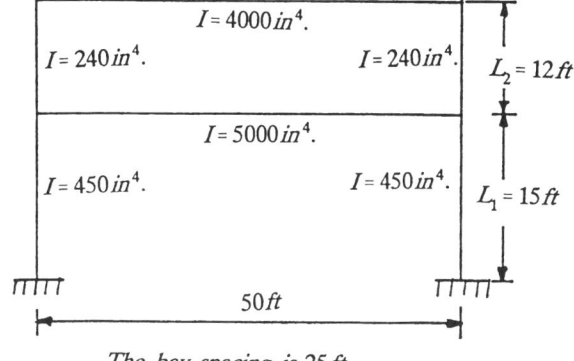

The bay spacing is 25 ft

Figure P6.9.

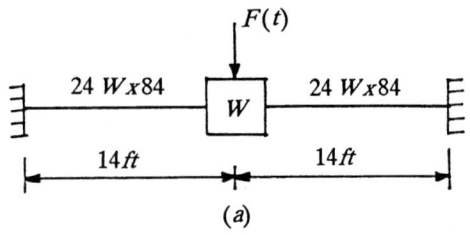

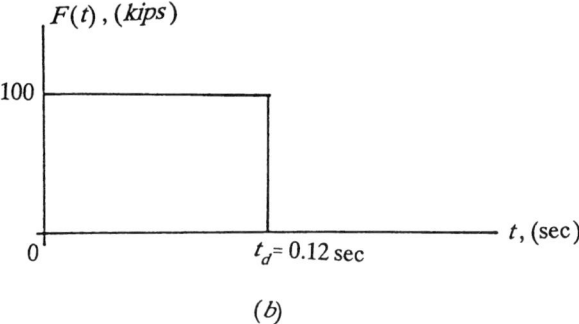

Figure P6.14.

P6.14b. By applying elastoplastic analysis, determine its maximum displacement due to $F(t)$ and the ductility ratio η, by using an appropriate idealized one-degree spring-mass system. The moment of inertia $I = 2{,}364.3 \text{ in.}^4$, the section modulus $S = 196.3 \text{ in.}^3$, the modulus $E = 30 \times 10^6 \text{ psi}$, and the plastic modulus $Z_p = 224 \text{ in.}^3$ Assume that the beam is weightless and neglect damping. The yield stress is equal to 30 ksi.

6.15 The fixed-fixed uniform beam in Fig. P6.15a supports the weight $W = 30$ kips at its center, and is also subjected to the dynamic force $F(t)$, located as shown. The time variation of $F(t)$ is illustrated in Fig. P6.15b. By applying elastoplastic analysis, determine the maximum displacement and ductility ratio η, by using an appropriate idealized one-degree spring-mass system. The moment of inertia $I = 2{,}096 \text{ in.}^4$, the modulus $E = 30 \times 10^6 \text{ psi}$, the plastic modulus $Z_p = 200.1 \text{ in.}^3$, and the yield stress $\sigma_y = 30$ ksi. Assume that the beam is weightless and neglect damping.

6.16 Rework the problem in Example 6.6 and select a column section so that the ductility ratio η is about 4. Compare the results.

6.17 Rework the problem in Example 6.6 and select a column section so that the ductility ratio η is about 6. Compare the results.

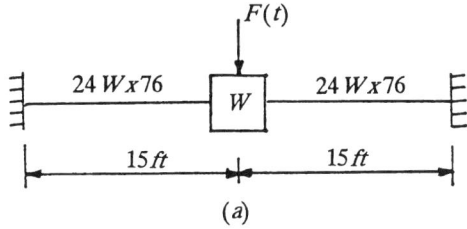

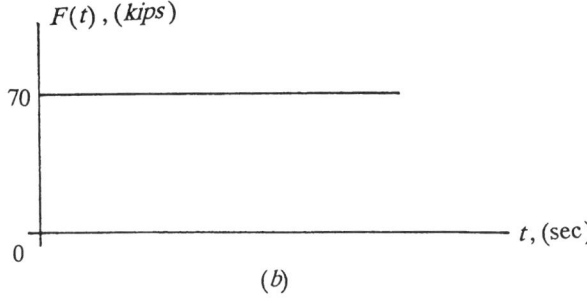

Figure P6.15.

6.18 Rework the problem in Example 6.6 and select a column section so that the ductility ratio η is about 2. Compare the results.

6.19 Rework the problem in Example 6.7 for a 10-KT typical surface burst and compare the results.

6.20 The required dimensions of an aboveground building are shown in Fig. P6.20. The building is constructed by using reinforced concrete walls and roof, and it houses delicate instrumentation and equipment that need to be protected, because it is expected to be subjected to a peak overpressure $P_{so} = 5.45$ psi, which is expected to be produced by a detonation of 1.2 KT. Determine and plot the time variations of the average front- and back-face pressures, and the average net horizontal pressure. Assume for convenience that $t_{od} = t_{o2}$ and $U_r = U_o$.

6.21 For the aboveground building in Problem 6.20, determine and plot the time variation of the average side and roof pressure.

6.22 Rework Problem 6.20 for a peak overpressure $P_{so} = 10.5$ psi, which results from a 10-KT detonation located at a distance of 0.4 mile (2,112 ft) from the building. Compare the results.

6.23 Rework Problem 6.21 for a peak overpressure $P_{so} = 10.5$ psi, which results from a 10-KT detonation located at a distance of 0.4 mile (2,112 ft) from the building. Compare the results.

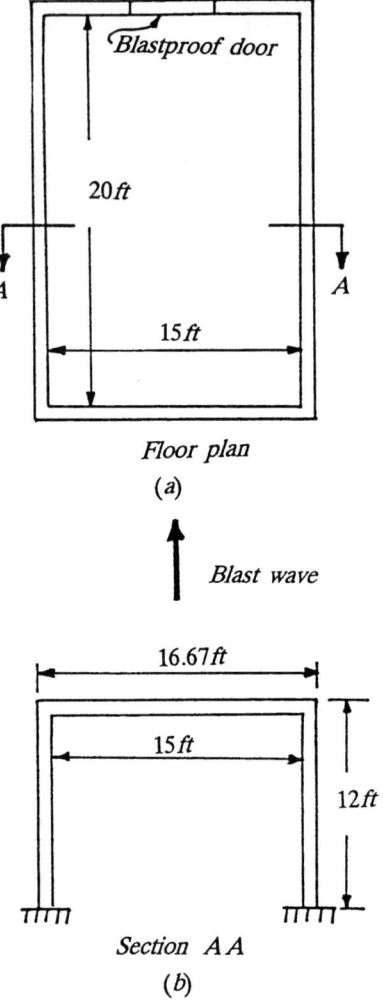

Figure P6.20.

7 Energy Concepts and Methods, Finite Element, and Finite Difference Methods

7.1 INTRODUCTION

In this chapter, energy concepts and methods of analysis are introduced because, although their use and existence can be traced back to the ancient Greek mathematicians and philosophers [6], their use and development reached unsurpassed heights during the 19th and 20th centuries. The discussion in this chapter includes the various types of strain energy produced by forces and moments, computation of deflections by equating external and internal work, the method of virtual work, and Castigliano's second theorem. Statically determinate, as well as statically indeterminate, problems are solved.

In addition, two numerical methods that became very popular during the second half of the 20th century, as well as the application of the author's method of the equivalent systems to plates of variable thickness, are discussed. The two numerical methods are the finite element and the finite difference methods. The use of these two numerical methods is based on the availability of computer hardware and computer software. The tremendous advancements in computer technologies made the utilization of these two numerical methods possible, because they can be effectively used for the analysis of very complex structural problems with very complex mathematical modeling.

7.2 CONCEPTS OF STRAIN ENERGY IN STRUCTURAL ANALYSIS

Saying that a structure is designed to withstand the effects of various types of loading conditions is tantamount to saying that it is designed to absorb well-defined levels of energy while still securing its safety and functional characteristics. The amount of strain energy that can be stored in a structure depends upon the types of internal forces acting at the cross sections of the component elements of the stucture. In general, the types of internal forces that store energy in a structure are axial forces, shear forces, torsional moments, and bending moments. Such internal forces are the result of externally applied and resisting forces, and of the dead weight of the structure. They can also be produced by temperature gradients, foundation settlements, operating machinery, and many other related causes.

Although most of the developments related to energy concepts and methods of analysis took place during the 19th and 20th centuries, their existence can be traced back to the ancient Greek mathematicians and philosophers [6]. Aristotle (384–322 B.C.), in order to prove the static equilibrium equation $\Sigma F_i x_i = 0$, introduced the principle of virtual work $\Sigma F_i \delta_i = 0$. Alexander of Afrodisias worked on the relation of motion to potential and kinetic energy, and he introduced the concept of conservation of energy as: "For the thing which, by causing them to pass from heaviness in potentia to heaviness in actu, causes them to be in a different state from the one they were before, is the cause of natural motion." In later years, Leonardo da Vinci (1452–1519) used the principle of virtual displacements to analyze various systems of pulleys and levers in hoisting devices. This shows that the importance of energy concepts and principles in the analysis of structural systems was realized many centuries ago.

In this section, strain energies produced by axial forces, shear forces, torsional moments, and bending moments are discussed, with applications to practical engineering problems.

7.2.1 Strain Energy Produced by Axial Loading

The columns of a building, the piers of a bridge, or the members of a truss structure are just a few practical applications where axially loaded members are involved. To determine the strain energy stored in axially loaded members, we consider the simple problems shown in Fig. 7.1a through 7.1c. Within certain limits, many structural materials are considered to be perfectly elastic, and they are assumed to obey Hooke's law. On this basis, the load-deflection curve for the prismatic bar in Fig. 7.1a would be a straight line, as shown in Fig. 7.1d. When the axial tensile load is increasing, the bar elongates, and the axial load produces work that is stored in the bar in the form of strain energy. For conservative systems, the work W performed by the load P is equal to the internal strain energy U stored in the bar.

We consider the condition where the load P_x in Fig. 7.1d reaches the value $P_x = P$. At this value of the axial tensile load, the elongation δ_x of the bar is equal to δ, as shown in Fig. 7.1d. The strain energy U stored in the bar is equal to the work W produced by the axial load $P_x = P$, and it is given by the expression

$$U = W = \frac{P\delta}{2} \quad (7.1)$$

Equation (7.1) represents the shaded area of the triangle OAB in Fig. 7.1d.

From Eq. (2.8) in Section 2.2, we have

$$\delta = \frac{PL}{AE} \quad (7.2)$$

CONCEPTS OF STRAIN ENERGY IN STRUCTURAL ANALYSIS   463

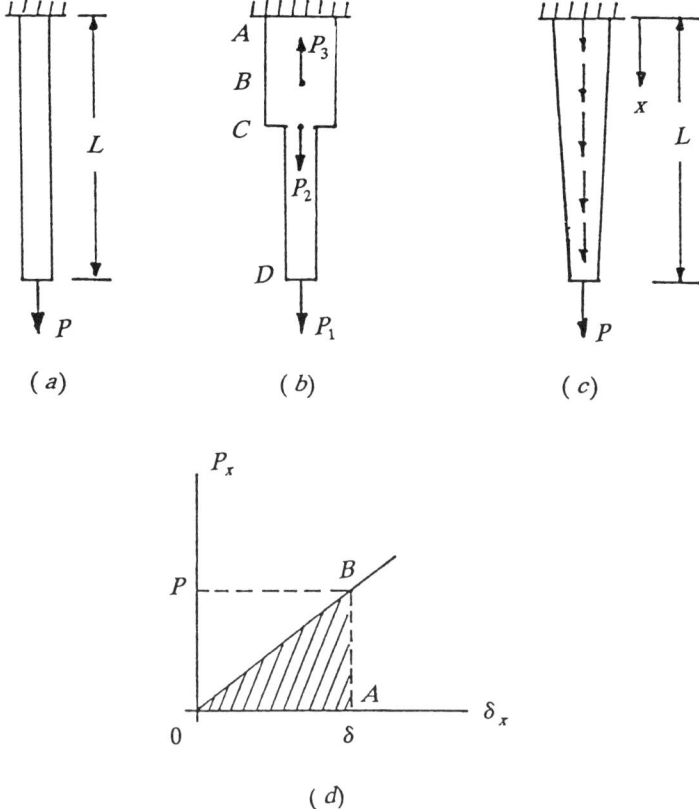

Figure 7.1. (*a*) Axially loaded prismatic bar. (*b*) Bar consisting of two prismatic portions. (*c*) Nonprismatic bar loaded by variable axial loading. (*d*) Load-deflection curve.

where L is the length of the member, A is its uniform cross-sectional area, and E is its modulus of elasticity. By substituting Eq. (7.2) into Eq. (7.1), we find

$$U = \frac{P^2 L}{2AE} \tag{7.3}$$

Equation (7.3) gives the total strain energy U stored in the bar in Fig. 7.1*a* as a function of the axial tensile load P.

The strain energy U may be also expressed as a function of the axial deformation δ. From Eq. (7.2), we find

$$P = \frac{\delta AE}{L} \tag{7.4}$$

By substituting Eq. (7.4) into Eq. (7.3), we obtain

$$U = \frac{AE\delta^2}{2L} \qquad (7.5)$$

which expresses U as a function of the axial displacement δ.

If the bar consists of two or more prismatic parts, and if it is subjected to axial loads as shown in Fig. 7.1b, then the total strain energy U may be obtained by applying Eq. (7.3) to the individual prismatic parts and summing up the results. For the problem in Fig. 7.1b, we apply Eq. (7.3) to parts AB, BC, and CD, and we sum up the results. On this basis, a general expression for the strain energy U may be written as follows:

$$U = \sum_{i=1}^{n} \frac{P_i^2 L_i}{2 A_i E_i} \qquad (7.6)$$

It should be noted that P_i in Eq. (7.6) is the axial force acting on the part L_i of the bar, and n is the number of the parts.

If the axial force $P(x)$ varies along the length of the member, say as a function of x, then the total strain energy U may be obtained by integrating the following expression:

$$U = \int_0^L \frac{[P(x)]^2 \, dx}{2A(x)E} \qquad (7.7)$$

In Eq. (7.7), $P(x)$ represents the expression of the variable axial force at a distance x, and $A(x)$ is the expression for the variable cross-sectional area at a distance x. Figure 7.1c illustrates a nonprismatic bar subjected to a distributed axial force $p(x)$ along the length of the bar, and to a concentrated axial force P at its free end. At any distance x from the fixed end, the variation $P(x)$ of the axial force may be obtained by using statics and appropriate free-body diagrams.

It should be noted, however, that the principle of superposition does not apply here. For example, in Fig. 7.1b, it would not be correct to first apply the load P_1 and determine U_1, then apply P_2 and determine U_2, and finally apply P_3 to determine U_3, and calculate the total strain energy U as the sum of U_1, U_2, and U_3. The total strain energy must be determined by using Eq. (7.6) and summing up the energies stored in parts AB, BC, and CD. Similar reasoning can be applied for the problems in Fig. 7.1c. We cannot superimpose the results obtained by applying individually the loads P and $p(x)$.

The following examples illustrate the application of the above theory.

Example 7.1: A circular stepped shaft made out of solid steel is subjected to three axial forces located as shown in Fig. 7.2a. Determine the total strain energy U stored in the shaft. The modulus of elasticity $E = 30 \times 10^3$ ksi.

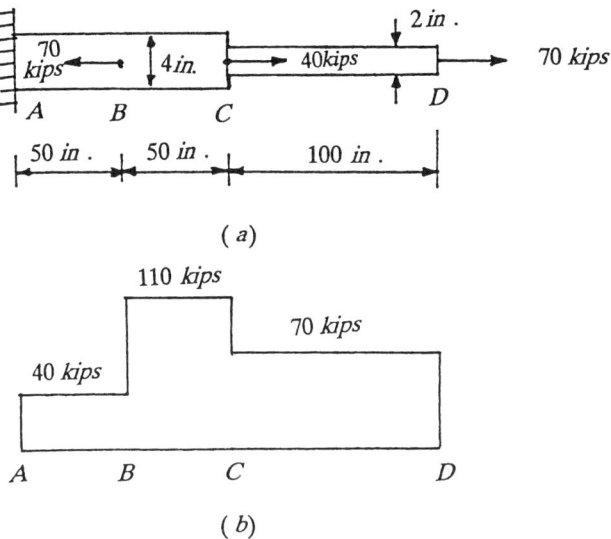

Figure 7.2. (a) Steel stepped shaft of circular cross section. (b) Axial force diagram of the shaft.

SOLUTION: By applying statics, the axial force diagram of the shaft is determined, and it is shown in Fig. 7.2b. By applying Eq. (7.6), we find

$$U = \sum_{i=1}^{3} \frac{P_i^2 L_i}{2A_i E} = \frac{P_{AB}^2 L_{AB}}{2A_{AB}E} + \frac{P_{BC}^2 L_{BC}}{2A_{BC}E} + \frac{P_{CD}^2 L_{CD}}{2A_{CD}E}$$

$$= \frac{(40)^2(50)}{2\pi(2)^2(30)(10)^3} + \frac{(110)^2(50)}{2\pi(2)^2(30)(10)^3} + \frac{(70)^2(100)}{2\pi(1)^2(30)(10)^3}$$

$$= 0.1061 + 0.8024 + 2.5995$$

$$= 3.508 \text{ kip-in.}$$

The value of 0.1061 kip-in. gives the energy stored in portion AB, 0.8024 kip-in. of energy is stored in portion BC, and 2.5995 kip-in. of energy is stored in portion CD. The results show that most of the strain energy is stored in portion CD, because this is the most severely stressed portion with a normal stress $\sigma = 22.28$ ksi. The normal stresses in portions AB and BC are 3.18 and 8.75 ksi, respectively.

Example 7.2: The simple steel truss in Fig. 7.3 is subjected to a vertical force $P = 10$ kN at joint C, as shown in the figure. Determine the total strain energy U stored in the truss. Each member of the truss has a cross-sectional area $A = 7.85 \times 10^{-3}$ m^2 and a modulus of elasticity $E = 210 \times 10^9$ Pa.

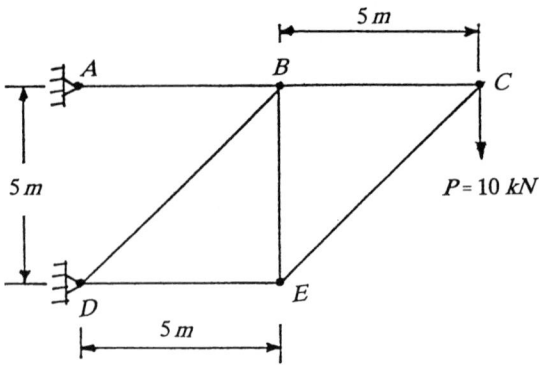

Figure 7.3. Simple steel truss loaded as shown.

SOLUTION: We assume here that the joints of the truss are frictionless and, consequently, each member of the truss will be subjected to either an axial compressive force, or to an axial tensile force. On this basis, the total strain energy U stored in the truss would be the sum of the individual strain energies stored in each member of the truss. Thus, in this case, the total strain energy U can be determined by applying Eq. (7.3) to each member of the truss and summing up the results. The axial forces F acting in the members of the truss are determined by using the method of joints, and they are as shown in the second column of Table 7.1. In the same table, the energy stored in each member of the truss is shown in the last column of the table. The total strain energy U is

$$U = \sum \frac{F^2 L}{2AE} = 1.9193 \text{ N} \cdot \text{m}$$

as shown in Table 7.1.

TABLE 7.1. Illustrates the Energy Stored in the Individual Members of the Truss and the Total Strain Energy U Stored in the Truss

(1) Member	(2) Axial Force F (kN)	(3) Length L (m)	(4) $(L/2AE) \times 10^{-9}$	(5) $F^2L/2AE$ (N·m)
AB	+20.00	5.00	1.5165	0.6066
BC	+10.00	5.00	1.5165	0.1517
DE	−10.00	5.00	1.5165	0.1517
DB	−14.14	7.07	2.1444	0.4288
BE	+10.00	5.00	1.5165	0.1517
EC	−14.14	7.07	2.1444	0.4288
		$U = \sum F^2 L/2AE = 1.9193$ N·m		

Example 7.3: The building concrete column in Fig. 7.4 supports the axial compressive load P that is transmitted to the column from the upper stories of the building. The column is tapered, the cross-sectional radius at the free end A is r_1, and it is r_2 at the fixed end B. Derive the expression for the total strain energy U stored in the column.

SOLUTION: At any distance x from the free end A, the radius $r(x)$ of the cross section is

$$r(x) = r_1 + \frac{x}{L}(r_2 - r_1) \tag{7.8}$$

Thus, the cross-sectional area $A(x)$ at any distance x is

$$A(x) = \pi[r(x)]^2 = \pi\left[r_1 + \frac{(r_2 - r_1)}{L}x\right]^2 \tag{7.9}$$

By applying Eq. (7.7), we obtain

$$U = \int_0^L \frac{P^2\,dx}{2A(x)E} = \frac{P^2}{2E\pi}\int_0^L \frac{dx}{[r_1 + [(r_2 - r_1)/L]x]^2} \tag{7.10}$$

By integrating Eq. (7.10), we find

$$\begin{aligned}U &= -\frac{P^2}{2E\pi}\left\{\frac{1}{[(r_2 - r_1)/L][r_1 + [(r_2 - r_1)/L]x]}\right\}_0^L \\ &= \frac{P^2 L}{2r_1 r_2 E\pi}\end{aligned} \tag{7.11}$$

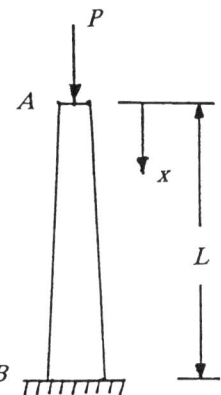

Figure 7.4. Tapered concrete building column subjected to an axial compressive load P.

7.2.2 Strain Energy Induced by Bending

We consider the prismatic member in Fig. 7.5a, which is subjected to pure bending in a principal plane. For linearly elastic systems, the angle of rotation θ produced by bending is proportional to the bending moment M, and this relationship is represented by the straight line in Fig. 7.5b. When the value of $M(x)$ is M, the rotation $\theta(x)$ is θ, and the strain energy of bending U is equal to the shaded triangular area OAB in Fig. 7.5b. This is also equal to the work W of M moving through the rotation θ. Thus,

$$U = W = \frac{M\theta}{2} \tag{7.12}$$

The deflection curve of the member in Fig. 7.5c is a circular arc of constant curvature $1/\rho$, where

$$\frac{1}{\rho} = -\frac{M}{EI} \tag{7.13}$$

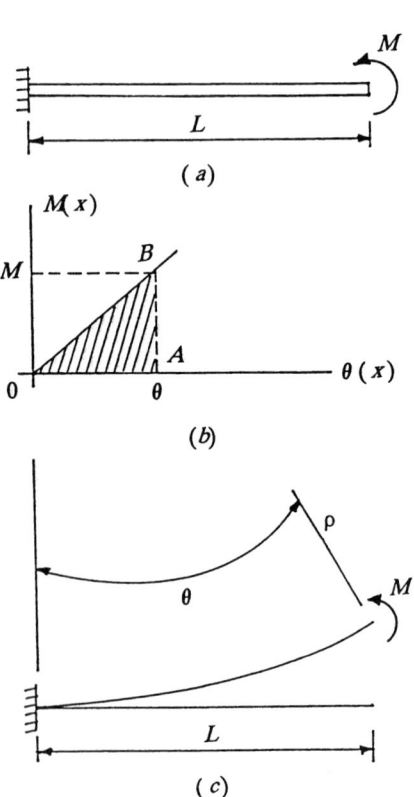

Figure 7.5. (a) Bar subjected to pure bending. (b) Curve of bending moment versus rotation θ. (c) Deflection configuration of the member.

CONCEPTS OF STRAIN ENERGY IN STRUCTURAL ANALYSIS 469

In Eq. (7.13), ρ is the radius of curvature, I is the cross-sectional moment of inertia, and E is the modulus of elasticity. Since $L = \rho\theta$, we have

$$\theta = \frac{L}{\rho} = \frac{ML}{EI} \tag{7.14}$$

Only absolute values are considered in the expression for θ represented by Eq. (7.14).

By substituting Eq. (7.14) into Eq. (7.12), we obtain

$$U = \frac{M^2 L}{2EI} \tag{7.15}$$

Equation (7.15) expresses the strain energy of bending in terms of the bending moment M. It may also be expressed in terms of the rotation θ by solving Eq. (7.14) for M and substituting into Eq. (7.12). This yields

$$U = \frac{EI\theta^2}{2L} \tag{7.16}$$

If the bending moment M and moment of inertia I are variable along the length of a member, then the strain energy U of bending may be determined by integration using the following equation:

$$U = \int_0^L \frac{M_x^2 \, dx}{2EI_x} \tag{7.17}$$

where M_x is the bending moment at any x, and I_x is the cross-sectional moment of inertia at any x.

The strain energy of bending may be also expressed in terms of the deflection y of the member. Thus, since

$$d\theta = \frac{M \, d_x}{EI} = \frac{d^2 y}{dx^2} \, dx \tag{7.18}$$

we have

$$dU = \frac{EI}{2} \left(\frac{d^2 y}{dx^2}\right)^2 dx \tag{7.19}$$

and

$$U = \int_0^L \frac{EI}{2} \left(\frac{d^2 y}{dx^2}\right)^2 dx \tag{7.20}$$

470   ENERGY CONCEPTS AND METHODS

Equation (7.20) gives the strain energy of bending in terms of the deflection y of the member. In this equation, both y and I may be variable along the length of a member.

The following example illustrates the application of the preceding methodology.

Example 7.4: The steel simply supported beam in Fig. 7.6a has a W21 × 62 wide-flange cross section, and it is loaded by a uniformly distributed load as shown in the same figure. Determine the total bending strain energy stored in the beam by the application of the applied distributed loading. The moment of inertia $I = 1{,}330 \text{ in.}^4$, and the modulus of elasticity $E = 30 \times 10^3$ ksi.

SOLUTION: In order to determine the total strain energy of bending U, we need to apply Eq. (7.17) for portions AB and BC and add the results. In order to accomplish this objective, we need to derive the required expression of M

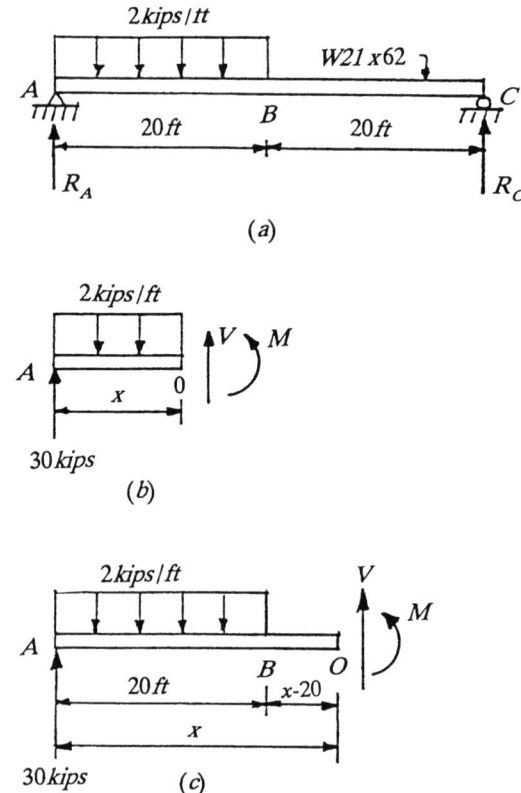

Figure 7.6. (a) Simply supported beam loaded as shown. (b) Free-body diagram for portion AB. (c) Free-body diagram for portion BC.

CONCEPTS OF STRAIN ENERGY IN STRUCTURAL ANALYSIS

for portions AB and BC. The reaction R_A in Fig. 7.6a is determined by applying statics, and is equal to 30 kips.

By using the free-body diagram in Fig. 7.6b and equating to zero the sum of the moments about point O, we find

$$-30x + (2)(x)\left(\frac{x}{2}\right) + M = 0$$

or

$$M = 30x - x^2 \qquad 0 \leqslant x \leqslant 20\,\text{ft} \qquad (7.21)$$

By using the free-body diagram in Fig. 7.6c and taking moments about point O, we find

$$-30x + (2)(20)(x - 10) + M = 0$$

or

$$M = 400 - 10x \qquad 20\,\text{ft} \leqslant x \leqslant 40\,\text{ft} \qquad (7.22)$$

By substituting Eqs. (7.21) and (7.22) into Eq. (7.17), we find

$$U = \frac{1}{2EI}\left[\int_0^{20\,\text{ft}} (30x - x^2)^2\,dx + \int_{20\,\text{ft}}^{40\,\text{ft}} (400 - 10x)^2\,dx\right]$$

$$= \frac{1}{2EI}\left[\int_0^{20\,\text{ft}} (900x^2 - 60x^3 + x^4)\,dx + \int_{20\,\text{ft}}^{40\,\text{ft}} 160{,}000 - 8{,}000x + 100x^2)\,dx\right]$$

$$= \frac{1}{2EI}\left[300x^3 - 15x^4 + \frac{x^5}{5}\right]_0^{20} + \frac{1}{2EI}\left[160{,}000x - 4{,}000x^2 + \frac{100}{3}x^3\right]_{20}^{40}$$

$$= \frac{453{,}333}{EI}$$

We know that

$$EI = (30)(10)^3(1{,}330) = 39.9 \times 10^6\,\text{kip-in.}^2$$
$$= 277{,}083.33\,\text{kip-ft}^2$$

On this basis,

$$U = \frac{453{,}333}{277{,}083.33} = 1.6361\,\text{kip-ft}$$
$$= 19.6331\,\text{kip-in.}$$

472   ENERGY CONCEPTS AND METHODS

7.2.3 Strain Energy Stored by Shear Forces

When a member is subjected to bending, shear forces are also developing at cross sections along the length of the member and, consequently, strain energy due to shear is stored in the member. This energy can be determined by using the following equation:

$$U = \int_0^L \frac{KV^2}{2GA} dx \qquad (7.23)$$

In Eq. (7.23), V is the shear force at cross sections along the length of the member, G is the shear modulus of the material, A is the cross-sectional area of the member, and K is the shear factor. For rectangular cross sections, $K = 1.2$, and it is equal to about 1.1 for wide-flange cross sections. Complete derivation of Eq. (7.23) may be found in [6, Section 1.5].

The following example illustrates the application of Eq. (7.23).

Example 7.5: For the beam in Example 7.4, determine the strain energy due to shear that is stored in the member, and compare the result with the energy that is obtained by considering bending only. The shear modulus $G = 11.5 \times 10^3$ ksi, and the cross-sectional area $A = 18.3$ in.² Assume $K = 1.1$.

SOLUTION: By using the free-body diagram in Fig. 7.6b and applying static equilibrium in the vertical direction, we find

$$V = 2x - 30 \qquad 0 \leqslant x \leqslant 20 \text{ ft} \qquad (7.24)$$

By using the free-body diagram in Fig. 7.6c and applying static equilibrium in the vertical direction, we find

$$V = 10 \text{ kips} \qquad 20 \text{ ft} \leqslant x \leqslant 40 \text{ ft} \qquad (7.25)$$

By substituting Eqs. (7.24) and (7.25) into Eq. (7.23), the strain energy U produced by shear is

$$U = \frac{K}{2GA} \left\{ \int_0^{20 \text{ ft}} (2x - 30)^2 \, dx + \int_{20 \text{ ft}}^{40 \text{ ft}} (10)^2 \, dx \right\}$$

$$= \frac{K}{2GA} \{6{,}666.67\}$$

But, we have

$$GA = (11.5)(10)^3(18.3) = 210.45 \times 10^3 \text{ kips}$$

CONCEPTS OF STRAIN ENERGY IN STRUCTURAL ANALYSIS 473

Thus, with $K = 1.1$, we find

$$U = \frac{(1.1)(6,666.67)}{(2)(210.45)(10)^3}$$

$$= 0.0174 \text{ kip-ft}$$

$$= 0.2091 \text{ kip-in.}$$

The preceding value of U for shear is very small compared to the value of U obtained for bending in Example 7.4. It is only 1.06 percent of the strain energy due to bending, which means that the shear contribution to the deformation of the member is small compared to the one produced by bending, and it is often neglected in practical applications.

7.2.4 Strain Energy Produced by Twisting Moments

For a uniform shaft, such as the one shown in Fig. 7.7a, the strain energy of torsion can be determined by using the diagram shown in Fig. 7.7b. This diagram illustrates the relationship between the torque T and the angular twist ϕ for linearly elastic systems. Again, at a value $T(x) = T$ and $\phi(x) = \phi$, the

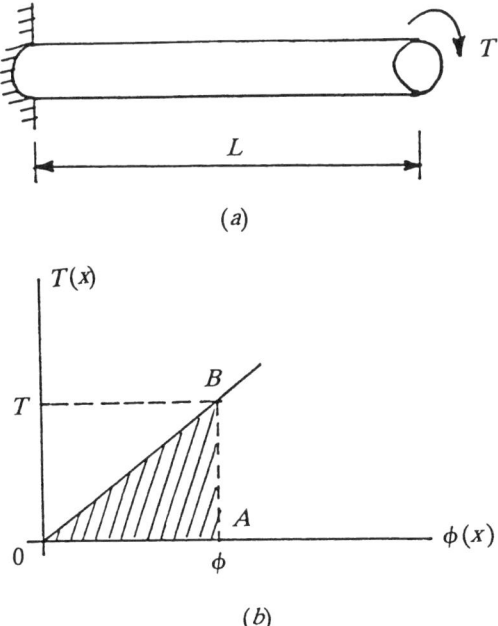

Figure 7.7. (a) Circular solid shaft subjected to torsion. (b) Diagram representing the variation of $T(x)$ with respect to $\phi(x)$.

work W done by the torque T is equal to the internal strain energy U stored in the member, and it is represented by the triangular area OAB in Fig. 7.7b. Thus,

$$U = W = \frac{T\phi}{2} \tag{7.26}$$

For members of circular and tubular cross sections, the angular twist ϕ is given by Eq. (2.30) in Section 2.2.2, and it is as follows:

$$\phi = \frac{TL}{JG} \tag{7.27}$$

where J is the cross-sectional polar moment of inertia and G is the shear modulus.

By substituting Eq. (7.27) into Eq. (7.26), we obtain

$$U = \frac{T^2 L}{2JG} \tag{7.28}$$

In terms of the angle of twist ϕ, the strain energy of torsion is written as

$$U = \frac{GJ\phi^2}{2L} \tag{7.29}$$

If the cross section of the twisted member is rectangular where b is the long side and h is the short one, the maximum angular twist ϕ may be determined by using Eq. (2.32) in Section 2.2.2, and it is as follows:

$$\phi = \frac{TL}{\beta b h^3 G} \tag{7.30}$$

where β is a numerical factor that depends on the ratio b/h. Values of β for various b/h are given in Table 2.1 of Section 2.2.2.

By substituting Eq. (7.30) into Eq. (7.26), we obtain

$$U = \frac{T^2 L}{2\beta b h^3 G} \tag{7.31}$$

In terms of ϕ, the strain energy of torsion is

$$U = \frac{\phi^2 \beta b h^3 G}{2L} \tag{7.32}$$

If the circular cross section of the member in Fig. 7.7a varies along its length and the applied load is also variable, then the internal strain energy of torsion stored in the member may be obtained from the equation

$$U = \int_0^L \frac{[T(x)]^2 \, dx}{2GJ(x)} \tag{7.33}$$

CONCEPTS OF STRAIN ENERGY IN STRUCTURAL ANALYSIS 475

or from the equation

$$U = \int_0^L \frac{GJ(x)}{2}\left(\frac{d^2\phi}{dx^2}\right)^2 dx \qquad (7.34)$$

The following examples illustrate the application of the preceding methodology.

Example 7.6: The steel stepped shaft in Fig. 7.8a has a circular cross section and it is loaded by two torsional moments as shown. Determine the strain energy of torsion that is stored in the shaft. The shear modulus $G = 10.0 \times 10^6$ psi.

SOLUTION: The twisting moment diagram for the stepped shaft is shown in Fig. 7.8b. By applying Eq. (7.28), we find

$$U = \frac{T_{AB}^2 L_{AB}}{2J_{AB}G} + \frac{T_{BC}^2 L_{BC}}{2J_{BC}G}$$

$$= \frac{(100)^2(60)}{(2)(\pi/2)(2)^4(11.5)(10)^3} + \frac{(-50)^2(40)}{(2)(\pi/2)(1)^4(11.5)(10)^3}$$

$$= 1.0380 + 2.7679$$

$$= 3.8059 \text{ kip-in.}$$

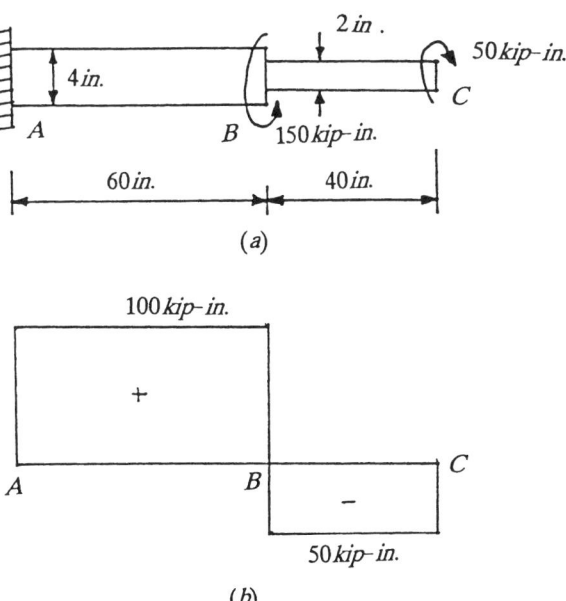

Figure 7.8. (a) Circular stepped shaft loaded as shown. (b) Twisting moment diagram.

476 ENERGY CONCEPTS AND METHODS

Note that 72.73 percent of the strain energy is stored in portion BC of the stepped shaft. In this portion, the maximum shear stress is equal to 31.83 ksi. In portion AB of the shaft, the maximum shear stress is 7.96 ksi.

Example 7.7: For the beam problem in Example 2.4, determine the torsional strain energy stored in the member for the two types of cross section and compare the results. The shear modulus $G = 10 \times 10^3$ ksi.

SOLUTION: For the circular cross section of radius $r = 1.7841$ in., Eq. (7.28) yields

$$U = \frac{T^2 L}{2JG} = \frac{(60{,}000)^2 (120)}{(2)(\pi/2)(1.7841)^4 (10.0)(10)^6}$$

$$= 1{,}357.24 \text{ lb-in.}$$

For the rectangular cross section, Eq. (7.31) yields

$$U = \frac{T^2 L}{2\beta b h^3 G} = \frac{(60{,}000)^2 (120)}{(2)(0.249)(5)(2)^3 (10.0)(10)^6}$$

$$= 2{,}168.67 \text{ lb-in.}$$

The value of $\beta = 0.249$, and it is obtained from Table 2.1 of Section 2.2.2.

The preceding results indicate that greater strain energy is stored in the member when its cross section is rectangular. From Example 2.4, we note that the maximum shear stress is 11,628 psi when the cross section is rectangular, and it is equal to 6,726 psi when the cross section is circular.

7.2.5 Total Strain Energy

If a member is subjected to a load combination consisting of axial forces, twisting moments, bending moment, and shear forces, then the total strain energy stored in the member is

$$U = \int_0^L \left(\frac{M^2}{2EI} + \frac{T^2}{2GC} + \frac{P^2}{2AE} + \frac{KV^2}{2GA} \right) dx \tag{7.35}$$

In Eq. (7.35), C is a torsional constant that is equal to the polar moment of inertia J for circular and tubular cross sections, and equal to $\beta b h^3$ for rectangular cross sections. The superposition of these types of strain energies is permissible here, because the stress-strain relations are assumed to be linear and to obey Hooke's law.

The following example illustrates a combination of such strain energies.

CONCEPTS OF STRAIN ENERGY IN STRUCTURAL ANALYSIS 477

Example 7.8: The semicircular steel arch in Fig. 7.9a is loaded by a uniformly distributed load w as shown. Determine the total strain energy stored in the arch by using Eq. (7.35). The rectangular cross section of the arch is shown in Fig. 7.9b. Assume that $w = 2\,\text{kips/ft}$, $r = 20\,\text{ft}$, $E = 30 \times 10^3\,\text{ksi}$, $G = 11.5 \times 10^3\,\text{ksi}$, $b = 10\,\text{in.}$, $h = 6\,\text{in.}$, and $K = 1.2$.

SOLUTION: The free-body diagram of a segment of the arch is shown in Fig. 7.9c. At cross section C, the internal resisting forces are the bending moment M_s, the shear force V_s, and the normal force P_s. On this basis, strain energies due to bending, shear, and axial force will be stored in the arch. By applying static equilibrium equations, we obtain the following general expressions for

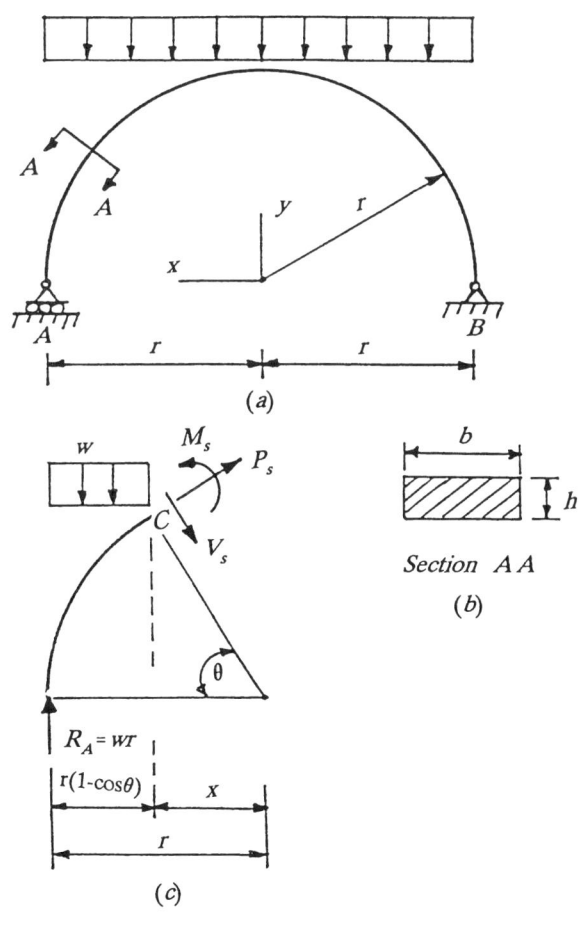

Figure 7.9. (a) Semicircular arch loaded as shown. (b) Cross-sectional shape of the arch. (c) Free-body diagram of a portion of the arch.

M_s, V_s, and P_s:

$$M_s = wr^2(1 - \cos\theta) - \frac{wr^2}{2}(1 - \cos\theta)^2 \tag{7.36}$$

$$P_s = -wr\cos^2\theta \tag{7.37}$$

$$V_s = wr\cos\theta\sin\theta \tag{7.38}$$

The strain energy U_1 stored in the arch by the bending moment M_s is

$$\begin{aligned}U_1 &= 2\int_0^{\pi/2} \frac{M_s^2}{2EI}\,d\theta \\ &= \frac{1}{EI}\int_0^{\pi/2}\left[wr^2(1-\cos\theta) - \frac{wr^2}{2}(1-\cos\theta)^2\right]^2 d\theta \\ &= \frac{1}{EI}\int_0^{\pi/2}\left[\frac{w^2r^4}{4} - \frac{w^2r^4}{2}\cos^2\theta + \frac{w^2r^4}{4}\cos^4\theta\right]d\theta \\ &= \frac{1}{EI}\left[\frac{w^2r^4\theta}{4} + \frac{w^2r^4}{2}\left(\frac{\theta}{2} + \frac{\sin 2\theta}{4}\right) + \frac{w^2r^4}{4}\left(\frac{3\theta}{8} + \frac{\sin 2\theta}{4} + \frac{\sin 4\theta}{32}\right)\right]_0^{\pi/2} \\ &= \frac{1}{EI}\left[\frac{w^2r^4\pi}{8} - \frac{w^2r^4}{2}\left(\frac{\pi}{4}\right) + \frac{w^2r^4}{4}\left(\frac{3\pi}{16}\right)\right] \tag{7.39} \\ &= \frac{3w^2r^4\pi}{64EI}\end{aligned}$$

By substituting in Eq. (7.39) the appropriate values for w, r, E, and I, we obtain

$$U_1 = \frac{(3)(2/12)^2(20\times 12)^4}{(64)(30)(10)^3[(10)(5)^3/12]} = 1.3824 \text{ kip-in.}$$

which provides the amount of strain energy stored in the arch as a result of bending.

The strain energy U_2 due to the axial force P_s is

$$\begin{aligned}U_2 &= 2\int_0^{\pi/2}\frac{P_s^2}{2AE}\,d\theta \\ &= \frac{1}{AE}\int_0^{\pi/2}[-wr\cos^2\theta]^2\,d\theta \\ &= \frac{1}{AE}\int_0^{\pi/2} w^2r^2\cos^2\theta\,d\theta \tag{7.40} \\ &= \frac{w^2r^2}{AE}\left[\frac{\theta}{2} + \frac{\sin 2\theta}{4}\right]_0^{\pi/2} \\ &= \frac{w^2r^2\pi}{4AE}\end{aligned}$$

Equation (7.40), by substitution, yields

$$U_2 = \frac{(2/12)^2(20 \times 12)^2 \pi}{(4)(10)(6)(30)(10)^3} = 0.0007 \text{ kip-in.} \tag{7.41}$$

This result shows that the strain energy stored in the arch by the axial force P_s is very small compared to the strain energy stored by the bending moment M_s.

The strain energy U_3 stored by the shear force V_s is

$$\begin{aligned} U_3 &= 2 \int_0^{\pi/2} \frac{K V_s^2}{2GA} d\theta \\ &= \frac{K}{GA} \int_0^{\pi/2} [wr \cos\theta \sin\theta]^2 \, d\theta \\ &= \frac{w^2 r^2 K}{GA} \int_0^{\pi/2} \cos^2\theta \sin^2\theta \, d\theta \\ &= \frac{w^2 r^2 K}{GA} \left[-\frac{\sin 4\theta}{32} + \frac{\theta}{8} \right]_0^{\pi/2} \\ &= \frac{w^2 r^2 K \pi}{16 GA} \end{aligned} \tag{7.42}$$

By substitution, we obtain

$$U_3 = \frac{(2/12)^2(20 \times 12)^2(1.2)\pi}{(16)(11.5)(10)^3(10)(6)} = 0.00055 \text{ kip-in.}$$

which shows that the strain energy stored in the arch by V_s is very small compared to the strain energy stored by M_s.

The total strain energy U stored in the arch is

$$\begin{aligned} U &= U_1 + U_2 + U_3 \\ &= 1.3824 + 0.0007 + 0.00055 \\ &= 1.38365 \text{ kip-in.} \end{aligned}$$

7.3 DEFLECTIONS BY EQUATING EXTERNAL AND INTERNAL WORK

For statically determinate structures that are subjected to an external force F, or an external moment M, a convenient way to determine the deflection in the direction of F, or the rotation in the direction of M, would be to equate the work done by F, or the work done by M, to the internal strain energy stored

480 ENERGY CONCEPTS AND METHODS

in the structure. In the preceding section, the expressions of strain energy that are developed for various cases of loadings may be used here for this purpose.

The following examples illustrate the application of the methodology, which is applicable to systems that obey the law of conservation of energy.

Example 7.9: For the uniform simply supported beam loaded as shown in Fig. 7.10a, determine the angular rotation at end A by equating the external work of the bending moment M_A to the internal strain energy stored in the member.

SOLUTION: By applying statics, we find that the reaction R_A is

$$R_A = \frac{M_A}{L} \tag{7.43}$$

The expression for the bending moment M at any distance x from support A is

$$M = M_A \left(1 - \frac{x}{L}\right) \tag{7.44}$$

The work W done by M_A moving through the rotation θ_A at end A of the member is

$$W = \frac{M_A \theta_A}{2} \tag{7.45}$$

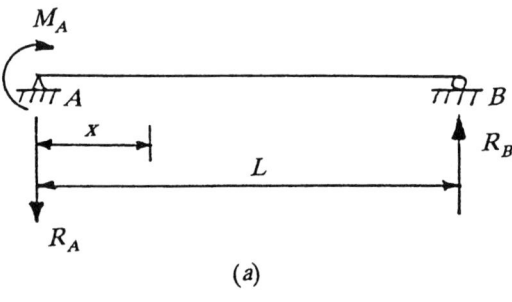

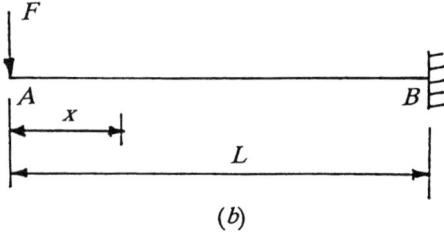

Figure 7.10. (*a*) Simply supported beam loaded as shown. (*b*) Cantilever beam loaded as shown.

By using Eq. (7.17) and Eq. (7.44), the internal strain energy U stored in the beam is

$$U = \int_0^L \frac{M^2\,dx}{2EI} = \int_0^L \frac{\{M_A[1-(x/L)]\}^2}{2EI}\,dx$$

$$= \frac{M_A^2}{2EI} \int_0^L \left(1 - \frac{x}{L}\right)^2 dx \qquad (7.46)$$

$$= \frac{M_A^2}{2EI} \left[x - \frac{x^2}{L} + \frac{x^3}{3L^2} \right]_0^L$$

$$= \frac{M_A^2 L}{6EI}$$

By applying the expression

$$W = U \qquad (7.47)$$

we find

$$\frac{M_A \theta_A}{2} = \frac{M_A^2 L}{6EI}$$

or

$$\theta_A = \frac{M_A L}{3EI} \qquad (7.48)$$

which is the required rotation at end A of the beam.

Example 7.10: Determine the vertical deflection y_A at end A of the uniform cantilever beam in Fig. 7.10b, by equating the work W done by the force F to the internal strain energy of bending and shear stored in the member ($W = U$).

SOLUTION: The work W done by the force F is

$$W = \frac{F y_A}{2} \qquad (7.49)$$

The internal strain energy U can be determined from the equation

$$U = \int_0^L \frac{M^2\,dx}{2EI} + \int_0^L \frac{KV^2}{2GA} \qquad (7.50)$$

482 ENERGY CONCEPTS AND METHODS

By applying statics, the expressions for M and V at any distance x from free end A are

$$M = -Fx \tag{7.51}$$

$$V = F \tag{7.52}$$

By substituting into Eq. (7.50), we find

$$\begin{aligned} U &= \int_0^L \frac{[-Fx]^2}{2EI} dx + \int_0^L \frac{KF^2}{2GA} dx \\ &= \frac{F^2}{2EI} \int_0^L x^2 \, dx + \frac{KF^2}{2GA} \int_0^L dx \\ &= \frac{F^2 L^3}{6EI} + \frac{KF^2 L}{2GA} \end{aligned} \tag{7.53}$$

By utilizing the expression $W = U$, we find

$$\frac{Fy_A}{2} = \frac{F^2 L^3}{6EI} + \frac{KF^2 L}{2GA}$$

or

$$y_A = \frac{FL^3}{3EI} + \frac{KFL}{GA} \tag{7.54}$$

The first term on the right-hand side of Eq. (7.54) is the deflection produced by the bending of the member, and the second term on the same side of the equation is the contribution of the shear.

Example 7.11: For the truss in Fig. 7.3, determine the vertical deflection y_C at the joint C of the truss by equating the work done by F to the internal strain energy stored in the truss ($W = U$). All required information is given in Example 7.2.

SOLUTION: The work W done by the vertical force F acting at the joint C of the truss is

$$W = \frac{Fy_C}{2} = 5{,}000 y_C \tag{7.55a}$$

The strain energy U stored in the truss is determined in Example 7.2, Table 7.1, and is as follows:

$$U = 1.9193 \text{ N} \cdot \text{m} \tag{7.55b}$$

From Eqs. (7.55a) and (7.55b), we find

$$5,000 y_C = 1.9193$$

or

$$y_C = \frac{1.9193}{5,000} = 0.000384 \text{ m}$$

$$= 0.384 \text{ mm}$$

7.4 THE METHOD OF VIRTUAL WORK WITH APPLICATIONS TO STATICALLY DETERMINATE AND STATICALLY INDETERMINATE PROBLEMS

The method of virtual work can be traced back to Aristotle (384–322 B.C.), who introduced the principle of virtual work $\Sigma F_i \delta_i = 0$ to prove the static equilibrium equation $\Sigma F_i x_i = 0$, and to Johann Bernoulli, who presented the principle of virtual velocities in 1717. On this basis, the principle of virtual work was formulated, which led to the development of a very powerful method that can be used for the computation of deflections and rotations in structures. The method is general, and it can be applied to trusses, beams, frames, and other structural components.

The principle of virtual work may be stated as follows:

> If a deformable structure is in equilibrium under the action of a given known load, or set of loads, and then is subjected to a virtual displacement by an additional action, the resulting external work of a given load, or loads, moving through the virtual displacement, is equal to the resulting internal work of the stresses produced by the given load, or set of loads.

It should be noted that the action producing the virtual displacement is independent of the system of loads acting on the structure just before the virtual displacement; the original given loads remain on the structure during the time the virtual displacement is taking place; and the original loads are the only ones producing the internal and external virtual work.

The original load system usually consists of a unit load, or a unit moment, that is applied to the structure at the point and direction where the required deflection, or required angular displacement, needs to be determined. For example, if it is required that we determine the vertical deflection at joint E of the truss in Fig. 7.3, we apply a vertical unit load at joint E, as shown in Fig. 7.11a. This is the original load system. The application of the unit vertical load at joint E will produce an axial force in each member of the truss, which we denote here as u.

484 ENERGY CONCEPTS AND METHODS

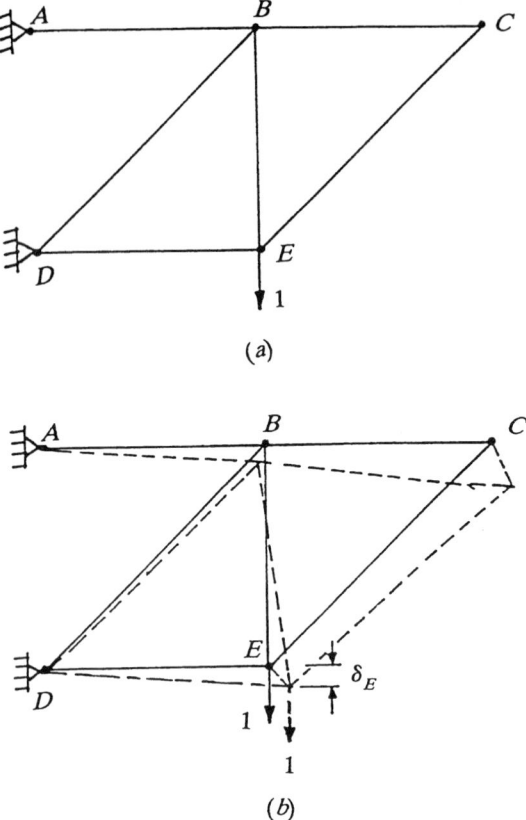

Figure 7.11. (a) Truss structure acted upon by the original load system consisting of a unit load applied as shown. (b) The same truss subjected to the deflection configuration created by the additional load P.

Figure 7.11b shows the same truss under the action of the additional vertical load P applied at joint C. The deflected configuration of the truss caused by the application of the load P is represented by the dashed lines shown in the same figure, and it took place from the original position where only the unit load was acting. On this basis, the deflection δ_E in the direction of the unit load is as shown in Fig. 7.11b. The external virtual work δW done by the unit load is the product of the unit load and δ_E.

In order to determine the internal virtual work stored in the truss, we note that each member of the truss changes in length by an amount equal to FL/AE, where F is the force in the member caused by the application of the additional load P, L is its length, A is its cross-sectional area, and E is its modulus of elasticity. The axial force u produced in each member of the truss moves through the length change FL/AE caused by the application of the additional

THE METHOD OF VIRTUAL WORK   485

load P. Thus, the work done by u in each member of the truss is equal to FLu/AE. Thus, the total internal virtual work δU of the truss is

$$\delta U = \sum \frac{FuL}{AE} \qquad (7.56)$$

By equating the external virtual work δW to the internal virtual work δU, we have

$$1 \times \delta_E = \sum \frac{FuL}{AE}$$

or

$$\delta_E = \sum \frac{FuL}{AE} \qquad (7.57)$$

The deflection at any other joint of the truss may be determined in a similar manner.

The application of the method is convenient, and it can be used for other types of structural problems. The following examples illustrate the application of the methodology to both statically determinate and statically indeterminate problems.

Example 7.12: For the steel truss of Example 7.2, determine the vertical displacement δ_E of joint E by using the method of virtual work.

SOLUTION: We apply first a unit load at joint E of the truss, as shown in Fig. 7.11a, and we calculate the force u in each member of the truss produced by the unit load. The calculated values of u are shown in the second column of Table 7.2. Next, we calculate the forces F in the members of the truss that are produced by the application of the load $P = 10\,\text{kN}$ at joint C. These forces are shown in the third column of Table 7.2. The last column of the table shows the value of FuL/AE for each member of the truss. The sum of the values of FuL/AE is also shown in the same table.

The vertical deflection δ_E of joint E may be obtained by equating external and internal work; that is,

$$1 \times \delta_E = \sum \frac{FuL}{AE}$$

or

$$\delta_E = 0.17675 \times 10^{-3}\,\text{m}$$
$$= 0.17675\,\text{mm}$$

486 ENERGY CONCEPTS AND METHODS

TABLE 7.2. Calculated Values of u, F, and FuL/AE

(1) Member	(2) u (N)	(3) F (kN)	(4) $A \times 10^{-3}$ (m²)	(5) $(L/AE) \times 10^{-9}$	(6) $(FuL/AE) \times 10^{-3}$ (m)
AB	+1.0000	+20.00	7.85	3.0330	0.06066
BC	0	+10.00	7.85	3.0330	0
DE	0	−10.00	7.85	3.0330	0
DB	−1.4142	−14.14	7.85	4.2888	0.08576
BE	+1.0000	+10.00	7.85	3.0330	0.03033
EC	0	−14.14	7.85	4.2888	0

$$\delta_E = \sum \frac{FuL}{AE} = 0.17675 \times 10^{-3} \text{ m}$$

$$= 0.1768 \text{ mm}$$

Example 7.13: For the steel truss of Example 7.12, determine the angular rotation θ_{BE} of member BE produced by the application of the vertical load $P = 10$ kN at joint C. Use the method of virtual work to obtain this result.

SOLUTION: In this case, the original system of loading will consist of a force equal to $1/L_{BE}$ applied at joint B, and a force equal to $1/L_{BE}$ applied at joint E, and directed as shown in Fig. 7.12. The internal forces u developed in the members of the truss are shown in the second column of Table 7.3. The internal forces F produced in the members by the application of the load $P = 10$ kN at joint C are shown in the third column of Table 7.3. The last column of the table gives the values of FuL/AE for each member of the truss.

The rotation θ_{BE} of member BE of the truss may be determined from the equation

$$1 \times \theta_{BE} = \sum \frac{FuL}{AE}$$

or

$$\theta_{BE} = 18.198 \times 10^{-6} \text{ rad}$$

Note that the original load system in Fig. 7.12 forms a moment couple whose magnitude is equal to unity.

Example 7.14: For the uniform statically indeterminate beam loaded as shown in Fig. 7.13a, determine the supporting reaction at end A by using the method of virtual work.

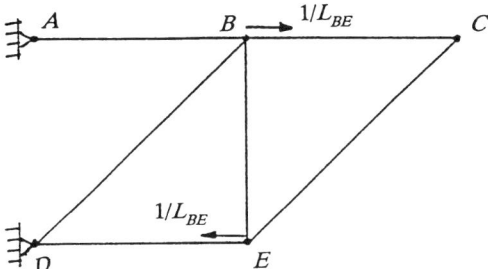

Figure 7.12. Truss structure showing the application of the original load system that is required to determine the rotation of member BE of the truss.

SOLUTION: The beam in Fig. 7.13a is statically indeterminate because we have three static equilibrium equations with four unknown reactive forces and moments at the supports. The horizontal reaction at end B is zero, since the applied load w does not have a component in the horizontal direction. By following the general rules for the solution of statically indeterminate problems, the reaction R_A at support A, or the moment M_B at the fixed support B, may be taken as the redundant quantity without destroying the static equilibrium of the member. Here we choose R_A to be the redundant quantity. On this basis, we have a statically determinate cantilever beam loaded with the distributed load w and the reaction R_A, as shown in Fig. 7.13b.

The method of virtual work can be easily used here to solve for the unknown reaction R_A. Since the purpose of the reaction R_A is to maintain zero deflection at support A, we choose the original load system to consist of a vertical unit load applied at end A of the cantilever beam, as shown in Fig. 7.13c.

TABLE 7.3. Calculated Values of u, F, and FuL/AE

(1) Member	(2) u (N)	(3) F (kN)	(4) $(L/AE) \times 10^{-9}$	(5) $(FuL/AE) \times 10^{-6}$ (rad)
AB	$1/L_{BE} = 0.20$	+20.00	3.0330	12.132
BC	0	+10.00	3.0330	0
DE	$-1/L_{BE} = -0.20$	−10.00	3.0330	6.066
DB	0	−14.14	4.2888	0
BE	0	+10.00	3.0330	0
EC	0	−14.14	4.2888	0

$$\theta = \sum \frac{FuL}{AE} = 18.198 \times 10^{-6} \text{ rad}$$

488 ENERGY CONCEPTS AND METHODS

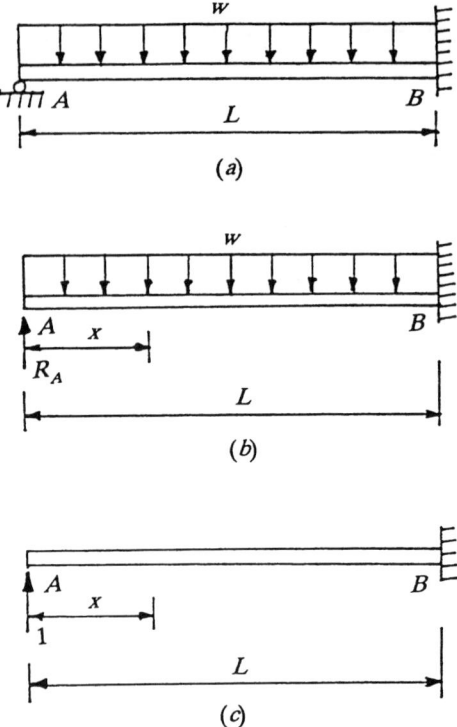

Figure 7.13. (*a*) Statically indeterminate single-span beam loaded as shown. (*b*) Cantilever beam loaded by the uniformly distributed load *w* and the redundant force R_A. (*c*) Original load system.

The expression for the bending moment M for the loading shown in Fig. 7.13*b* is

$$M = R_A x - \frac{wx^2}{2} \qquad (7.58)$$

The expression for the bending moment m caused by the unit load is

$$m = x \qquad (7.59)$$

If the vertical deflection at end A of the cantilever beam in Fig. 7.13*b* is denoted as y_A, then the work δW done by the unit load in Fig. 7.13*c* moving through the displacement y_A is

$$\delta W = 1 \times y_A \qquad (7.60)$$

By neglecting the effect of shear, the internal virtual work δU stored in the member may be determined from the expression

$$\delta U = \int_0^L \frac{Mm}{EI} dx \qquad (7.61)$$

By substituting Eqs. (7.58) and (7.59) into Eq. (7.61) and integrating, we find

$$\begin{aligned}\delta U &= \frac{1}{EI}\int_0^L \left(R_A x - \frac{wx^2}{2}\right)(x)\,dx \\ &= \frac{1}{EI}\int_0^L \left(R_A x^2 - \frac{wx^3}{2}\right)dx \\ &= \frac{R_A L^3}{3EI} - \frac{wL^4}{8EI}\end{aligned} \qquad (7.62)$$

By using the identity

$$\delta W = \delta U$$

we find

$$y_A = \frac{R_A L^3}{3EI} - \frac{wL^4}{8EI} \qquad (7.63)$$

Since the purpose of the reaction R_A is to maintain zero deflection at the end support A, we have $y_A = 0$, and Eq. (7.63) yields

$$\frac{R_A L^3}{3EI} - \frac{wL^4}{8EI} = 0$$

or

$$R_A = \frac{3wL}{8} \qquad (7.64)$$

Equation (7.64) yields the value of the reaction R_A at support A of the statically indeterminate member in Fig. 7.13a. The three reactions at support B, consisting of vertical and horizontal reactions and a bending moment, may be determined by using the three equations of equilibrium from statics. From $\Sigma F_x = 0$, we note that the horizontal reaction at B is equal to zero.

Example 7.15: For the semicircular arch loaded as shown in Fig. 7.9a, determine the horizontal displacement at support A by using the method of virtual work. Assume that $r = 20$ ft, $w = 2$ kips/ft, $E = 30 \times 10^3$ ksi, $b = 10$ in., and $h = 6$ in. Consider only the effect of bending.

490 ENERGY CONCEPTS AND METHODS

SOLUTION: Since we are interested in determining the horizontal displacement of support A, the original force system would consist of a horizontal unit load applied at support A, as shown in Fig. 7.14a. The expression for the bending moment m that is produced at cross sections of the arch may be determined from the free-body diagram in Fig. 7.14b, and it is as follows:

$$m = r \sin \theta \tag{7.65}$$

The moment M produced by the distributed load w in Fig. 7.9a is determined in Example 7.8, Eq. (7.36), and it is as follows:

$$M = wr^2(1 - \cos \theta) - \frac{wr^2}{2}(1 - \cos \theta)^2 \tag{7.66}$$

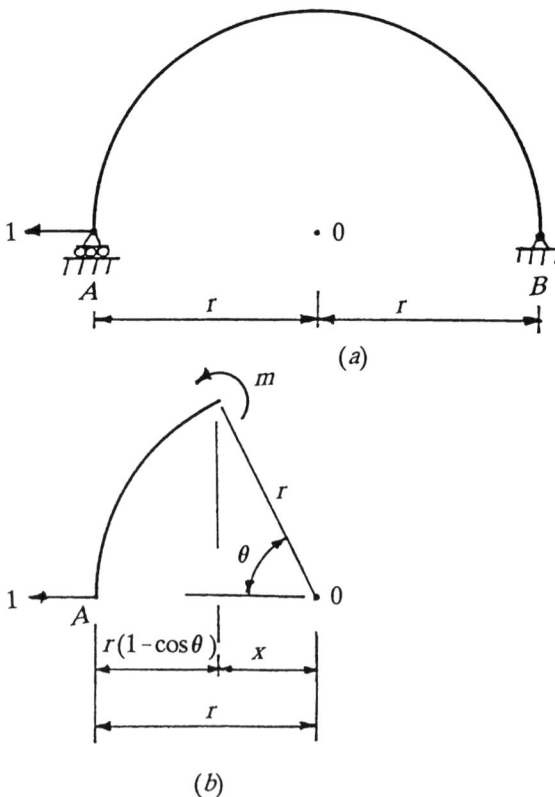

Figure 7.14. (a) Original load system consisting of a horizontal unit load located as shown. (b) Free-body diagram of a segment of the arch.

If the horizontal displacement of the arch at support A is denoted as δ_A, the external work δW done by the unit load is

$$\delta W = 1 \times \delta_A \tag{7.67}$$

The internal work δU stored in the arch is

$$\delta U = \int_0^\pi \frac{Mm}{EI} d\theta$$
$$= 2 \int_0^{\pi/2} \frac{Mm}{EI} d\theta \tag{7.68}$$

By substituting Eqs. (7.65) and (7.66) into Eq. (7.68) and integrating, we find

$$\delta U = \frac{2}{EI} \int_0^{\pi/2} \left[wr^2(1 - \cos\theta) - \frac{wr^2}{2}(1 - \cos\theta)^2 \right] [r \sin\theta] \, d\theta$$
$$= \frac{wr^3}{EI} \int_0^{\pi/2} \sin^3\theta \, d\theta \tag{7.69}$$
$$= \frac{wr^3}{EI} \left[-\frac{1}{3} \cos\theta(\sin^2\theta + 2) \right]_0^{\pi/2}$$
$$= \frac{2wr^3}{3EI}$$

From the relation $\delta W = \delta U$, we find

$$1 \times \delta_A = \frac{2wr^3}{3EI}$$

or

$$\delta_A = \frac{2wr^3}{3EI} \tag{7.70}$$

which is the horizontal displacement at support A of the arch in Fig. 7.9a.
By substituting the appropriate values for w, r, E, and I, Eq. (7.70) yields

$$\delta_A = \frac{(2)(2/12)(20 \times 12)^3}{(3)(30)(10)^3[(10)(6)^3]/12}$$
$$= 0.284 \text{ in.}$$

7.5 CASTIGLIANO'S SECOND THEOREM WITH APPLICATIONS TO STATICALLY DETERMINATE AND STATICALLY INDETERMINATE STRUCTURES

Alberto Castigliano (1847–1884) of Italy developed his famous energy thorems in 1870, and they are still used extensively today by many mathematicians, physicists, and engineers. The second theorem of Castigliano is discussed in this section. For information regarding his first theorem, the reader may refer to Reference [6]. In words, this theorem may be expressed as follows:

> If a linearly elastic structure is subjected to a system of loads, which includes both forces and moments, the displacement in the direction of one of these loads is equal to the partial derivative, or rate of change, of the total strain energy with respect to that load.

If the total strain energy is designated as U, and we need to determine the deflection δ in the direction of a concentrated load P, then, in accordance with this theorem, we have

$$\delta = \frac{\partial U}{\partial P} \tag{7.71}$$

If we need to determine the angular rotation θ, in radians, in the direction of a concentrated moment M, then we have

$$\theta = \frac{\partial U}{\partial M} \tag{7.72}$$

Equations (7.71) and (7.72) may be used to determine deflections and rotations in both statically determinate and statically indeterminate structures. Their application, however, requires us to express U in terms of P, if the deflection in the direction of P is required, or in terms of M, if the rotation in the direction of M is required, and to take the partial derivative with respect to P, or with respect to M. Methods of determining the strain energy U for bending, shear, torsion, and axial force are discussed in detail in the preceding sections of this chapter.

The methodology is general, and it can be applied to both statically determinate and statically indeterminate problems. If we need to determine the deflection, or rotation, at a point of a structure where there is no P or M, then a fictitious load P_0, or a fictitious moment M_0, may be applied. In such a case, the strain energy U would be in terms of P_0 or M_0. The partial derivative of U with respect to P_0 yields the deflection in the direction of P_0, and the partial derivative of U with respect to M_0 yields the rotation in the sense of M_0. After the partial derivatives are carried out, then we can set P_0 or M_0 equal to zero, because they are fictitious.

The following examples illustrate the methodology to both statically determinate and statically indeterminate problems.

Example 7.16: For the simple truss in Example 7.2, Fig. 7.3, determine the vertical deflection δ_C in the direction of the applied load P at joint C, by using Castigliano's second theorem.

SOLUTION: The total strain energy U of the truss, in terms of the applied load P, is determined as shown in Table 7.4, and it is as follows:

$$U = \sum \frac{F^2 L}{2AE} = 19.1929 P^2 (10)^{-9} \text{ N} \cdot \text{m}$$

In accordance with Eq. (7.71), the deflection δ_C in the direction of the load P is

$$\delta_C = \frac{\partial U}{\partial P} = (2)(19.1929) P (10)^{-9}$$

$$= (38.3858)(10,000)(10)^{-9}$$

$$= 0.000384 \text{ m}$$

$$= 0.384 \text{ mm}$$

Total derivative of U with respect to P could be also used here, because U is only in terms of P.

Note that δ_C is identical to the value y_c obtained in Example 7.11.

Example 7.17: For the simply supported beam in Example 7.9, Fig. 7.10a, determine the angular rotation θ_A at end A, by using Castigliano's second theorem. Consider only strain energy of bending.

TABLE 7.4. Energy Stored in the Truss by the Application of Load P at Joint C of the Truss

(1) Member	(2) Axial Force F (N)	(3) Length L (m)	(4) $(L/2AE) \times 10^{-9}$	(5) $(F^2 L/2AE) \times 10^{-9}$ (N·m)
AB	$+2P$	5.00	1.5165	$6.066 P^2$
BC	$+P$	5.00	1.5165	$1.5165 P^2$
DE	$-P$	5.00	1.5165	$1.5165 P^2$
DB	$-1.4142 P$	7.07	2.1444	$4.2887 P^2$
BE	$+P$	5.00	1.5165	$1.5165 P^2$
EC	$-1.4142 P$	7.07	2.1444	$4.2887 P^2$

$$U = \sum \frac{F^2 L}{2AE} = 19.1929 P^2$$

494   ENERGY CONCEPTS AND METHODS

SOLUTION: From Example 7.9, the total internal strain energy U, in terms of M_A, stored in the member, is

$$U = \frac{M_A^2 L}{6EI}$$

In accordance with Eq. (7.72), the rotation θ_A at end A is

$$\theta_A = \frac{\partial U}{\partial M_A} = \frac{M_A L}{3EI}$$

which is identical to Eq. (7.48) obtained in Example 7.9.

Example 7.18: For the simple truss in Example 7.2, Fig. 7.3, determine the vertical deflection δ_E of joint E by using Castigliano's second theorem.

SOLUTION: In this case we do not have a vertical load at joint E. In order to make the application of the method possible, we apply a fictitious vertical load P_0 at joint E, as shown in Fig. 7.15.

The second column in Table 7.5 shows the axial force F in each member of the truss, resulting from the application of the loads $P = 10\,\text{kN}$ and P_0. The last column of the table gives the internal strain energy stored in each member of the truss by the application of P and P_0. Thus, the total strain energy U stored in the truss, is

$$U = \sum \frac{F^2 L}{2AE}$$

$$= 0.7322 + (1{,}187.0013 + 176.7523 P_0 + 7.3218 P_0^2) 10^{-3}$$

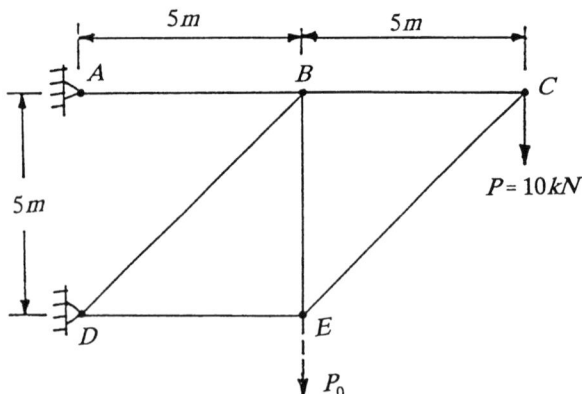

Figure 7.15. Simple truss loaded with the load $P = 10\,\text{kN}$ and the fictitious load P_0, as shown.

TABLE 7.5. Energy Stored in the Truss by the Application of the Loads $P = 10$ kN and P_0

(1) Member	(2) Axial Force F (kN)	(3) Length L (m)	(4) $(L/2AE) \times 10^{-9}$	(5) $(F^2L/2AE)$ (N·m)
AB	$20.00 + P_0$	5.00	1.5165	$(20 + P_0)^2(1.5165)(10)^{-3}$
BC	10.00	5.00	1.5165	0.1517
DE	-10.00	5.00	1.5165	0.1517
DB	$-14.14 - 1.4142P_0$	7.07	2.1444	$(14.14 + 1.4142P_0)^2(2.1444)(10)^{-3}$
BE	$10.00 + P_0$	5.00	1.5165	$(10 + P_0)^2(1.5165)(10)^{-3}$
EC	-14.14	7.07	2.1444	0.4288

$$\sum \frac{F^2 L}{2AE} = 0.7322 + 10^{-3}(1{,}187.0013 + 176.7523P_0 + 7.3218P_0^2)$$

496 ENERGY CONCEPTS AND METHODS

By applying Eq. (7.71), we can determine the vertical deflection δ_E of joint E of the truss as follows:

$$\delta_E = \frac{\partial U}{\partial P_0} = 176.7523(10)^{-3} + 14.6436 P_0 (10)^{-3}$$

Since P_0 is a fictitious force, its magnitude is zero and, consequently,

$$\delta_E = 176.7523(10)^{-3} \text{ m}$$
$$= 0.17675 \text{ mm}$$

which is identical to the value obtained in Example 7.12.

Example 7.19: For the two-span continuous beam loaded as shown in Fig. 7.16a, determine the reaction R_C at support C by using Castigliano's second theorem. The stiffness EI of the beam is constant and equal to 60×10^6 kip-in^2.

SOLUTION: The continuous beam in Fig. 7.16a is statically indeterminate. We consider the vertical reaction R_C at support C as this redundant quantity. This is also convenient, since we need to determine R_C. On this basis, the problem is reduced to a simply supported beam with an overhang, and loaded with the uniformly distributed load w and reaction R_C, as shown in Fig. 7.16b.

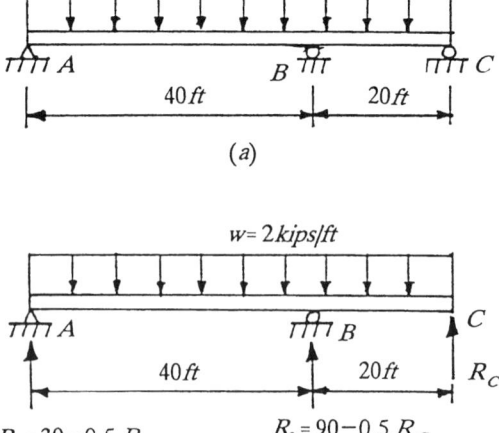

Figure 7.16. (a) Two-span continuous beam loaded as shown. (b) Simply supported beam with overhang loaded by w and R_C, as shown.

CASTIGLIANO'S SECOND THEOREM 497

By using statics, the expressions of the bending moment M for portions AB and BC, with support A as the origin, are as follows:

$$M = 30x - 0.5R_C x - x^2 \qquad 0 \leqslant x \leqslant 40\,\text{ft} \qquad (7.73)$$

$$M = 3{,}600 - 20R_C - 120x + R_C x - x^2 \qquad 40\,\text{ft} \leqslant x \leqslant 60\,\text{ft} \qquad (7.74)$$

We neglect in this problem the effect of shear as being negligibly small.
The strain energy U of bending is

$$U = \int \frac{M^2\,dx}{2EI}$$

$$= \frac{1}{2EI}\int_0^{40}(30x - 0.5R_C x - x^2)^2\,dx$$

$$+ \frac{1}{2EI}\int_{40}^{60}(3{,}600 - 20R_C - 120x + R_C x + x^2)^2\,dx$$

$$= \frac{1}{2EI}\left[30x^3 - 10R_C x^3 - 15x^4 + \frac{0.25R_C^2 x^3}{3} + \frac{R_C x^4}{4} + \frac{x^5}{5}\right]_0^{40}$$

$$+ \frac{1}{2EI}\left[12.96(10)^6 x - 144(10)^3 R_C x - 432(10)^3 x^2 + 6(10)^3 R_C x^2\right.$$

$$+ 7.2(10)^3 x^3 + 400R_C^2 x - 20R_C^2 x^2 - \frac{280 R_C x^3}{3} - 60x^4$$

$$\left.+ \frac{R_C^2 x^3}{3} + 0.5R_C x^4 + \frac{x^5}{5}\right]_{40}^{60}$$

$$= \frac{1}{2EI}[-15.56(10)^6 + 133.333(10)^3 R_C + 24(10)^3 R_C^2]$$

Knowing that the deflection δ_C at support C of the original problem in Fig. 7.16a is zero, the reaction R_C at support C can be determined by equating to zero the partial derivative of the strain energy U with respect to R_C. It yields

$$\delta_C = \frac{\partial U}{\partial R_C} = \frac{1}{2EI}[133.333(10)^3 + 48(10)^3 R_C] = 0$$

or, by solving for R_C,

$$R_C = -2.777\,\text{kips}$$

The minus sign indicates that R_C is directed downward. The assumed upward direction in Fig. 7.16b is not correct.

7.6 INTRODUCTION TO THE FINITE ELEMENT METHOD

The basic purpose in the finite element method is to subdivide a given structure into a finite number of simple elements, and then to solve the complex differential equations by using the simple elements. We transform the problem into a linear algebra problem that can be solved rather conveniently by using computer hardware and computer software. On this basis, we can solve very complex structural and mechanical problems to determine stress and deflection characteristics throughout the structure, print tables of deflection and stress, and perform important research analysis and design configurations, which would be extremely time consuming if performed manually by an engineer.

In order to perform a finite element analysis, we must idealize the physical structural system and prepare a mathematical model that includes simple Euclidean geometry, boundary conditions, material behavior, displacement continuity, and stress equilibrium. The displacement boundary conditions prescribe displacement or rotation components at each node of the structure, and the forces are idealized as point forces, uniformly distributed, or as forces whose intensity may vary linearly or quadratically along the boundary.

The structural system is subdivided into pieces, and a particular element type is associated with each of the structural pieces. The subdivision process is completed when we assign numerical values to the element stiffness matrix, and to the equivalent loading vector coefficients for all element models of the structural system. The most important responsibility of the engineer performing finite element analysis is to interpret numerical results. The engineer has to make sure that the idealization of the subdivision process used for the structural system, as well as any computer-induced errors, do not make the analytical results irrelevant.

Historically, the basis of the continuum discretization may be attributed to the Halzer–Gumbel–Tolle–van den Dungen method and the transfer matrix method [5, 37, 39, 76]. The origin, however, of the finite element method, as such, may be traced back to the 1943 paper by Courant [77], who suggested the division of a continuum problem into discrete triangles; the development may be attributed to the intense interest of the aircraft industry. The method became a reality in a paper by Turner et al. [78], published in 1956. Clough [79, 80], one of Turner's coauthors, later coined the term "finite element method" in his 1960 paper. Argyris is another major contemporary in the development of the finite element method [81], but he did not use the term "finite element." His work was published in the *Journal of Aircraft Engineering* in 1954 and 1955. In 1966, a conference on matrix methods in structural mechanics [82] was held at Wright-Paterson Air Force Base in Ohio, and the implementation of the method on the computer had started. Under a NASA project, the computer program NASTRAN (NASA Structural Analysis Program) was developed.

Fundamental aspects and applications of the finite element method are discussed in this section. Extensive discussions and applications of the finite element method may be found in References [3, 5, 6, 83–85], and in the more than 8,000 journal papers and 200 textbooks and monographs published on this subject [86].

7.6.1 Stiffness Matrix of an Axial or Truss Bar Element with Application to Trusses

We derive here the stiffness matrix for an axial or truss bar element that can be used for the analysis of trusses and members that are subjected to axial forces. Since certain members of truss structures are inclined, or have different orientations, a global coordinate system will be also established that is common to all members.

Consider the truss bar element shown in Fig. 7.17a, where f_1 and f_2 are the axial forces at nodes 1 and 2, respectively, and u_1 and u_2 are the respective displacements in the x direction. By using Eq. (2.8) from Section 2.2.1, we can write the following equation for the displacement u_2:

$$u_2 = \frac{f_2 L}{AE} + u_1 \qquad (7.75)$$

where A is the cross-sectional area of the element and E is the modulus of elasticity.

By applying static equilibrium to the element in Fig. 7.17a, we find

$$f_1 = -f_2 \qquad (7.76)$$

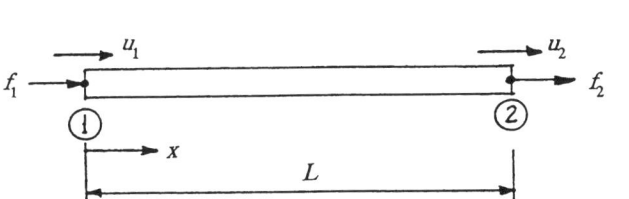

(a)

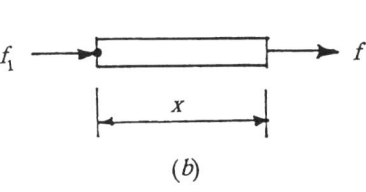

(b)

Figure 7.17. (a) Truss bar element. (b) Free-body diagram of a portion of the element.

By substituting Eq. (7.76) into Eq. (7.75) and solving for f_1, we obtain

$$f_1 = \frac{AE}{L}(u_1 - u_2) \qquad (7.77)$$

and

$$f_2 = -\frac{AE}{L}(u_1 - u_2) \qquad (7.78)$$

Equations (7.77) and (7.78) may be written in matrix form as follows:

$$\begin{bmatrix} f_1 \\ f_2 \end{bmatrix} = \frac{AE}{L} \begin{bmatrix} 1 & -1 \\ -1 & 1 \end{bmatrix} \begin{bmatrix} u_1 \\ u_2 \end{bmatrix} \qquad (7.79)$$

In a more compact form, we may write Eq. (7.79) as

$$\{f\} = [K]\{u\} \qquad (7.80)$$

where

$$\{f\} = \begin{bmatrix} f_1 \\ f_2 \end{bmatrix} \qquad (7.81)$$

$$\{u\} = \begin{bmatrix} u_1 \\ u_2 \end{bmatrix} \qquad (7.82)$$

are, respectively, the force and displacement vectors, and

$$[K] = \frac{AE}{L} \begin{bmatrix} 1 & -1 \\ -1 & 1 \end{bmatrix} \qquad (7.83)$$

is the stiffness matrix for the truss element. In Eq. (7.83), the term AE/L is the axial stiffness of the truss bar, which indicates that the bar responds like a spring of spring constant equal to AE/L and units of force per unit length.

We consider now the case where a member of a truss has a different orientation, such as the one shown in Fig. 7.18. In this figure, we choose the x, y axes to be the global system coordinates, and we also choose this global coordinate system to be common for all members of a truss. In Fig. 7.18a, the angle of inclination is designated as α, and we relate the components $\bar{u}_1, \bar{v}_1, \bar{u}_2$, and $\bar{v}_2$ of the displacements u_1 and u_2 along the x and y global coordinates by the following equations:

$$u_1 = \bar{u}_1 \cos \alpha + \bar{v}_1 \sin \alpha \qquad (7.84)$$

$$u_2 = \bar{u}_2 \cos \alpha + \bar{v}_2 \sin \alpha \qquad (7.85)$$

INTRODUCTION TO THE FINITE ELEMENT METHOD 501

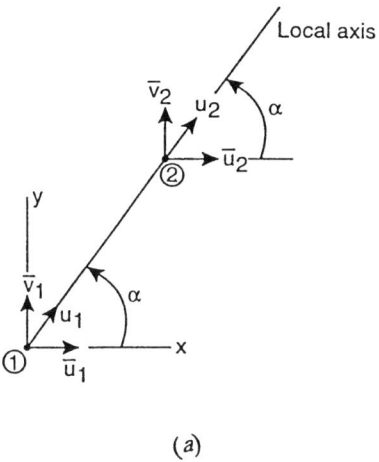

(a)

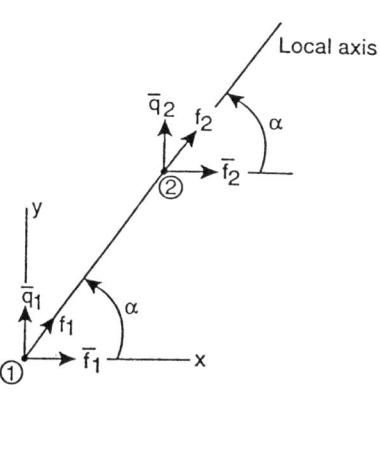

(b)

Figure 7.18. (a) Components of u_1 and u_2 in the global coordinates. (b) Components of f_1 and f_2 in the global coordinates.

In matrix form, we write Eqs. (7.84) and (7.85) as

$$\begin{bmatrix} u_1 \\ u_2 \end{bmatrix} = \begin{bmatrix} \cos\alpha & \sin\alpha & 0 & 0 \\ 0 & 0 & \cos\alpha & \sin\alpha \end{bmatrix} \begin{bmatrix} \bar{u}_1 \\ \bar{v}_1 \\ \bar{u}_2 \\ \bar{v}_2 \end{bmatrix} \quad (7.86)$$

or, in a more compact form,
$${s} = [T]{\bar{s}} \tag{7.87}$$
where
$$[T] = \begin{bmatrix} \cos \alpha & \sin \alpha & 0 & 0 \\ 0 & 0 & \cos \alpha & \sin \alpha \end{bmatrix} \tag{7.88}$$

is known as the transformation matrix.

In a similar manner, we can write the global components $\bar{f}_1$, $\bar{q}_1$, $\bar{f}_2$, and $\bar{q}_2$ of the forces f_1 and f_2 in Fig. 7.18b, as follows:

$$\bar{f}_1 = f_1 \cos \alpha \tag{7.89}$$
$$\bar{q}_1 = f_1 \sin \alpha \tag{7.90}$$
$$\bar{f}_2 = f_2 \cos \alpha \tag{7.91}$$
$$\bar{q}_2 = f_2 \sin \alpha \tag{7.92}$$

Equations (7.89) through (7.92) are written in matrix form as

$$\begin{bmatrix} \bar{f}_1 \\ \bar{q}_1 \\ \bar{f}_2 \\ \bar{q}_2 \end{bmatrix} = \begin{bmatrix} \cos \alpha & 0 \\ \sin \alpha & 0 \\ 0 & \cos \alpha \\ 0 & \sin \alpha \end{bmatrix} \begin{bmatrix} f_1 \\ f_2 \end{bmatrix} \tag{7.93}$$

or, in a more compact form,
$${\bar{S}} = [T^T]{S} \tag{7.94}$$
where
$$[T^T] = \begin{bmatrix} \cos \alpha & 0 \\ \sin \alpha & 0 \\ 0 & \cos \alpha \\ 0 & \sin \alpha \end{bmatrix} \tag{7.95}$$

is the transformation matrix for the forces f_1 and f_2.

By substituting Eq. (7.86) into Eq. (7.79) and performing the matrix multiplication, we obtain

$$\begin{bmatrix} f_1 \\ f_2 \end{bmatrix} = \frac{AE}{L} \begin{bmatrix} \cos \alpha & \sin \alpha & -\cos \alpha & -\sin \alpha \\ -\cos \alpha & -\sin \alpha & \cos \alpha & \sin \alpha \end{bmatrix} \begin{bmatrix} \bar{u}_1 \\ \bar{v}_1 \\ \bar{u}_2 \\ \bar{v}_2 \end{bmatrix} \tag{7.96}$$

or

$$\{\bar{f}\} = [B]\{\bar{s}\} \tag{7.97}$$

where

$$[B] = [K][T] \tag{7.98}$$

On this basis, we can use Eq. (7.96) to determine the truss bar forces from the global coordinate displacement components.

Now we first premultiply both sides of Eq. (7.96) by the transpose of the transformation matrix, Eq. (7.88), then replace the left-hand side of the resulting equation by Eq. (7.93). If we carry out the matrix multiplication on the right-hand side of the resulting equation, we obtain

$$\begin{bmatrix} \bar{f}_1 \\ \bar{q}_1 \\ \bar{f}_2 \\ \bar{q}_2 \end{bmatrix} = \frac{AE}{L} \begin{bmatrix} \cos^2\alpha & \cos\alpha\sin\alpha & -\cos^2\alpha & -\cos\alpha\sin\alpha \\ \cos\alpha\sin\alpha & \sin^2\alpha & -\cos\alpha\sin\alpha & -\sin^2\alpha \\ -\cos^2\alpha & -\cos\alpha\sin\alpha & \cos^2\alpha & \cos\alpha\sin\alpha \\ -\cos\alpha\sin\alpha & -\sin^2\alpha & \cos\alpha\sin\alpha & \sin^2\alpha \end{bmatrix} \begin{bmatrix} \bar{u}_1 \\ \bar{v}_1 \\ \bar{u}_2 \\ \bar{v}_2 \end{bmatrix}$$

$$\tag{7.99}$$

which in compact form is written as

$$\{\bar{S}\} = [\bar{K}]\{\bar{s}\} \tag{7.100}$$

where

$$[\bar{K}] = \frac{AE}{L} \begin{bmatrix} \cos^2\alpha & \cos\alpha\sin\alpha & -\cos^2\alpha & -\cos\alpha\sin\alpha \\ \cos\alpha\sin\alpha & \sin^2\alpha & -\cos\alpha\sin\alpha & -\sin^2\alpha \\ -\cos^2\alpha & -\cos\alpha\sin\alpha & \cos^2\alpha & \cos\alpha\sin\alpha \\ -\cos\alpha\sin\alpha & -\sin^2\alpha & \cos\alpha\sin\alpha & \sin^2\alpha \end{bmatrix} \tag{7.101}$$

is the truss stiffness matrix in the global coordinate system, and Eq. (7.100) is the force-displacement equation in the global coordinates. Note that the local axis in Fig. 7.18 is directed from endpoint 1 of the member to its endpoint 2. The angle α is measured counterclockwise from the global x axis to the local axis.

The following example illustrates the element assembly and the application of the methodology.

Example 7.20: For the three-bar truss loaded as shown in Fig. 7.19a, determine the forces acting in the three bars of the truss. The inclined bar has a cross-sectional area of 20 in.^{2}, and each of the other two bars has a cross-sectional area of 10 in.^{2} The modulus of elasticity $E = 30 \times 10^3$ ksi.

504   ENERGY CONCEPTS AND METHODS

SOLUTION: We begin the solution by numbering the elements as shown in Fig. 7.19b. We note that joint 2 has a horizontal degree of freedom r_1 and a vertical degree of freedom r_2. Joint 3 has a horizontal degree of freedom r_3. The lengths L_1, L_2, and L_3 of the three bars 1, 2, and 3, respectively, are

$$L_1 = 15 \text{ ft}$$
$$L_2 = 30 \text{ ft}$$
$$L_3 = 33.54 \text{ ft}$$

In the global coordinates, the element force-displacement expressions may be determined from Eq. (7.99). If we designate by x_1 and y_1 the global coordinates of the first end of the element, and by x_2 and y_2 the global coordinates of the second end, the values of $\cos \alpha$, $\sin \alpha$, and L can be obtained from the following general expressions:

$$\cos \alpha = \frac{x_2 - x_1}{L} \tag{7.102}$$

$$\sin \alpha = \frac{y_2 - y_1}{L} \tag{7.103}$$

$$L = \sqrt{(x_2 - x_1)^2 + (y_2 - y_1)^2} \tag{7.103a}$$

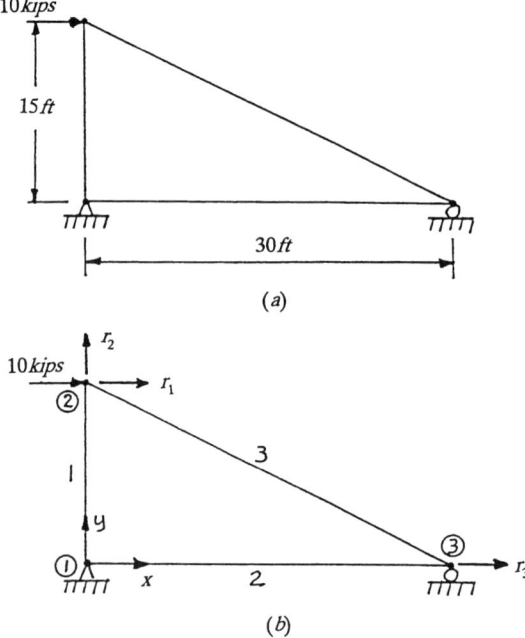

Figure 7.19. (a) Three-bar truss loaded as shown. (b) Degrees of freedom of the truss.

INTRODUCTION TO THE FINITE ELEMENT METHOD   505

For element 1, Eqs. (7.102) and (7.103) yield

$$\cos \alpha = \frac{0-0}{15} = 0 \tag{7.104}$$

$$\sin \alpha = \frac{15-0}{15} = 1 \tag{7.105}$$

By using Eqs. (7.104) and (7.105), the force-displacement expressions for element 1 can be determined from Eq. (7.99), and they are as follows:

$$\begin{bmatrix} \bar{f}_1 \\ \bar{q}_1 \\ \bar{f}_2 \\ \bar{q}_2 \end{bmatrix}^1 = \begin{matrix} \mathbf{0} & \mathbf{0} & \mathbf{r_1} & \mathbf{r_2} \\ \begin{bmatrix} 0 & 0 & 0 & 0 \\ 0 & 1{,}666.67 & 0 & -1{,}666.67 \\ 0 & 0 & 0 & 0 \\ 0 & -1{,}666.67 & 0 & 1{,}666.67 \end{bmatrix} \end{matrix} \begin{bmatrix} 0 \\ 0 \\ r_1 \\ r_2 \end{bmatrix} \tag{7.106}$$

For element 2, we have

$$\cos \alpha = \frac{30-0}{30} = 1 \tag{7.107}$$

$$\sin \alpha = \frac{0-0}{30} = 0 \tag{7.108}$$

Thus, from Eq. (7.99), we find

$$\begin{bmatrix} \bar{f}_1 \\ \bar{q}_1 \\ \bar{f}_2 \\ \bar{q}_2 \end{bmatrix}^2 = \begin{matrix} \mathbf{0} & \mathbf{0} & \mathbf{r_3} & \mathbf{0} \\ \begin{bmatrix} 833.33 & 0 & -833.33 & 0 \\ 0 & 0 & 0 & 0 \\ -833.33 & 0 & 833.33 & 0 \\ 0 & 0 & 0 & 0 \end{bmatrix} \end{matrix} \begin{bmatrix} 0 \\ 0 \\ r_3 \\ 0 \end{bmatrix} \tag{7.109}$$

For element 3, we have

$$\cos \alpha = \frac{30-0}{33.54} = 0.8945 \tag{7.110}$$

$$\sin \alpha = \frac{-15-0}{33.54} = -0.4472 \tag{7.111}$$

and from Eq. (7.99) we obtain

$$\begin{bmatrix} \bar{f}_1 \\ \bar{q}_1 \\ \bar{f}_2 \\ \bar{q}_2 \end{bmatrix}^3 = \begin{bmatrix} \overset{r_1}{1{,}192.80} & \overset{r_2}{-596.33} & \overset{r_3}{-1{,}192.80} & \overset{0}{596.33} \\ -596.33 & 298.13 & 596.33 & -298.13 \\ -1{,}192.80 & 596.33 & 1{,}192.80 & -596.33 \\ 596.33 & -298.13 & -596.33 & 298.13 \end{bmatrix} \begin{bmatrix} r_1 \\ r_2 \\ r_3 \\ 0 \end{bmatrix} \quad (7.112)$$

We are ready now to begin the assembly process of the elements by satisfying equilibrium conditions in terms of the structure stiffness matrix and external force, or load vector. Since we have three degrees of freedom, we have a 3 × 3 structure stiffness matrix. There is only one externally applied load of 10 kips acting at node 2 in the positive horizontal direction. The other two terms of the load vector are zero. Each element of the structure stiffness matrix is obtained by adding algebraically the contributing terms from the element stiffness matrices in Eqs. (7.106), (7.109), and (7.112). On this basis, the structure stiffness matrix represents three equilibrium equations associated with the three degrees of freedom, which are put in matrix form.

The structure stiffness matrix is as follows:

$$\begin{matrix} 1 \\ 2 \\ 3 \end{matrix} \begin{bmatrix} \overset{r_1}{1{,}192.80} & \overset{r_2}{-596.33} & \overset{r_3}{-1{,}192.80} \\ -596.33 & (1{,}666.67 + 298.13) & 596.33 \\ -1{,}192.80 & 596.33 & (833.33 + 1{,}192.80) \end{bmatrix} \begin{bmatrix} r_1 \\ r_2 \\ r_3 \end{bmatrix} = \begin{bmatrix} 10 \\ 0 \\ 0 \end{bmatrix} \quad (7.113)$$

In the first row of Eq. (7.113), the first element is the algebraic sum of the third element of the third row of Eq. (7.106), which is zero in this case, and the first element of the first row of Eq. (7.112), which is equal to 1,192.80. The second term of the same row is the algebraic sum of the fourth element of the third row in Eq. (7.106), which is zero, and the second element of the first row in Eq. (7.112), yielding -596.33. The third element of the row is $-1,192.80$, which is the third element of the first row in Eq. (7.112). The completion of the first row in Eq. (7.113) establishes the requirement for equilibrium in the r_1 direction. In a similar manner, the elements of the remaining rows of the structural stiffness matrix in Eq. (7.113) are determined.

The solution of the three expressions represented by Eq. (7.113) yields the following values for the displacements r_1, r_2, and r_3:

$$\begin{bmatrix} r_1 \\ r_2 \\ r_3 \end{bmatrix} = \begin{bmatrix} 0.02188 \\ 0.00300 \\ 0.01200 \end{bmatrix} \quad (7.114)$$

The forces in the bars of the truss may be determined by using Eq. (7.96), because we know the displacements r_1, r_2, and r_3. For element 1, Eq. (7.96) yields

$$\begin{bmatrix} f_1 \\ f_2 \end{bmatrix}^1 = \frac{(10)(30)(10)^3}{(15)(12)} \begin{bmatrix} 0 & 1 & 0 & -1 \\ 0 & -1 & 0 & 1 \end{bmatrix} \begin{bmatrix} 0 \\ 0 \\ 0.02188 \\ 0.00300 \end{bmatrix} \quad (7.115)$$

$$= \begin{bmatrix} -5.00 \\ 5.00 \end{bmatrix}$$

For element 2, we find

$$\begin{bmatrix} f_1 \\ f_2 \end{bmatrix}^2 = \frac{(10)(30)(10)^3}{(30)(12)} \begin{bmatrix} 1 & 0 & -1 & 0 \\ -1 & 0 & 1 & 0 \end{bmatrix} \begin{bmatrix} 0 \\ 0 \\ 0.012 \\ 0 \end{bmatrix}$$

$$= \begin{bmatrix} -10.00 \\ 10.00 \end{bmatrix}$$

For element 3, we have

$$\begin{bmatrix} f_1 \\ f_2 \end{bmatrix}^3 = \frac{(10)(30)(10)^3}{(33.54)(12)} \begin{bmatrix} 0.8945 & -0.4472 & -0.8945 & 0.4472 \\ -0.8945 & 0.4472 & 0.8945 & -0.4472 \end{bmatrix} \begin{bmatrix} 0.02188 \\ 0.00300 \\ 0.01200 \\ 0 \end{bmatrix}$$

$$= \begin{bmatrix} 11.1747 \\ -11.1747 \end{bmatrix} \quad (7.117)$$

7.7 FINITE DIFFERENCE METHOD FOR RECTANGULAR PLATES OF UNIFORM AND VARIABLE THICKNESS WITH APPLICATIONS

Plates are two-dimensional straight flat structural elements, with a thickness that is considered to be very small in comparison with the other dimensions of the plate. Common practical geometric plate shapes are rectangular, triangular, circular, folded, continuous, and so on, with free, simply supported, and fixed boundary conditions. Their boundary conditions may also include elastic supports and elastic restraints. The loads that are applied to plates may be static, or dynamic, and the loads act predominately in a direction perpendicular to the surface of the plate.

Plates are extensively used by practicing design engineers, because their two-dimensional structural action results in lighter structures and in numerous economic advantages. When folded or curved plates are used as structural elements, the economic advantages are even greater, because they combine the advantages of both plates and shells. Plates are used extensively in architectural structures, hydraulic structures, pavements, bridges, airplanes, missiles, instruments, ships, and in numerous other practical applications.

Plates are often categorized as stiff plates, membranes, flexible plates, and thick plates. Stiff plates are considered to be thin, and the applied loads are resisted by internal bending and torsional moments, and by transverse shears. Membranes are considered to be thin plates without flexural rigidity, and the external loads are resisted by internal axial and central shear forces. Flexible plates are thin plates representing a combination of stiff plates and membranes. The externally applied loads are resisted by the combined action of internal moments, transverse and central shear forces, and axial forces. They are used extensively by the aerospace industry. Finally, thick plates are structural elements with internal stress condition resembling the one that is developing in three-dimensional continua.

A short history regarding the mathematical solution of thin plates is given in Section 1.3.4. More on the history associated with the analysis of plates may be found in References [6, 13].

The thickness of a plate may be either uniform or variable. Most of the analytical work in the past has concentrated on uniform thickness plates. For rectangular plates of variable thickness, the first analytical solution was attempted by R. G. Olsson [87] in 1934, by considering linearly varying rigidity and a special load. An improvement on this solution was later made by E. Reissner [88] and R. G. Olsson [89], and an exponentially varying rigidity was tested by R. G. Olsson [90] and H. D. Conway [91]. Energy and numerical methods have also been used by various investigators [92–94] to solve variable thickness rectangular plates. H. Favre and B. Gilg [95], in 1952, assumed a linear variation of thickness and used a parameter method to expand the differential equation of the plate into an infinite series of equations.

During recent years, the author and his collaborators have developed unique linear and nonlinear methods for the solution of beams, plates, and frames of variable thickness [3, 5, 6, 37, 39, 42, 47, 96–98]; as a group these approaches are known as the method of the equivalent systems. For rectangular and circular plates of variable thickness, the solution is given in the form of an equivalent system of flat plates that replaces the original variable thickness plate. The sum of the deflections of the flat plates is equal to the deflection of the original variable thickness plate.

Numerical methods of analysis, such as the finite difference and finite element methods, are used by many researchers for the solution of the plate problem. The bulk of the work, however, is concentrated on uniform thickness plates. In this section, the finite difference method is used for the solution of rectangular thin flat plates of uniform and variable thickness. Extensive work on the general subject of plates may be found in References [3, 6, 71, 99].

7.7.1 Differential Equations of Thin Rectangular Plates

The derivation of the differential equations of thin rectangular plates is based on the assumptions that the applied loads are normal to the x–y plane of the plate; the edges of the plate are free to move in the plane of the plate, so that the reactive forces at the boundaries are normal to the x–y plane. Thus, any strain in the middle plane of the plate can be neglected during bending; the influence of the shear forces and the compressive forces due to the loading is neglected; and the material of the plate is homogeneous and isotropic, and its elastic limit is not exceeded.

The differential element of a plate, showing the shear forces and moments acting on it, is shown in Fig. 7.20. By using the static force equilibrium conditions in the vertical z direction, we find the following equilibrium equation:

$$\frac{\partial V_x}{\partial x} + \frac{\partial V_y}{\partial y} + q = 0 \qquad (7.118)$$

where q is the applied distributed load on the upper surface of the plate.

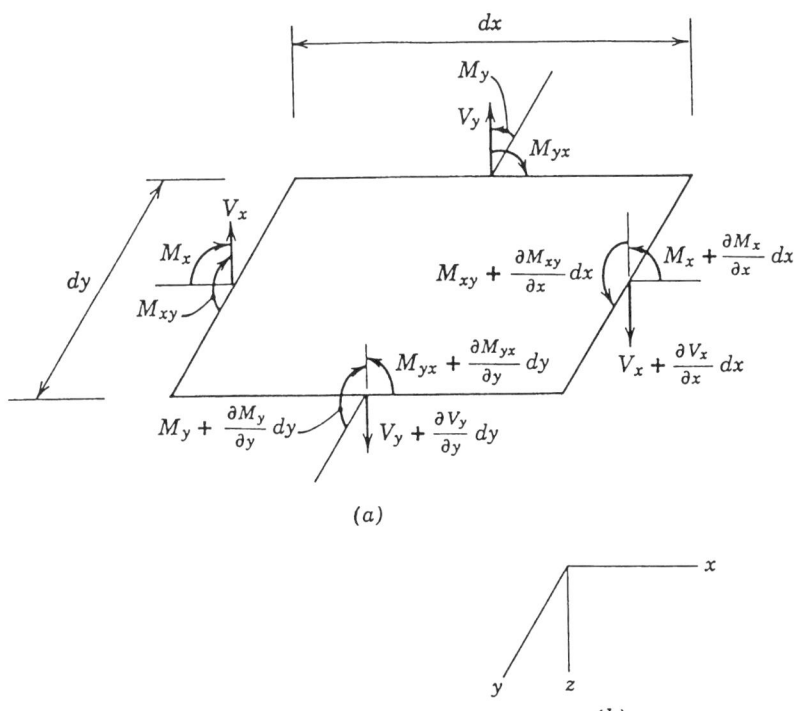

Figure 7.20. (a) Free-body diagram of an element of a plate with sides dx and dy. (b) Positive directions of the x, y, and z axes.

By applying static moment equilibrium about the x and y axes and neglecting higher order terms, we obtain the following two equations of equilibrium:

$$\frac{\partial M_{xy}}{\partial x} - \frac{\partial M_y}{\partial y} + V_y = 0 \tag{7.119}$$

$$\frac{\partial M_{yx}}{\partial y} + \frac{\partial M_x}{\partial x} - V_x = 0 \tag{7.120}$$

Thus, the three equations of equilibrium of the plate element, which define its equilibrium state, are Eqs. (7.118)–(7.120).

By solving Eqs. (7.119) and (7.120) for V_x and V_y and substituting into Eq. (7.118), we obtain

$$\frac{\partial^2 M_x}{\partial x^2} + \frac{\partial^2 M_y}{\partial y^2} + \frac{\partial^2 M_{yx}}{\partial x\, \partial y} - \frac{\partial^2 M_{xy}}{\partial x\, \partial y} = -q \tag{7.121}$$

In Eq. (7.121) we have $M_{yx} = -M_{xy}$ because $\tau_{yx} = \tau_{xy}$. Thus, Eq. (7.121) yields

$$\frac{\partial^2 M_x}{\partial x^2} + \frac{\partial^2 M_y}{\partial y^2} - 2\frac{\partial^2 M_{xy}}{\partial x\, \partial y} = -q \tag{7.122}$$

which is the partial differential equation of a rectangular plate in terms of the bending moments M_x and M_y about the x and y axes, respectively, and the twisting moment M_{xy}.

Equation (7.122) may be also written in terms of the vertical displacement w of the plate. From the theory of plates [6, 71], we have

$$M_x = -D\left(\frac{\partial^2 w}{\partial x^2} + v\frac{\partial^2 w}{\partial y^2}\right) \tag{7.123}$$

$$M_y = -D\left(\frac{\partial^2 w}{\partial y^2} + v\frac{\partial^2 w}{\partial x^2}\right) \tag{7.124}$$

$$M_{xy} = -M_{yx} = D(1 - v)\frac{\partial^2 w}{\partial x\, \partial y} \tag{7.125}$$

where D is the plate rigidity given by the expression

$$D = \frac{Eh^3}{12(1 - v^2)} \tag{7.126}$$

v is the Poisson ratio, and h is the thickness of the plate. Note that D and h may be a function of both x and y.

By differentiating Eqs. (7.123) through (7.125), we obtain

$$\frac{\partial^2 M_x}{\partial x^2} = -D\left(\frac{\partial^4 w}{\partial x^4} + v\frac{\partial^4 w}{\partial x^2 \partial y^2}\right) - 2\frac{\partial D}{\partial x}\left(\frac{\partial^3 w}{\partial x^3} + v\frac{\partial^3 w}{\partial x \partial y^2}\right)$$
$$- \frac{\partial^2 D}{\partial x^2}\left(\frac{\partial^2 w}{\partial x^2} + v\frac{\partial^2 w}{\partial y^2}\right) \quad (7.127)$$

$$\frac{\partial^2 M_y}{\partial y^2} = -D\left(\frac{\partial^4 w}{\partial y^4} + v\frac{\partial^4 w}{\partial x^2 \partial y^2}\right) - 2\frac{\partial D}{\partial y}\left(\frac{\partial^3 w}{\partial y^3} + v\frac{\partial^3 w}{\partial x^2 \partial y}\right)$$
$$- \frac{\partial^2 D}{\partial y^2}\left(\frac{\partial^2 w}{\partial y^2} + v\frac{\partial^2 w}{\partial x^2}\right) \quad (7.128)$$

$$\frac{\partial^2 M_{xy}}{\partial x \partial y} = (1-v)\left(D\frac{\partial^4 w}{\partial x^2 \partial y^2} + \frac{\partial D}{\partial y}\frac{\partial^3 w}{\partial x^2 \partial y} + \frac{\partial D}{\partial x}\frac{\partial^3 w}{\partial x \partial y^2} + \frac{\partial^2 D}{\partial x \partial y}\frac{\partial^2 w}{\partial x \partial y}\right)$$
$$(7.129)$$

By substituting Eqs. (7.127)–(7.129) into Eq. (7.122), we find the following differential equation for the rectangular plate, which is in terms of the deflection w:

$$D\nabla^4 w + 2\frac{\partial D}{\partial x}\frac{\partial}{\partial x}\nabla^2 w + 2\frac{\partial D}{\partial y}\frac{\partial}{\partial y}\nabla^2 w + \nabla^2 D \nabla^2 w$$
$$- (1-v)\left(\frac{\partial^2 D}{\partial x^2}\frac{\partial^2 w}{\partial y^2} - 2\frac{\partial^2 D}{\partial x \partial y}\frac{\partial^2 w}{\partial x \partial y} + \frac{\partial^2 D}{\partial y^2}\frac{\partial^2 w}{\partial x^2}\right) = q(x,y) \quad (7.130)$$

where

$$\nabla^2 = \frac{\partial^2}{\partial x^2} + \frac{\partial^2}{\partial y^2} \quad (7.131)$$

$$\nabla^4 = (\nabla^2)^2 = \left(\frac{\partial^4}{\partial x^4} + 2\frac{\partial^4}{\partial x^2 \partial y^2} + \frac{\partial^4}{\partial y^4}\right) \quad (7.132)$$

The plate rigidity D in Eq. (7.130) is given by Eq. (7.126).

If the thickness variation of the plate is restricted to the y direction only, Eq. (7.130) yields

$$D\nabla^4 w + 2\frac{\partial D}{\partial y}\frac{\partial}{\partial y}\nabla^2 w + \frac{\partial^2 D}{\partial y^2}\left(\frac{\partial^2 w}{\partial y^2} + v\frac{\partial^2 w}{\partial x^2}\right) = q(x,y) \quad (7.133)$$

Equations (7.130) and (7.133) are linear differential equations of the fourth order with variable coefficients.

Equation (7.133) may be also written in a more compact form as follows:

$$\nabla^2(D\nabla^2 w) - (1-v)\frac{\partial^2 D}{\partial y^2}\frac{\partial^2 w}{\partial x^2} = q(x,y) \qquad (7.134)$$

All preceding equations describe linear systems and, consequently, the principle of superposition may be used in their solution.

The geometric boundary conditions of the rectangular plate may be prescribed in terms of the lateral deflection and the rotation. For example, if the edges of the rectangular plate are simply supported, the vertical deflection w and the moments M_x and M_y along the edges are zero. This, however, is a mixed boundary condition, since it involves both displacement and moment. If the edges are fixed, then the deflection and rotation at the edges are zero; see Fig. 7.21. On the other hand, the edge of a plate may be free, which is a statical

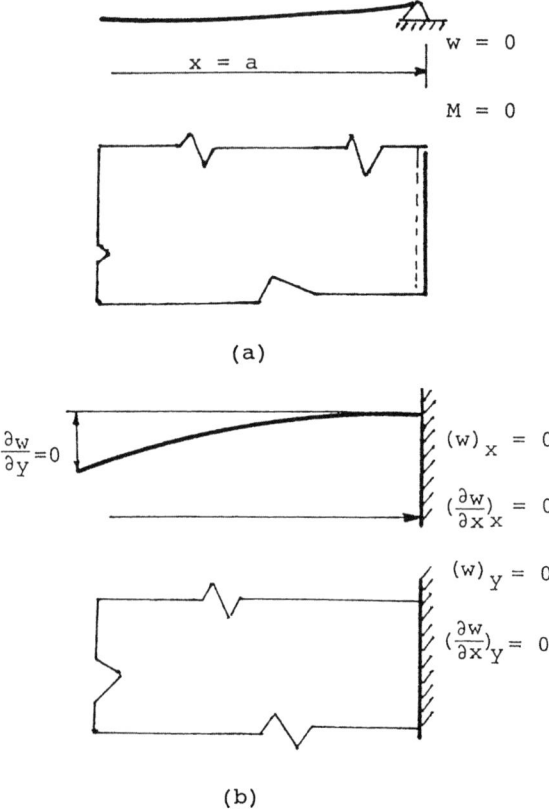

Figure 7.21. (a) Boundary conditions of a simply supported rectangular plate. (b) Boundary conditions of a rectangular plate with fixed edges.

boundary condition. In such a case, we can say that the bending moment and the transverse shear force are both zero at this edge. Another case of mixed boundary condition is when the edge of a plate is elastically supported, say by vertical springs, or if a simply supported edge is subjected to rotational elastic restraints. Such boundary conditions must be taken into consideration in the analysis of rectangular plates.

7.7.2 Finite Difference Equations for Rectangular Plates of Variable Thickness in the y Direction

The general purpose in the finite difference method is to replace the actual total or partial differential equation and the equations defining the boundary conditions with finite difference equations. In our particular problem in this section, we wish to replace the derivatives of the partial differential equation

$$D\nabla^4 w + 2\frac{\partial D}{\partial y}\frac{\partial}{\partial y}\nabla^2 w + \frac{\partial^2 D}{\partial y^2}\left(\frac{\partial^2 w}{\partial y^2} + v\frac{\partial^2 w}{\partial x^2}\right) = q(y) \qquad (7.135)$$

which is the same as Eq. (7.133), with finite difference quantities, thus forming a finite difference equation. The finite difference equation is then used for selected points located at the nodes of a rectangular network, such as the one in Fig. 7.22, known as the finite difference mesh.

It is an approximate numerical method, and its accuracy improves with finer mesh. Central differences are used in the derivation of the finite difference equations, because they are more suitable for such types of problems and the boundary conditions they represent.

We start the derivation of finite differences by considering the function $y = f(x)$, shown plotted in Fig. 7.23. The derivative dy/dx of a function $y = f(x)$ at some point i, where $x = x_i$, is, by definition,

$$\left(\frac{dy}{dx}\right)_i = \lim_{\Delta x \to 0} \frac{f(x_i + \Delta x) - f(x_i - \Delta x)}{2\Delta x}$$

or, by using the notation in Fig. 7.23 where r means one point on the right-hand side of point i, and l means one point on the left-hand side of point i, we write

$$\left(\frac{dy}{dx}\right)_i = \frac{f_r - f_l}{2\Delta x} = \frac{y_r - y_l}{2\Delta x} \qquad (7.136)$$

In Eq. (7.136), f_r and f_l are the values of $f(x)$ at points r and l, respectively, and y_r and y_l are the corresponding values of y at points r and l. If Δx is small,

514 ENERGY CONCEPTS AND METHODS

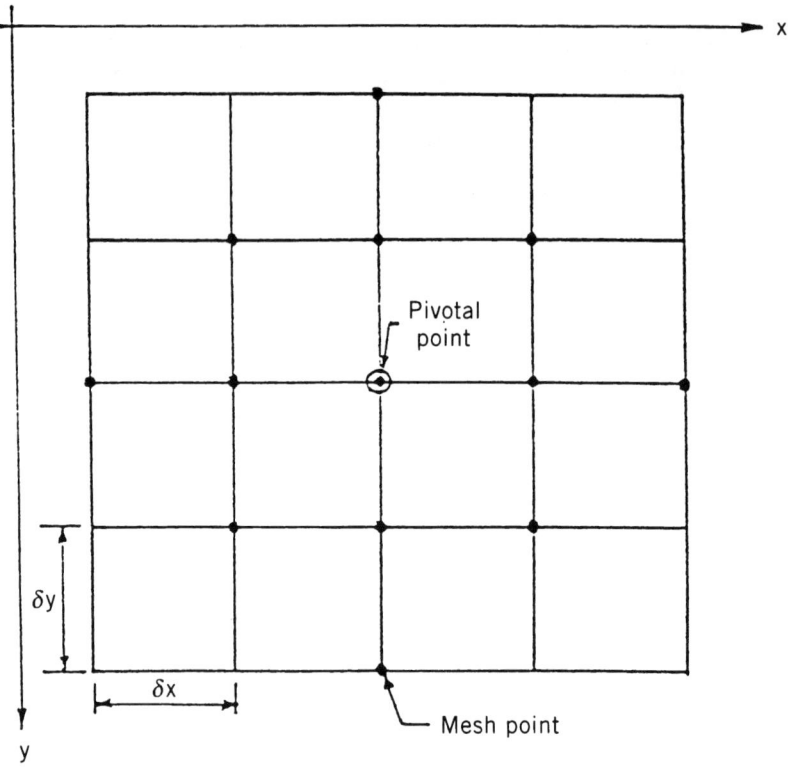

Figure 7.22. Finite difference mesh.

the derivative dy/dx is approximately equal to $\Delta y/\Delta x$. Therefore, we have

$$\left(\frac{dy}{dx}\right)_i = \frac{y_r - y_l}{2\Delta x} \qquad (7.137)$$

In a similar manner, we can write the second derivative with respect to x as

$$\left(\frac{d^2y}{dx^2}\right)_i = \frac{d}{dx}\left(\frac{dy}{dx}\right)_i$$
$$= \frac{y_r - 2y_i + y_l}{(\Delta x)^2} \qquad (7.138)$$

In Eq. (7.138), the second central difference is derived by taking the forward difference of point i minus the forward difference of point l. This is considered to be a better approximation of the second central difference.

FINITE DIFFERENCE METHOD FOR RECTANGULAR PLATES

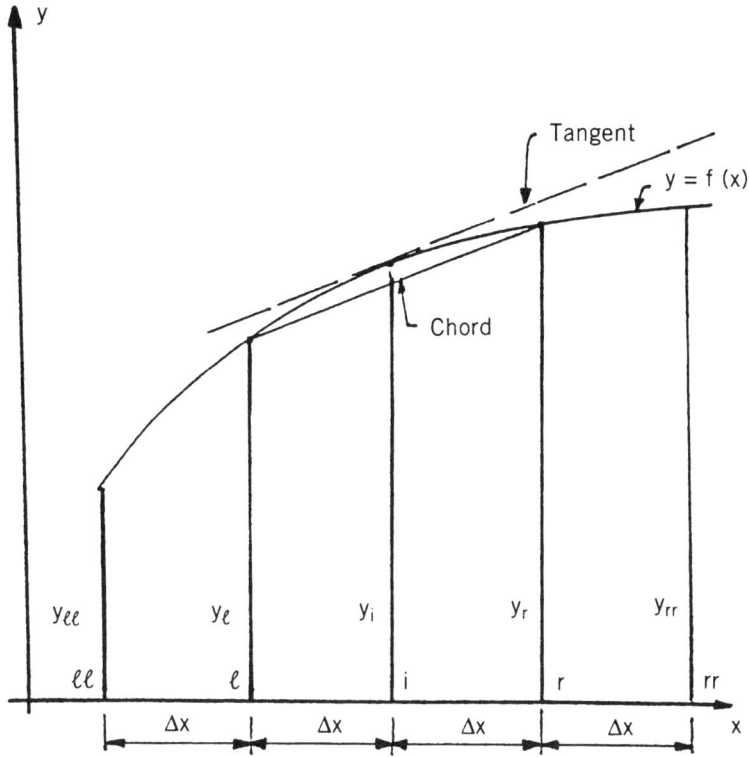

Figure 7.23. Plot of function $y = f(x)$ and indicated notation.

Similarly,

$$\left(\frac{d^3y}{dx^3}\right)_i = \frac{y_{rr} - 2y_r + 2y_l - y_{ll}}{2(\Delta x)^3} \tag{7.139}$$

$$\left(\frac{d^4y}{dx^4}\right)_i = \frac{y_{rr} - 4y_r + 6y_i - 4y_l + 4y_{ll}}{(\Delta x)^4} \tag{7.140}$$

In Eqs. (7.139) and (7.140), the subscript rr means the value of y at the second point on the right-hand side of point i, and the subscript ll indicates the value of y at the second point on the left-hand side of point i.

By following the preceding definitions used for the development of finite differences for ordinary derivatives, we can extend the procedure to include partial derivatives, since this is the purpose in this section. For this purpose, we consider a rectangular area with uniform grid spacing δx and δy, as shown in Fig. 7.24. We wish to determine finite differences for point i. The notations r, rr, l, and ll have the same meaning as in the preceding discussion. In addition,

516   ENERGY CONCEPTS AND METHODS

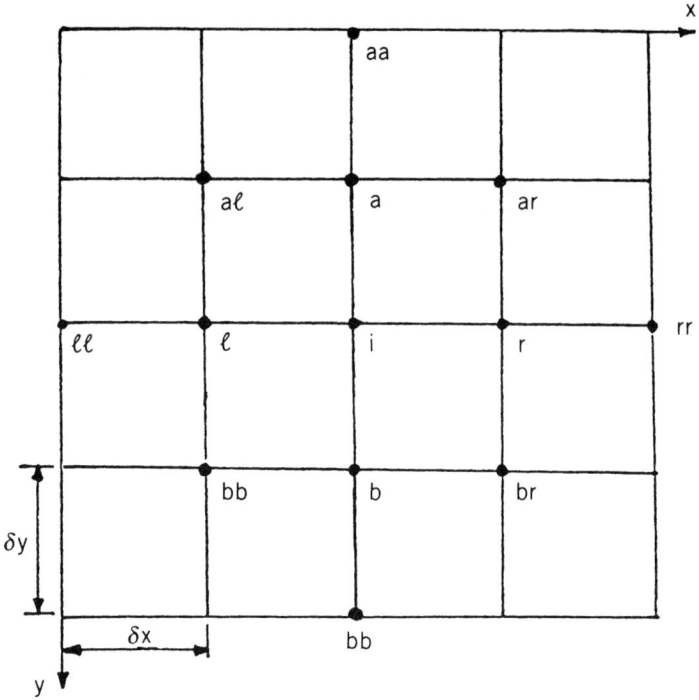

Figure 7.24. Grid pattern for finite difference representation.

for the y direction, we introduce the notation a and b, which means the first point above point i and the first point below point i, respectively. The notations aa and bb are used to denote the second point above and below point i, respectively.

On this basis, the finite differences for the partial derivatives with respect to x are

$$\left(\frac{\partial w}{\partial x}\right)_i = \frac{w_r - w_l}{2\delta x} \tag{7.141}$$

$$\left(\frac{\partial^2 w}{\partial x^2}\right)_i = \frac{w_r - 2w_i + w_l}{(\delta x)^2} \tag{7.142}$$

$$\left(\frac{\partial^3 w}{\partial x^3}\right)_i = \frac{w_{rr} - 2w_r + 2w_l - w_{ll}}{2(\delta x)^3} \tag{7.143}$$

$$\left(\frac{\partial^4 w}{\partial x^4}\right)_i = \frac{w_{rr} - 4w_r + 6w_i - 4w_l + w_{ll}}{(\delta x)^4} \tag{7.144}$$

where w represents the vertical deflection of a rectangular plate.

In a similar manner, for the y direction, we have

$$\left(\frac{\partial w}{\partial y}\right)_i = \frac{w_a - w_b}{2\delta y} \tag{7.145}$$

$$\left(\frac{\partial^2 w}{\partial y^2}\right)_i = \frac{w_a - 2w_i + w_b}{(\delta y)^2} \tag{7.146}$$

$$\left(\frac{\partial^3 w}{\partial y^3}\right)_i = \frac{w_{aa} - 2w_a + 2w_b - w_{bb}}{2(\delta y)^3} \tag{7.147}$$

$$\left(\frac{\partial^4 w}{\partial y^4}\right)_i = \frac{w_{aa} - 4w_a + 6w_i - 4w_b + w_{bb}}{(\delta y)^4} \tag{7.148}$$

We also have the identity

$$2\frac{\partial D}{\partial y}\frac{\partial}{\partial y}\nabla^2 w = 2\frac{\partial D}{\partial y}\left(\frac{\partial^3 w}{\partial y\,\partial x^2} + \frac{\partial^3 w}{\partial y^3}\right) \tag{7.149}$$

where

$$D = \frac{Eh^3}{12(1-v^2)} \tag{7.150}$$

We assume here that the variation of the depth h in the y direction is given by the equation

$$h = h_0 + ky \tag{7.151}$$

where h_0 is the reference value and k is a constant. By substituting Eq. (7.151) in Eq. (7.150), we find

$$D = \frac{E(h_0 + ky)^3}{12(1-v^2)} \tag{7.152}$$

Thus,

$$2\frac{\partial D}{\partial y} = \frac{2E}{12(1-v^2)}(3h_0^2 k + 6k^2 h_0 y + 3k^3 y^2) \tag{7.153}$$

$$\frac{\partial^2 D}{\partial y^2} = \frac{E}{12(1-v^2)}(6k^2 h_0 + 6k^3 y) \tag{7.154}$$

We introduce the symbol F_2 to represent the identity

$$F_2 = \frac{E}{12(1-v^2)}(6k^2 h_0 + 6k^3 y) \tag{7.155}$$

518   ENERGY CONCEPTS AND METHODS

By substituting Eq. (7.153) into Eq. (7.149), we obtain

$$\frac{2E}{12(1-v^2)}(3h_0^2 k + 6k^2 h_0 y + 3k^3 y^2)\left(\frac{\partial^3 w}{\partial y\, \partial x^2} + \frac{\partial^3 w}{\partial y^3}\right) \quad (7.156)$$

We use the symbol F_1 to represent the identity

$$F_1 = \frac{2E}{12(1-v^2)}(3h_0^2 k + 6k^2 h_0 y + 3k^3 y^2) \quad (7.157)$$

In molecule form, the preceding equations are written as shown in Table 7.6.

TABLE 7.6. Schematic Representation of Central Differences

$\dfrac{\partial w}{\partial x}$	$\boxed{-1\ \vert\ 0\ \vert\ 1}$	$\dfrac{w}{2\delta x}$
$\dfrac{\partial^2 w}{\partial x^2}$	$\boxed{1\ \vert\ -2\ \vert\ 1}$	$\dfrac{w}{(\delta x)^2}$
$\dfrac{\partial^3 w}{\partial x^3}$	$\boxed{-1\ \vert\ 2\ \vert\ 0\ \vert\ -2\ \vert\ -1}$	$\dfrac{w}{2(\delta x)^3}$
$\dfrac{\partial^4 w}{\partial x^4}$	$\boxed{1\ \vert\ -4\ \vert\ 6\ \vert\ -4\ \vert\ 1}$	$\dfrac{w}{(\delta x)^4}$
Point	$ll \quad l \quad i \quad r \quad rr$	

$\dfrac{\partial w}{\partial y}$	$\dfrac{\partial^2 w}{\partial y^2}$	$\dfrac{\partial^3 w}{\partial y^3}$	$\dfrac{\partial^4 w}{\partial y^4}$
		1	1
		-2	-4
1	1	0	6
0 $\dfrac{w}{2\delta y}$	-2 $\dfrac{w}{(\delta y)^2}$	2 $\dfrac{w}{2(\delta y)^3}$	-4 $\dfrac{w}{(\delta y)^4}$
-1	1	-1	1

FINITE DIFFERENCE METHOD FOR RECTANGULAR PLATES

For the mixed derivatives, we can also obtain finite difference expressions in a similar manner. They are as follows:

$$\frac{\partial^2 w}{\partial x\,\partial y} = \frac{\partial}{\partial x}\left(\frac{\partial w}{\partial y}\right) = \frac{w_{ar} - w_{al} - w_{br} + w_{bl}}{4\delta y\,\delta x} \quad (7.158)$$

$$\frac{\partial^3 w}{\partial x^2\,\partial y} = \frac{w_{ra} - 2w_a - 2w_{la} - w_{rb} + 2w_b - w_{lb}}{2\delta y(\delta x)^2} \quad (7.159)$$

$$\frac{\partial^3 w}{\partial x\,\partial y^2} = \frac{w_{ar} - 2w_r + w_{lr} - w_a + 2w_b - w_{lb}}{2\delta x(\delta y)^2} \quad (7.160)$$

$$\frac{\partial^4 w}{\partial x^2\,\partial y^2} = \frac{w_{ar} - 2w_a + w_{al} - 2w_r + 4w_i - 2w_l + w_{br} - 2w_b + w_{bl}}{(\delta x)^2(\delta y)^2} \quad (7.161)$$

In molecule form, the preceding four mixed drivatives are written as shown in Table 7.7.

TABLE 7.7 Equations (7.158) through (7.161) Arranged in a Molecule Form

$\dfrac{\partial^2 w}{\partial x\,\partial y} =$

-1	0	1
0	0	0
1	0	-1

$\dfrac{w}{4\delta x\,\delta y}$

$\dfrac{\partial^3 w}{\partial x^2\,\partial y} =$

1	-2	1
0	0	0
-1	2	-1

$\dfrac{w}{2(\delta x)^2 \delta y}$

$\dfrac{\partial^3 w}{\partial y^2\,\partial x} =$

-1	0	1
2	0	-2
-1	0	1

$\dfrac{w}{2(\delta y)^2 \delta x}$

$2\dfrac{\partial^4 w}{\partial x^2\,\partial y^2} =$

1	-2	1
-2	4	-2
1	-2	1

$\dfrac{w}{(\delta x)^2(\delta y)^2}$

Based on the preceding finite differences and molecule arrangements, the partial differential equation given by Eq. (7.135) may be represented as shown in Table 7.8. The stiffness equation may be represented in the form of a finite difference operator, as shown in Fig. 7.25, where A, B, C, D, E, F, G, H, P, Q, R, S, and T represent corresponding terms in Table 7.8. For example, if we wish to write a finite difference equation for a point i using the notation given in Fig. 7.25, we have

$$[Fw_{aa} + Gw_a + Tw_{ar} + Ew_{rr} + Dw_r + Qw_{bl} + Pw_{bb} + Hw_b + Cw_i + Rw_{br} + Bw_l + Sw_{al} + Aw_{ll}] = q_i(y) \quad (7.162)$$

Such equations can be written for every node of the selected finite difference mesh. The simultaneous solution of the resulting algebraic finite difference equations yields the values of the vertical deflection w at the node points of the mesh.

7.7.3 Boundary Conditions

The solution of Eq. (7.135) by using the finite difference method requires proper finite difference representation of the boundary conditions of the rectangular plate. The utilization of central differences makes it mandatory to consider fictitious points outside the edges of the rectangular plate, as shown in Fig. 7.26. The edges of the actual plate are shown by the four heavy lines in the same figure. For example, points $7a$, $15a$, $27b$, $21b$, $14l$, $23r$, and so on, are fictitious points located outside the boundary lines of the rectangular plate. This means that the finite difference equations generated for such nodes would be in terms of unknown internal deflections w_i and unknown external deflections w_e. On this basis, the operations must be applied to boundary nodes in order to generate additional equations that can be used for the solution of the external unknown displacements.

In the case of simply supported or fixed edges of the rectangular plate, the external nodes are represented in terms of internal points, and they are added in their respective location to the stiffness operator or matrix. The cases with fixed and simply supported edges are shown in Figs. 7.27a and 7.27b, respectively. Note that $w_i = -w_e$, or $w_i = w_e$, since the vertical deflection w at the boundaries of simply supported edges and the slope at fixed edges are, respectively, zero. For example, for the fixed boundaries shown in Fig. 7.27a, we have for points m, n the following:

$$\left(\frac{\partial w}{\partial x}\right)_{m,n} = \frac{1}{2\Delta x}(w_i - w_e) = 0 \quad (7.163)$$

or, in schematic representation,

$$\left(\frac{\partial w}{\partial x}\right)_{m,n} = \boxed{-1 \;|\; 0 \;|\; 1} \; \frac{w}{2\Delta x} = 0 \quad (7.164)$$

TABLE 7.8. Schematic Representation of the Differential Equation (7.135)

$$\frac{w}{(\delta x)^4} D \quad = q(y)$$

Stencil (cross-shaped) coefficients:

Top cell:
$$\frac{w}{(\delta x)^4} D$$

Upper row (three cells):

Left:
$$\frac{2w}{(\delta x)^2(\delta y)^2} D + \frac{w}{2(\delta x)^2 \delta y} F_1$$

Center:
$$\left[-\frac{4w}{(\delta x)^4} - \frac{4w}{(\delta x)^2(\delta y)^2} \right] D + \frac{vw}{(\delta x)^2} F_2$$

Right:
$$\frac{2w}{(\delta x)^2(\delta y)^2} D - \frac{w}{2(\delta x)^2 \delta y} F_1$$

Middle row (five cells):

Far left:
$$\frac{w}{(\delta y)^4} D + \frac{w}{2(\delta y)^3} F_1$$

Left:
$$\left[-\frac{4w}{(\delta y)^4} - \frac{4w}{(\delta x)^2(\delta y)^2} \right] D + \left[-\frac{2w}{2(\delta y)^3} - \frac{2w}{2(\delta x)^2 \delta y} \right] F_1 + \frac{w}{(\delta y)^2} F_2$$

Center:
$$\left[\frac{6w}{(\delta x)^4} + \frac{6w}{6 y^4} + \frac{8w}{(\delta x)^2(\delta y)^2} \right] D - 2 \left[\frac{w}{(\delta x)^2} + \frac{vw}{(\delta y)^2} \right] F_2$$

Right:
$$\left[-\frac{4w}{(\delta y)^4} - \frac{4w}{(\delta x)^2(\delta y)^2} \right] D + \left[\frac{2w}{2(\delta y)^3} + \frac{2w}{2(\delta x)^2 \delta y} \right] F_1 + \frac{w}{(\delta y)^2} F_2$$

Far right:
$$\frac{w}{(\delta y)^4} D - \frac{w}{2(\delta y)^3} F_1$$

Lower row (three cells):

Left:
$$\frac{2w}{(\delta x)^2(\delta y)^2} D + \frac{w}{2(\delta x)^2 \delta y} F_1$$

Center:
$$\left[-\frac{4w}{(\delta x)^4} - \frac{4w}{(\delta x)^2(\delta y)^2} \right] D + \frac{vw}{(\delta x)^2} F_2$$

Right:
$$\frac{2w}{(\delta x)^2(\delta y)^2} D - \frac{w}{2(\delta x)^2 \delta y} F_1$$

Bottom cell:
$$\frac{w}{(\delta x)^4} D$$

522 ENERGY CONCEPTS AND METHODS

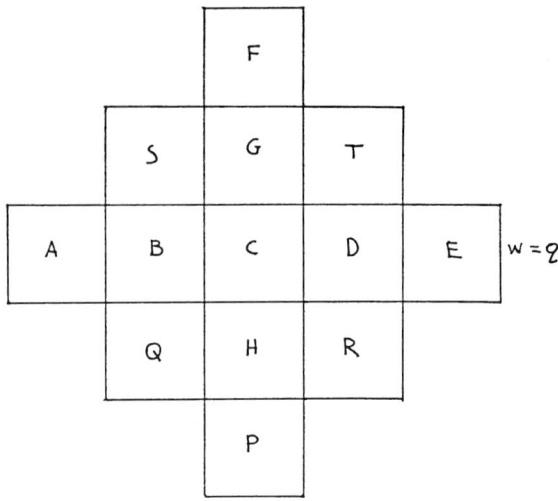

Figure 7.25. Finite difference operator for the finite difference equation shown in Table 7.8.

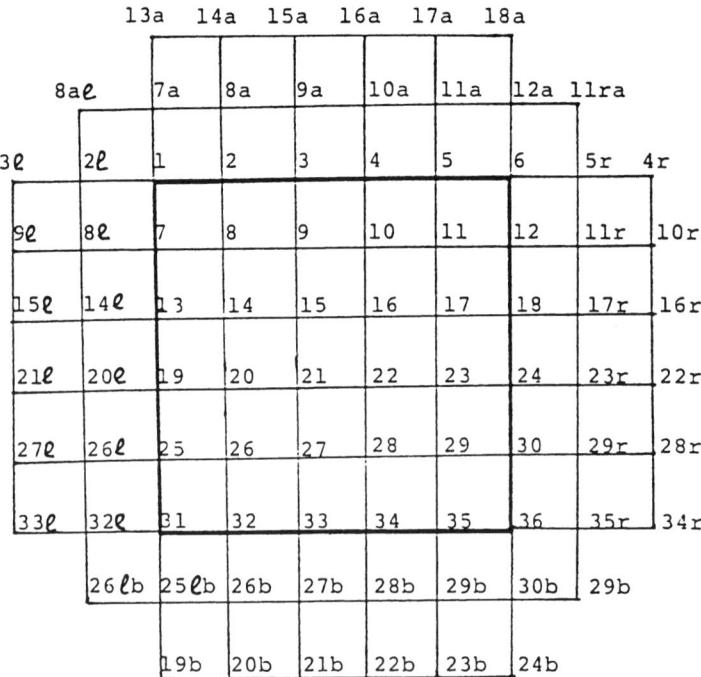

Figure 7.26. Grid system of a rectangular plate with imaginary points outside its boundaries.

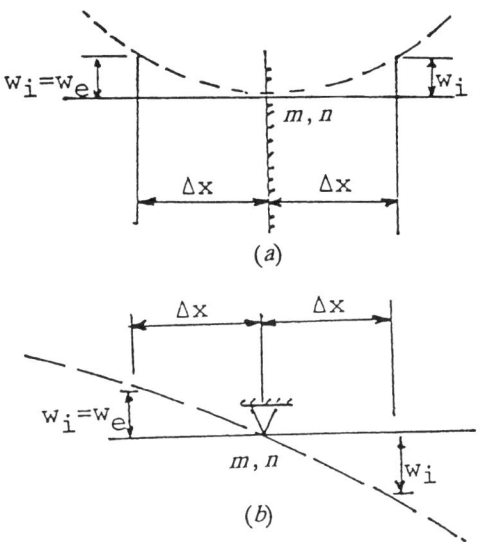

Figure 7.27. Boundary conditions. (*a*) Fixed edges. (*b*) Simply supported edges.

Thus, we have $w_i = w_e$. For simply supported edges, the second finite difference of $\partial^2 w/\partial x^2$ is zero at the boundary points, which again leads to the identity $w_i = -w_e$.

As a final note regarding this section, I wish to point out that the preceding finite difference equations were derived for rectangular plates with linear variation of thickness in the y direction, as illustrated by Eq. (7.151). Finite difference equations for other thickness variations of the plate may be derived in a similar manner. The preceding finite difference equations apply also to flat thin plates of uniform thickness when we set $k = 0$ in Eq. (7.151) and in subsequent derivations.

7.8 APPLICATION OF THE FINITE DIFFERENCE METHOD TO UNIFORM AND VARIABLE THICKNESS RECTANGULAR PLATES

The application of the finite difference method to uniform and variable thickness plates in the y direction is discussed in this section.

7.8.1 Rectangular Plates of Uniform Thickness

We assume here that the thickness h of the rectangular plate is uniform, and that $\delta x = \delta y = \lambda$. On this basis, the finite difference equation for a node point

524 ENERGY CONCEPTS AND METHODS

i is reduced to the following equation:

$$\frac{D}{\lambda^4}[20i - 8(w_r + w_l + w_a + w_b) + 2(w_{ra} + w_{la} + w_{rb} + w_{lb})$$

$$+ w_{rr} + w_{ll} + w_{aa} + w_{bb}] = (q_y)_i \quad (7.165)$$

The application of Eq. (7.165) becomes very convenient if we put it in the schematic form shown in Fig. 7.28. Every time we apply Eq. (7.165) for a node point i, 12 additional points around point i are affected. The indicated values at each point of the diagram in Fig. 7.28 give the factor, or influence, for the deflection w at each point.

The following examples illustrate the application of Eq. (7.165) to practical rectangular plate problems of uniform thickness.

Example 7.21: A uniform simply supported square plate is loaded by a uniformly distributed vertical load q over the entire surface of the plate. The sides of the plate are equal to a. By using a 4×4 mesh, determine the deflection at the node points of the mesh.

SOLUTION: The 4×4 mesh of the uniform square plate is shown in Fig. 7.29. Because of the double symmetry, only points 1, 2, and 3 need to be used in the analysis. The boundary conditions for simply supported plates require

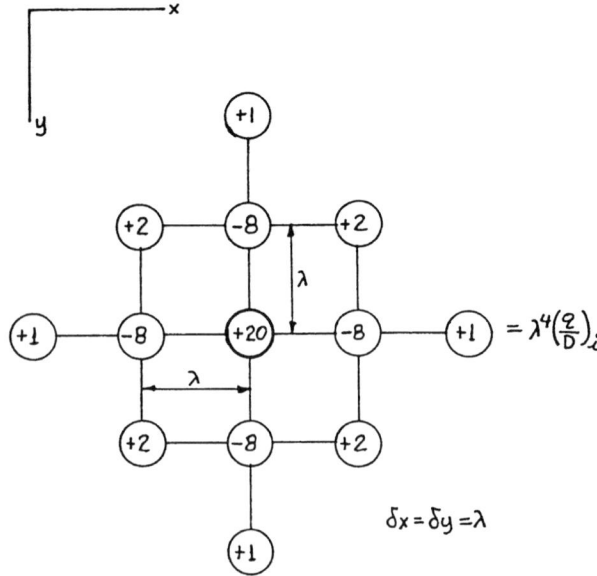

Figure 7.28. Schematic representation of Eq. (7.165).

APPLICATION OF THE FINITE DIFFERENCE METHOD 525

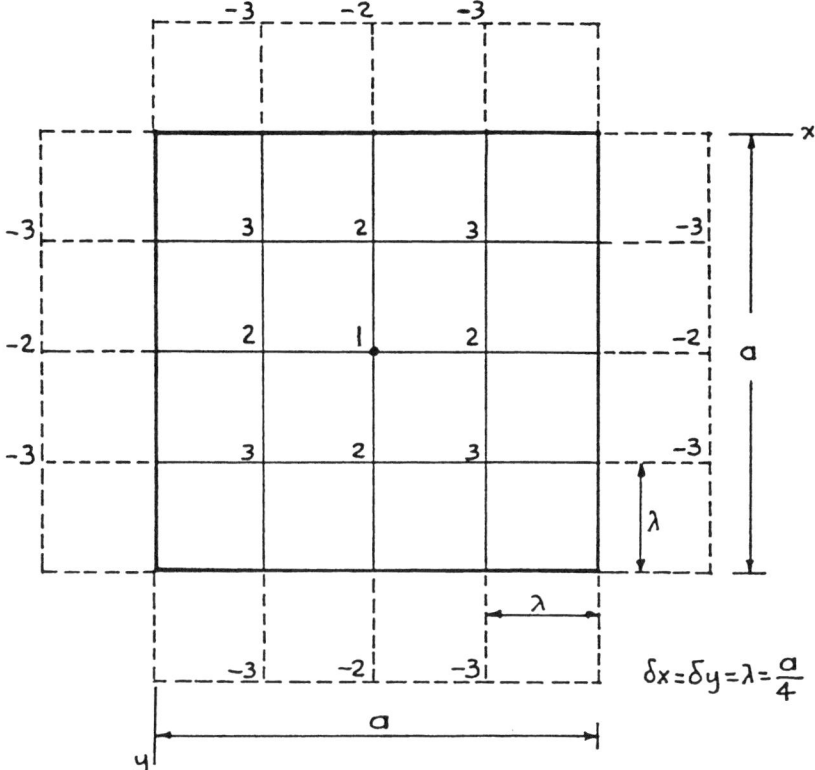

Figure 7.29. 4 × 4 mesh of the uniform square plate.

$w_e = -w_i$ and, consequently, the points outside the boundaries are indicated as negative. The mesh size $\lambda = a/4$, where a is the side of the square plate. By using the schematic representation shown in Fig. 7.28, the finite difference equations for points 1, 2, and 3 are as follows:

Point 1

$$20w_1 - 32w_2 + 8w_3 = \frac{qa^4}{256D} \tag{7.166}$$

Point 2

$$20w_2 - 16w_3 - 8w_1 + 4w_2 + w_2 - w_2 = \frac{qa^4}{256D}$$

or (7.167)

$$-8w_1 + 24w_2 - 16w_3 = \frac{qa^4}{256D}$$

526 ENERGY CONCEPTS AND METHODS

Point 3

$$20w_3 - 16w_2 + 2w_1 + w_3 - w_3 - w_3 = \frac{qa^4}{256D}$$

or   (7.168)

$$2w_1 - 16w_2 + 20w_3 = \frac{qa^4}{256D}$$

Simultaneous solution of Eqs. (7.166) through (7.168), yields

$$w_1 = 0.004028 \frac{qa^4}{D} \qquad (7.169)$$

$$w_2 = 0.00293 \frac{qa^4}{D} \qquad (7.170)$$

$$w_3 = 0.002136 \frac{qa^4}{D} \qquad (7.171)$$

We note that the maximum value of the deflection w is w_1, which is located at the center of the plate. For this case, the correct value of $w_1 = 0.004029 qa^4/D$. Because of the nature of the problem and the symmetry involved, the 4×4 mesh produced very accurate results. For other cases, however, a finer mesh is usually required for an accurate solution of the problem.

Example 7.22: Repeat the problem in Example 7.21 by assuming that all the sides of the square plate are fixed. Also, by using a 6×6 mesh, write the finite difference equations for each node of the mesh.

SOLUTION: The 4×4 mesh for the uniform square plate with fixed edges is shown in Fig. 7.30a. Because of symmetry for both loading and geometry, only points 1, 2, and 3 need to be considered. The boundary conditions for fixed boundary require $w_e = w_i$ for the outside points, as shown in the figure. By using the schematic representation shown in Fig. 7.28 and $\delta x = \delta y = \lambda = a/4$, the finite difference equations for points 1, 2, and 3 are as follows:

Point 1

$$20w_1 - 32w_2 + 8w_3 = \frac{qa^4}{256D} \qquad (7.172)$$

Point 2

$$20w_2 - 16w_3 - 8w_1 + 4w_2 + w_2 + w_2 = \frac{qa^4}{256D}$$

or   (7.173)

$$-8w_1 + 26w_2 - 16w_3 = \frac{qa^4}{256D}$$

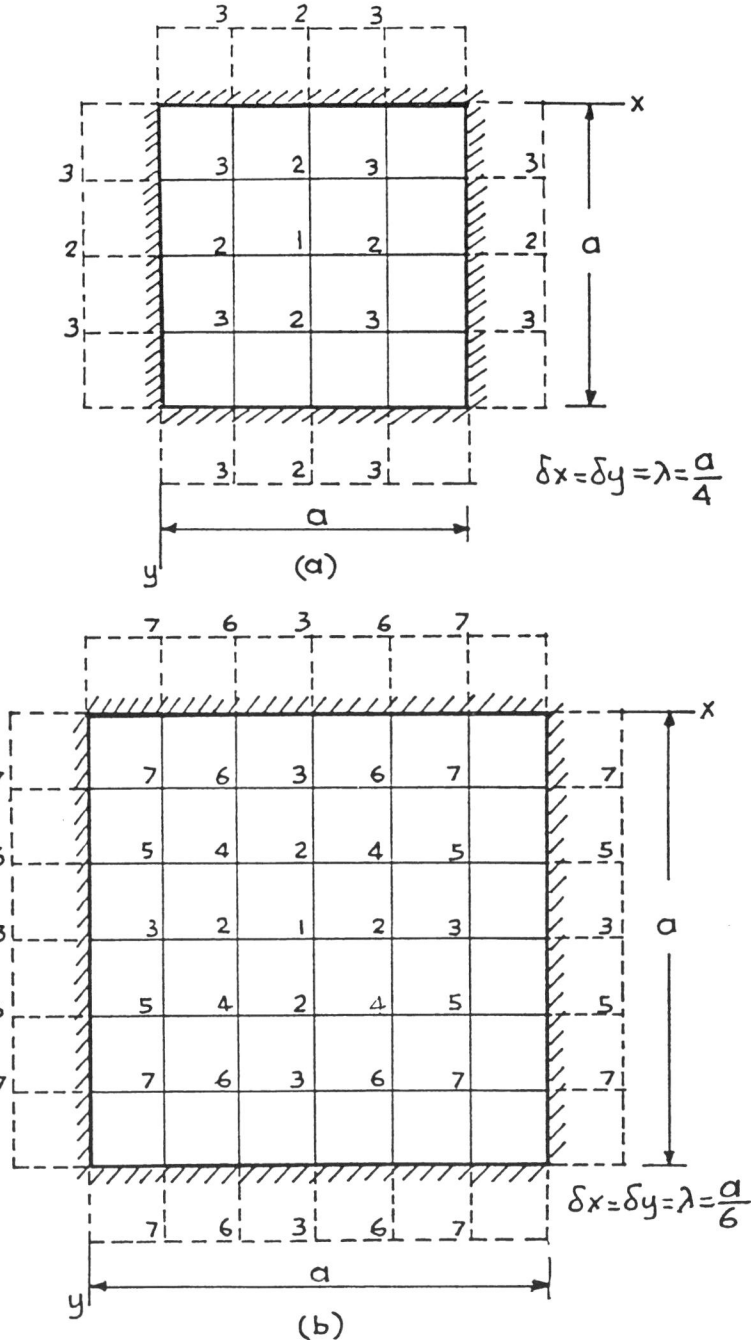

Figure 7.30. (a) 4 × 4 mesh for the uniform square plate with fixed edges. (b) 6 × 6 mesh for the same plate.

528 ENERGY CONCEPTS AND METHODS

Point 3

$$20w_3 - 16w_2 + 2w_1 + w_3 + w_3 + w_3 + w_3 = \frac{qa^4}{256D}$$

or (7.174)

$$2w_1 - 16w_2 + 24w_3 = \frac{qa^4}{256D}$$

The simultaneous solution of Eqs. (7.172) through (7.174) yields

$$w_1 = 0.0018 \frac{qa^4}{D} \qquad (7.175)$$

$$w_2 = 0.001207 \frac{qa^4}{D} \qquad (7.176)$$

$$w_3 = 0.000817 \frac{qa^4}{D} \qquad (7.177)$$

We note that the maximum value of the deflection is $w_1 = 0.0018qa^4/D$, and that it occurs at the center of the plate. The correct value [71] for w_1 is

$$w_1 = 0.001264 \frac{qa^4}{D} \qquad (7.178)$$

Thus, the solution using a 4 × 4 mesh yields 42.4 percent error, indicating that a finer mesh is required for an accurate solution of the problem.

For the 6 × 6 mesh shown in Fig. 7.30b, the finite difference equations for points 1 through 7 are as follows:

Point 1

$$20w_1 - 32w_2 + 8w_4 + 4w_3 = \frac{qa^4}{1{,}296D} \qquad (7.179)$$

Point 2

$$-8w_1 - 25w_2 - 8w_3 - 16w_4 + 4w_5 + 2w_6 = \frac{qa^4}{1{,}296D} \qquad (7.180)$$

Point 3

$$w_1 - 8w_2 - 21w_3 + 4w_4 - 16w_5 + 2w_7 = \frac{qa^4}{1{,}296D} \qquad (7.181)$$

Point 4

$$2w_1 - 16w_2 + 4w_3 + 22w_4 - 8w_5 - 8w_6 + 2w_7 = \frac{qa^4}{1{,}296D} \quad (7.182)$$

Point 5

$$3w_2 - 8w_3 - w_4 + 22w_5 + 2w_6 - 8w_7 = \frac{qa^4}{1{,}296D} \quad (7.183)$$

Point 6

$$3w_2 - 8w_3 - 8w_4 + 2w_5 + 22w_6 - 8w_7 = \frac{qa^4}{1{,}296D} \quad (7.184)$$

Point 7

$$2w_3 + 2w_4 - 8w_5 - 8w_6 + 22w_7 = \frac{qa^4}{1{,}296D} \quad (7.185)$$

The simultaneous solution of Eqs. (7.179) through (7.185) should yield much more accurate results. This task, however, requires the use of a computer software in order to make the solution convenient. Such software are easily available.

7.8.2 Rectangular Plates with Linearly Varying Thickness in the y Direction

In order to illustrate the procedure, the square simply supported plate in Fig. 7.31 is considered. The thickness of the square plate may vary linearly in either the x or the y direction, as shown in the figure, and a vertical uniformly distributed load $q = 130$ psi over the entire surface of the plate is applied to the plate. The modulus of elasticity $E = 29 \times 10^6$ psi, and the Poisson ratio $v = 0.3$.

The analysis of the plate is performed by considering the finite difference equations developed in Section 7.7, and implementing them using a computer program. A linear system of equations was developed by using the finite difference operator shown in Fig. 7.25. One finite difference equation was written for every node point of the selected mesh, and the boundary conditions were satisfied as explained in the preceding section. The developed equations were solved simultaneously by using a computer program that uses the Gaus elimination method. The mesh size selected here is 7×7. The calculated values of the vertical deflection w at points 1, 2, and 3 in Fig. 7.31, are $0.00223qa^4/D_0$, $0.00270qa^4/D_0$, and $0.00183qa^4/D_0$, respectively. Better accuracy may be obtained if a finer mesh is used. A very accurate closed-form solution of this

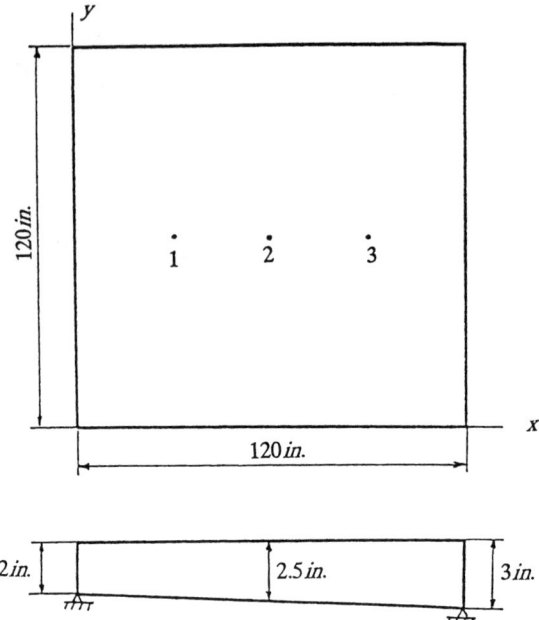

Figure 7.31. Simply supported square plate with linearly varying thickness.

problem is given in the following section of this chapter by using the author's method of the equivalent systems for rectangular plates.

7.9 RIGOROUS SOLUTION OF THE VARIABLE THICKNESS RECTANGULAR PLATE BY USING EQUIVALENT SYSTEMS

We consider here the case of a rectangular plate with variable thickness in the y direction only. In the x direction the thickness is constant. The procedure, however, although not the purpose in this section, is also applicable if the thickness of the plate varies in both x and y directions.

For a rectangular plate with thickness variation in both x and y directions, the partial differential equation, in terms of the deflection w, is given by the following equation, as shown in Section 7.7:

$$D\nabla^4 w + 2\frac{\partial D}{\partial x}\frac{\partial}{\partial x}\nabla^2 w + 2\frac{\partial D}{\partial y}\frac{\partial}{\partial y}\nabla^2 w + \nabla^2 D \nabla^2 w$$
$$-(1-v)\left[\frac{\partial^2 D}{\partial x^2}\frac{\partial^2 w}{\partial y^2} - 2\frac{\partial^2 D}{\partial x \partial y}\frac{\partial^2 w}{\partial x \partial y} + \frac{\partial^2 D}{\partial y^2}\frac{\partial^2 w}{\partial x^2}\right] = q(x, y) \quad (7.186)$$

where

$$\nabla^2 = \frac{\partial^2}{\partial x^2} + \frac{\partial^2}{\partial y^2} \tag{7.187}$$

$$\nabla^4 = (\nabla^2)^2 = \frac{\partial^4}{\partial x^4} + 2\frac{\partial^4}{\partial x^2 \partial y^2} + \frac{\partial^4}{\partial y^4} \tag{7.188}$$

$$D = \frac{Eh^3}{12(1-v^2)} \tag{7.189}$$

and v is the Poisson's ratio.

Equation (7.186) is the fourth order partial differential equation of the plate in two dimensions with variable coefficients. In Eq. (7.189) the depth h, as well as the modulus of elasticity E, may be variable quantities. We assume however, that there are no abrupt variations in thickness. On this basis, the equations for the bending and the twisting of flat plates can also be used with sufficient accuracy for the elastic analysis of variable thickness plates.

If the thickness of a rectangular plate varies in one direction only, say the y direction, Eq. (7.186) yields

$$D\nabla^4 w + 2\frac{\partial D}{\partial y}\frac{\partial}{\partial y}\nabla^2 w + \frac{\partial^2 D}{\partial y^2}\left[\frac{\partial^2 w}{\partial y^2} + v\frac{\partial^2 w}{\partial x^2}\right] = q(x,y) \tag{7.190}$$

The solution of either Eq. (7.186) or Eq. (7.190) is very difficult, and approximate closed-form or numerical methods of analysis are often used. The method of the equivalent systems is used here for the elastic anlysis of variable thickness plates in the y direction. Mathematically exact equivalent systems consisting of flat plates are derived, which can be used to simplify the solution of such complex problems.

By assuming that the rectangular plate has variable thickness h in the y direction and constant E, we write

$$h = h_0[1 + \lambda f(y)] \tag{7.191}$$

where h_0 is the constant reference thickness, $f(y)$ describes the thickness variation, and λ is a constant that keeps the plate thin enough to fall within the range of the thin plate theory. By substituting Eq. (7.191) into Eq. (7.189), we find

$$D = D_0[1 + \lambda f(y)]^3 \tag{7.192}$$

where

$$D_0 = \frac{Eh_0^3}{12(1-v^2)} \tag{7.193}$$

532 ENERGY CONCEPTS AND METHODS

By substituting Eq. (7.192) into Eq. (7.190) and performing the required differentiations, we obtain

$$\nabla^4 w = \frac{q}{D_0[1+\lambda f]^3} - \frac{6\lambda}{[1+\lambda f]}\frac{\partial f}{\partial y}\frac{\partial}{\partial y}\nabla^2 w - \frac{3\lambda}{[1+\lambda f]}\frac{\partial^2 f}{\partial y^2}\left[\frac{\partial^2 w}{\partial y^2} + v\frac{\partial^2 w}{\partial x^2}\right]$$

$$- \frac{6\lambda^2}{[1+\lambda f]^2}\left[\frac{\partial f}{\partial y}\right]^2\left[\frac{\partial^2 w}{\partial y^2} + v\frac{\partial^2 w}{\partial x^2}\right] \quad (7.194)$$

For the elastic case, it can be proven that a variable thickness plate represented by Eq. (7.194) may be replaced by a set of plates of uniform rigidity D_0, which are represented by the following series of differential equations [3, 6]:

$$\nabla^4 w_0 = \frac{q}{D_0} \quad (7.195a)$$

$$\lambda\nabla^4 w_1 = \lambda\left\{(-3f)\frac{q}{D_0} - 6\frac{\partial f}{\partial y}\frac{\partial}{\partial y}\nabla^2 w_0 - 3\frac{\partial^2 f}{\partial y^2}\left[\frac{\partial^2 w_0}{\partial y^2} + v\frac{\partial^2 w_0}{\partial x^2}\right]\right\} \quad (7.195b)$$

$$\lambda^2\nabla^4 w_2 = \lambda^2\left\{(6f^2)\frac{q}{D_0} + 6\frac{\partial f}{\partial y}f\frac{\partial}{\partial y}\nabla^2 w_0 + 3\frac{\partial^2 f}{\partial y^2}f\left[\frac{\partial^2 w_0}{\partial y^2} + v\frac{\partial^2 w_0}{\partial x^2}\right]\right.$$

$$\left. - 6\left[\frac{\partial f}{\partial y}\right]^2\left[\frac{\partial^2 w_0}{\partial y^2} + v\frac{\partial^2 w_0}{\partial x^2}\right] - 6\frac{\partial f}{\partial y}\frac{\partial}{\partial y}\nabla^2 w_1 - 3\frac{\partial^2 f}{\partial y^2}\left[\frac{\partial^2 w_1}{\partial y^2} + v\frac{\partial^2 w_1}{\partial x^2}\right]\right\} \quad (7.195c)$$

$$\lambda^3\nabla^4 w_3 = \lambda^3\left\{-10f^3\frac{q}{D_0} - 6\frac{\partial f}{\partial y}f^2\frac{\partial}{\partial y}\nabla^2 w_0\right.$$

$$- \left[3\frac{\partial^2 f}{\partial y^2}f^2 - 12\left(\frac{\partial f}{\partial y}\right)^2 f\right]\left[\frac{\partial^2 w_0}{\partial y^2} + v\frac{\partial^2 w_0}{\partial x^2}\right]$$

$$+ 6\frac{\partial f}{\partial y}f\frac{\partial}{\partial y}\nabla^2 w_1 + \left[3\frac{\partial^2 f}{\partial y^2}f - 6\left(\frac{\partial f}{\partial y}\right)^2\right]\left[\frac{\partial^2 w_1}{\partial y^2} + v\frac{\partial^2 w_1}{\partial x^2}\right]$$

$$\left. - 6\frac{\partial f}{\partial y}\frac{\partial}{\partial y}\nabla^2 w_2 - 3\frac{\partial^2 f}{\partial y^2}\left[\frac{\partial^2 w_2}{\partial y^2} + v\frac{\partial^2 w_2}{\partial x^2}\right]\right\} \quad (7.195d)$$

$$\lambda^m\nabla^4 w_m = \cdots \quad (7.195e)$$

The first of the preceding set of equations, Eq. (7.195a), represents a flat plate with constant rigidity D_0 and loading q that is identical to the load applied on the original variable thickness plate. The remaining equations in this set represent flat plates with different loads. These loads can be determined once the deflection from the preceding equation is determined. Thus, Eq. (7.195) is the set of equations describing an equivalent system of flat plates that replaces the original variable thickness plate. The solution applies to all boundary conditions and all continuous thickness variations with continuous first and second derivatives. For practical applications, an accurate solution may be obtained by using only the first two or three equations from the set of equations given by Eq. (7.195); that is

$$w = w_0 + \lambda w_1 + \lambda^2 w_2 + \cdots \tag{7.196}$$

where the parameter λ is considered to be smaller than one.

The moments M_x and M_y in the x and y directions, respectively, as well as the maximum stresses σ_x and σ_y in the same directions, are [3, 6]:

$$M_x = -D_0(1 + \lambda f)^3 \left\{ \left(\frac{\partial^2 w_0}{\partial x^2} + v \frac{\partial^2 w_0}{\partial y^2} \right) + \lambda \left(\frac{\partial^2 w_1}{\partial x^2} + v \frac{\partial^2 w_1}{\partial y^2} \right) \right. \\ \left. + \lambda^2 \left(\frac{\partial^2 w_2}{\partial x^2} + v \frac{\partial^2 w_2}{\partial y^2} \right) + \lambda^3 \left(\frac{\partial^2 w_3}{\partial x^2} + v \frac{\partial^2 w_3}{\partial y^2} \right) \right\} \tag{7.197}$$

$$M_y = -D_0(1 + \lambda f)^3 \left\{ \left(\frac{\partial^2 w_0}{\partial y^2} + v \frac{\partial^2 w_0}{\partial x^2} \right) + \lambda \left(\frac{\partial^2 w_1}{\partial y^2} + v \frac{\partial^2 w_1}{\partial x^2} \right) \right. \\ \left. + \lambda^2 \left(\frac{\partial^2 w_2}{\partial y^2} + v \frac{\partial^2 w_2}{\partial x^2} \right) + \lambda^3 \left(\frac{\partial^2 w_3}{\partial y^2} + v \frac{\partial^2 w_3}{\partial x^2} \right) \right\} \tag{7.198}$$

$$\sigma_x = -\frac{Eh_0(1 + \lambda f)}{2(1 - v^2)} \left\{ \left(\frac{\partial^2 w_0}{\partial x^2} + v \frac{\partial^2 w_0}{\partial y^2} \right) + \lambda \left(\frac{\partial^2 w_1}{\partial x^2} + v \frac{\partial^2 w_1}{\partial y^2} \right) \right. \\ \left. + \lambda^2 \left(\frac{\partial^2 w_2}{\partial x^2} + v \frac{\partial^2 w_2}{\partial y^2} \right) + \lambda^3 \left(\frac{\partial^2 w_3}{\partial x^2} + v \frac{\partial^2 w_3}{\partial y^2} \right) \right\} \tag{7.199}$$

$$\sigma_y = -\frac{Eh_0(1 + \lambda f)}{2(1 - v^2)} \left\{ \left(\frac{\partial^2 w_0}{\partial y^2} + v \frac{\partial^2 w_0}{\partial x^2} \right) + \lambda \left(\frac{\partial^2 w_1}{\partial y^2} + v \frac{\partial^2 w_1}{\partial x^2} \right) \right. \\ \left. + \lambda^2 \left(\frac{\partial^2 w_2}{\partial y^2} + v \frac{\partial^2 w_2}{\partial x^2} \right) + \lambda^3 \left(\frac{\partial^2 w_3}{\partial y^2} + v \frac{\partial^2 w_3}{\partial x^2} \right) \right\} \tag{7.200}$$

They are derived by substituting Eqs. (7.192) and (7.196) into the following plate equations and rearranging terms:

$$M_x = -D \left(\frac{\partial^2 w}{\partial x^2} + v \frac{\partial^2 w}{\partial y^2} \right) \tag{7.201}$$

$$M_y = -D \left(\frac{\partial^2 w}{\partial y^2} + v \frac{\partial^2 w}{\partial x^2} \right) \tag{7.202}$$

$$\sigma_x = -\frac{Ez}{1 - v^2} \left(\frac{\partial^2 w}{\partial x^2} + v \frac{\partial^2 w}{\partial y^2} \right) \tag{7.203}$$

$$\sigma_y = -\frac{Ez}{1 - v^2} \left(\frac{\partial^2 w}{\partial y^2} + v \frac{\partial^2 w}{\partial x^2} \right) \tag{7.204}$$

534 ENERGY CONCEPTS AND METHODS

where z is the distance from the median plane of the plane. The maximum stress occurs at the location $z = h/2$.

The following examples illustrate the application of the methodology to practical rectangular plate problems.

Example 7.23: Consider the rectangular plate in Fig. 7.32a, which has a linear thickness variation in the y direction as shown in Fig. 7.32b. The plate is subjected to a vertical uniformly distributed load q applied throughout its

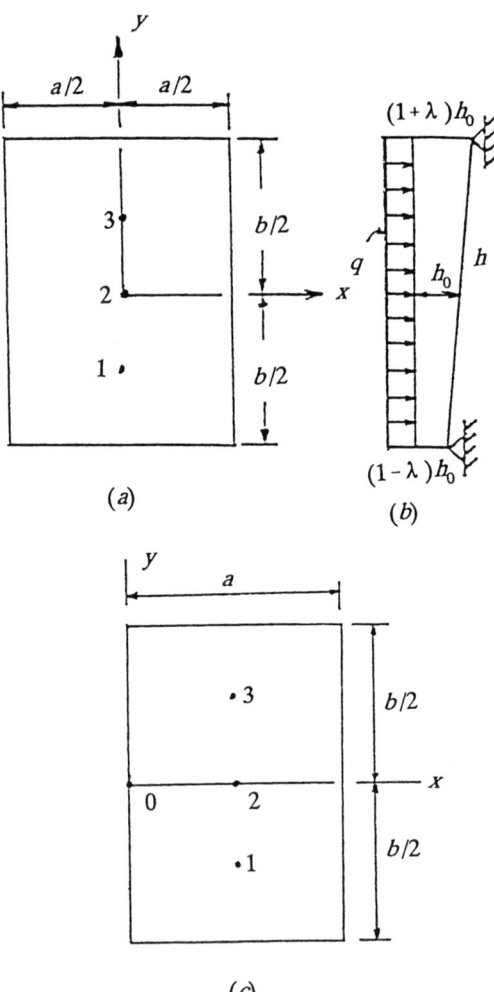

Figure 7.32. (a) Rectangular plate with linear thickness variation along the y direction. (b) Load q acting on the plate. (c) Coordinate axes used in the solution of the simplified equivalent system represented by Eq. (7.208).

upper surface. By using Eq. (7.195), determine the set of equivalent plates of constant rigidity D_0 representing the variable thickness plate. Also, determine a simplified equivalent system consisting of only one partial differential equation that simplifies the solution of the variable thickness plate. The modulus of elasticity E is constant.

SOLUTION: The variation in thickness h along the y axis is given by the following expression:

$$h = h_0 \left[1 + \lambda \left(\frac{2y}{b} \right) \right] \qquad (7.205)$$

where h_0 is the reference thickness value located as shown in Fig. 7.32b. By comparing Eq. (7.205) with Eq. (7.191), we note that

$$f(y) = \frac{2y}{b} \qquad (7.206a)$$

$$\frac{\partial f}{\partial y} = \frac{2}{b} \qquad (7.206b)$$

$$\frac{\partial^2 f}{\partial y^2} = 0 \qquad (7.206c)$$

By substituting into Eq. (7.195), we obtain the following set of differential equations representing equivalent flat plates of constant rigidity D_0 and uniform thickness h_0:

$$\nabla^4 w_0 = \frac{q}{D_0} \qquad (7.207a)$$

$$\lambda \nabla^4 w_1 = \lambda \left[-\frac{6y}{b} \frac{q}{D_0} - \frac{12}{b} \frac{\partial}{\partial y} \nabla^2 w_0 \right] \qquad (7.207b)$$

$$\lambda^2 \nabla^4 w_2 = \lambda^2 \left[\frac{24 y^2}{b^2} \frac{q}{D_0} + \frac{24y}{b^2} \frac{\partial}{\partial y} \nabla^2 w_0 - \frac{24}{b^2} \left(\frac{\partial^2 w_0}{\partial y^2} + v \frac{\partial^2 w_0}{\partial x^2} \right) - \frac{12}{b} \frac{\partial}{\partial y} \nabla^2 w_1 \right] \qquad (7.207c)$$

$$\lambda^3 \nabla^4 w_3 = \lambda^3 \left[-\frac{80 y^3}{b^3} \frac{q}{D_0} - \frac{48 y^2}{b^3} \frac{\partial}{\partial y} \nabla^2 w_0 + \frac{96y}{b^3} \left(\frac{\partial^2 w_0}{\partial y^2} + v \frac{\partial^2 w_0}{\partial x^2} \right) \right.$$
$$\left. + \frac{24y}{b^2} \frac{\partial}{\partial y} \nabla^2 w_1 - \frac{24}{b^2} \left(\frac{\partial^2 w_1}{\partial y^2} + v \frac{\partial^2 w_1}{\partial x^2} \right) - \frac{12}{b} \frac{\partial}{\partial y} \nabla^2 w_2 \right] \qquad (7.207d)$$

$$\vdots$$

$$\lambda^m \nabla^4 w_m = \qquad (7.207e)$$

536 ENERGY CONCEPTS AND METHODS

The solution of each differential equation given in Eq. (7.207) can be obtained either by using rigorous solutions, if such solutions could be available, or by using approximate and numerical methods of analysis, such as finite difference or finite element method.

A simplified equivalent system consisting of only one differential equation may be also used with sufficient accuracy for practical applications. This equation is as follows:

$$\nabla^4 w = \frac{q}{D_0}\left[1 - \frac{6\lambda y}{b} + \frac{24\lambda^2 y^2}{b^2} - \frac{80\lambda^3 y^3}{b^3}\right] - \frac{12\lambda}{b}\frac{\partial}{\partial y}\nabla^2 w_0 \quad (7.208)$$

Equation (7.208) combines Eqs. (7.207a) and (7.207b), and also includes the first term of Eq. (7.207c) and the first term of Eq. (7.207d). The selection of the terms other than the ones included in Eqs. (7.207a) and (7.207b) is based on the idea that the terms that converge the least rapidly are associated with the known applied load. On this basis, the solution of Eq. (7.208) will yield an accuracy that would be close to the one obtained by using the set of the three equations given by Eqs. (7.207a), (7.207b), and (7.207c).

The values of the term

$$\frac{\partial}{\partial y}\nabla^2 w_0 \quad (7.209)$$

in Eq. (7.208) may be obtained from the solution of the differential equation

$$\nabla^4 w_0 = \frac{q}{D_0} \quad (7.210)$$

which can be found in texts that deal with the theory of plates [6, 71, 99].

Simplified equivalent systems for other cases of rectangular plate problems may be obtained in a similar manner.

Example 7.24: By using the simplified equivalent system given by Eq. (7.208), determine the vertical deflection w at the quarter points 1, 2, and 3, along the y direction in Fig. 7.32a. Consider the cases where $b/a = 1.0$ and 1.5.

SOLUTION: By using the coordinate system of axes shown in Fig. 7.32c, the complete solution of Eq. (7.208) is derived in References [3, 6, 39], and it is as follows:

$$w = \frac{qa^4}{D_0}\sum_{n=1,3,\ldots}(A_n\cosh\alpha_n y + B_n\alpha_n y\sinh\alpha_n y$$
$$+ C_n\sinh\alpha_n y + D_n\alpha_n y\cosh\alpha_n y + K_0 + K_1 y \quad (7.211)$$
$$+ K_2 y^2 + K_3 y^3 + K_4 y^2\sinh\alpha_n y)\sin\alpha_n x$$

RIGOROUS SOLUTION OF THE VARIABLE THICKNESS RECTANGULAR PLATE 537

where

$$K_0 = \frac{4}{n^5\pi^5}\left(1 + \frac{96\lambda^2 a^2}{n^2\pi^2 b^2}\right) \tag{7.212}$$

$$K_1 = -\frac{24\lambda}{n^5\pi^5 b}\left(1 + \frac{120\lambda^2 a^2}{n^2\pi^2 b^2}\right) \tag{7.213}$$

$$K_2 = \frac{96\lambda^2}{n^5\pi^5 b^2} \tag{7.214}$$

$$K_3 = -\frac{320\lambda^3}{n^5\pi^5 b^3} \tag{7.114a}$$

$$K_4 = -\frac{6\lambda}{n^4\pi^4 ab \cosh \gamma_n} \tag{7.215}$$

$$\alpha_n = \frac{n\pi}{a} \tag{7.216}$$

$$\gamma_n = \frac{n\pi b}{2a} \tag{7.217}$$

$$A_n = -\frac{4}{n^5\pi^5 \cosh \gamma_n}\left[1 + 6\lambda^2 + \frac{96\lambda^2 a^2}{n^2\pi^2 b^2}\right] - B_n \gamma_n \tanh \gamma_n \tag{7.218}$$

$$B_n = \frac{2}{n^5\pi^5 \cosh \gamma_n}\left[1 + 6\lambda^2 + \frac{48\lambda^2 a^2}{n^2\pi^2 b^2}\right] \tag{7.219}$$

$$C_n = \frac{\lambda}{n^5\pi^5 \sinh \gamma_n}\left[12 + 3\gamma_n \tanh \gamma_n + 40\lambda^2 + \frac{(6)(240)\lambda^2 a^2}{n^2\pi^2 b^2}\right] - D_n \gamma_n \coth \gamma_n \tag{7.220}$$

$$D_n = -\frac{4\lambda}{n^5\pi^5 \sinh \gamma_n}\left[5\lambda^2 - \frac{3a^2}{n^2\pi^2 b^2}(\gamma_n \tanh \gamma_n - 20\lambda^2)\right] \tag{7.221}$$

Square Plate with $\lambda = 0.2$: For a square plate we have $b = a$. Thus, by substituting into Eqs. (7.212) through (7.221), we find:

For $n = 1$:

$$K_0 = 0.018157 \qquad \gamma_{n=1} = \frac{\pi}{2}$$

$$K_1 = \frac{-0.023314}{b} \qquad A_{n=1} = -0.013868$$

$$K_2 = \frac{0.012548}{b^2} \qquad B_{n=1} = 0.003736$$

538   ENERGY CONCEPTS AND METHODS

$$K_3 = \frac{-0.008365}{b^3} \qquad C_{n=1} = 0.006757$$

$$K_4 = \frac{-0.004910}{b^2} \qquad D_{n=1} = -0.000006$$

$$\alpha_{n=1} = \frac{\pi}{a}$$

For $n = 3$:

$$K_0 = 0.000056 \qquad \gamma_{n=3} = \frac{3\pi}{2}$$

$$K_1 = \frac{-0.000068}{b} \qquad A_{n=3} = -0.000006$$

$$K_2 = \frac{0.000052}{b^2} \qquad B_{n=3} = 0.000001$$

$$K_3 = \frac{-0.000034}{b^3} \qquad C_{n=3} = 0.000001$$

$$K_4 = \frac{-0.000003}{b^2} \qquad D_{n=3} = -0.00000001312$$

$$\alpha_{n=3} = \frac{3\pi}{a}$$

By substituting into Eq. (7.211), we obtain the following expression for the vertical deflection w of the square plate:

$$\begin{aligned} w = \frac{qa^4}{D_0} \Bigg[&-0.013868 \cosh\frac{\pi y}{a} + 0.003736 \left(\frac{\pi y}{a}\right) \sinh\left(\frac{\pi y}{a}\right) \\ &+ 0.006757 \sinh\left(\frac{\pi y}{a}\right) - 0.000006 \left(\frac{\pi y}{a}\right) \cosh\left(\frac{\pi y}{a}\right) + 0.018157 \\ &- 0.023314 \frac{y}{b} + 0.012548 \frac{y^2}{b^2} - 0.008365 \frac{y^3}{b^3} \\ &- 0.004910 \left(\frac{y^2}{b^2}\right) \sinh\left(\frac{\pi y}{a}\right) \Bigg] \sin\frac{\pi x}{a} + \frac{qa^4}{D_0} \Bigg[-0.000006 \cosh\left(\frac{3\pi y}{a}\right) \\ &+ 0.000001 \left(\frac{3\pi y}{a}\right) \sinh\left(\frac{3\pi y}{a}\right) + 0.000001 \sinh\left(\frac{3\pi y}{a}\right) \\ &- 0.00000001312 \left(\frac{3\pi y}{a}\right) \cosh\left(\frac{3\pi y}{a}\right) + 0.000056 \\ &- 0.000068 \frac{y}{b} + 0.000052 \frac{y^2}{b^2} - 0.000034 \frac{y^3}{b^3} \end{aligned} \qquad (7.222)$$

$$-0.000003\frac{y^2}{b^2}\sinh\left(\frac{3\pi y}{a}\right)\bigg]\sin\left(\frac{3\pi x}{a}\right)$$

At $x = a/2$ and $y = 0$, which is point 2 in Fig. 7.32c, Eq. (7.222) yields

$$w_{x=a/2, y=0} = \frac{qa^4}{D_0}(0.004289) - \frac{qa^4}{D_0}(0.000050)$$
$$= 0.004239\frac{qa^4}{D_0} \qquad (7.223)$$

At $x = a/2$ and $y = b/4$, which is point 3 in Fig. 7.32c, Eq. (7.222) yields

$$w_{x=a/2, y=b/4} = \frac{qa^4}{D_0}(0.002757) - \frac{qa^4}{D_0}(0.000025)$$
$$= 0.002732\frac{qa^4}{D_0} \qquad (7.224)$$

For point 1 in Fig. 7.32c, we have $x = a/2$, $y = -b/4$, and Eq. (7.222) yields

$$w_{x=a/2, y=-b/4} = \frac{qa^4}{D_0}(0.003439) - \frac{qa^4}{D_0}(0.000053)$$
$$= 0.003386\frac{qa^4}{D_0} \qquad (7.225)$$

Note that an accurate solution is obtained by using only the terms for $n = 1$ and $n = 3$ in Eq. (7.211). If better accuracy is required, terms corresponding to $n = 5, 7 \ldots$ may be added to the solution. Practical applications usually require only few values of n for a satisfactory solution.

Rectangular Plate with $\lambda = 0.2$ ***and*** $b/a = 1.5$: By substituting into Eqs. (7.212) through (7.221), we find:

For $n = 1$:

$$K_0 = 0.015331 \qquad \gamma_{n=1} = \frac{\pi b}{a}$$

$$K_1 = \frac{-0.019076}{b} \qquad A_{n=1} = -0.007240$$

$$K_2 = \frac{0.012548}{b^2} \qquad B_{n=1} = 0.001629$$

$$K_3 = \frac{-0.008365}{b^3} \qquad C_{n=1} = 0.002887$$

$$K_4 = \frac{-0.002314}{0.666667b^2} \qquad D_{n=1} = 0.000002$$

$$\alpha_{n=1} = \frac{\pi}{a}$$

For $n = 3$:

$$K_0 = 0.000055 \qquad \gamma_{n=3} = \frac{3\pi b}{2a}$$

$$K_1 = \frac{-0.000066}{b} \qquad A_{n=3} = -0.000001$$

$$K_2 = \frac{0.000052}{b^2} \qquad B_{n=3} = 0.000000057$$

$$K_3 = \frac{-0.000034}{b^3} \qquad C_{n=3} = 0.0000001744$$

$$K_4 = -\frac{0.000000259}{0.666667b^2} \qquad D_{n=3} = -0.00000000194$$

$$\alpha_{n=3} = 3\pi/a$$

By using the above values and Eq. (7.211), we find the following results for the deflection w at points 1, 2, and 3 in Fig. 7.32c.

For point 2, we have $x = a/2$ and $y = 0$. Thus,

$$w_{x=a/2, y=0} = \frac{qa^4}{D_0}(0.008091) - \frac{qa^4}{D_0}(0.000054)$$
$$= 0.008037 \frac{qa^4}{D_0} \tag{7.226}$$

For point 3, we have $x = a/2$ and $y = b/4$. Thus,

$$w_{x=a/2, y=b/4} = \frac{qa^4}{D_0}(0.005092) - \frac{qa^4}{D_0}(0.000025)$$
$$= 0.005067 \frac{qa^4}{D_0} \tag{7.227}$$

For point 1, we have $x = a/2$ and $y = -b/4$. Thus,

$$w_{x=a/2, y=-b/4} = \frac{qa^4}{D_0}(0.007034) - \frac{qa^4}{D_0}(0.000059)$$
$$= 0.006975 \frac{qa^4}{D_0} \tag{7.228}$$

The deflection at any other point of the plate may be obtained by substituting the appropriate values for the x and y coordinates in Eq. (7.211).

The results obtained by using the simplified equivalent system given by Eq. (7.208) are in very close agreement with the results obtained by using the first three equations of the set given in Eq. (7.207), which are Eqs. (7.207a), (7.207b), and (7.207c). The error is within 2 percent.

PROBLEMS

7.1 For the axially loaded bars in Fig. P7.1, determine the expression for the strain energy U stored in each bar.

7.2 For the truss loaded as shown in Fig. P7.2, determine the total strain energy stored in the truss. All members of the truss have the same cross-sectional area $A = 4 \text{ in.}^2$, and the modulus of elasticity $E = 30 \times 10^3$ ksi. The calculated forces in the members of the truss are

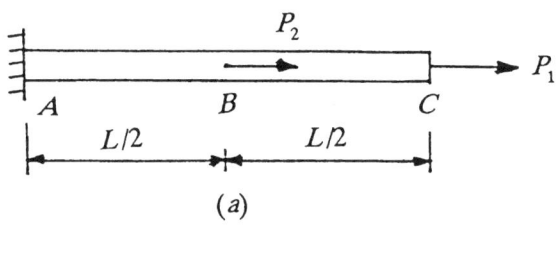

(a)

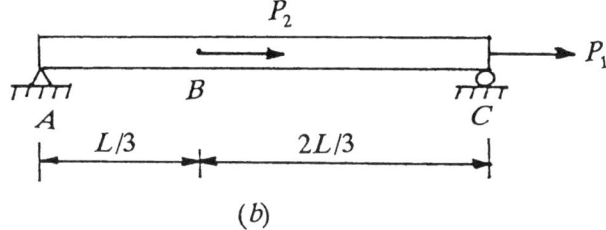

(b)

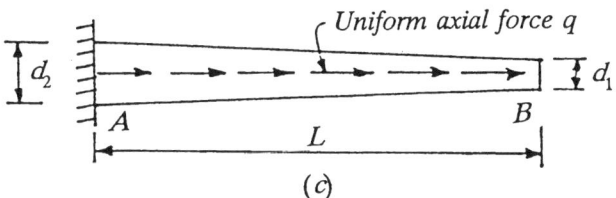

(c)

Figure P7.1.

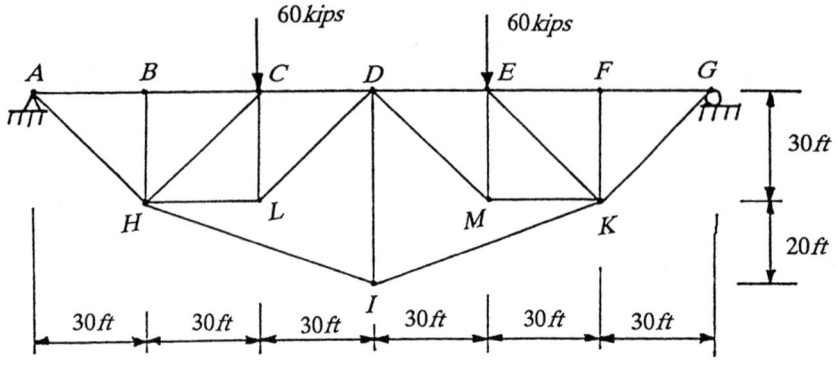

Figure P7.2.

as follows:

Member	Force (kips)	Member	Force (kips)
AB	−50.0	AH	+70.7
BC	−50.0	HI	+56.9
CD	−82.0	IK	+56.9
DE	−62.0	KG	+56.6
EF	−40.0	HL	+28.0
FG	−40.0	MK	+8.0
BH	0	HC	−45.2
CL	−28.0	LD	+39.6
DI	−36.0	MD	+11.3
EM	−8.0	KE	−31.1
FK	0		

7.3 For the steel truss loaded as shown in Fig. P7.3, determine the total strain energy stored in the truss. All members of the truss have the same cross-sectional area $A = 4$ in.^{2}, and the modulus of elasticity $E = 30 \times 10^3$ ksi. The values of the axial force in each member of the truss, in units of kips, are shown in Fig. P7.3, where plus signs indicate tension and minus signs are used for compression.

7.4 For the simple steel truss loaded as shown in Fig. P7.4, determine the strain energy U stored in the truss. The cross-sectional area for each member of the truss is 12 cm^2 and the modulus of elasticity $E = 200$ GPA.

7.5 For the loaded beam in Fig. P7.5, determine the bending strain energy

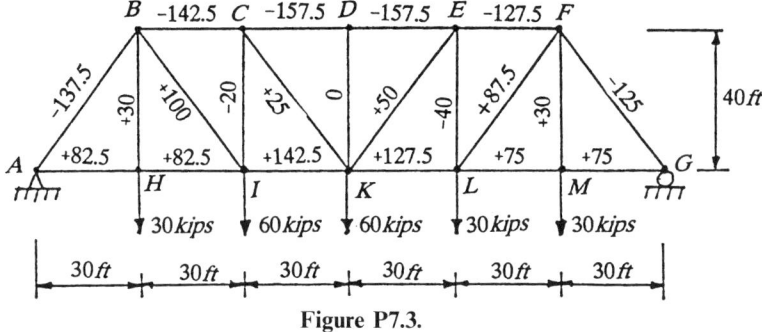

Figure P7.3.

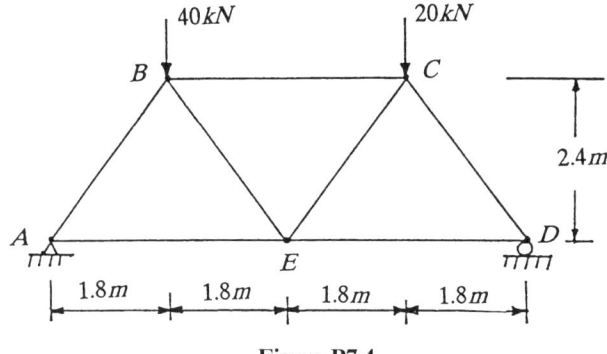

Figure P7.4.

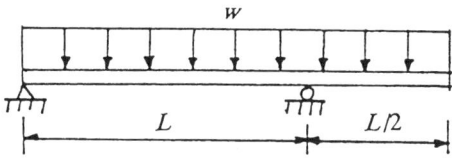

Figure P7.5.

stored in the member. The constant stiffness $EI = 60 \times 10^6$ kip-in.^{2}, the length $L = 40$ ft, and the distributed load $w = 2$ kips/ft.

7.6 For the beam problems in Fig. 7.6, determine in each case the bending strain energy stored in the member. For the members in Figs. P7.6a, P7.6b, P7.6c, and P7.6d, the bending stiffness EI is constant.

7.7 For the beam in Problem 7.5, determine the shear strain energy stored in the member. Assume that the shear modulus $G = 11.5 \times 10^3$ ksi, the cross-sectional area $A = 18.3$ in.^{2}, and shear factor $K = 1.1$.

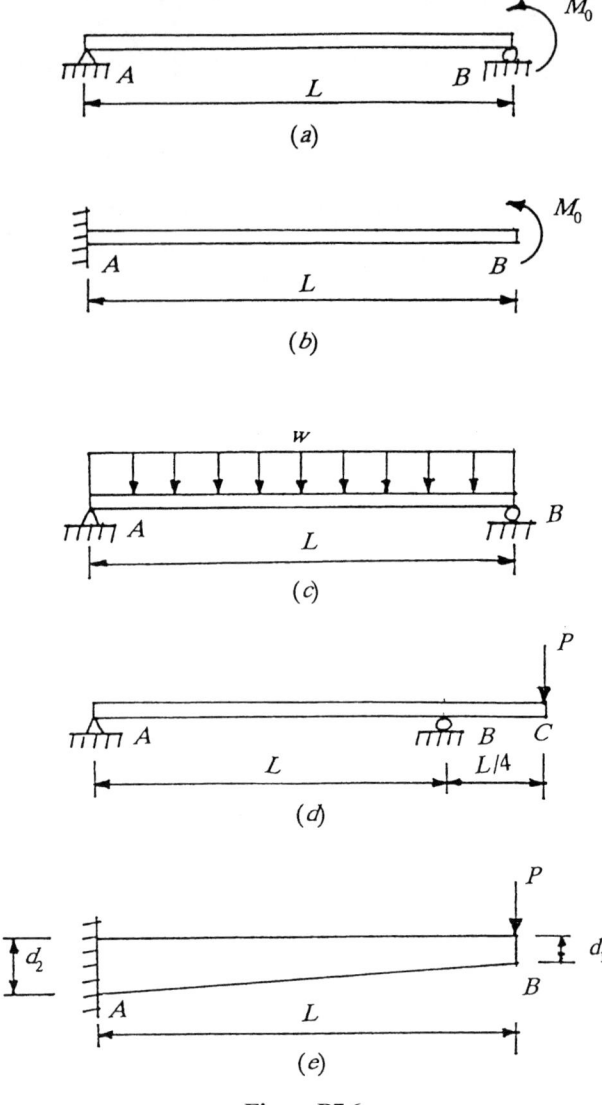

Figure P7.6.

7.8 For each member in Fig. P7.6, derive an expression for the strain energy stored in the member by shear.

7.9 The cantilever circular bar in Fig. P7.9 is subjected to a uniformly distributed torque q, as shown. Determine the strain energy of torsion stored in the bar.

7.10 Repeat Problem 7.9 by assuming that the distributed torque q varies

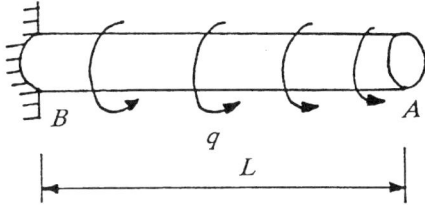

Figure P7.9.

linearly between ends A and B of the bar. At end A, the torque $q = 0$, and at end B, the torque $q = q_B$. Compare the results.

7.11 A steel bar of solid circular cross section is subjected to pure torsion by a torque T. Calculate the strain energy stored in the bar when the maximum shear stress is equal to 70 MPa. The length L of the bar is 5 m, its diameter $d = 150$ mm, and the shear modulus $G = 80$ GPa.

7.12 Solve the problem in Example 7.8 by assuming that the cross section is circular of area $A = 60$ in.^{2}, which is equal to the rectangular area in Fig. 7.9b. Compare the results.

7.13 For the beam problem in Fig. 7.6a, determine the rotation of end B by equating the external work done by the bending moment M_0 to the internal strain energy of bending.

7.14 For the beam problem in Fig. 7.6b, determine the rotation of the free end B by equating the external work to the internal strain energy of bending.

7.15 For the beam problem in Fig. P7.6d, determine the vertical deflection at end C by equating the external work to the internal strain energy of bending.

7.16 For the beam problem in Fig. P7.6e, determine the vertical deflection at end B by equating the external work to the internal strain energy of both bending and shear. Discuss the results.

7.17 For the uniform simply supported beam loaded as shown in Fig. P7.17,

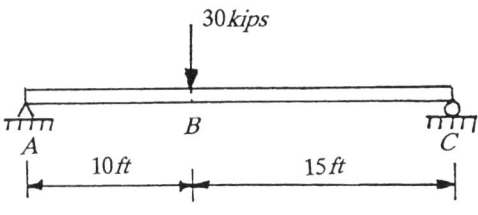

Figure P7.17.

546 ENERGY CONCEPTS AND METHODS

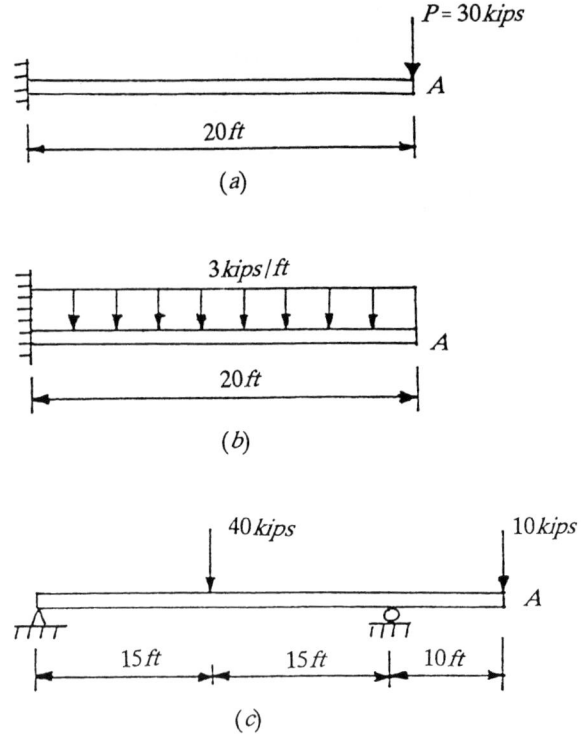

Figure P7.18.

determine the vertical deflection at point B by equating the external work to the internal strain energy of bending. The moment of inertia $I = 1{,}000$ in.^{4}, and the modulus of elasticity $E = 29 \times 10^6$ psi.

7.18 For the beam problems in Fig. P7.18, determine in each case the vertical deflection at the free end A by using the method of virtual work. For the beams in Figs. P7.18a and P7.18b, the uniform moment of inertia $I = 5{,}000$ in.^{4}, while $I = 1{,}000$ in.^{4} for the beam in Fig. P7.18c. The modulus of elasticity $E = 29 \times 10^6$ psi.

7.19 For the frame loaded as shown in Fig. P7.19, determine the horizontal displacement of point D by using the method of virtual work. The modulus of elasticity $E = 29 \times 10^3$ ksi.

7.20 For the beam problems in Problem 7.18, determine in each case the angular rotation θ_A at point A by using the method of virtual work.

7.21 For the uniform simply supported beam loaded as shown in Fig. P7.21, determine the angular rotation θ_C at point C by using the method of virtual work. The stiffness EI is constant.

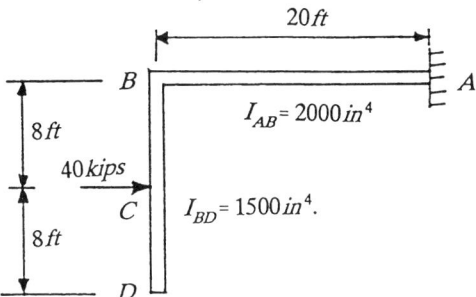

Figure P7.19.

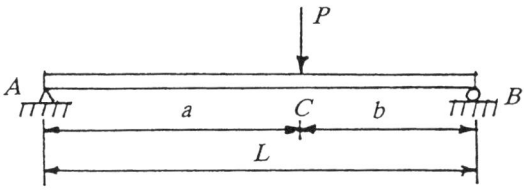

Figure P7.21.

7.22 For the simple steel truss in Problem 7.4, Fig. P7.4, determine the vertical deflection δ_E of joint E by using the method of virtual work.

7.23 For the simple truss loaded as shown in Fig. P7.23, determine the vertical deflection δ_I of joint I by using the method of virtual work. The cross-sectional areas of all members, in square inches, are shown in parentheses in the figure. The modulus of elasticity E for each member is 10.5×10^3 ksi.

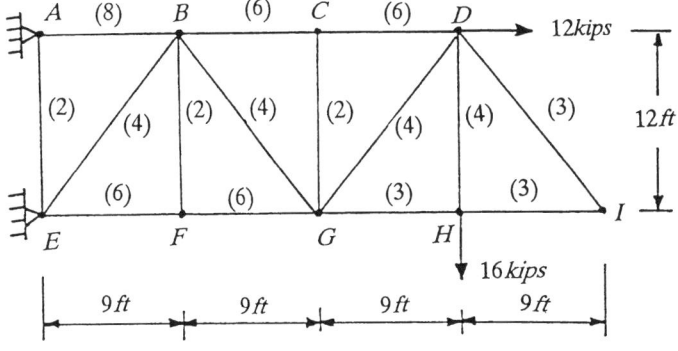

Figure P7.23.

7.24 Repeat Example 7.13 in order to determine the angular rotation θ_{EC} of member EC.

7.25 Repeat Example 7.13 in order to determine the angular rotation θ_{BC} of member BC.

7.26 Repeat Example 7.14 and determine the fixed-end moment M_{F_B} at the fixed end B.

7.27 Assume that the arch structure shown in Fig. 7.9a is hinged at both supports A and B. By using the method of virtual work, determine the horizontal reaction H_A at support A. All required information regarding load, dimensions, and so on, is given in Example 7.8.

7.28 Repeat Example 7.15 in order to determine the angular displacement at support A.

7.29 For the statically indeterminate beam problems shown in Fig. P7.29, determine in each case the reaction R_B at support B by applying the method of virtual work. The stiffness EI is constant.

7.30 For the statically indeterminate problems in Figs. P7.29a and P7.29b, determine in each case the bending moment M_A at the fixed support A by applying the method of virtual work. The stiffness EI is constant.

7.31 Solve Problem 7.22 by using Castigliano's second theorem.

7.32 Solve Problem 7.23 by using Castigliano's second theorem.

7.33 Solve Problem 7.19 by using Castigliano's second theorem.

7.34 Solve Problem 7.18 by using Castigliano's second theorem.

7.35 For the beam problems in Fig. P7.18, determine in each case the rotation of point A by using Castigliano's second theorem.

7.36 Solve Problem 7.29 by using Castigliano's second theorem.

7.37 Solve Problem 7.30 by using Castigliano's second theorem.

7.38 Repeat the problem in Example 7.20 by assuming that the truss is subjected to an horizontal external force of 10 kips at node 2 and to an external vertical downward force of 15 kips at the same node.

7.39 By following the methodology explained in Section 7.6, determine the forces in the members of the truss in Fig. 7.3. Each member has a cross-sectional area $A = 7.85 \times 10^{-3} \, m^2$, and a modulus of elasticity $E = 210 \times 10^9$ Pa. The truss is loaded by a vertical force $P = 10 \, kN$ applied at joint C, as shown.

7.40 The simple truss in Fig. P7.40 is loaded at joint D as shown. By following the methodology explained in Section 7.6, determine the forces acting in

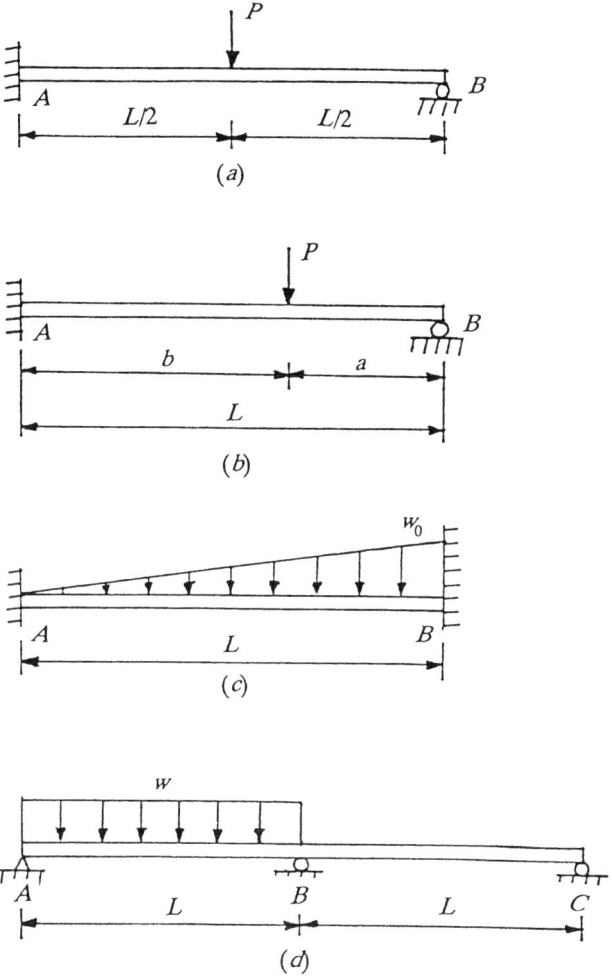

Figure P7.29.

the members of the truss. All bars of the truss have a cross-sectional area $A = 4 \text{ in.}^2$, and a modulus of elasticity $E = 30 \times 10^3$ ksi.

7.41 Rework the plate problem in Example 7.21 by assuming that the applied distributed load has a linear variation starting with zero at the left-hand edge, and increasing to the value q_0 at the right-hand edge.

7.42 Rework the plate problem in Example 7.21 by assuming that the applied distributed load has a linear variation starting with zero at the left-hand edge, reaching its maximum value q_0 at $x = a/2$, and then reducing linearly to zero at the right-hand edge.

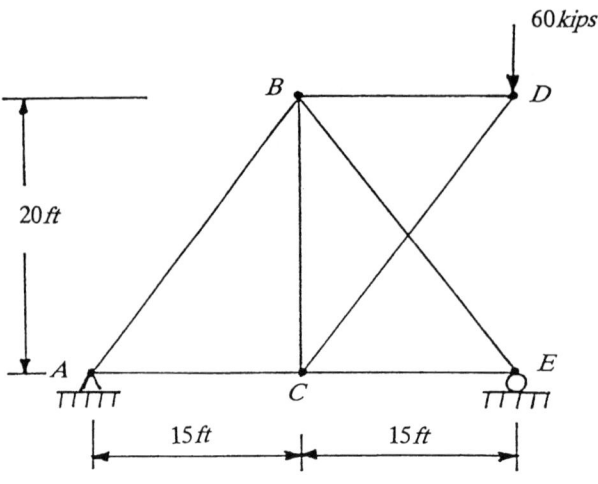

Figure P7.40.

7.43 Rework the plate problem in Example 7.22 by assuming that the applied distributed load has a linear variation starting with zero at the left-hand edge, reaching its maximum value q_0 at $x = a/2$, and then reducing linearly to zero at the right-hand edge.

7.44 Rework the plate problem in Example 7.21 by assuming that the applied distributed load is parabolic with zero values at the left-hand and right-hand edges, and a maximum value q_0 at the center line of the plate.

7.45 Repeat Problem 7.44 by assuming that all sides of the square plate are fixed.

7.46 Derive the finite difference equation for a simply supported plate with quadratic thickness variation in the y direction, which is represented by the equation

$$h = \left[1 + \alpha \left(\frac{2y}{b}\right)^2\right] h_0$$

where h_0 is the plate thickness at its center, and α is a constant.

7.47 By assuming that a simply supported square plate has a quadratic thickness variation, such as the one in Problem 7.46, and that it is loaded by a uniformly distributed load q over its entire surface, determine the maximum vertical deflection at its center by using the finite difference method and an 8×8 mesh. Assume $\alpha = 0.2$.

7.48 Repeat Problem 7.47 by assuming that $\alpha = 0.4$. Compare the results.

7.49 Repeat the square plate problem in Example 7.24 by assuming $\lambda = 0.3$ and compare the results.

7.50 Repeat the rectangular plate problem in Example 7.24 by assuming $\lambda = 0.3$ and $b/a = 2$. Compare the results.

APPENDIX A
Formulas of Vertical Deflection of Uniform Single-Span Beams Subjected to Various Types of Loadings

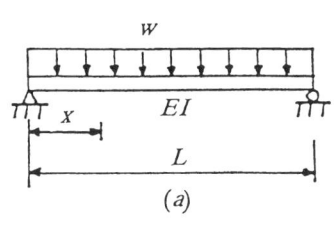

(a) $\quad y_x = \dfrac{wx}{24EI}[L^3 - 2Lx^2 + x^3]$

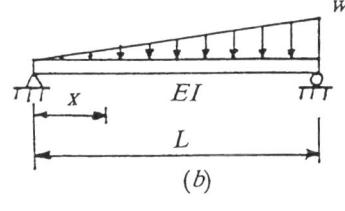

(b) $\quad y_x = \dfrac{w_0 x}{360EIL}[3x^4 - 10L^2x^2 + 7L^4]$

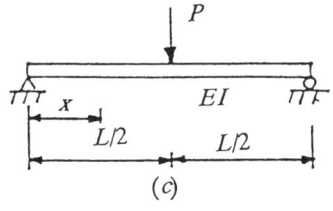

(c) $\quad y_x = \dfrac{Px}{48EI}[3L^2 - 4x^2] \qquad x \leqslant \dfrac{L}{2}$

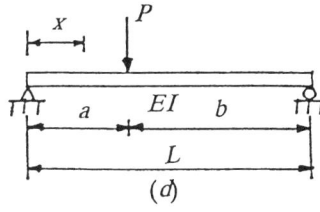

(d) $\quad y_x = \dfrac{Pbx}{6EIL}[L^2 - b^2 - x^2] \qquad x \leqslant a$

(e) $\quad y_x = \dfrac{Px}{96EI}[3L^2 - 5x^2] \qquad x \leqslant \dfrac{L}{2}$

$\quad y_x = \dfrac{P}{96EI}[x - L]^2[11x - 2L] \qquad x > \dfrac{L}{2}$

553

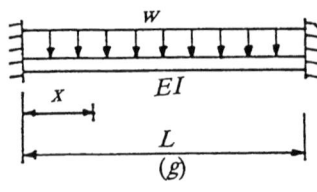
(f)

$$y_x = \frac{Pb^2x}{12EIL^3}[3aL^2 - 2Lx^2 - ax^2] \quad x \leq a$$

$$y_x = \frac{Pa}{12EIL^3}[L - x]^2(3L^2x - a^2x - 2a^2L)$$
$$x > a$$

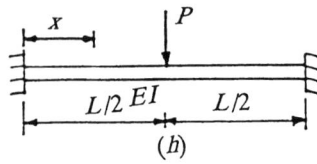
(g)

$$y_x = \frac{wx^2}{24EI}[L - x]^2$$

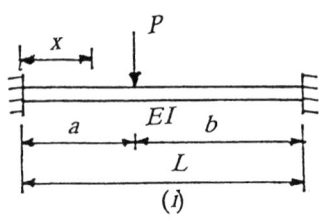
(h)

$$y_x = \frac{Px^2}{48EI}[3L - 4x] \quad x \leq \frac{L}{2}$$

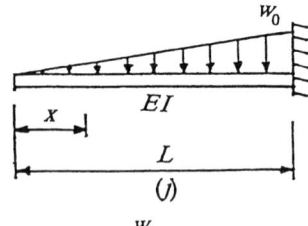
(i)

$$y_x = \frac{Pb^2x^2}{6EIL^3}[3aL - 3ax - bx] \quad x \leq a$$

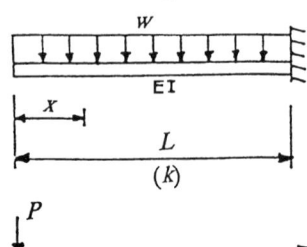
(j)

$$y_x = \frac{w_0}{120EIL}[x^5 - 5L^4x + 4L^5]$$

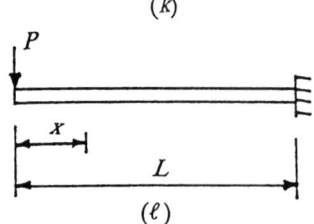
(k)

$$y_x = \frac{w}{24EI}[x^4 - 4L^3x + 3L^4]$$

(ℓ)

$$y_x = \frac{P}{6EI}[2L^3 - 3L^2x + x^3]$$

APPENDIX B
Simpson's One-Third Rule

Simpson's one-third rule is one of the most commonly used methods to approximate integration. Consider, for example, the integral

$$\lambda = \int_a^b f(x)\, dx \tag{B.1}$$

between the limits a and b. If we divide the integral between the limits $x = a$ and $x = b$ into n equal parts, where n is an even number, and if $y_0, y_1, y_2, \ldots, y_{n-1}, y_n$ are the ordinates of the curve $y = f(x)$, as shown in Fig. B.1, then, according to Simpson's one-third rule we have

$$\int_a^b f(x)\, dx = \frac{\Delta}{3}(y_0 + 4y_1 + 2y_2 + 4y_3 + \cdots + 2y_{n-2} + 4y_{n-1} + y_n) \tag{B.2}$$

where

$$\Delta = \frac{b - a}{n} \tag{B.3}$$

Simpson's rule provides reasonably accurate results for practical applications.

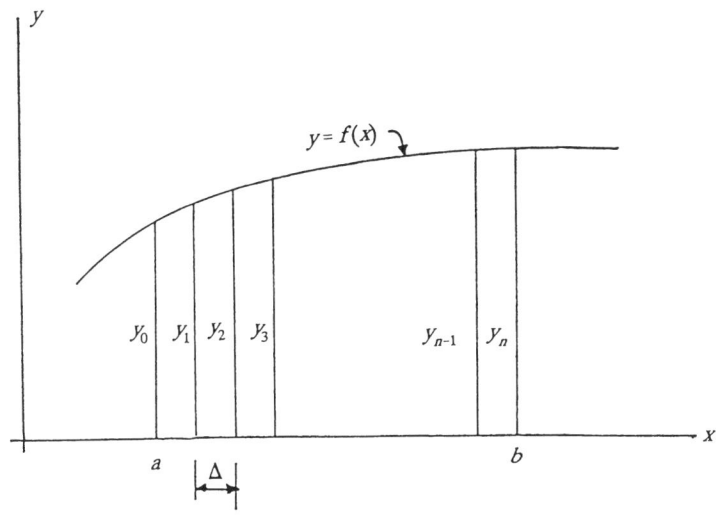

Figure B.1. Plot of a function $y = f(x)$.

APPENDIX B

Let it be assumed that it is required to determine the value λ of the integral

$$\lambda = \int_0^L x^2 \, dx \tag{B.4}$$

We divide the length L into 10 equal segments, yielding $\Delta = 0.1L$. By applying Simpson's rule, given by Eq. (B.2), we find

$$\lambda = \frac{0.1L}{3}[(1)(0)^2 + (4)(0.1)^2 + (2)(0.2)^2 + (4)(0.3)^2 + (2)(0.4)^2$$
$$+ (4)(0.5)^2 + (2)(0.6)^2 + (4)(0.7)^2 + (2)(0.8)^2 + (4)(0.9)^2 + (1)(1)^2]L^2$$
$$= \frac{L^3}{3}$$

In this case, the exact value of the integral is obtained.

As a second example, let it be assumed that it is required to find the value λ of the integral

$$\lambda = \int_0^L x^3 \, dx \tag{B.5}$$

Again, we subdivide length L into 10 equal segments, yielding $\Delta = 0.1L$. In this case, the Simpson's one-third rule yields

$$\lambda = \frac{0.1L}{3}[(1)(0)^3 + (4)(0.1)^3 + (2)(0.2)^3 + (4)(0.3)^3 + (2)(0.4)^3 + (4)(0.5)^3$$
$$+ (2)(0.6)^3 + (4)(0.7)^3 + (2)(0.8)^3 + (4)(0.9)^3 + (1)(1)^3]L^3$$
$$= \frac{0.75L^4}{3} = \frac{L^4}{4}$$

The exact value of the integral is obtained again in this case.

More complicated integrals may also be evaluated in a similar manner.

APPENDIX C
Integration of the Euler–Bernoulli Nonlinear Differential Equation

This appendix provides the methodology used for the integration of the nonlinear Euler–Bernoulli differential equation for elastic analysis discussed in Chapter 5. We write this equation as follows:

$$\frac{y''}{[1+(y')^2]^{3/2}} = -\frac{M_x}{E_x I_x} \tag{C.1}$$

or

$$\frac{y''}{[1+(y')^2]^{3/2}} = -\frac{M_x}{EI_1 f(x) g(x)} \tag{C.2}$$

where $f(x)$ is the moment of inertia function representing the variation of I_x with I_1 as the reference. The modulus of elasticity E is assumed constant, and on this basis the modulus function $g(x) = 1$.

We integrate Eq. (C.2) by making changes in the variables. We let $y' = p$ and then $y'' = p'$. Thus, from Eq. (C.1), we obtain

$$\frac{p'}{[1+p^2]^{3/2}} = \lambda(x) \tag{C.3}$$

where

$$\lambda(x) = \frac{M_x}{E_x I_x} \tag{C.4}$$

Now we write Eq. (C.3) as follows:

$$\frac{dp/dx}{[1+p^2]^{3/2}} = \lambda(x) \tag{C.5}$$

By multiplying both sides of Eq. (C.5) by dx and integrating once, we find

$$\int \frac{dp}{[1+p^2]^{3/2}} = \int \lambda(x)\, dx \tag{C.6}$$

557

We can integrate Eq. (C.6) by making the following substitutions:

$$p = \tan\theta \tag{C.7}$$

$$dp = \sec^2\theta \, d\theta \tag{C.8}$$

On this basis, Eqs. (C.7) and (C.8) yield

$$\cos\theta \frac{1}{[1+p^2]^{1/2}} \tag{C.9}$$

$$\sin\theta \frac{p}{[1+p^2]^{1/2}} \tag{C.10}$$

By substituting into Eq. (C.6), we find

$$\int \cos\theta \, d\theta = \int \lambda(x) \, dx \tag{C.11}$$

Integration of Eq. (C.11) yields

$$\sin\theta = \varphi(x) + C \tag{C.12}$$

Equation (C.12) may be written in terms of p and y' as follows:

$$\frac{p}{[1+p^2]^{1/2}} = \varphi(x) + C \tag{C.13}$$

$$\frac{y'}{[1+(y')^2]^{1/2}} = \varphi(x) + C \tag{C.14}$$

where C is the constant of integration, which can be determined from the boundary conditions of the given problem. If we solve for $y'(x)$ and $y(x)$, we obtain the following two equations:

$$y'(x) = \frac{\varphi(x) + C}{\sqrt{1 - [\varphi(x) + C]^2}} \tag{C.15}$$

$$y(x) = \int_0^x \frac{\varphi(\eta) + C}{\sqrt{1 - [\varphi(\eta) + C]^2}} \, d\eta \tag{C.16}$$

This shows that when $M_x/E_x I_x$ is known and it is integrable, then the Euler–Bernoulli equation may be solved directly for $y'(x)$. The required integrations may be carried out more efficiently if they are expressed in a nondimensional form.

APPENDIX D
Computer Program That Can Be Used to Carry Out the Inelastic Analysis of Prismatic and Nonprismatic Cantilever Beams

The following computer program is used to carry out the inelastic analysis in Sections 5.9 and 5.10 of Chapter 5, and to determine ultimate loads for cantilever beams made out of mild steel, monel, and aluminum alloy. The computer program can be easily modified for use with simply supported beams, or beams of other boundary conditions. The beam may have a constant thickness, or its thickness may vary linearly along its length.

```
c    ************************************************************
c    *                                                          *
c    *                                                          *
c    *                                                          *
c    *                                                          *
c    *                                                          *
c    *                                                          *
c    *                                                          *
c    *                                                          *
c    *                                                          *
c    *                                                          *
c    *          A General Beam Analysis Program                 *
c    *                      Using                               *
c    *                Equivalent System                         *
c    *                                                          *
c    *         ****************    ****************            *
c    *                                                          *
c    *                   Version 2.0                            *
c    *                   May, 1995                              *
c    *                                                          *
c    ************************************************************
c
c
c TABLE OF VARIABLES
c
        implicit real*8   (a-h, o-z)
        implicit integer (i-n)
        real*8   lyield
        real*8   inert
        character*8 outname
```

559

560 APPENDIX D

```fortran
      PARAMETER  (NXMAX=22,NYMAX=40,NLINE=2)
      dimension hx(0:NXMAX),inert(0:NXMAX),
     +    act(0:NXMAX),acb(0:NXMAX),dt(0:NXMAX),
     +    dy(0:NXMAX),ct(0:NXMAX),xl(0:NXMAX),yn(0:NXMAX),
     +    yy(0:NYMAX),xgm(0:NYMAX),epc(0:NYMAX),
     +    g(0:NXMAX),ee(0:NXMAX),
     +    rm(0:NXMAX),Me(0:NXMAX),v1(0:NXMAX),v2(0:NXMAX)
      DATA n,m/20,20/

c---------------------------------
      DATA hb,hs,b/12.,8.,6./       ! hb -- fixed end height
                                    ! hs -- free end height
c---------------------------------
c     DATA e1,ep1/22000.,.0021818182/,p/0./     ! 3-line Monel
c     DATA e1,ep1/26000.,.001923076923/,p/0./ ! 2-line Monel
c     DATA e1,ep1/22000.,.0021818182/,p/0./     ! E/P Monel

      DATA e1,ep1/9436.6,3.55e-3/,p/0./         ! 4-line steel
c     DATA e1,ep1/9436.6,3.55e-3/,p/0./         ! E/P Steel
c---------------------------------------------------------------
c     DATA e1,ep1/9500.,,8.398947e-3/,p/0./     ! 2-line aluminum
c     DATA e1,ep1/9510.,6.302839e-3/,p/0./      ! 3-line aluminum
c     DATA e1,ep1/9490.,6.0e-3/,p/0./           ! 6-line aluminum
c---------------------------------------------------------------
c
      write(*,*) 'please key-in output filename:'   ! message
      read(*,'(\a)') outname                ! read in the output file name
      open(3,file=outname,status='unknown')  ! output data file 2
      write(*,*)'SpanLength ='              ! length of beam
      read(*,*)  tl                         !
      write(*,531)                          !
c---------------------------------------------------------------
      tdx=tl/n                              !
      do 111 i=0,n                          !
        xl(i)=i*tdx                         ! x(i) for cantilever beam
        hx(i)=hb-xl(i)/tl*(hb-hs)           ! height for taped beam section
        inert(i)=b*hx(i)**3/12.             ! inertia of moment
        act(i)=p/b/hx(i)/e1                 ! initial trial top strain
        acb(i)=act(i)                       ! initial trial bot strain
111   continue                              !
c---------------------------------------!
      w0=4.*e1*inert(0)*ep1/hx(0)/tl**2    !
      pp=.70                               !
      wint=pp*w0                           !
      wincre=(1.-pp)*w0                    !
      dytemp=0.                            !
        write(*,9877)0.,0.,0.              !
        write(3,6453) 0.,0.,0.             !
9001  w=wint+wincre                        !
```

```
c-------------------------------------------------------------------
365     eer = e1                              !
        serror0 = 1.                          !
        ndf = 0                               !
        consl = w/24./e1/inert(0)             !
        do 320 i = 1,n
          dy(i) = cons1*((tl-xl(i))**4-4.*tl**3        ! initial curve
     *      *(tl − xl(i)) + 3.*tl**4)  ! initial       ! and
          ct(i) = (dy(i) − dy(i−1))/(xl(i) − xl(i − 1))  ! rotation
320     continue
        dy(0) = 0.
        ct(0) = 0.
        end_moment = 0.
c--------------------------------------------------
9       dttemp = 0.                           !
        do 121 i = 0,n                        ! start of the main loop
          strs0 = p/b/hx(i)                   ! initial stress by p only
          ac0 = strs0/e1                      ! initial strain by p only
          acc0 = ac0                          !
          xmnt = .5*w*(tl − xl(i))**2-end_moment
          acb(i) = ac0 + .5*xmnt*hx(i)/e1/inert(i)
          act(i) = 2.*ac0 − acb(i)
c--------------- solving P,M ------------------
        if(abs(xmnt).le.1.e − 16)  goto 901
        nm = 0
66      np = 0
77      dt(i) = acb(i) − act(i)
        r = hx(i)/dt(i)
        yn(i) = r*acb(i)
c--------------- solving P' ------------------
        ddy = hx(i)/m
        do 670 jj = 0,m
          yy(jj) = yn(i) − hx(i)*jj/m
          epc(jj) = yy(jj)/r
          xgm(jj) = strs(epc(jj))
670     continue
        s1 = 0.
        do 671 jj = 0,m−1
          s1 = s1 + .5*(xgm(jj) + xgm(jj + 1))*ddy
671     continue
        s1 = s1*b
c----------trial-and-error for P ----------------
55      if(np.eq.0)then
          xb = acb(i−1)
          yb = 0.
        else
          xb = xa
          yb = ya
        endif
```

562 APPENDIX D

```
            xa = acb(i)
            if(p.eq.0.)then
                ya = s1
            else
                ya = s1/p - 1.
            endif
            if(abs(ya).gt.1.e-4.and.abs(yb-ya).gt.1.e-16)then
                np = np + 1
                acb(i) = xa - ya/(yb-ya)*(xb-xa)
                goto 77
            endif
909         continue
c--------- trial-and-error for M ----------------
            sm2 = 0.
            sm1 = .5*w*(tl-xl(i))**2 - p*dy(i) - end_moment
            sm1 = sm1 - p*yn(i)
c
            do 621 jj = 0,m-1
                sm2 = sm2 + .5*(xgm(jj)*epc(jj) + xgm(jj+1)*epc(jj+1))*
     +              (epc(jj+1) - epc(jj))
621         continue
            sm2 = -b*r*r*sm2
            if(nm.eq.0)then
                tb = act(i-1)
                qb = .1
            else
                tb = ta
                qb = qa
            endif
            ta = act(i)
            qa = sm2/sm1 - 1.
            if(abs(qa).gt.1.e-4.and.abs(qb-qa).gt.1.e-16)then
                act(i) = ta - qa/(qb-qa)*(tb-ta)
                acc0 = .5*(act(i) + acb(i))
                nm = nm + 1
                goto 66
            endif
211         continue
c----------------------------------------------------------
901         if(abs(xmnt).1e.1.e-16)then
                ee(i) = 0.
                rm(i) = 0.
                g(i) = 0.
                v1(i) = 0.
                v2(i) = 0.
            else
                    ee(i) = r/inert(i)*sm2
                    rm(i) = sm2
                    g(i) = ee(i)/e1
```

```
              if(abs(g(i)).1e.1.e-16)then
              endif
              v1(i)=.5*hx(i)-yn(i)
              if(abs(act(i)).gt.ep1.and.abs(acb(i)).gt.ep1)then
                 v2(i)=hx(i)/dt(i)*p/b/hx(i)/ee(i)
              else
                 v2(i)=hx(i)/dt(i)*p/b/hx(i)/e1
              endif
           endif
           if(abs(dttemp).lt.abs(dt(i)))then
              dttemp=dt(i)
              itemp=i
           endif
121     continue
c------------------------------
        do 999 i=0,n
           xmnt=.5*w*(tl-xl(i))**2-end_moment
           ta=inert(i)/inert(n)
           if(abs(xmnt).lt.1.e-16)then
              Me(i)=0.
           else
              Me(i)=rm(i)/g(i)/ta
           endif
999     continue
        serror=0.
        error_endcita=0.
        do 321 i=1,n
           ct(i)=0.
           dy(i)=0.
           dy(0)=0.
           do 322 j=1,i
              ct(i)=ct(i)+.5*(xl(j)-xl(j-1))*(Me(j)+Me(j-1))/inert(i)
322        continue
           ct(i)=ct(i)/e1
           ct(0)=0.
           do 323 j=1,i
              dy(i)=dy(i)+.5*(xl(j)-xl(j-1))*(Me(j)+Me(j-1))
     +              *(xl(i)-.5*(xl(j-1)+xl(j)))/inert(i)
323        continue
           dy(i)=dy(i)/e1
           serror=serror+dy(i)**2
321     continue
        serror=sqrt(serror)
        if(abs(serror/serror0-1.).gt.1.e-5)then
           serror0=serror
           ndf=ndf+1
           goto 9
        endif
c------------------------------
```

```
              cvt=2.*ep1
              xx1=0.
              xx2=0.
              do 3033 i=1,n
                if(abs(dt(i)).ge.cvt.and.abs(dt(i-1)).lt.cvt)then
                  if(abs(dt(i+1)).lt.cvt)
     *              xx2=xl(i)+(dt(i)-cvt)*(xl(i+1)-xl(i))/(dt(i)-dt(i+1))
                    xx1=xl(i)-(dt(i)-cvt)*(xl(i)-xl(i-1))/(dt(i)-dt(i-1))
                  else if(abs(dt(i)).ge.cvt.and.abs(dt(i+1)).lt.cvt)then
                    xx2=xl(i)+(dt(i)-cvt)*(xl(i+1)-xl(i))/(dt(i)-dt(i+1))
                  endif
3033          continue
1009          continue
              lyield=xx2-xx1
              if(dttemp-cvt.lt.0.)then
                write(*,9877)w,dy(n),lyield
                write(3,6453)w,dy(n),lyield
              else
                write(*,9877)w,dy(n),lyield
                write(3,6453)w,dy(n),lyield
              endif
              if(wint.eq.0.)then
                stt=dy(n)/w
                sttemp=100000.
              else
                stt=(dy(n)-dytemp)/(w-wint)
              endif
c-------------------------------------!
              if(stt.gt.1.1*sttemp)then           !
                wincre=wincre*.70                 !
                if(wincre.lt..001)wincre=.001     !
              endif                               !
c-------------------------------------!
              dytemp=dy(n)
              wint=w
              sttemp=stt
              if(dy(n).le..40*tl)goto 9001
c----------------------------------------------------------------------
531           format(
```

```
c      èëëëëëëëëëëëëëëëëëëëëëëëëëëëëëëëëëëëëëëëëëëëëëëëëëëë£
c      □                                                          □
c      □        ÉÉÉÉ      ÉÉÉÉÉÉÉ    ÉÉÉ    ÉÉÉÉÉÉÉ    ÉÉÉÉÉ       □
c      □      ßÚÚÚÚÚÚÏ   ÚÚÚÚÚÚÚÚ  ßÚÚÚÚÏ  ÚÚÚÚÚÚÚ   ÚÚÚÚÚÚÚ      □
c      □      ÚÚÏ  ßÚÚ   ÚÚÚ  ÚÚÚ  ÚÚÚÚÚÚ  ÚÚÚ      ßÚÚ  ÔÔ       □
c      □      ßÚÚ  ÉÉÉÉÉ ÚÚÚÉÉÚÚÔ  ßÚÚ ÚÚÏ ÚÚÚÚÚÚÚ  ÚÚÚÚÚÚÉ      □
c      □      ßÚÚ  ÔÚÚÚÔ ÚÚÚÔÔÚÚÚ  ÚÚÚÚÚÚÚÚ ÚÚÚÔÔÔÔ  ÔÔÔÚÚÚÏ     □
c      □      ÚÚÏ  ÚÚÚ   ÚÚÚ  ÚÚÚ  ßÚÚÏ ßÚÚÏ ÚÚÚÚÚÚ   ÉÉ  ßÚÚÏ    □
c      □      ßÚÚÚÚÚÚÚ   ÚÚÚÚÚÚÚÚ  ÚÚÚ   ÚÚÚ ÚÚÚÚÚÚ   ÚÚÚÚÚÚÚ     □
c      □       ÔÔÔÔÔÔ    ÔÔÔÔÔÔÔ   ÔÔÔ   ÔÔÔ ÔÔÔÔÔÔ    ÔÔÔÔÔ      □
c      □                                                          □
c      □          A General Beam Analysis Program                 □
c      □                        Using                             □
c      □                   Equivalent System                      □
c      □                                                          □
c      □           ááááááááááááááá   ááááááááááááááá              □
c      □                                                          □
c      □                     Version 2.0                          □
c      □                      May, 1995                           □
c      □                                                          □
c      àëëëëëëëëëëëëëëëëëëëëëëëëëëëëëëëëëëëëëëëëëëëëëëëëëëëY
c      ─────────────────────────────────────────────────────────
c
566    format(/,'START WITH THE ELASTIC'
      +          ' DEFLECTION without AXIAL FORCE')
567    format(/,'START WITH A GIVEN DEFLECTION' )
871    format(/,'FINAL END MOMENT =',f9.2,'(kips-in)',//)
980    format(//,'TRIAL Number: ',i3,',   Axial Force = ',f10.4)
931    format(//,
      +' xi      hi      fxi     strain1  strain2    xigma     Ee     g=Ee/E1',
      +   '     Mreq        Me',/,
      +'(in)    (in)            *.001    *.001',
      +    '    (in-lb)    (in-lb)',/)
981    format(f4.0,1x,f4.1,1x,f5.3,3(1x,f7.4),3x,f7.4,3(1x,f7.4))
982    format(4x,f4.0,3x,f9.4,3x,f8.4,3(3x,f7.4))
983    format(//,'   TOTAL TRIAL LOOPS =',i3,/,
      +       '   FINAL AXIAL FORCE =',f7.2,'(kips)',/,)
984    format(//,'      xl(i),  Me(i)*1000  cita*.001  deflection')
9876   format(4x,'w =',f7.3,'(kips)   ',4x,'f =',f10.6,'  (in)')
9878   format(4x,'w =',f7.3,'(kips)*',4x,'f =',f10.6,
      +       '  (in)',3x,f9.6,4x,i2)
6453   format(2x,f11.5,4x,f11.5,4x,f11.5)
9877   format(4x,'w =',f7.3,'   (kips)  ',4x,'y =',f8.4,'   (in)',4x,
      +        'Ly =',f10.6,'  (in)')
c─────────────────────────────────────────────────────────────
       close(1)
       close(2)
       close(3)
       stop
       end


       FUNCTION strs(strn)
       implicit real*8  (a-h,o-z)
       dimension stif(1:4),strncr(1:4),stress(1:4)
       integer j

       DATA mmm/4/,stif/9436.6,282.89,37.85,-79.71/, ! + +
```

```
c         +          strncr/0.,3.55e-3,.076356273,.234876748/  ! 4-line mild steel
c                    DATA mmm/2/,stif/9436.6,.01/,strncr/0,3.55e-3/ ! elastic-plastic steel
c-----------------------------------------------------------------------
c                    DATA mmm/2/,stif/22000.,.1/,strncr/0.,2.18182e-3/ ! E/P monel
c                    DATA mmm/3/,stif/22000.,504.,125./,       ! ++
c         +          strncr/0.,2.18182e-3,.0240072/            ! 3-line monel
c                    DATA mmm/2/,stif/26000.,53./,strncr/0.,1.92308e-3/ ! 2-line monel
c-----------------------------------------------------------------------
c                    DATA mmm/2/,stif/9500.,200./,strncr/0.,8.398947e-3/ ! 2-line aluminum
c                    DATA mmm/3/,stif/9510.,8630.,190./,       ! ++
c         +          strncr/0.,6.302839117e-3,8.589050e-3/     ! 3-line aluminum
c                    DATA mmm/6/,stif/9490.,8650.,4990.,1240.,500.,199./, ! ++
c         +          strncr/0.,6.e-3,7.848555e-3,8.848555e-3,  ! ++
c         +          .010058232,.011058232/                    ! 6-line aluminum
c-----------------------------------------------------------------------
                    mit=2    ! mit: 0 -- unlimited, 1 -- 1-0,  2 -- 1-2-0,  ...
c-----------------------------------------------------------------------
                    stress(1)=0.
                    do 23 i=2,mmm
                      stress(i)=stress(i-1)+(strncr(i)-strncr(i-1))*stif(i-1)
23                  continue

                    do 16 j=1,mmm ! mmm -- number of lines in stress-strain fig.
                      if(abs(strn).gt.strncr(mmm))goto 31
                        if(abs(strn).ge.strncr(j).and.
                          abs(strn).le.strncr(j+1))then
                          strstemp=stress(j)+stif(j)*(abs(strn)-strncr(j))
                          goto 111
                        endif
16                  continue
31                  strstemp=stress(mmm)+stif(mmm)*(abs(strn)-strncr(mmm))

111                 if(mit.ne.0)then
                      if(abs(strstemp.gt.stress(mit+1))then
                        strstemp=stress(mit+1)
                      endif
                    endif

                    if(strn.ge.0.)then
                      strs=strstemp
                    else
                      strs=-strstemp
                    endif
                    return
                    END
```

APPENDIX E
The Superposition Principle

In many cases, the response of structural systems is described by linear equations. Such linear structures are often, but not always, assumed to obey the principle of superposition, which may be stated as follows:

> With respect to a coordinate system of axes, components of stress at a point in an elastic body, or its displacements, may be algebraically added in any sequence.

For example, if a member is acted upon by forces $F_1, F_2, F_3, \ldots, F_n$, producing a total displacement y at a point C of this member, then

$$y = y_1 + y_2 + y_3 + \cdots + y_n \tag{E.1}$$

where $y_1, y_2, y_3, \ldots, y_n$ are the displacements produced individually by the application of $F_1, F_2, F_3, \ldots, F_n$, respectively. The forces $F_1, F_2, F_3, \ldots, F_n$ may be applied to the structure in any sequence and the corresponding displacements $y_1, y_2, y_3, \ldots, y_n$ may be algebraically added in any order.

This is one of the most important principles in the linear theory of mechanics, and many of the methods that are developed for the study of such systems are based, at least in part, on the assumption that this principle is valid. The application of this principle to structural problems presupposes that all members of the structural system are constructed of linearly elastic, continuous, homogeneous and isotropic material, and that the deformations are small so that the geometry of the loading system is not appreciably altered.

The principle of superposition is not valid when the deformation of a structural system resulting from any one of the forces is affected by the deformation due to one of the other forces. For example, in the case of slender flexible members subjected to large axial loads, experience has shown that axial forces may appreciably alter the shape of the elastic curve. Under these conditions, the transverse displacements are not of the magnitude that permits us to neglect their influence on the equilibrium conditions for a deformed body. As a result, deflections are no longer proportional to the transverse loading, and the principle of superposition is not directly applicable.

APPENDIX F
Formulas and Mode Shapes of Free Vibrations of Single Span Uniform Beams [101]

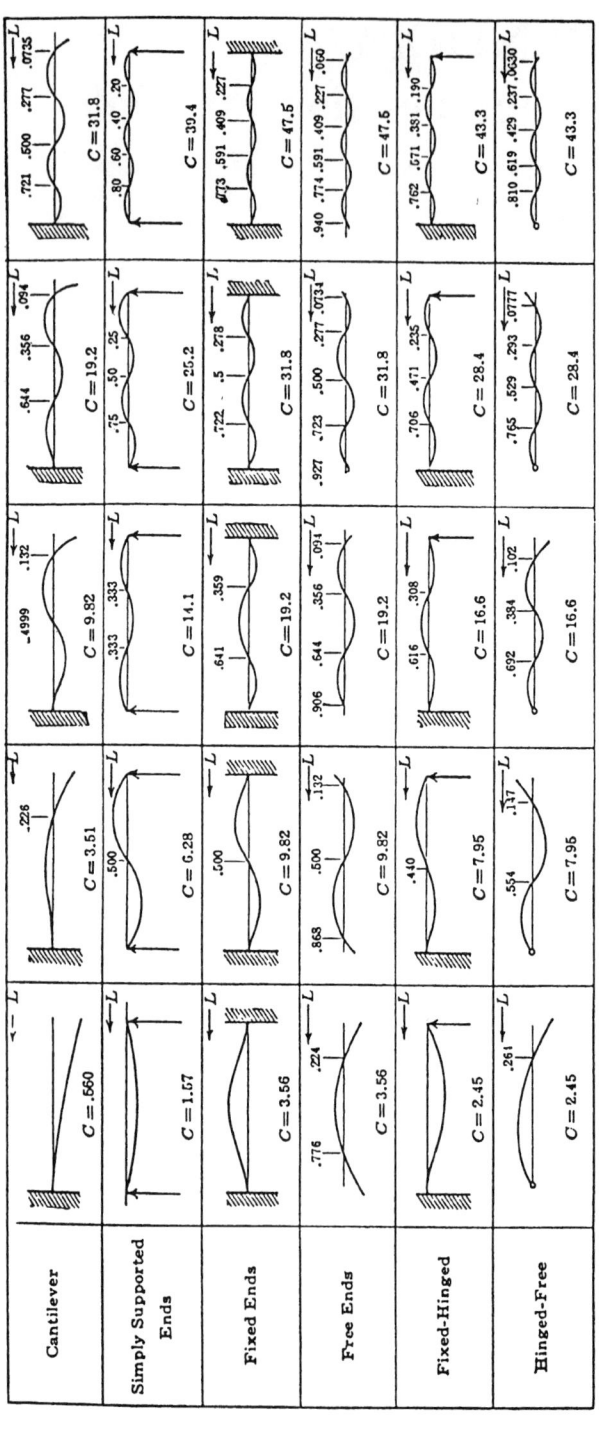

$$f_n = C\sqrt{\frac{EI}{mL^4}}$$

f_n = natural frequency in cycles/sec.
C = constant from above table
E = modulus of elasticity in lb./in.^{2}
I = sectional moment of inertia in in.^{4}
m = mass of unit length beam in lb. sec./in.^{2}
L = beam length in inches

APPENDIX G
Charts of Elastoplastic Response for Undamped One-Degree Spring-Mass Systems

The results shown in the graphs of Figs. G.1 through G.4 may be used for one-degree undamped spring-mass systems, and they are based on a bilinear resistance function R. The initial motion of the spring-mass system is assumed to be zero. These charts provide maximum response only, and they apply to

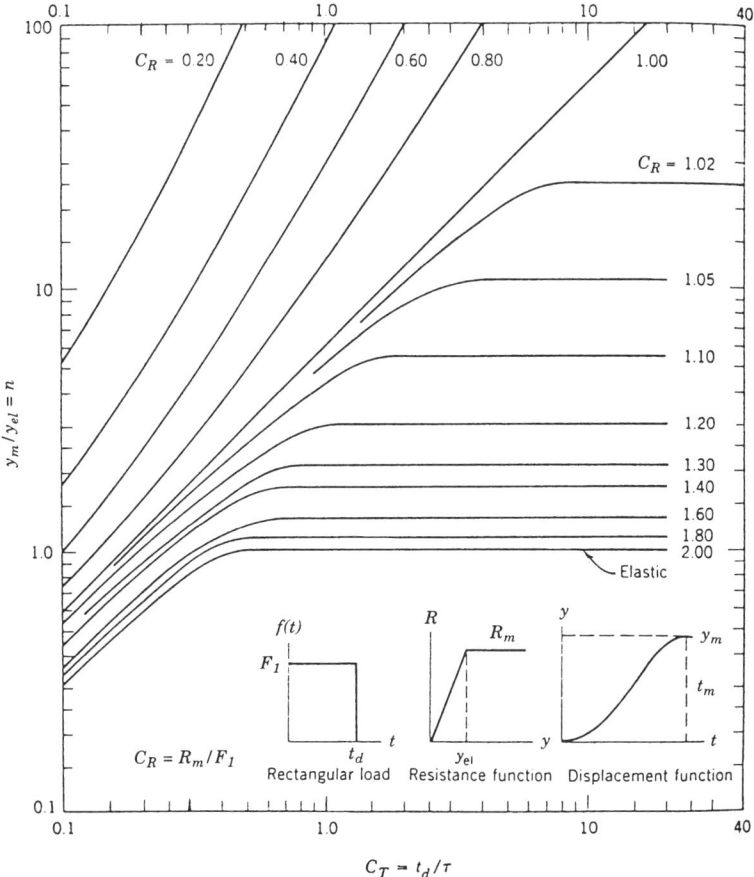

Figure G.1. The y_m/y_{el} curves for elastoplastic system, rectangular load [40].

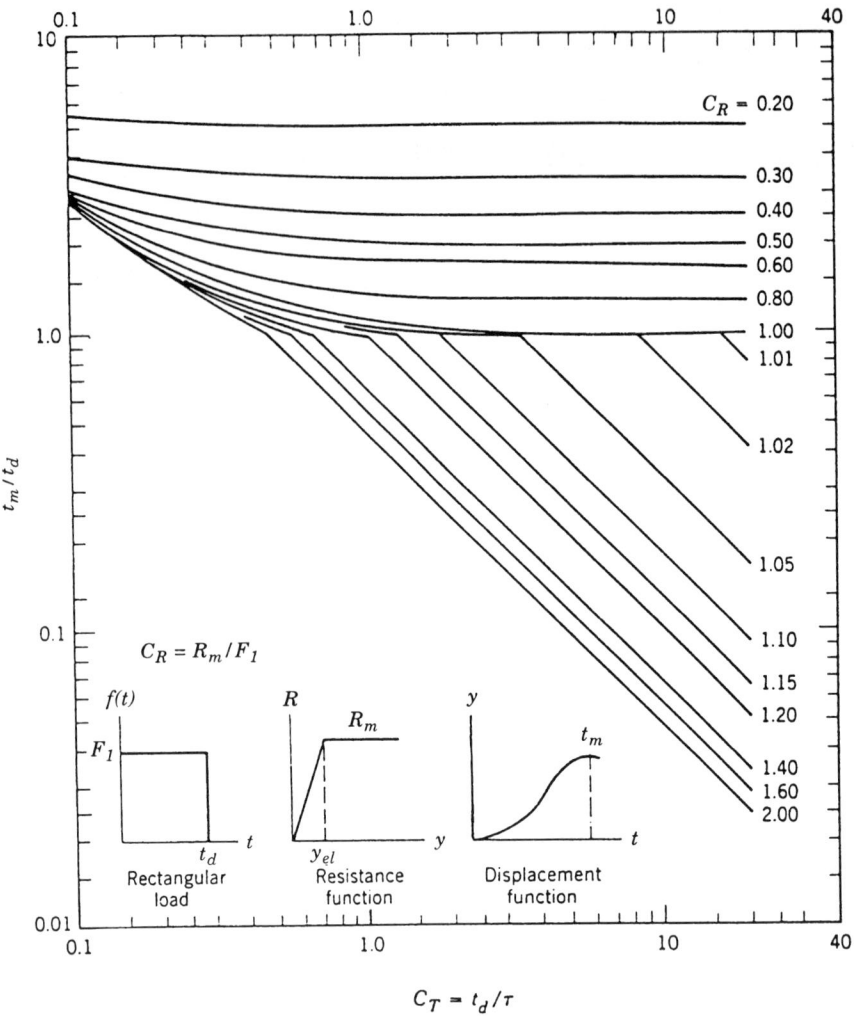

Figure G.2. The t_m/t_d curves for elastoplastic system, rectangular load [40].

cases where the design requirements are based on maximum displacement. Although they are limited to two types of load-time variations, many practical situations can be approximated by a rectangular or a triangular time variation. The explanation of the symbols used in these graphs is given in the figures, with the exception of τ, which denotes the undamped free period of vibration of the one-degree spring-mass system. Note that the response for $R_m/F_1 = 2$ is completely elastic, and it is inelastic for curves with R_m/F_1 larger than 2.

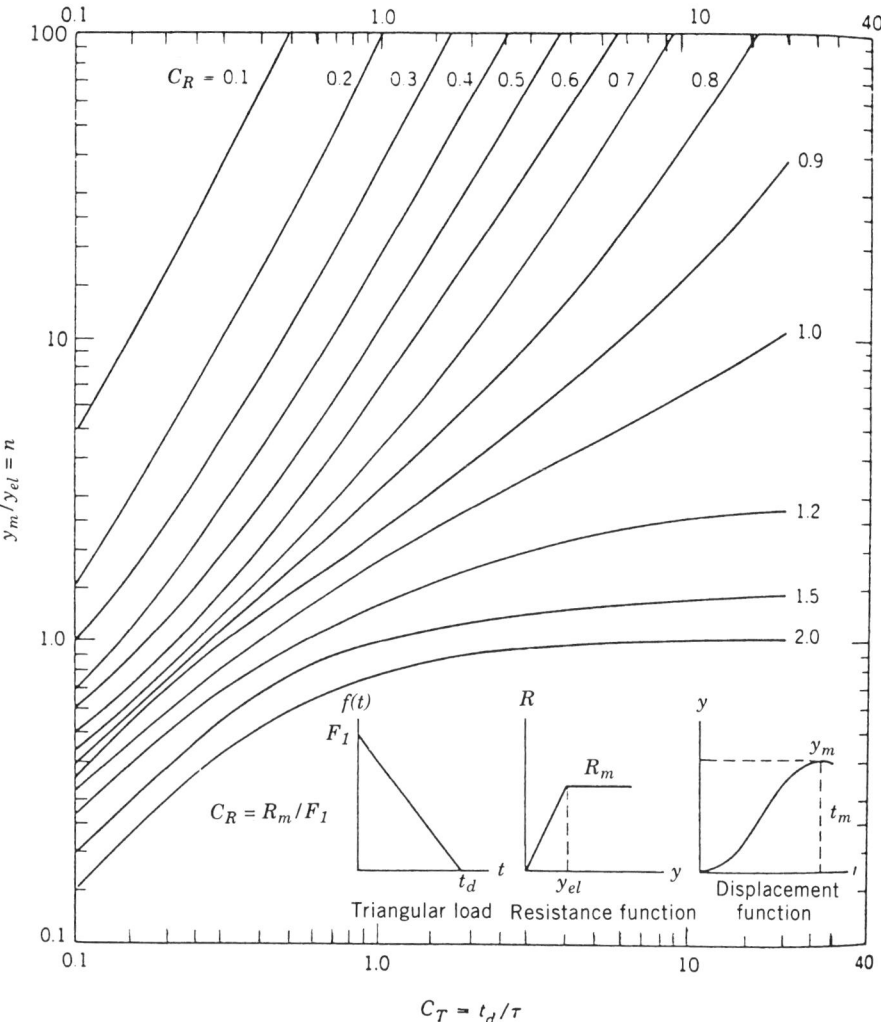

Figure G.3. The y_m/y_{el} curves for elastoplastic system, triangular load [40].

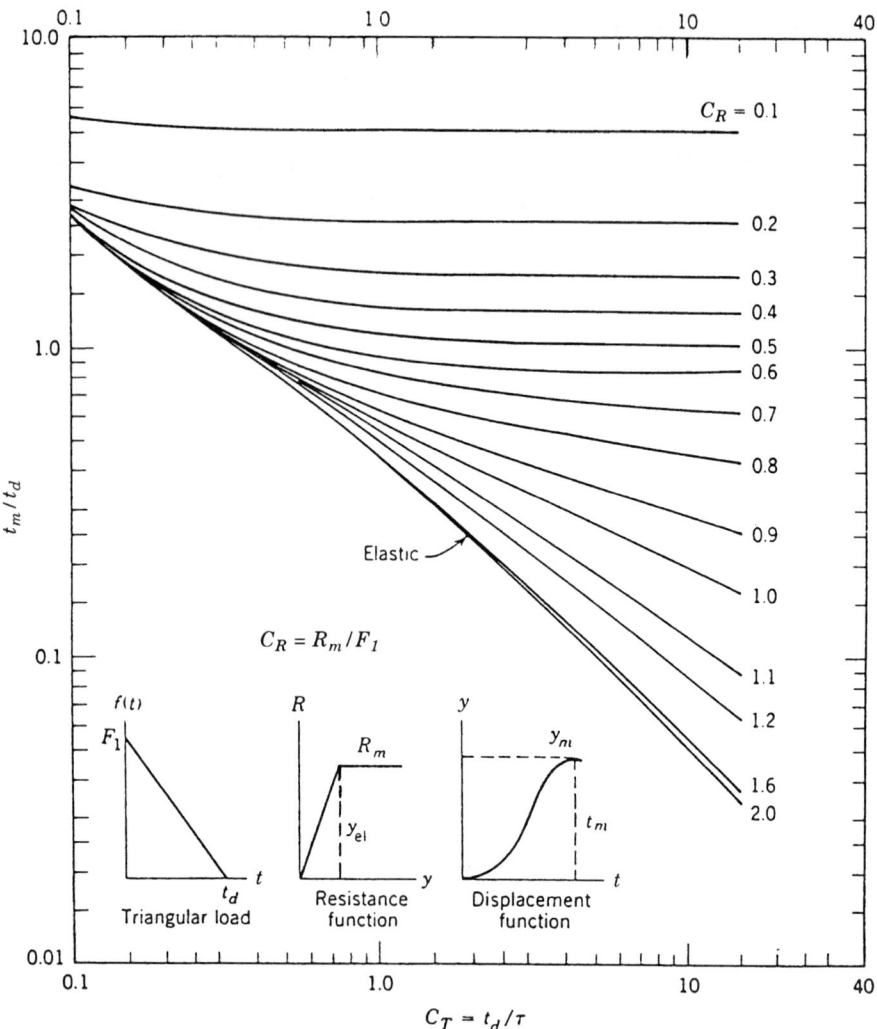

Figure G.4. The t_m/t_d curves for elastoplastic system, triangular load [40].

APPENDIX H
Graphs of Maximum Magnification Factors and Times of Maximum Response for Undamped One-Degree Spring-Mass Systems

The graphs in Figs. H.1 through H.4 contain, in each case, a plot of the maximum magnification factor Γ_{max} versus the ratio of t_d/τ, and a plot of the

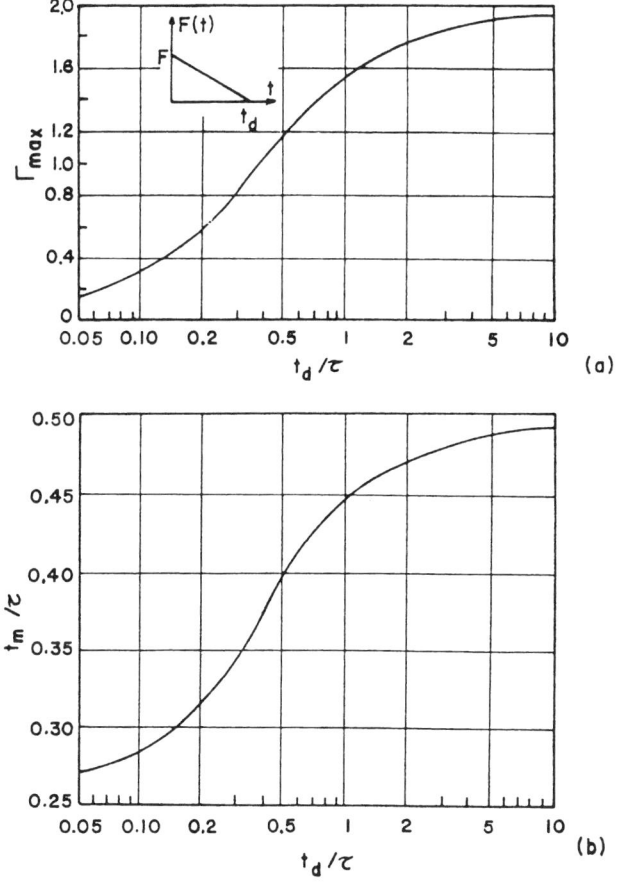

Figure H.1. Reference [40].

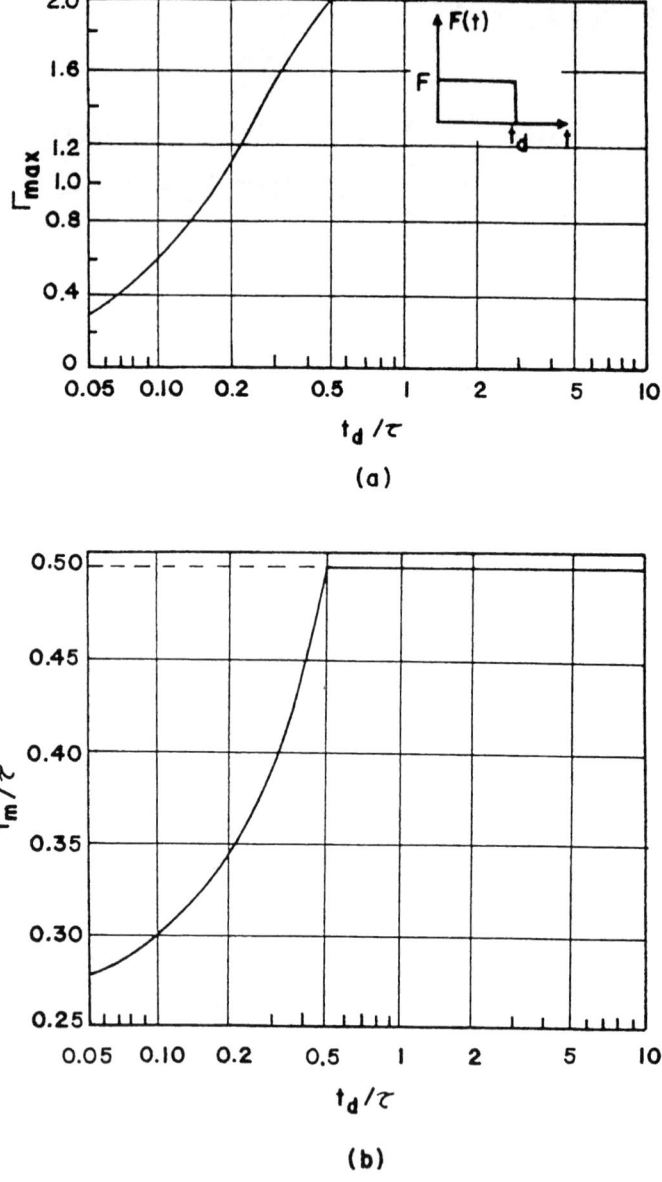

Figure H.2. Reference [40].

ratio t_m/τ versus the ratio t_d/τ, for undamped one-degree spring-mass systems, where τ is the period of vibration of the spring-mass system, t_d is the time duration of the dynamic force $F(t)$, and t_m is the time of maximum response. These graphs illustrate the great importance of the ratio t_d/τ regarding the

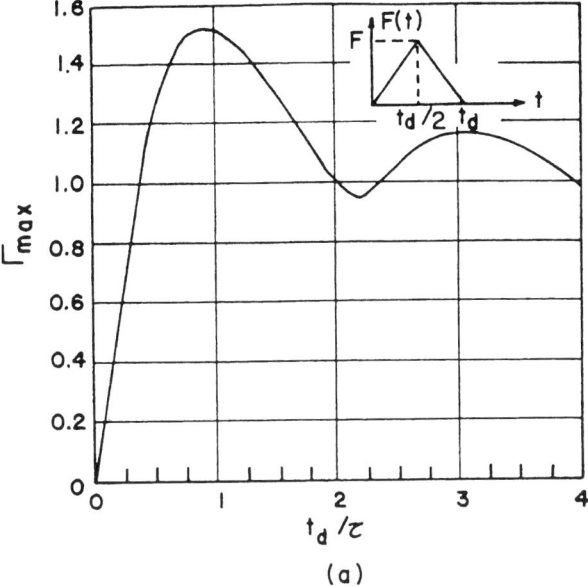

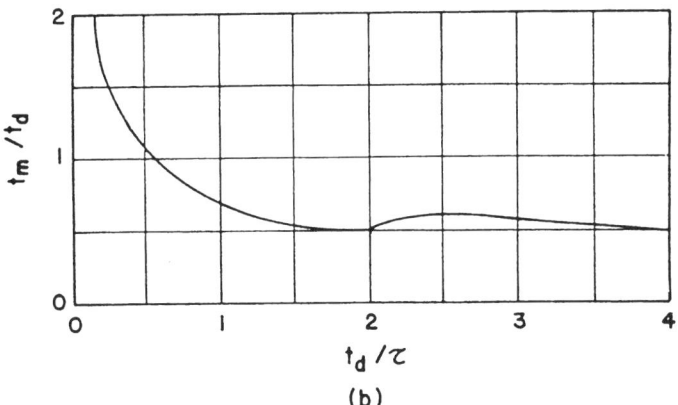

Figure H.3. Reference [40].

dynamic response of a structural system and, consequently, they help the design engineer to make wise decisions regarding the dynamic design of a structure. For a given duration of the dynamic force $F(t)$, the structure could be designed for the most favorable t_d/τ ratio. They are also a convenient way to determine Γ_{max} and t_m, and such factors are applicable for displacements, stresses, shear forces, bending moments, and other related quantities.

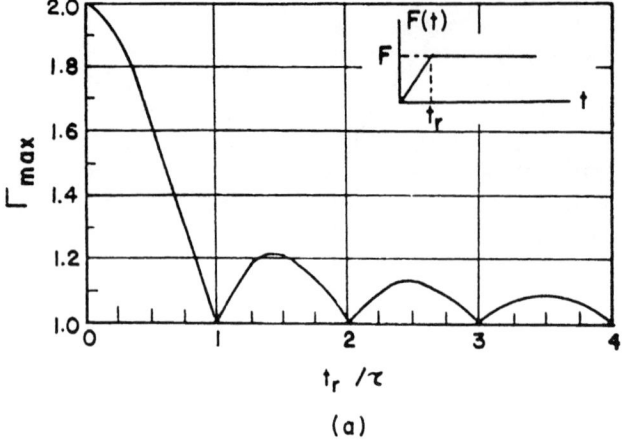

(a)

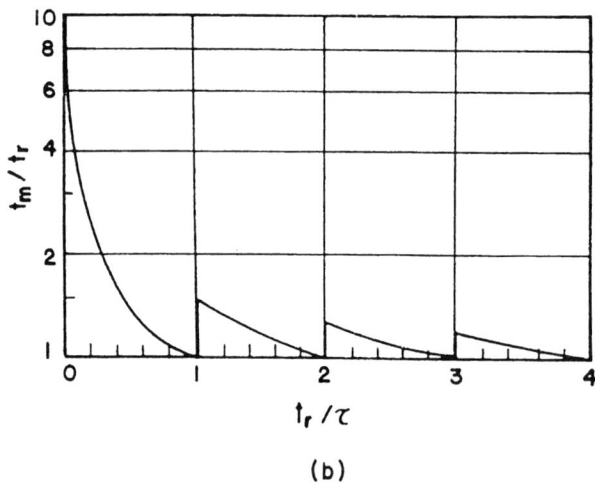

(b)

Figure H.4. Reference [40].

APPENDIX I
Computer Program Using the Acceleration Impulse Extrapolation Method (AIEM) for the Elastoplastic Analysis in Example 6.6

This short computer program was used to carry out the computation in Example 6.6 of Section 6.5.2 of Chapter 6, which involves the elastoplastic analysis of a two-story building subjected to the earthquake ground displacements shown in Fig. 6.11.

```
10          PRINT 100
20      100 FORMAT(' INPUT IX, SX, AND ZX FOR CHOSEN COLUMN')
30          READ *, XI,SX,ZX
40          EM1=0.22526
50          EM2=0.14451
60          EK1=0.07144*XI
70          EK2=27.*EK1/8.
71          EKM=(EK1+EK2)/(EM1)+EK2/EM2
72          EKK=4.*EK1*EK2/(EM1*EM2)
73          W12=0.50*EKM+0.50*SQRT(EKM**2-EKK)
74          W22=0.50*EKM-0.50*SQRT(EKM**2-EKK)
75          TAU1=2.*3.141593/SQRT(W12)
76          TAU2=2.*3.141593/SQRT(W22)
80          Y1EL=9.331*SX/XI
90          Y2EL=4.*Y1EL/9.
100         R1M=0.3333*ZX
110         R2M=0.5000*ZX
120         PRINT 110,EK1,EK2,Y1EL,Y2EL,R1M,R2M
130     110 FORMAT(' K1 =',F7.3,' K2=',F7.3,' Y1EL =',F7.4,' Y2EL=',F7.4,
140        $' R1M=', F7.3,' R2M =',F7.3)
141         PRINT 115,TAU1,TAU2
142     115 FORMAT(' TAU1 =',F7.3,' TAU2 =',F7.3)
150         J=0
160         XO1=0.0
170         XO2=0.0
180         PRINT 120
190     120 FORMAT(' ENTER TIME INCREMENT TO BE USED')
200         READ *,DT
220         DT2=DT**2
```

```
230         PRINT 130
240     130 FORMAT(' ENTER 1ST GROUND DISPLACEMENT' )
250      10 READ *,U
260         IF(J.GT.1) GO TO 50
270         CN=EK2*DT2/(6.*EM2)
280         C1A=CN/(1.+CN)
290         EK1M=EK1*DT2/(6.*EM1)
300         EK2M=EK2*DT2/(6.*EM1)
310         X1 = EK1M*U/(1.+EK2M*(1.−C1A)+EK1M)
320         X2=C1A*X1
330      50 X1U=X1−U
340         R1=X1U*EK1
350         IF(ABS(R1).LE.R1M) GO TO 80
351         IF(R1.LT.0.0) R1 = −R1M
352         IF(R1.GT.R1M) R1=R1M
360      80 X21=X2−X1
370         R2=X21*EK2
380         IF(ABS(R2).LE.R2M) GO TO 90
381         IF(R2.LT.0.0) R2 = −R2M
382         IF(R2.GT.R2M) R2=R2M
390      90 X1A=(R2−R1)/EM1
400         X1AT2=X1A*DT2
410         X2A=−R2/EM2
420         X2AT2=X2A*DT2
422         ETA1=ABS(X1/Y1EL)
423         ETA2=ABS(X21/Y2EL)
430         PRINT 150, X1U,R1,X21,R2,X1A,X1AT2,X1,ETA1,X2A,X2AT2,X2,ETA2
440     150 FORMAT(' X1−U=',F8.4,' R1=',F8.2,',X2−X1=',F8.4,' R2=',
450        $F8.2/'  X1A=',F8.2,' X1AT2=',F8.3,' X1=',F8.4,'ETA1=',F5.2/
460        $'     X2A=',F8.2,' X2AT2=',F8.3,' X2=',F8.4,' ETA2=',F5.2)
470         CX1=X1
480         CX2=X2
490         X1=2.*X1−XO1+X1AT2
500         X2=2.*X2−XO2+X2AT2
510         XO1=OX1
520         XO2=OX2
530         J=2
540         GO TO 10
550         END
```

Answers to Selected Problems

Chapter 1

1.1 At left support A:

$$H_A = 10 \text{ kips} \leftarrow$$
$$V_A = 4.33 \text{ kips} \leftarrow$$

At right support B:

$$V_B = 15.67 \text{ kips} \uparrow$$

1.2
$$F_{CD} = -45.00 \text{ kips}$$
$$F_{CF} = 5.16 \text{ kips}$$
$$F_{EF} = 42.10 \text{ kips}$$

1.4 For Fig. P1.3a, members CB, HD, FG, and FE. For Fig. P1.3c, members KL and EO.

1.5 For Fig. P1.5b:

$$V_A = 45 \text{ kips} \uparrow \qquad V_B = 56.33 \text{ kips} \downarrow$$
$$H_A = 10 \text{ kips} \leftarrow \qquad V_C = 11.33 \text{ kips} \uparrow$$

For Fig. P1.5d:

$$V_A = 47.36 \text{ kips} \uparrow \qquad H_A = 28.68 \text{ kips} \rightarrow$$
$$V_B = 29.81 \text{ kips} \uparrow \qquad V_C = 27.83 \text{ kips} \uparrow$$
$$H_C = 1.32 \text{ kips} \rightarrow$$

1.6 For Fig. P1.6a:

$$F_{GD} = 30 \text{ kips} \quad \text{(compression)}$$
$$F_{EB} = 40 \text{ kips} \quad \text{(compression)}$$

All other members in Fig. P1.6a are zero-force members. For Fig. P1.6c:

580   ANSWERS TO SELECTED PROBLEMS

$$V_A = 15 \text{ kips} \uparrow \qquad H_B = 0$$
$$H_A = 20 \text{ kips} \rightarrow \qquad H_C = 10 \text{ kips} \rightarrow$$
$$V_B = 15 \text{ kips} \downarrow \qquad V_D = 10 \text{ kips} \uparrow$$

1.8 For Fig. P1.8*b*:

$$A_z = 37 \text{ kips} \uparrow \qquad C_x = 0$$
$$C_z = 30 \text{ kips} \uparrow \qquad B_x = 0$$
$$B_z = 27 \text{ kips} \downarrow \qquad C_y = 0$$

For Fig. P1.8*e*:

$$A_x = 80 \text{ kips} \rightarrow \qquad C_z = 4.33 \text{ kips} \downarrow$$
$$A_z = 83.33 \text{ kips} \uparrow \qquad D_z = 19 \text{ kips} \uparrow$$
$$B_y = 0 \qquad D_x = 40 \text{ kips} \rightarrow$$

1.9 For Fig. P1.9*a*, $\Sigma F_x \neq 0$, beam is free to move horizontally. For Fig. P1.9*h*, $\Sigma F_x \neq 0$, truss is free to move horizontally at roller.

1.11 Axial force $P = 13.66$ kips (compressive); shear force $V = 3.66$ kips $\nwarrow$; bending moment $M = 73.20$ kip-ft $\circlearrowright$.

1.12 For Fig. P1.12*c*: axial force $P = 0$; shear force $V = 7.5$ kips $\uparrow$; bending moment $M = 250$ kip-ft $\circlearrowleft$. For Fig. P1.12*f*: axial force $P = 10.58$ kips (tensile); shear force $V = 1.78$ kips $\downarrow$; bending moment $M = 2,280$ kip-ft $\circlearrowleft$.

1.13 For Fig. P1.12*a*:

$$\text{Shear force} \qquad V = 11.25 - \frac{3x^2}{80}$$

$$\text{Bending moment} \qquad M = \frac{x^3}{80} - 11.25x$$

and x is measured from support B. For Fig. P1.12*f*:

Bending moment $\quad M = 1{,}875 \sin\theta + 93.75 \cos\theta - 937.5 \sin^2\theta - 93.75,$
$$0 \leqslant \theta \leqslant 90°$$

Shear force $\quad V = -75 \cos\theta + 75 \sin\theta \cos\theta + 3.75 \sin\theta,$
$$0 \leqslant \theta \leqslant 90°$$

Axial force $\quad P = \dfrac{3.75}{\cos\theta} + 75 \sin\theta - 75 \sin^2\theta - 3.75 \dfrac{\sin^2\theta}{\cos\theta},$
$$0 \leqslant \theta \leqslant 90°$$

ANSWERS TO SELECTED PROBLEMS   581

θ is zero at the left support and 90° at the middle of the arch. Similar equations may be written for the right half of the arch. For Fig. P1.12d:

Bending moment $M = 2{,}400 \, (\sin\theta + 2\cos\theta) + 60(1 - \cos\theta) - 4{,}800,$
$$0 \leqslant \theta \leqslant 90°$$

Axial force $P = 60 \, (\sin\theta + 2\cos\theta), \qquad 0 \leqslant \theta \leqslant 90°$

Shear force $V = \dfrac{60}{\cos\theta} - \dfrac{60\sin\theta}{\cos\theta}(\sin\theta + 2\cos^2\theta), \quad 0 \leqslant \theta \leqslant 90°$

θ is zero at the left support and 90° at the center of the arch where the hinge is located.

1.16 $M_B = 365 \text{ kip-ft}$
$M_C = 95 \text{ kip-ft}$

Chapter 2

2.1   In Portion BC, $\sigma_{\max} = 1{,}666.67$ psi. Total elongation $\delta = 0.0505$ in.

2.3 $A = 6.668 \times 10^{-6} \text{ m}^2$
$= 6.668 \text{ mm}^2$

2.5 $\delta = \dfrac{PL}{Ebh}\log(2) = 0.693\dfrac{PL}{Ebh}$

2.7 $T = 229.73$ N·m.

2.8 $T = 376.99$ N·m.

2.10 $T = 257.77$ N·m.

2.12 $T = 41{,}280$ in.·lb for rectangular cross section. $T = 71{,}362$ in.·lb for circular solid section of equal cross-sectional area.

2.14 $T = 15{,}710$ in.·lb.

2.17   For Fig. P2.17a:

$$y_x = \frac{w_0}{120EIL}(x^5 - 5L^4 x + 4L^5)$$

$$y_{x=0} = y_B = \frac{w_0 L^4}{30EI}$$

$$\theta_x = \frac{dy_x}{dx} = \frac{w_0}{120EIL}(5x^4 - 5L^4)$$

$$\theta_{x=0} = \theta_B = -\frac{w_0 L^3}{24EI}$$

For Fig. P2.17d:

$$y_x = \frac{w_0 x}{960 EIL}(5L^2 - 4x^2)^2, \qquad x \leq \frac{L}{2}$$

$$y_{x=L/2} = y_C = \frac{w_0 L^4}{120 EI}$$

$$\theta_x = \frac{dy_x}{dx} = \frac{w_0}{960 EIL}(25L^4 - 120L^2 x^2 + 80x^4), \qquad x \leq \frac{L}{2}$$

$$\theta_{x=L/2} = \theta_C = 0$$

$$\theta_{x=0} = \theta_A = \Theta_B = \frac{w_0 L^3}{38.4 EI}$$

For Fig. P2.17g:

$$y_x = \frac{wx}{48 EIL}(L^4 - 3L^2 x^2 + 2Lx^3) \quad \text{(between supports)}$$

$$y_{x_1} = \frac{wx_1}{24 EI}\left(2L^3 + \frac{3L^2 x_1}{2} - 2Lx_1^2 + x_1^3\right) \quad \text{(for overhang)}$$

x_1 is zero at B and $L/2$ at D.

$$y_C = y_{x=L/2} = \frac{wL^4}{192 EI}$$

$$y_{x_1 = L/2} = y_D = \frac{19 wL^4}{384 EI}$$

2.19 1.741×10^7 Pa.

2.21 $P = 1{,}837$ lb (compression governs).

2.24
$\sigma_A = 1{,}100$ psi (tension)
$\sigma_B = 900$ psi (compression)
$\sigma_C = 1{,}500$ psi (compression)
$\sigma_D = 500$ psi (tension)

2.26 $r = 1.318$ in.

2.28 $\tau_{max} = 1.305 \times 10^5$ Pa.

2.29
$\tau_{max} = 4{,}150$ psi
$\sigma_1 = 6{,}740$ psi (tension)
$\sigma_2 = 1{,}557$ psi (compression)

2.30 $P = 724.08$ N (bending governs).

2.32 $P_{max} = 10{,}689$ lb.

2.35 $\sigma_y = 30.54$ MPa; $\tau_{yx} = 15.73$ MPa.

Chapter 3

3.1 For Fig. P3.1a:

$$k = \frac{3EI}{L^3} \qquad W = W \qquad F = F(t)$$

For Fig. P3.1c:

$$k = \frac{81EI}{4L^3} \qquad W = W \qquad F = F(t)$$

For Fig. P3.1e:

$$k = \frac{k_1 k_3}{k_1 + k_3}, \qquad \text{where } k_3 = \frac{192 k_2 EI}{4 k_2 L^3 + 48EI}, \qquad W = W, \qquad F = F(t)$$

3.3 For Fig. P3.3a:

$$k = 42{,}450.62 \text{ lb/in.}$$
$$W = 222{,}577.5 \text{ lb}$$
$$F = F(t)$$

For Fig. P3.3b:

$$k = 36{,}568.34 \text{ lb/in.}$$
$$W = 160{,}070 \text{ lb}$$
$$F = F(t)$$

3.5 For Fig. P3.1a:

$$\omega = 12.066 \text{ rps}$$
$$\tau = 0.521 \text{ sec}$$

For Fig. P3.1c:

$$\omega = 31.347 \text{ rps}$$
$$\tau = 0.2004 \text{ sec}$$

For Fig. P3.1e:

$$\omega = 27.15 \text{ rps}$$
$$\tau = 0.2314 \text{ sec}$$

3.7 For Fig. P3.1a:

$\omega_d = 11.93$ rps (15 percent damping)
$\omega_d = 11.68$ rps (25 percent damping)
$\omega_d = 11.30$ rps (35 percent damping)
$\tau_d = 0.5267$ sec (15 percent damping)
$\tau_d = 0.5379$ sec (25 percent damping)
$\tau_d = 0.5560$ sec (35 percent damping)

For Fig. P3.1e:

$\omega_d = 26.84$ rps (15 percent damping)
$\omega_d = 26.29$ rps (25 percent damping)
$\omega_d = 25.43$ rps (35 percent damping)
$\tau_d = 0.2341$ sec (15 percent damping)
$\tau_d = 0.2390$ sec (25 percent damping)
$\tau_d = 0.2471$ sec (35 percent damping)

3.8 For Fig. P3.3a:

$$\omega = 8.58 \text{ rps}$$
$$\tau = 0.7323 \text{ sec}$$

For Fig. P3.3b:

$$\omega = 9.395 \text{ rps}$$
$$\tau = 0.6688 \text{ sec}$$

3.9 For Fig. P3.3a:

$$\omega_d = 8.537 \text{ rps}$$
$$\tau_d = 0.736 \text{ sec}$$
$$\delta = 0.6315$$

For Fig. P3.3b:

$$\omega_d = 9.348 \text{ rps}$$
$$\tau_d = 0.6721 \text{ sec}$$
$$\delta = 0.6315$$

3.11 For Fig. P3.1a:

$$y_{max} = 0.1217 \text{ m}$$
$$\sigma_{max} = 228.19 \text{ MPa} \quad \text{(at the fixed end)}$$

For Fig. P3.1c:

$$y_{max} = 0.002051 \text{ m} \quad \text{(at free end)}$$
$$\sigma_{max} = 8.65 \text{ MPa} \quad \text{(at the right-hand support)}$$

For Fig. P3.1e:

$$y_{max} = 0.002983 \text{ m}$$
$$\sigma_{max} = 6.8829 \text{ MPa}$$

3.12 For Fig. P3.1a:

$$y_{max} = 0.05603 \text{ m}$$
$$\sigma_{max} = 105.0563 \text{ MPa}$$

For Fig. P3.1c:

$$y_{max} = 0.002044 \text{ m} \quad \text{(at free end)}$$
$$\sigma_{max} = 8.6231 \text{ MPa} \quad \text{(at right-hand support)}$$

3.15 For Fig. P3.1a:

$$y_{max} = 0.05998 \text{ m}$$
$$\sigma_{max} = 112.4625 \text{ MPa}$$

For Fig. P3.1c:

$$y_{max} = 0.008885 \text{ m}$$
$$\sigma_{max} = 37.4835 \text{ MPa}$$

3.16 For Fig. P3.3a:

$$x_{max} = -0.1721 \text{ in.}$$
$$\sigma_{1max} = 4{,}829.4 \text{ psi} \quad \text{(first column)}$$
$$\sigma_{2max} = 2{,}359.67 \text{ psi} \quad \text{(second column)}$$
$$\sigma_{3max} = 4{,}888.23 \text{ psi} \quad \text{(third column)}$$
$$\sigma_{4max} = 2{,}444.12 \text{ psi} \quad \text{(fourth column)}$$

3.17 For Fig. P3.3a: assume $\omega_f = 8$ rps. Then:

$$x_{max} = 2.6859 \text{ in.}$$
$$\sigma_{1max} = 75{,}368 \text{ psi} \quad \text{(first column)}$$
$$\sigma_{2max} = 36{,}825 \text{ psi} \quad \text{(second column)}$$
$$\sigma_{3max} = 76{,}286 \text{ psi} \quad \text{(third column)}$$
$$\sigma_{4max} = 38{,}143 \text{ psi} \quad \text{(fourth column)}$$

3.18 For Fig. P3.3a: assume $\omega_f = 8$ rps. Then:

$$x_{max} = 1.54868 \text{ in.}$$
$$\sigma_{1max} = 43{,}457 \text{ psi} \quad \text{(first column)}$$
$$\sigma_{2max} = 21{,}233 \text{ psi} \quad \text{(second column)}$$
$$\sigma_{3max} = 43{,}987 \text{ psi} \quad \text{(third column)}$$
$$\sigma_{4max} = 21{,}993 \text{ psi} \quad \text{(fourth column)}$$

3.19 For Fig. P3.3a: assume $\omega_f = \omega = 8.5846$ rps. Then:

$$x_{max} = 1.76676 \text{ in.}$$
$$\sigma_{1max} = 49{,}576 \text{ psi} \quad \text{(first column)}$$
$$\sigma_{2max} = 24{,}223 \text{ psi} \quad \text{(second column)}$$
$$\sigma_{3max} = 50{,}180 \text{ psi} \quad \text{(third column)}$$
$$\sigma_{4max} = 25{,}090 \text{ psi} \quad \text{(fourth column)}$$

3.21 For Fig. P3.1a:

$$y_{max} = 0.023992 \text{ m}$$
$$\sigma_{max} = 44.9859 \text{ MPa} \quad \text{(at the fixed end)}$$

ANSWERS TO SELECTED PROBLEMS   587

3.22 For Fig. P3.1a:

$$y_{max} = 0.020737 \text{ m}$$
$$\sigma_{max} = 38.88 \text{ MPa}$$

3.25 For Fig. P3.3a:

$$x_{max} = 0.7067 \text{ in.}$$
$$\sigma_{1_{max}} = 19{,}830.55 \text{ psi} \quad \text{(first column)}$$
$$\sigma_{2_{max}} = 9{,}689.29 \text{ psi} \quad \text{(second column)}$$
$$\sigma_{3_{max}} = 20{,}072.12 \text{ psi} \quad \text{(third column)}$$
$$\sigma_{4_{max}} = 10{,}036.07 \text{ psi} \quad \text{(fourth column)}$$

3.26 For Fig. P3.3a:

$$x_{max} = 0.611028 \text{ in.}$$
$$\sigma_{1_{max}} = 17{,}145.85 \text{ psi} \quad \text{(first column)}$$
$$\sigma_{2_{max}} = 8{,}377.53 \text{ psi} \quad \text{(second column)}$$
$$\sigma_{3_{max}} = 17{,}354.71 \text{ psi} \quad \text{(third column)}$$
$$\sigma_{4_{max}} = 8{,}677.36 \text{ psi} \quad \text{(fourth column)}$$

3.29 The answers for Figs. P3.1a, P3.1c, and P3.1e are similar to the answers in Problem 3.11 for the same three figures.

3.30 The answers for Figs. P3.1a and P3.1c are similar to the answers in Problem 3.12 for the same two figures.

3.32 The answer for Fig. P3.3a is similar to the answer in Problem 3.16 for the same figure.

3.33 The answer for Fig. P3.3a is similar to the answer in Problem 3.18 for the same figure.

3.34 The answer for Fig. P3.1a is similar to the answer in Problem 3.21 for the same figure.

3.35 The answer for Fig. P3.1a is similar to the answer in Problem 3.22 for the same figure.

3.38 The answer for Fig. P3.3a is similar to the answer in Problem 3.25 for the same figure.

3.39 The answer for Fig. P3.3a is similar to the answer in Problem 3.26 for the same figure.

588   ANSWERS TO SELECTED PROBLEMS

3.42
$$\omega = 13.70 \text{ rps}$$
$$f = \frac{\omega}{2\pi} = 2.18 \text{ Hz}$$

3.44
$$\omega = 10.56 \text{ rps}$$
$$f = \frac{\omega}{2\pi} = 1.68 \text{ Hz}$$

3.46 Results should be closely identical.

3.47 Results should be closely identical.

3.48
$$\omega = 15.75 \text{ rps}$$
$$f = \frac{\omega}{2\pi} = 2.50 \text{ Hz}$$

3.50
$$F_1 = 159.30 \text{ kips}$$
$$F_2 = 318.61 \text{ kips}$$
$$F_3 = 238.96 \text{ kips}$$
$$\text{First-story shear} = 716.87 \text{ kips}$$
$$\text{Second-story shear} = 557.57 \text{ kips}$$
$$\text{Third-story shear} = 238.96 \text{ kips}$$

Chapter 4

4.1 For Fig. P4.1b:

$$\text{Exact } M_e = \frac{L^3(24x - x^2)}{2(L + x)^3} \quad \text{(in units of kip-ft)}$$

EI_1 is constant with $I_1 = bh_1^3/12$.

4.4 For Fig. P4.1b:

$$y_A = 0.042 \text{ in.} \qquad y_B = 0.0351 \text{ in.}$$
$$\theta_A = 170.948(10)^{-6} \text{ rad} \qquad \theta_B = -267.797(10)^{-6} \text{ rad}$$

4.5 For Fig. P4.5a:

ANSWERS TO SELECTED PROBLEMS   589

x(ft)	f(x)	M_x(kip-ft)	M_e(kip-ft)
0	1	0	0
3	1	−4.50	−4.50
6	1	−18.00	−18.00
9	1	−40.50	−40.50
12	1	−72.00	−72.00
15	1	−112.50	−112.50
18	1.331	−48.75	−36.63
21	1.728	6.00	3.47
24	2.197	51.75	23.55
27	2.744	88.50	32.25
30	3.375	116.25	34.44
33	4.096	135.00	32.96
36	4.913	144.75	29.46
39	5.832	145.50	24.95
42	6.859	137.25	20.01
45	8.000	120.00	15.00

x is measured from the free end A.

4.6   For Fig. P4.5a, $y_A = 0.2069$ in.

4.12 For Fig. P4.12a, deflection at center = $y_C = 0.10205$ in.

4.16 For Fig. P4.16a:

$$\theta_A = 7.5483(10)^{-6} \text{ rad} \qquad \theta_B = 6.0386(10)^{-6} \text{ rad}$$
$$\theta_C = 3.0193 \text{ rad} \qquad y_C = 0.031137 \text{ m}$$

For Fig. P4.16d:

$$y_D = -3.75(10)^{-4} \text{ m} \qquad y_C = 4.932(10)^{-3} \text{ m}$$

4.17 Results for Figs. P4.16a and P4.16d are identical to the results of Problem 4.16 for the same two figures.

4.18 For Fig. P4.16a, with end A fixed $y_C = 0.027677$ m.

4.19 Results for Fig. P4.16a with end A fixed are identical to the results of Problem 4.18 for the same figure.

4.20 For Fig. P4.20a:

$$\theta_B = -5.12(10)^{-3} \text{ rad} \qquad y_C = 0.6912 \text{ in.}$$

4.21 For Fig. P4.20a:

$$\theta_B = -5.12(10)^{-3} \text{ rad} \qquad y_C = 0.6912 \text{ in.}$$

4.22 $F_B = 800 \text{ lb}$ $\qquad y_B = 460{,}800/EI_0 \text{ in.} \qquad y_D = 144{,}000/EI_0 \text{ in.}$

4.23 $F_B = 800 \text{ lb}$ $\qquad y_B = 460{,}800/EI_0 \text{ in.} \qquad y_D = 144{,}000/EI_0 \text{ in.}$

4.26
$$M_A = M_D = 0 \qquad M_B = 164.27 \text{ kip-ft}$$
$$M_C = 73.36 \text{ kip-ft} \qquad R_A = 11.79 \text{ kips} \uparrow$$
$$R_B = 45.48 \text{ kips} \uparrow \qquad R_C = 11.84 \text{ kips} \uparrow$$
$$R_D = 0.89 \text{ kips} \uparrow \qquad y_C = 0.935 \text{ in.}$$

4.28 Because of symmetry they are both equal to 582.58 kip-ft.

4.29 $y_C = 0.935 \text{ in.}$

4.30 Because of symmetry they are both equal to 584.0 kip-ft.

Chapter 5

5.2 $\delta_A = 689.58 \text{ in.,} \qquad \theta_A = 66.87°.$

5.4 $\delta_A = 747.33 \text{ in.,} \qquad \theta_A = 81.70°.$

5.5 $\Delta_A = 621.12 \text{ in.}$

5.6 $\Delta_A = 550.56 \text{ in.}$

5.8
$$y'(x) = \frac{\Phi(x)}{\sqrt{1 - [\Phi(x)]^2}}$$

$$\Phi(x) = \frac{w}{12EI}[2x^3 + 3\Delta x^2 + (L - \Delta)^3 - 3L(L - \Delta)^2]$$

x is measured from the free end A of the beam.

5.10 $\delta_A = 522.40 \text{ in., } \theta_A = 43.41°, \Delta_A = 173.60 \text{ in.}$

5.11 $\delta_A = 506.63 \text{ in., } \theta_A = 40.75°, \Delta_A = 160.34 \text{ in.}$

5.12 $\delta_A = 513.28 \text{ in., } \theta_A = 41.75°, \Delta_A = 165.34 \text{ in.}$

5.13 $\delta_A = 514.98 \text{ in., } \theta_A = 41.75°, \Delta_A = 166.80 \text{ in.}$

5.14 $\delta_D = 390.50 \text{ in., } \theta_B = 83.34°.$

5.15 $\delta_A = 466 \text{ in., } \theta_A = 42.12°, \Delta_A = 108 \text{ in.}$

5.18 For Fig. P5.18a, $P = 36{,}322$ lb, $\sigma_{BC} = 9{,}204$ psi (compressive), $\sigma_{AC} = 6{,}509$ psi (tensile).

5.22 W14 × 30.

5.32 $\quad E_r = 3.23 \times 10^6$ psi $\quad M_{req} = 4.50 \times 10^6$ lb-in.
$\quad g(x) = 0.34 \quad\quad\quad M_e = 2.469 \times 10^6$ lb-in.

5.33 $\quad E_r = 3.14 \times 10^6$ psi $\quad M_{req} = 4.50 \times 10^6$ lb-in.
$\quad g(x) = 0.33 \quad\quad\quad M_e = 2.544 \times 10^6$ lb-in.

Chapter 6

6.1 For Fig. P6.1a,

$$m_1 \ddot{y}_1 + k_1 y_1 - k_2(y_2 - y_1) + c\dot{y}_1 = F_1(t)$$
$$m_2 \ddot{y}_2 + k_3 y_2 + k_2(y_2 - y_1) = F_2(t)$$

6.3 $\quad X_{1\max} = 0.526$ in. $\quad X_{2\max} = 0.747$ in.

First story:

$$(\sigma_H)_{\max} = 8.90 \text{ ksi} \quad \text{(Fixed-hinged column)}$$
$$(\sigma_F)_{\max} = 17.79 \text{ ksi} \quad \text{(Fixed-fixed column)}$$

Second story:

$$\sigma_{\max} = 9.68 \text{ ksi}$$

6.6 $\quad X_{1\max} = 1.2906$ in. $\quad X_{2\max} = 1.4508$ in.
$\quad \sigma_{1\max} = 25.20$ ksi (First-story columns)
$\quad \sigma_{2\max} = 6.97$ ksi (Second-story columns)

6.7 $\quad X_{1\max} = 1.1445$ in. $\quad X_{2\max} = 1.2877$ in.
$\quad \sigma_{1\max} = 22.34$ ksi (First-story columns)
$\quad \sigma_{2\max} = 6.27$ ksi (Second-story columns)

6.8 $\quad u_{1\max} = -0.61236$ in. $\quad$ at $t = 0.85$ sec
$\quad (u_2 - u_1)_{\max} = -0.07790$ in. $\quad$ at $t = 0.85$ sec
$\quad \sigma_{1\max} = 14.387$ ksi (First-story columns)
$\quad \sigma_{2\max} = 4.118$ ksi (Second-story columns)

ANSWERS TO SELECTED PROBLEMS

6.9 $\Lambda_1 = 0.784$ $\Lambda_2 = 0.215$
$V_{1\,\text{max}} = -58.6 \text{ kips}$ at $t = 0.6$ sec
$V_{2\,\text{max}} = 41.7 \text{ kips}$ at $t = 0.87$ sec

Note: Two cycles of support motion are used in the computation of $V_{1\,\text{max}}$ and $V_{2\,\text{max}}$.

6.14 $y_{\text{max}} = 0.9371 \text{ in.}$ at $t = 0.11$ sec
$\eta = 2.80$

6.15 $y_{\text{max}} = 0.6187 \text{ in.}$ at time $t = 0.07$ sec
$\eta = 1.547$

6.20 $P_r = 12.56 \text{ psi}$ $(P_{\text{back}})_{\text{max}} = 3.75 \text{ psi}$
$P_{sb} = 4.47 \text{ psi}$ $t_{02} = 0.33 \text{ sec}$

6.21 $P_m = 4.90 \text{ psi}$ $\dfrac{L}{U_0} = 0.017 \text{ sec}$

Chapter 7

7.5 $U = 27.65$ kip-in.

7.6 For Fig. P7.6a:
$$U = \frac{M_0^2 L}{6EI}$$

For Fig. P7.6b:
$$U = \frac{M_0^2 L}{2EI}$$

For Fig. P7.6c:
$$U = \frac{w^2 L^5}{240 EI}$$

For Fig. P7.6d:
$$U = \frac{5 P^2 L^3}{384 EI}$$

7.9 $$U = \frac{q^2 L^3}{6GJ}$$

7.10 $$U = \frac{q_B^2 L^3}{40GJ}$$

7.11 $U = 1{,}352.97 \text{ N} \cdot \text{m}$.

7.13 $$\theta_B = \frac{M_0 L}{3EI}$$

7.14 $$\theta_B = \frac{M_0 L}{EI}$$

7.15 $$y_C = \frac{5PL^3}{192EI}$$

7.17 $y_B = 0.536$ in.

7.18 For Fig. P7.18a, $y_A = 0.955$ in. For Fig. P7.18b, $y_A = 0.714$ in. For Fig. P7.18c, $y_A = 0.546$ in. (upward).

7.19 $\delta_D = 3.74$ in.

7.20 For Fig. P7.18a, $\theta_A = 5.9586 \times 10^{-3}$ rad. For Fig. P7.18b, $\theta_A = 47.669 \times 10^{-3}$ rad.

7.21 $$\theta_C = \frac{P}{9EIL}(2b^3 - a^3)$$

7.22 $\delta_E = 1.092$ mm.

7.23 $\delta_I = 0.807$ in.

7.29 For Fig. P7.29a:

$$R_B = \frac{5P}{16}$$

For Fig. P7.29b:

$$R_B = \frac{Pb^2}{2L^3}(a + 2L)$$

For Fig. P7.29d:

$$R_B = \frac{5wL}{8}$$

7.30 For Fig. P7.29a:

$$M_A = \frac{3PL}{16}$$

For Fig. P7.29b:

$$M_A = \frac{Pab}{2L^2}(a + L)$$

7.38 For Bar 1, $F_1 = -10$ kips. For Bar 2, $F_2 = 10$ kips. For Bar 3, $F_3 = -11.18$ kips.

7.40
$F_{AC} = 0$ $F_{BE} = -75$ kips
$F_{AB} = 0$ $F_{BC} = 60$ kips
$F_{BD} = 45$ kips $F_{CE} = 45$ kips

7.42
$$w_1 = 0.00293 \frac{q_0 a^4}{D} \quad \text{(maximum)}$$

$$w_2 = 0.00214 \frac{q_0 a^4}{D}$$

$$w_3 = 0.00202 \frac{q_0 a^4}{D}$$

$$w_4 = 0.00148 \frac{q_0 a^4}{D}$$

7.47
$$w_{max} = 0.00349 \frac{qa^4}{D_0}$$

7.48
$$w_{max} = 0.00298 \frac{qa^4}{D_0}$$

References

1. K. R. White, J. Minor, and K. N. Derucher, *Bridge Maintenance Inspection and Evaluation*, Marcel Dekker, New York, 1992.
2. "FHWA Proposes NICET Certification for Bridge Inspectors," *Professional Engineering News*, National Society of Professional Engineers, Alexandria, Virginia, Fall 1987.
3. D. G. Fertis, *Nonlinear Mechanics*, CRC Press, Boca Raton, Florida, 1993.
4. P. A. Bosela, D. G. Fertis, and F. J. Shaker, "A New Preloaded Beam Geometric Stiffness Matrix with Full Rigid Body Capabilities," *Int. J. Comput. Struct.*, Vol. 45, No. 1, 1992.
5. D. G. Fertis, *Mechanical and Structural Vibrations*, John Wiley, New York, 1995.
6. D. G. Fertis, *Advanced Mechanics of Structures*, Marcel Dekker, New York, 1996.
7. S. P. Timoshenko, *Theory of Structures*, 2nd ed., McGraw-Hill, New York, 1965.
8. G. R. Kirchhoff, *J. Math.* (Crelle), Vol. 40, 1850.
9. J. W. S. Lord Rayleigh, "On the Infinitesimal Bending of Surfaces of Revolution," *London Math. Soc. Proc.*, Vol. 13, 1882.
10. J. W. S. Lord Rayleigh, *The Theory of Sound*, Dover Publications, New York, 1945.
11. S. Germaine, *Recherches sur la Théorie de Surfaces Élastiques*, Paris, 1821.
12. D. J. Dowrick, *Earthquake Resistant Design*, John Wiley, New York, 1977.
13. S. P. Timoshenko, *History of Strength of Materials*, McGraw-Hill, New York, 1953.
14. D. G. Fertis, "Safety of Long Span Highways Bridges Based on Dynamic Response," Bridges and Transmission Line Structures, Proceedings Structures Congress 1987/ST Div/ASCE, August 17–20, 1987.
15. S. R. Bender, "Dynamic Response of Single Span Highway Bridges," M.S. Thesis, University of Akron, Akron, Ohio, 1976.
16. C. E. Inglis, *A Mathematical Treatise on Vibration in Railway Bridges*, Cambridge University Press, London, 1934.
17. A. Hillerborg, "Dynamic Influences of Smoothly Running Loads on Simply Supported Girders," Institute of Structural Engineering and Bridge Building of the Royal Institute of Technology, Stockholm, Sweden, 1951.
18. J. M. Biggs, H. S. Suer, and J. M. Louw, "The Vibration of Simple Span Highway Bridges," *ASCE Trans.*, Vol. 124, 1959.

19. T. P. Tung, L. E. Goodman, T. Y. Chen, and N. M. Newmark, "Highway Bridge Impact Problems," HRB Bulletin 124, National Academy of Science, Washington, D.C., 1955.

20. L. S. Jacobsen and R. S. Ayre, *Engineering Vibrations*, McGraw-Hill, New York, 1958.

21. The AASHO Road Test, Highway Research Board, Special Report 71, "Dynamic Studies of Bridges on the AASHO Road Test," National Academy of Sciences, National Research Council, Washington, D.C., 1962, Publ. 968.

22. A. S. Veletsos and T. Huang, "Analysis of Dynamic Response of Highway Bridges," *J. Eng. Mechanics Division, Proc. ASCE*, Vol. 96, No. EM.5, October, 1970.

23. L. Fryba, *Vibration of Solids and Structures Under Moving Loads*, Noordhoff International, Groningen, The Netherlands, 1972.

24. R. Willis, "Preliminary Essay to the Appendix B: Experiments for Determining the Effects Produced by Causing Weights to Travel over Bars with Different Velocities," Report to the Commissioners Appointed to Inquire into the Application of Iron to Railway Structures, W. Clowes and Sons, London, 1849.

25. G. C. Stokes, "Discussions of a Differential Equation Relating to the Breaking of Railway Bridges," *Trans., Cambridge Philosoph. Soc.*, 1869.

26. A. N. Krylov, *Mathematical Collection of Papers of the Academy of Sciences*, Vol. 61, Petersburg, 1905.

27. S. Timoshenko and D. H. Young, *Vibration Problems in Engineering*, 3rd ed., Van Nostrand, New York, 1955.

28. C. F. Scheffey, "Dynamic Load Analysis and Design of Highway Bridges," HRB Bulletin 124, National Academy of Science, Washington, D.C., 1955.

29. R. K. Wen and A. S. Veletsos, "Dynamic Behavior of Simple-Span Highway Bridges," HRB Bulletin 315, National Academy of Science, Washington, D.C., 1962.

30. R. K. Wen, "Dynamic Response of Beams Traversed by Two-Axle Loads," *J. Eng. Mechanics Division, Proc. ASCE*, Vol. 86, No. EM.5, October, 1960.

31. J. M. Hayes and J. A. Sbarounis, "Vibration Study of Three-Span Continuous I-Beam Bridge," HRB Bulletin 124, National Academy of Science, Washington, D.C., 1955.

32. G. M. Foster and L. T. Oehler, "Vibration and Deflection of Rolled Beams and Plate-Girder-Bridges," HRB Bulletin 124, National Academy of Science, Washington, D.C., 1955.

33. V. Kolousek, *Dynamics in Engineering Structures*, John Wiley, New York, 1973.

34. J. H. Hodgson, *Earthquake and Earth Structure*, Prentice-Hall, Englewood Cliffs, New Jersey, 1964.

35. J. A. Blume, N. M. Newmark, and L. H. Corning, "Design of Multistory Reinforced Concrete Buildings for Earthquake Motions," Portland Cement Association, Chicago, Illinois, 1961.

36. D. G. Fertis, "Proposed Method for the Design of Earthquake Resistant Buildings," M.S. Thesis, Michigan State University, 1955.

37. D. G. Fertis, *Dynamics and Vibration of Structures*, John Wiley, New York, 1973.

38. J. M. Gere and S. P. Timoshenko, *Mechanics of Materials*, 3rd ed., PWS-KENT Publishing, Boston, 1990.

39. D. G. Fertis, *Dynamics and Vibration of Structures*, revised ed., Robert E. Krieger Publishing, Malabar, Florida, 1984.

40. U.S. Army Corps of Engineers, "Design of Structures to Resist the Effects of Atomic Weapons."
    Manual EM-1110-345-413, Weapon Effects Data. Published July 1, 1959.
    Manual EM-1110-345-414, Strength of Materials and Structural Elements. Published March 15, 1957.
    Manual EM-1110-345-415, Principles of Dynamic Analysis and Design. Published March 15, 1957.
    Manual EM-1110-345-416, Structural Elements Subjected to Dynamic Loads. Published March 15, 1957.

41. D. G. Fertis, "Research Report," Michigan Department of Transportation, 1957.

42. D. G. Fertis and H. Cunningham, "Equivalent Systems for Shear Deflections of Variable Thickness Members," *Industrial Math.*, Vol. 12, Part 1, The Industrial Mathematics Society, Detroit, Michigan, 1962.

43. "Standard Specifications for Highway Bridges," American Association of State Highway and Transportation Officials, 12th ed., Washington, D.C., 1977.

44. L. J. Ritter, Jr. and R. J. Paquette, *Highway Engineering*, 3rd ed. Ronald Press, New York, 1967.

45. H. Cross, "Analysis of Continuous Frames by Distributing Fixed End Moments," *Trans. Amer. Soc. Civil Eng. (ASCE)*, Vol. 96, 1–10, 1932.

46. L. Euler, "Methodus Inveniedi Lineas Curvas," 1744.

47. D. G. Fertis and A. O. Afonta, "Equivalent Systems for Large Deformation of Beams of Any Stiffness Variation," *Europ. J. Mechanics, A/Solids*, Vol. 10, No. 3, 265–293, 1991.

48. A. Chajes, *Principles of Structural Stability Theory*, Prentice-Hall, Englewood Cliffs, New Jersey, 1974.

49. S. P. Timoshenko and J. M. Gere, *Theory of Elastic Stability*, 2nd ed., McGraw-Hill, New York, 1961.

50. D. G. Fertis and Q. Lu, "Inelastic Analysis of Members of Different Materials and Various Cross-Sectional Shapes," Proceedings of ESDA 96 Engineering Systems Design and Analysis, Nonlinear Systems Symposium, Montpellier, France, July 1–4, 1996.

51. S. Timoshenko, *Strength of Materials, Part II, Advanced Theory and Problems*, 3rd ed., Robert E. Krieger Publishing, Huntington, New York, 1976.

52. D. G. Fertis and F. R. Schubert, "Inelastic Analysis of Prismatic and Nonprismatic Aluminum Members," *Int. J. Comput. Structures*, Vol. 52, No. 2, 287–295, 1994.

53. E. Simiu and R. H. Scanlan, *Wind Effects on Structures*, John Wiley, New York, 1986.

54. C. Scruton, "Experimental Investigation of Aerodynamic Stability of Suspension Bridges with Special Reference to Proposed Seven Bridge," *Proc. Inst. Civ. Eng.*, Vol. 1, Par 1, No. 2, 189–222, March, 1952.

55. A. G. Davenport, "The Use of Taut Strip Models in the Prediction of the Response of Long-Span Bridges to Turbulent Wind Flow-Induced Structural Vibrations," *Proceedings of the IUTAM-IAHR Symposium on Flow-Induced Structural Vibrations*, Karlsruhe, West Germany, 1972, Springer-Verlag, Berlin, 1974, pp. 373–382.

56. Y. Nakamura and T. Yoshimura, "Binary Flutter of Suspension Bridge Deck Sections," *J. Eng. Mech. Div. ASCE*, Vol. 102, No. EM.4, 685–700, August, 1976.

57. F. Leonhardt, "Latest Developments of Cable-Stayed Bridges for Long Spans," *Bygningsstakiske Medd.*, Vol. 45, No. 4, 89–143, 1974.

58. M. Ito and Y. Nakamura, "Aerodynamic Stability of Structures in Wind," IABSE Surveys S-20/82, International Association for Bridge and Structural Engineering, ETH-Hönggerberg, Zürich, 1982.

59. D. Berreby, "The Great Bridge Controversy," *Discover*, pp. 26–33, February 1992.

60. R. C. Fuller, D. A. Zollman, and T. C. Campbell, "The Puzzle of the Tacoma Narrows Bridge Collapse," The American Association of Physics Teachers, 1982. Also videodisc, available from John Wiley, New York, 1982.

61. H. Petroski, *Design Paradigms; Case Histories of Error and Judgment in Engineering*, Cambridge University Press, 1994.

62. J. Gotchy, "Bridging the Narrows," The Peninsula Historical Society, 1990.

63. A. H. Flax and L. Goland, "Dynamic Effect in Rotor Blade Bending," *J. Aerospace Sci.*, Vol. 18, No. 12, 813–829, December, 1951.

64. J. D. Hobolt and G. W. Brooks, "Differential Equations of Motion for Combined Flapwise Bending, Chordwise Bending, and Torsion of Twisted Non-uniform Rotor Blades," Report No. 1346, Langley Aeronautical Laboratory, Langley Field, Virginia, 1956.

65. A. H. Flax, "The Bending of Rotor Blades," *J. Aerospace Sci.*, Vol. 14, No. 1, 42–50, January, 1947.

66. R. C. DiPrima and G. H. Handelman, "Vibration of Twisted Beams," *Quart. Appl. Math.*, Vol. XII, No. 3, 241–259, October, 1954.

67. D. G. Fertis, "Dynamic Response of Nonuniform Rotor Blades," *J. Aircraft*, Vol. 14, No. 5, 417–418, May, 1977.

68. D. G. Fertis, "Airfoil Design Concept that Increases Lift, Reduces Drag, and Improves Stall," Proceedings of the American Institute of Aeronautics and Astronautics, 29th AIAA/ASME/ASCE/AHS SMD Structures, Structural Dynamics and Materials Conference, 1988.

69. D. G. Fertis, "Airfoil Design Concept that Increases Lift, Reduces Drag, and Improves Stall," *J. Aerospace Eng., ASCE*, Vol. 7, No. 3, 328–339, 1994.

70. J. A. Blume, "Design of High-Rise Building under the New Code Requirements for Ductile Reinforced Concrete," Session No. 4, ASCE Seminar on Buildings in Earthquake and Wind, Seattle, Washington, 1967b.

71. S. Timoshenko, *Theory of Plates and Shells*, McGraw-Hill, New York, 1940.

72. R. L. Wiegel, Coordinating Ed., *Earthquake Engineering*, Prentice-Hall, Englewood Cliffs, New Jersey, 1970.

73. "The Effects of Nuclear Weapons," prepared by the U.S. Department of Defense and published by the U.S. Atomic Energy Commission, April, 1962.

74. "NSWC Technical Report," Report No. NSWC/WOL/TR 75-134, 26 March, 1976.

75. "Structures to Resist the Effects of Accidental Explosions," U.S. Department of the Army, the Navy, and the Airforce, Technical Manual No. TM 5-1300, NAVFAC P-397, AFM 88-22, June, 1969.

76. E. C. Pestel and F. A. Leckie, *Matrix Methods in Elastomechanics*, McGraw-Hill, New York, 1963.

77. R. Courant, "Variational Method for the Solution of Problems of Equilibrium and Vibrations," *Bull. Amer. Math. Soc.*, Vol. 49, 1943.

78. M. J. Turner, R. W. Clough, H. L. Martin, and L. J. Topp, "Stiffness and Deflection Analysis of Complex Structures," *J. Aeronaut. Sci.*, Vol. 23, No. 9, 1956.

79. R. Clough, "The Finite Element Method in Plane Stress Analysis," Proceedings of the Second Conference on Electronic Computations, ASCE, pp. 345–377, New York, 1960.

80. R. Clough and J. Penzien, *Dynamics of Structures*, McGraw-Hill, New York, 1975.

81. J. H. Argyris and S. Kelsey, *Energy Theorems in Structural Analysis*, Butterworth, London, 1960.

82. "Proceedings of the Matrix Method in Structural Mechanics," AFFDL-TR-66-80, Wright-Patterson Air Force Base, Ohio, 1966.

83. R. J. Melosh, *Structural Engineering Analysis by Finite Element Method*, Prentice-Hall, Englewood Cliffs, New Jersey, 1990.

84. R. D. Cook, *Concepts and Applications of Finite Element Analysis*, John Wiley, New York, 1981.

85. O. C. Zienkiewicz, *Finite Element Method*, 3rd ed., McGraw-Hill, London, 1977.

86. A. K. Noor, "Textbooks and Monographs on Finite Element Technology," *J. Finite Elements*, February 1985.

87. R. G. Olsson, "Biegung der Rechteckplatte bei Linear Veramderlicher Biegungssteifigkeit," *Ing. Arch.*, Vol. 5, 363, 1934.

88. E. Reisner, "Remark on the Theory of Bending of Plates of Variable Thickness," *J. Math. Phys.*, Vol. 16, No. 43, 1937.

89. R. G. Olsson, "Biengung der Rechteckplatte bei Linear Veranderlicher Steifigkeit und Beliebiger Belastung," *Bauingenier*, Vol. 22, No. 10, 1941.

90. R. G. Olsson, "Biengung der Rechteckplatte Von Exponential Veranderlicher Dicke," *Bauingenier*, Vol. 24, 230, 1940.

91. H. D. Conway, "Flexure of Infinite Rectangular Plates of Variable Thickness," *Ing. Arch.*, Vol. 26, 143–145, 1958.

92. Y. C. Fung, "Bending of Thin Plastic Plates of Variable Thickness," *J. Aeronaut. Sci.*, Vol. 20, 455–463, 1953.

93. B. Klein, "Shear Buckling of Simply Supported Rectangular Plates Tapered in Thickness," *Franklin Inst. J.*, Vol. 263, 537–541, 1957.

94. R. J. Melosh, "A Stiffness Matrix for the Analysis of Thin Plates in Bending," *J. Aerospace Sci.*, Vol. 28, 34–42, 1961.

95. H. Favre and B. Gilg, "La Plaque Rectangulaire Flechie D'epaisseur Linearement Variable," *Z. Ange. Math. Mech.*, Vol. 3, 354, 1952.

96. D. G. Fertis and M. E. Keene, "Elastic and Inelastic Analysis of Nonprismatic Members," *J. Struct. Eng., ASCE*, Vol. 116, No. 2, 1990.

97. D. G. Fertis and C. T. Lee, "Elastic and Inelastic Analysis of Variable Thickness Plates by Using Equivalent Systems," *Int. J. Mech. Struct. Mach.*, Vol. 21, No. 2, 1993.

98. D. G. Fertis and A. O. Afonta, "Small Vibrations of Flexible Bars by Using the Finite Element Method with Equivalent Uniform Stiffness and Mass Methodology," *Int. J. Sound Vibration*, Vol. 163, No. 2, 1993.

99. R. Szilard, *Theory and Analysis of Plates, Classical and Numerical Methods*, Prentice-Hall, Englewood Cliffs, New Jersey, 1974.

100. N. S. Grigg, *Infrastructure Engineering and Management*, John Wiley, New York, 1988.

101. G. L. Rogers, *Dynamics of Framed Structures*, John Wiley, New York, 1959.

Index

Acceleration impulse extrapolation method, 51, 53, 185–194
Angle of incidence, 439
Arches, 16–18, 476–479, 489–491
  three-hinge, 28, 29
  types of, 18
Average loading caused by blast:
  back face, 441, 442
  front face, 440, 441
  net horizontal loading, 443, 452
  roof and sides, 443, 444
Axial force, 31–38

Beams, 8, 147
  curved, 8, 17
  design of, 147
  dynamic analysis of, 147, 157
  idealized, 149–153
  slender, 13
  vertical deflection formulas for, 553, 554
  vibration formulas for, 568
Bending moment, 1, 31–38
  diagrams, 31–38
  equations, 31–38
Bending stiffness, 91
Blast, 350, 391, 432–437, 439–452
Bridge vehicle system, 48–54
Bridges, 1, 147
  collapsed suspension, 340–350
  composite, 196–200
  continuous highway, 200–208, 273–277
  dynamic behavior of suspension, 341–344
  dynamic characteristics of, 48–54
  dynamic testing of, 49, 54, 342
  equivalent systems for highway, 230, 234, 273–277
  flexible, 2
  highway, 2
  inspection of, 2
  long-span, 2, 51
  low fundamental frequency, 51
  natural frequencies of, 2, 196–208
  noncomposite, 196–200
  resonant vibration of, 50, 51
  specifications for, 230, 231
  suspension, 340–350
  vibration of, 48–54, 194–208
Buildings, 1, 147, 153–157, 169–191, 208–212, 283–291, 398–417, 424–432, 439–452
  blast loadings on, 439–452
  idealized, 153–157
  multistory, 424–432
  practical earthquake design for, 208–212
  response to harmonic excitations, 169–191
  with members of variable stiffness, 283–291
Bursts, 434
  types of, 434

Castigliano's second theorem, 461, 492–497
Coefficient:
  deflection, 200
  drag, 441
  influence, 200
  seismic, 209
Columns, 303
  buckling of, 326–336
  critical load of, 327, 329
  eccentrically loaded, 331–336
  elastically supported beam, 338–340
  Euler, 328–331
  free vibration of, 336–338
  inelastic, 333
Combined loading, 105–113
Computer programs, 559–566, 577, 578
Conjugate-beam method, 219, 258–266
  statements of, 259
Conservative systems, 462
Convolution integral, 180

Critical damping factor, 163
Critical divergence velocity, 343
Crust, 54
Culverts, 270–272
Curvature, 91

Damping:
  coefficient of viscous, 149
  coulomb, 149
  critical, 163
  dry friction, 149
  hysteresis, 149
  light, 164
  overdamping, 163
  solid damping, 149
  underdamping, 164
  viscous, 148, 157, 395–397
Damping ration, 164
Deformations:
  axial, 72–77
  bending, 88–105
  large, 219, 303–326
  torsional, 77, 78
Design loads:
  ultimate, 364–383
Differential equations:
  closed-form solution of, 170–178
  homogeneous, 160
  numerical solution of, 185–194
  solution of, 159–185
Double integration method, 96–98, 101–105
Ductility ration, 423
Duhamel's integral, 180
Dynamic equilibrium, 210

Earthquakes, 1, 20, 58–62, 147, 303, 350, 391
  design codes for, 57, 147, 208
  elastoplastic analysis for, 417–432
  intensity of, 55
  modal analysis for, 409–417
  practical design for, 208–212
  practical design considerations for, 57–61
  response spectra for, 55
  tectonic, 54
  volcanic, 54
Earthquake-resistant structures, 57

Elastic line, 88, 91
Elastica theory, 304
Elastoplastic response:
  chards of, 569–572
Energy, 461–479
  concept of conservation of, 462
  concepts of, 461–479
  historical notes on concepts of, 462
  methods, 479–497
  strain energy, 461–479
    axial loading, 462–467
    bending, 468–471
    shear, 472, 473
    total, 476–479
    twisting moment, 473–476
Epicenter, 54
Epicentral zone, 54
Equations:
  transformation, 122, 499–507
Equilibrium conditions, 71
Equilibrium of, 21–31
  equations of, 21, 26, 27
  particles, 21–26
  rigid bodies or structures, 26–31
Equivalent systems, 219–248, 283–291
  approximate method of the, 224–230
  for frames, 283–291
  for highway bridges, 230–234, 273–277
  for inelastic analysis and design, 350–383
  for nonlinear response, 303–326, 350–383
  for rectangular plates, 530–541
  pseudolinear, 306–308, 311–323
  simplified nonlinear, 304, 308, 309, 323–326
  theorem of equivalent systems, 220
  theory and method of the, 220–224
Euler-Bernoulli law of deformation, 220, 557, 558
  integration of, 557, 558
Explosions, 432–452
  conventional, 437–439
  nuclear, 432–437, 439–452

Fault blocks, 54
Fault plane, 54
Finite difference method, 219, 461, 507–530

thin rectangular plates, 509–530
Finite element method, 19, 219, 461, 498–507
Footings, 19–21
Forces, 147
   centrifugal, 150
   complanar concurrent, 21
   damping, 147–149
   drag, 342
   general type, 147, 178–194
   harmonic, 147, 169–178
   impulsive, 148
   inertia, 201, 210
   lift, 343
   periodic nonharmonic, 147
   three-dimensional concurrent, 21
Foundations, 19–21, 147
   critical dynamic excitations for, 20
   depth of, 20
   machine, 147
   types of, 20
Frames, 13–16, 34–38, 147, 283–291
   concrete, 13
   design of, 147
   dynamic analysis of, 147, 283–291, 402–417, 424–432
   equations of equilibrium for, 15
   idealized, 153–157, 283–291, 402–417, 424–432
   monolithic, 14
   statically indeterminate, 13
   three-dimensional, 13–16
   two-dimensional, 13
   types of, 15
   variable thickness, 14, 283–291
Frequencies of vibration, 4–8
   damped, 162–169
   forced, 6
   free, 6, 162
   resonant, 6, 195
Friction:
   coefficient of kinetic, 149

Generalized coordinates, 391–395

Hooke's law, 79, 90, 328, 333, 462

Idealized systems, 147–194, 391

Inelastic analysis and design, 219, 350–383
Inertia:
   rotatory, 338–340
   shear, 338–340
Infrastructure, 2, 147–212
   dynamic aspects of, 47–54, 147–212
   mechanics, 2, 3
Infrastructure systems, 1–3, 303–383
   component 1–3
      dynamic aspects of, 1, 47–54, 147–212
   comprehensive master plan for, 3
   inelastic response of, 303, 350–383
   instabilities of, 303, 326–340
   major, 1–3
   nonlinearities of, 303–326
   optimization parameters for, 3
Instabilities:
   aerodynamic, 341
   aeroelastic, 343
   flutter, 338–340
   static, 338–340
   torsional divergence, 343
Integration, 93–105
   method of, 93–105

Kern County earthquake, 56

Lagrange's equation, 391–397, 398
Large deformations, 303–326
Liquefaction, 20
Logarithmic decrement, 166–169
Los Angeles earthquake, 61, 62

Magnification factor, 51, 573–576
   graphs of maximum, 573–576
Members:
   axially loaded, 72–77
   bending of, 88–105
   circular and tubular, 80
   combined loading on, 105–113
   elastically supported variable stiffness, 241–248
   flexible, 303–326
   highway bridges of variable stiffness, 230–234, 273–277
   nonlinear analysis of, 303–326, 350–383

Members (cont'd)
  rectangular, 80
  statically indeterminate variable stiffness, 234–248
  torsion of, 77–88
  ultimate design loads of, 364–383
  ultimate load-carrying capacity of, 357, 364–383
  variable stiffness, 219–248, 303–326, 350–383
Method of vibration analysis:
  finite difference, 200
  finite element, 200
  Myklestad's, 200
  Rayleigh's, 200
  Stodola's, 200–208
  transfer matrix, 200
Modal analysis, 398–417
Modal displacement, 399
Modal participation factor, 412
Modulus function, 221
Mohr's circle of stress, 128–132
Moment-area method, 219, 248–258, 263–266, 362
  theorems of, 248–251
Moment diagrams, 31–38
Moment distribution method, 266–277
  for bridge girders of variable stiffness, 273–277
  step-by-step procedure for, 269
Moment equations, 31–38
Moment of inertia:
  polar, 79
Moment of inertia function, 221

Nonlinear analysis, 303–383
Nonlinear integral equations, 306, 313, 319

Overpressure:
  peak, 434–437
  peak dynamic, 437, 444
  reflected, 441, 447

Plane stress, 122
Plates, 18, 19, 507–541
  boundary condition for rectangular, 512, 520–523
  buckling of, 18
  differential equations for rectangular, 509–513
  equivalent systems for rectangular, 530–541
  finite difference equations for rectangular, 513–523
  history of, 19
  rigorous solution of variable thickness rectangular, 530–541
  types of, 18
  variable thickness rectangular, 513–523, 529–541
Principal angles, 124
Principal planes, 124
Probability, 57
Protective structures, 437–439

Quasi resonance, 51

Rectangular structures, 439–452
  dynamic loading on closed, 439–452
Reduced modulus, 354–364
References, 595–600
Relations between load, shear, and moment, 38–47
Resistance of structures, 417–432
Response spectra, 55
Reynolds number, 343
Rift, 54

Scour, 20
Selected problems:
  answers to, 579–594
Shear factor, 472
Shear force, 1, 31–38
  diagrams for, 31–38
  equation for, 31–38
Shear modulus, 79
Shells, 18, 19
  history of, 19
  thin, 19
  types of, 19
Shock front, 434
Shock wave, 434
Sign conventions, 92
Simpson's rule, 314, 316, 555, 556
Slip, 54

Soil-structure interaction, 20
Sound velocity, 441
Spring-mass systems, 149–194
    differential equations of motion for, 157–185
    dynamic analysis of, 157–194
    idealized spring-mass systems, 149–194
    practical applications of, 149–194
Stochastic methods, 56
Strain, 71
    axial, 71–77
    fundamentals of, 71–105
    longitudinal, 90
    maximum shear, 78
Stresses, 71–132
    allowable bending, 98–101
    average normal, 125
    axial, 71–77
    bending, 88–105, 169, 177
    dynamic, 169–194
    fundamentals of, 71–132
    maximum and minimum normal, 121–127
    maximum and minimum shear, 121–127
    maximum shear, 80
    normal, 90
    principal, 123
    shear, 113–121
    torsional, 77–88
Structural elements and loading, 1, 3–21
    arches, 16–18
    beam-columns, 4–8
    beams, 4–8
    columns, 4–8
    footings, 19–21
    foundations, 19–21
    frames, 8, 13–16
    plates, 18, 19
    shells, 18, 19
    trusses, 8–13
Superposition principle, 567
Support motion, 6
    harmonic, 6

Tacoma narrows suspension bridge, 340, 341, 344–348
Three-moment equation, 277–283
Tire hop, 50
Trusses, 8–13, 499–507
    buckling of, 10
    complex, 10
    compound, 10
    finite elements for, 499–507
    plane, 8
    simple, 10
    space, 8, 10, 12
    statically determinate, 10
    statically indeterminate, 10
    types of, 10
Tubes, 87, 88
    thin-walled, 87, 88

Variable stiffness frames:
    dynamic response of, 283–291
Vibrations:
    forced, 169–185
    formulas of beam, 562
    free undamped, 159–162
    highway bridge, 48–54, 194–208
    random, 55, 56
    resonant, 50, 51, 55
    viscously damped, 162–169
Virtual work, 461, 483–491
    method of, 483–491
    principle of, 462

Young's modulus, 91